UNSATURATED
SOIL MECHANICS

UNSATURATED SOIL MECHANICS

NING LU
Colorado School of Mines

and

WILLIAM J. LIKOS
University of Missouri–Columbia

JOHN WILEY & SONS, INC.

This book is printed on acid-free paper. ∞

Copyright © 2004 by John Wiley & Sons, Inc. All rights reserved.

Published by John Wiley & Sons, Inc., Hoboken, New Jersey
Published simultaneously in Canada.

No part of this publication may be reproduced, stored in a retrieval system, or transmitted in any form or by any means, electronic, mechanical, photocopying, recording, scanning, or otherwise, except as permitted under Section 107 or 108 of the 1976 United States Copyright Act, without either the prior written permission of the Publisher, or authorization through payment of the appropriate per-copy fee to the Copyright Clearance Center, Inc., 222 Rosewood Drive, Danvers, MA 01923, (978) 750-8400, fax (978) 750-4470, or on the web at www.copyright.com. Requests to the Publisher for permission should be addressed to the Permissions Department, John Wiley & Sons, Inc., 111 River Street, Hoboken, NJ 07030, (201) 748-6011, fax (201) 748-6008, e-mail: permcoordinator@wiley.com.

Limit of Liability/Disclaimer of Warranty: While the publisher and author have used their best efforts in preparing this book, they make no representations or warranties with respect to the accuracy or completeness of the contents of this book and specifically disclaim any implied warranties of merchantability or fitness for a particular purpose. No warranty may be created or extended by sales representatives or written sales materials. The advice and strategies contained herein may not be suitable for your situation. You should consult with a professional where appropriate. Neither the publisher nor author shall be liable for any loss of profit or any other commercial damages, including but not limited to special, incidental, consequential, or other damages.

For general information on our other products and services or for technical support, please contact our Customer Care Department within the United States at (800) 762-2974, outside the United States at (317) 572-3993 or fax (317) 572-4002.

Wiley also publishes its books in a variety of electronic formats. Some content that appears in print may not be available in electronic books. For more information about Wiley products, visit our web site at www.wiley.com.

Library of Congress Cataloging-in-Publication Data:

Lu, Ning, 1960–
  Unsaturated soil mechanics / by Ning Lu and William J. Likos.
    p. cm.
  Includes bibliographical references and index.
  ISBN 978-0-471-44731-3 (cloth)
  1. Soil mechanics. 2. Soil moisture. I. Likos, William J. II. Title.
  TA710.L74 2004
  624.1'5136—dc22
                                                          2004004218

10 9 8 7 6 5 4 3 2

To
Vivian, Connie, Benton, Jonas, Holly, and Shemin

CONTENTS

FOREWORD	xvii
PREFACE	xix
SYMBOLS	xxi
INTRODUCTION	1
1 STATE OF UNSATURATED SOIL	**3**

    1.1 Unsaturated Soil Phenomena / 3
          1.1.1 Definition of Unsaturated Soil Mechanics / 3
          1.1.2 Interdisciplinary Nature of Unsaturated Soil Mechanics / 4
          1.1.3 Classification of Unsaturated Soil Phenomena / 6
    1.2 Scope and Organization of Book / 8
          1.2.1 Chapter Structure / 8
          1.2.2 Geomechanics and Geo-environmental Tracks / 11
    1.3 Unsaturated Soil in Nature and Practice / 12
          1.3.1 Unsaturated Soil in Hydrologic Cycle / 12
          1.3.2 Global Extent of Climatic Factors / 12
          1.3.3 Unsaturated Zone and Soil Formation / 13
          1.3.4 Unsaturated Soil in Engineering Practice / 18
    1.4 Moisture, Pore Pressure, and Stress Profiles / 20
          1.4.1 Stress in the Unsaturated State / 20

1.4.2 Saturated Moisture and Stress Profiles: Conceptual Illustration / 21
1.4.3 Unsaturated Moisture and Stress Profiles: Conceptual Illustration / 22
1.4.4 Illustrative Stress Analysis / 23
1.5 State Variables, Material Variables, and Constitutive Laws / 26
1.5.1 Phenomena Prediction / 26
1.5.2 Head as a State Variable / 28
1.5.3 Effective Stress as a State Variable / 30
1.5.4 Net Normal Stresses as State Variables / 33
1.6 Suction and Potential of Soil Water / 34
1.6.1 Total Soil Suction / 34
1.6.2 Pore Water Potential / 35
1.6.3 Units of Soil Suction / 38
1.6.4 Suction Regimes and the Soil-Water Characteristic Curve / 39
Problems / 43

I FUNDAMENTAL PRINCIPLES 45

2 MATERIAL VARIABLES 47

2.1 Physical Properties of Air and Water / 47
2.1.1 Unsaturated Soil as a Multiphase System / 47
2.1.2 Density of Dry Air / 48
2.1.3 Density of Water / 50
2.1.4 Viscosity of Air and Water / 53
2.1.5 Flow Regimes / 55
2.2 Partial Pressure and Relative Humidity / 57
2.2.1 Relative Humidity in Unsaturated Soil Mechanics / 57
2.2.2 Composition and Partial Pressure of Air / 57
2.2.3 Equilibrium between Free Water and Air / 59
2.2.4 Equilibrium between Pore Water and Air / 62
2.2.5 Relative Humidity / 63
2.2.6 Dew Point / 64
2.3 Density of Moist Air / 65
2.3.1 Effect of Water Vapor on Density of Air / 65
2.3.2 Formulation for Moist Air Density / 66
2.4 Surface Tension / 73

2.4.1　Origin of Surface Tension / 73
　　　2.4.2　Pressure Drop across an Air-Water Interface / 76
　2.5　Cavitation of Water / 80
　　　2.5.1　Cavitation and Boiling / 80
　　　2.5.2　Hydrostatic Atmospheric Pressure / 82
　　　2.5.3　Cavitation Pressure / 84
　　　Problems / 86

3　INTERFACIAL EQUILIBRIUM　　　　　　　　　　　　　　　89

　3.1　Solubility of Air in Water / 89
　　　3.1.1　Henry's Law / 89
　　　3.1.2　Temperature Dependence / 91
　　　3.1.3　Volumetric Coefficient of Solubility / 92
　　　3.1.4　Henry's Law Constant and Volumetric Coefficient of Solubility / 93
　　　3.1.5　Vapor Component Correction / 94
　　　3.1.6　Mass Coefficient of Solubility / 95
　3.2　Air-Water-Solid Interface / 96
　　　3.2.1　Equilibrium between Two Water Drops / 96
　　　3.2.2　Equilibrium at an Air-Water-Solid Interface / 97
　　　3.2.3　Contact Angle / 99
　　　3.2.4　Air-Water-Solid Interface in Unsaturated Soil / 101
　3.3　Vapor Pressure Lowering / 104
　　　3.3.1　Implications of Kelvin's Equation / 104
　　　3.3.2　Derivation of Kelvin's Equation / 106
　　　3.3.3　Capillary Condensation / 111
　3.4　Soil-Water Characteristic Curve / 114
　　　3.4.1　Soil Suction and Soil Water / 114
　　　3.4.2　Capillary Tube Model / 115
　　　3.4.3　Contacting Sphere Model / 118
　　　3.4.4　Concluding Remarks / 124
　　　Problems / 124

4　CAPILLARITY　　　　　　　　　　　　　　　　　　　　　　128

　4.1　Young-Laplace Equation / 128
　　　4.1.1　Three-Dimensional Meniscus / 128
　　　4.1.2　Hydrostatic Equilibrium in a Capillary Tube / 131
　4.2　Height of Capillary Rise / 133

4.2.1 Capillary Rise in a Tube / 133
4.2.2 Capillary Finger Model / 136
4.2.3 Capillary Rise in Idealized Soil / 137
4.2.4 Capillary Rise in Soil / 139
4.3 Rate of Capillary Rise / 140
4.3.1 Saturated Hydraulic Conductivity Formulation / 140
4.3.2 Unsaturated Hydraulic Conductivity Formulation / 142
4.3.3 Experimental Verification / 145
4.4 Capillary Pore Size Distribution / 147
4.4.1 Theoretical Basis / 147
4.4.2 Pore Geometry / 150
4.4.3 Computational Procedures / 153
4.5 Suction Stress / 160
4.5.1 Forces between Two Spherical Particles / 160
4.5.2 Pressure in the Water Lens / 162
4.5.3 Effective Stress due to Capillarity / 163
4.5.4 Effective Stress Parameter and Water Content / 165
Problems / 168

II STRESS PHENOMENA 171

5 STATE OF STRESS 173

5.1 Effective Stress in Unsaturated Soil / 173
5.1.1 Macromechanical Conceptualization / 173
5.1.2 Micromechanical Conceptualization / 174
5.1.3 Stress between Two Spherical Particles with Nonzero Contact Angle / 175
5.1.4 Pore Pressure Regimes / 181
5.2 Hysteresis / 182
5.2.1 Hysteresis Mechanisms / 182
5.2.2 Ink-Bottle Hysteresis / 184
5.2.3 Contact Angle Hysteresis / 186
5.2.4 Hysteresis in the Soil-Water Characteristic Curve / 187
5.2.5 Hysteresis in the Effective Stress Parameter / 187
5.2.6 Hysteresis in the Suction Stress Characteristic Curve / 191
5.3 Stress Tensor Representation / 191
5.3.1 Net Normal Stress, Matric Suction, and Suction Stress Tensors / 191

          5.3.2 Stress Tensors in Unsaturated Soil: Conceptual Illustration / 195
    5.4 Stress Control by Axis Translation / 201
          5.4.1 Rationale for Axis Translation / 201
          5.4.2 Equilibrium for an Air-Water-HAE System / 202
          5.4.3 Equilibrium for an Air-Water-HAE-Soil System / 203
          5.4.4 Characteristic Curve for HAE Material / 204
          5.4.5 Controlled Stress Variable Testing / 204
    5.5 Graphical Representation of Stress / 207
          5.5.1 Net Normal Stress and Matric Suction Representation / 207
          5.5.2 Effective Stress Representation / 213
          Problems / 218

6 SHEAR STRENGTH   220

    6.1 Extended Mohr-Coulomb (M-C) Criterion / 220
          6.1.1 M-C for Saturated Soil / 220
          6.1.2 Experimental Observations of Unsaturated Shear Strength / 221
          6.1.3 Extended M-C Criterion / 229
          6.1.4 Extended M-C Criterion in Terms of Principal Stresses / 232
    6.2 Shear Strength Parameters for the Extended M-C Criterion / 233
          6.2.1 Interpretation of Triaxial Testing Results / 233
          6.2.2 Interpretation of Direct Shear Testing Results / 236
    6.3 Effective Stress and the M-C Criterion / 238
          6.3.1 Nonlinearity in the Extended M-C Envelope / 238
          6.3.2 Effective Stress Approach / 241
          6.3.3 Measurements of χ at Failure / 242
          6.3.4 Reconciliation between ϕ^b and χ_f / 244
          6.3.5 Validity of Effective Stress as a State Variable for Strength / 247
    6.4 Shear Strength Parameters for the M-C Criterion / 248
          6.4.1 Interpretation of Direct Shear Testing Results / 248
          6.4.2 Interpretation of Triaxial Testing Results / 250
    6.5 Unified Representation of Failure Envelope / 252
          6.5.1 Capillary Cohesion as a Characteristic Function for Unsaturated Soil / 252

      6.5.2   Determining the Magnitude of Capillary Cohesion / 256
      6.5.3   Concluding Remarks / 261
  Problems / 265

7 SUCTION AND EARTH PRESSURE PROFILES   267

7.1 Steady Suction and Water Content Profiles / 267
      7.1.1   Suction Regimes in Unsaturated Soil / 267
      7.1.2   Analytical Solutions for Profiles of Matric Suction / 270
      7.1.3   Hydrologic Parameters for Representative Soil Types / 272
      7.1.4   Profiles of Matric Suction for Representative Soil Types / 273
      7.1.5   Profiles of Water Content for Representative Soil Types / 275

7.2 Steady Effective Stress Parameter and Stress Profiles / 280
      7.2.1   Profiles of the Effective Stress Parameter χ / 280
      7.2.2   Profiles of Suction Stress and Their Solution Regimes / 282
      7.2.3   Profiles of Suction Stress for Representative Soil Types / 289
      7.2.4   Concluding Remarks / 292

7.3 Earth Pressure at Rest / 294
      7.3.1   Extended Hooke's Law / 294
      7.3.2   Profiles of Coefficient of Earth Pressure at Rest / 296
      7.3.3   Depth of Cracking / 297

7.4 Active Earth Pressure / 301
      7.4.1   Mohr-Coulomb Failure Criteria for Unsaturated Soil / 301
      7.4.2   Rankine's Active State of Failure / 302
      7.4.3   Active Earth Pressure Profiles for Constant Suction Stress / 306
      7.4.4   Active Earth Pressure Profiles for Variable Suction Stress / 308
      7.4.5   Active Earth Pressure Profiles with Tension Cracks / 310

7.5 Passive Earth Pressure / 312
      7.5.1   Rankine's Passive State of Failure / 312
      7.5.2   Passive Earth Pressure Profiles for Constant Suction Stress / 315
      7.5.3   Passive Earth Pressure Profiles for Variable Suction Stress / 318

CONTENTS **xiii**

      7.5.4 Concluding Remarks / 320
    Problems / 322

III FLOW PHENOMENA   323

8 STEADY FLOWS   325

  8.1 Driving Mechanisms for Water and Airflow / 325
      8.1.1 Potential for Water Flow / 325
      8.1.2 Mechanisms for Airflow / 326
      8.1.3 Regimes for Pore Water Flow and Pore Airflow / 326
      8.1.4 Steady-State Flow Law for Water / 328
  8.2 Permeability and Hydraulic Conductivity / 329
      8.2.1 Permeability versus Conductivity / 329
      8.2.2 Magnitude, Variability, and Scaling Effects / 331
  8.3 Hydraulic Conductivity Function / 333
      8.3.1 Conceptual Model for the Hydraulic Conductivity Function / 333
      8.3.2 Hysteresis in the Hydraulic Conductivity Function / 336
      8.3.3 Relative Conductivity / 336
      8.3.4 Effects of Soil Type / 338
  8.4 Capillary Barriers / 341
      8.4.1 Natural and Engineered Capillary Barriers / 341
      8.4.2 Flat Capillary Barriers / 342
      8.4.3 Dipping Capillary Barriers / 345
  8.5 Steady Infiltration and Evaporation / 349
      8.5.1 Horizontal Infiltration / 349
      8.5.2 Vertical Infiltration and Evaporation / 352
  8.6 Steady Vapor Flow / 359
      8.6.1 Fick's Law for Vapor Flow / 359
      8.6.2 Temperature and Vapor Pressure Variation / 359
      8.6.3 Vapor Density Gradient / 361
  8.7 Steady Air Diffusion in Water / 363
      8.7.1 Theoretical Basis / 363
      8.7.2 Air Diffusion in an Axis Translation System / 366
    Problems / 367

9 TRANSIENT FLOWS   369

  9.1 Principles for Pore Liquid Flow / 369
      9.1.1 Principle of Mass Conservation / 369

        9.1.2 Transient Saturated Flow / 371
        9.1.3 Transient Unsaturated Flow / 372
  9.2 Rate of Infiltration / 376
        9.2.1 Transient Horizontal Infiltration / 376
        9.2.2 Transient Vertical Infiltration / 380
        9.2.3 Transient Moisture Profile for Vertical Infiltration / 384
  9.3 Transient Suction and Moisture Profiles / 386
        9.3.1 Importance of Transient Soil Suction and Moisture / 386
        9.3.2 Analytical Solution of Transient Unsaturated Flow / 386
        9.3.3 Numerical Modeling of Transient Unsaturated Flow / 389
  9.4 Principles for Pore Gas Flow / 396
        9.4.1 Principle of Mass Conservation for Compressible Gas / 396
        9.4.2 Governing Equation for Pore Airflow / 397
        9.4.3 Linearization of the Airflow Equation / 398
        9.4.4 Sinusoidal Barometric Pressure Fluctuation / 400
  9.5 Barometric Pumping Analysis / 402
        9.5.1 Barometric Pumping / 402
        9.5.2 Theoretical Framework / 403
        9.5.3 Time Series Analysis / 404
        9.5.4 Determining Air Permeability / 407
  Problems / 412

IV MATERIAL VARIABLE MEASUREMENT AND MODELING   415

10 SUCTION MEASUREMENT   417

  10.1 Overview of Measurement Techniques / 417
  10.2 Tensiometers / 420
        10.2.1 Properties of High-Air-Entry Materials / 420
        10.2.2 Tensiometer Measurement Principles / 421
  10.3 Axis Translation Techniques / 424
        10.3.1 Null Tests and Pore Water Extraction Tests / 424
        10.3.2 Pressure Plates / 425
        10.3.3 Tempe Pressure Cells / 427
  10.4 Electrical/Thermal Conductivity Sensors / 429
  10.5 Humidity Measurement Techniques / 431
        10.5.1 Total Suction and Relative Humidity / 431

10.5.2 Thermocouple Psychrometers / 432
10.5.3 Chilled-Mirror Hygrometers / 438
10.5.4 Polymer Resistance/Capacitance Sensors / 441
10.6 Humidity Control Techniques / 443
10.6.1 Isopiestic Humidity Control / 444
10.6.2 Two-Pressure Humidity Control / 445
10.7 Filter Paper Techniques / 449
10.7.1 Filter Paper Measurement Principles / 449
10.7.2 Calibration and Testing Procedures / 451
10.7.3 Accuracy, Precision, and Performance / 452
Problems / 459

11 HYDRAULIC CONDUCTIVITY MEASUREMENT 462

11.1 Overview of Measurement Techniques / 462
11.2 Steady-State Measurement Techniques / 463
11.2.1 Constant-Head Method / 463
11.2.2 Constant-Flow Method / 466
11.2.3 Centrifuge Method / 472
11.3 Transient Measurement Techniques / 476
11.3.1 Hydraulic Diffusivity / 476
11.3.2 Horizontal Infiltration Method / 477
11.3.3 Outflow Methods / 480
11.3.4 Instantaneous Profile Methods / 484
Problems / 493

12 SUCTION AND HYDRAULIC CONDUCTIVITY MODELS 494

12.1 Soil-Water Characteristic Curve Models / 494
12.1.1 SWCC Modeling Parameters / 495
12.1.2 Brooks and Corey (BC) Model / 497
12.1.3 van Genuchten (VG) Model / 499
12.1.4 Fredlund and Xing (FX) Model / 505
12.2 Hydraulic Conductivity Models / 506
12.2.1 Empirical and Macroscopic Models / 509
12.2.2 Statistical Models / 516
Problems / 527

REFERENCES 531

INDEX 547

FOREWORD

Although a significant portion, if not the majority, of conditions encountered in geotechnical engineering practice involves unsaturated soils, the traditional analysis and design approach has been to assume the limiting conditions represented by either completely dry or completely saturated soils. The primary motivation for this assumption is that measuring the properties of soils containing only one fluid phase (i.e., either air or water) is vastly easier than that of soils containing two fluid phases (i.e., both air and water). The primary justification for the assumption is that the approach usually is conservative. For example, the shear strength of a water-saturated soil is lower than the shear strength of the same soil at the same void ratio under unsaturated conditions. However, several considerations within the past decade or so warrant a reassessment of this approach.

First, the assumption of saturated soil conditions is simply not appropriate in some applications, such as in evaluating the heave of foundations on swelling or expansive soils. Second, advances in technology continue to improve our ability to measure, characterize, and predict the properties, behavior, and performance of unsaturated soils. Third, the ever-increasing costs associated with construction make the continued reliance on conservatism less economically appealing. As a result, the motivation for applying the principles of unsaturated soil mechanics to geotechnical engineering problems where unsaturated soil conditions prevail is increasing.

Unfortunately, education and training of practitioners in the area of unsaturated soil mechanics currently is limited. This limitation is due, in part, to the lack of instructors educated in the area of unsaturated soil mechanics, the paucity of formal courses that are being offered in the area of unsaturated soil mechanics, and the dearth of formal textbooks emphasizing the principles of unsaturated soil mechanics.

Unsaturated Soil Mechanics has been written largely in response to both the increasing demand for geotechnical engineers who are knowledgeable in unsaturated soil mechanics and the current limitations in research and education in unsaturated soil mechanics. In writing *Unsaturated Soil Mechanics*, the authors have focused the presentation of the material on principles rather than applications because a fundamental knowledge based on principles is more likely to be retained and is more useful in terms of the depth and breadth of applications that subsequently can be addressed. The book offers a critical assessment of the state of the art with respect to the stress in and strength of unsaturated soils. Both the microscopic physical basis and the macroscopic thermodynamic framework for water retention and the state of stress in unsaturated soils are covered. The author's comprehensive treatment of measurement and modeling techniques not only enhances an understanding of the principles but also represents a valuable resource for future consultation. The overall result is that *Unsaturated Soil Mechanics* represents a thorough and comprehensive treatment of the subject that is written clearly and effectively and should remain a valuable textbook and reference source for many years to come.

CHARLES D. SHACKELFORD

Professor, Department of Civil Engineering
Colorado State University
Fort Collins, Colorado

PREFACE

The principal aim of this book is to provide a thorough grounding in unsaturated soil mechanics principles from three fundamental perspectives: thermodynamics, mechanics, and hydrology. The book is written to guide a first course on the subject and is primarily intended for undergraduate seniors, graduate students, and researchers with backgrounds in the more general fields of geotechnical engineering, soil science, environmental engineering, and groundwater hydrology.

In formulating this book, we have maintained the opinion that a first course in any branch of mechanics should emphasize the fundamental principles that govern the phenomena of interest. A principles-based approach to learning is most beneficial to the general reader and is particularly appropriate for the subject of unsaturated soil mechanics as it remains a young, dynamic, and rapidly emerging field of research and practice. Our general viewpoint towards the pursuit of understanding is reflected by Thomas Henry Huxley's (1825–1895) statement: "The known is finite, the unknown infinite; intellectually we stand on an islet in the midst of an illimitable ocean of inexplicability. Our business in every generation is to reclaim a little more land." We hope that this book will provide the necessary background and motivation for those who desire to explore and reclaim the ocean of unsaturated soil mechanics problems that nature and society continue to present.

A comprehensive introductory account of unsaturated soil mechanics is presented in Chapter 1 to provide readers with a road map for the remainder of the book. This includes a general introduction to unsaturated soil phenomena (Section 1.1), a formulation for the scope of the book (Section 1.2), a discussion of the role of unsaturated soil mechanics in nature and engineering practice (Section 1.3), a discussion of some essential differences between unsaturated soil mechanics and classical (saturated) soil mechanics (Section 1.4), an introduction to the state and material variables and constitutive laws forming the language of unsaturated soil mechanics (Section 1.5), and an introduction to suction and pore water potential concepts for unsaturated soil (Section 1.6).

The remainder of the book is presented as four progressive and interrelated parts. Part I examines the *fundamental principles* applicable to unsaturated soil mechanics. Parts II and III illustrate application of these principles to *stress and flow phenomena* in unsaturated soil, respectively. Finally, Part IV describes, illustrates, and evaluates the major *measurement and modeling* techniques used to quantify the state and material variables required to describe these stress and flow phenomena.

In formulating the first three parts of the text, we offer a perspective that unites the microscopic physical basis and the macroscopic thermodynamic framework for pore water retention and the state of stress in unsaturated soil. Two constitutive relationships are needed to describe unsaturated flow phenomena, namely, the soil-water characteristic curve and the hydraulic conductivity characteristic curve. For unsaturated stress phenomena, we contend that an additional relationship referred to as the suction stress characteristic curve is required.

The materials covered in this book have been an outgrowth of unsaturated soil mechanics courses taught at the Colorado School of Mines and University of Missouri–Columbia for graduating seniors and graduate students over the past four years. The book contains sufficient material for a one-semester, laboratory-supplemented course tailored along either a geomechanics or geoenvironmental track. Problems are provided at the end of each chapter with solutions available from the publisher's web site at www.wiley.com.

While many colleagues have been helpful in making the book possible in its present form, any error, bias, or inaccuracy remains ours. We are grateful to the following people who generously provided insightful reviews for at least one chapter: Jiny Carrera, Mandar M. Dewoolkar, Susan Eustes, Shemin Ge, Jonathan W. Godt, D. Vaughan Griffiths, Laureano R. Hoyos, Jr., Nasser Khalili, K.K. (Muralee) Muraleetharan, Harold W. Olsen, Paul M. Santi, Charles D. Shackelford, Radhey S. Sharma, Alexandra Wayllace, and Changfu Wei.

NING LU
WILLIAM J. LIKOS

SYMBOLS

Symbol	Description	Units
A	Hamaker's constant	N · m
A	area; cross-sectional area	m²
A'	projection of cross-sectional area	m²
A_n	atmospheric amplitude constant	—
a	SWCC modeling constant	—
a	shear strength parameter	kPa
B	footing width	m
B_2, B_3	viral coefficient for osmotic pressure	—
C	molar concentration of solute	mol/L or mol/m³
$C(h)$	specific moisture capacity as function of head	1/m
C_n	air pressure amplitude	kPa
C_r	SWCC modeling constant	Dimensionless
$C(\psi)$	specific moisture capacity as function of suction	1/kPa
$C(\psi)$	SWCC modeling correction function	Dimensionless
c	cohesion	kPa
c'	effective cohesion	kPa
c''	capillary cohesion	kPa
D	diameter of sphere	m
D	footing depth	m
D	electric displacement	C/m²
D	diffusivity	m²/s

D_v	diffusion coefficient for water vapor	m²/s
D_0	diffusion coefficient for free air	m²/s
D_{10}	10% finer particle diameter	m
D_{50}	50% finer particle diameter	m
d	thickness of boundary layer at air-water interface	m
d	diameter of capillary tube	m
d	representative pore size; pore diameter	m
d_{sc}	maximum pore diameter in simple cubic packing	m
d_{th}	maximum pore diameter in tetrahedral close packing	m
E	free energy per unit mass	N · m/kg
E	Young's modulus	kPa
e	void ratio	Dimensionless
e_{max}	void ratio in loosest state	Dimensionless
e_{min}	void ratio in densest state	Dimensionless
F	resultant force; reaction force	N
f	van der Waals potential factor	Dimensionless
G	dimensionless deformability variable for earth pressure	Dimensionless
G_s	specific gravity of soil solids	Dimensionless
g	gravimetric acceleration	m/s²
H_i	mass coefficient of solubility for gaseous species i	kg/kg
H_v	absolute humidity of solution	kg/m³
$H_{v,sat}$	absolute humidity of free water	kg/m³
h	suction head (absolute value of matric suction head)	m
h_a	air-entry head	m
h_{avg}	average suction head	m
h_c	maximum height of capillary rise	m
h_e	elevation head	m
h_i	suction head at wetting front	m
h_i	volumetric coefficient of solubility for gaseous species i	L/L
h_m	matric suction head (negative suction head)	m
h_o	osmotic suction head	m
h_p	pressure head	m
h_t	total head	m
h_0	suction head behind wetting front	m
i	hydraulic gradient	Dimensionless
i, j, m, s	series indices	Dimensionless

Symbol	Description	Units
K	intrinsic permeability	m^2
K	normalized unsaturated hydraulic conductivity	Dimensionless
K_a	coefficient of active earth pressure	Dimensionless
K_{au}	coefficient of active earth pressure, unsaturated condition	Dimensionless
K_{H_i}	Henry's law constant for gaseous species i	mol/L · bar
K_p	coefficient of passive earth pressure	Dimensionless
K_{pu}	coefficient of passive earth pressure, unsaturated condition	Dimensionless
K_0	coefficient of earth pressure at rest	Dimensionless
k	hydraulic conductivity	m/s
k_a	hydraulic conductivity of air	m/s
k_{ra}	relative hydraulic conductivity of air	m/s
k_{rw}	relative hydraulic conductivity of water	m/s
k_s, k_{sw}	saturated hydraulic conductivity	m/s
k_{sa}	saturated air conductivity	m/s
k_w	hydraulic conductivity of water	m/s
k_x, k_y, k_z	hydraulic conductivity in the x, y, and z directions	m/s
k_ψ	unsaturated hydraulic conductivity	m/s
L	diversion width for capillary barrier	m
L	length of soil specimen	m
M	shear strength parameter	Dimensionless
M	molar mass of solute	mol/L
M_i	mass of species i	kg
m	series index	Dimensionless
N	index variable	Dimensionless
N_A	Avogadro's number, 6.02×10^{23}	mol^{-1}
N_c, N_q, N_γ	bearing capacity parameters	Dimensionless
n	porosity	%
n	SWCC modeling constant	Dimensionless
n	series index	Dimensionless
n_a	air-filled porosity	%
n_i	molar quantity of gaseous species i	mol
p	air pressure squared	kPa^2
p	mean stress	kPa
pF	suction unit based on logarithm of head in cm of water	Dimensionless
p'	mean effective stress	kPa
Q	dimensionless flow variable	Dimensionless
Q	diversion capacity for capillary barrier	m^2/s

Q	infiltration displacement	m
Q	volumetric flow rate	m³/s
q	total fluid flux rate	m³/s
q	fluid discharge velocity	m/s
R	universal gas constant	J/mol·K
R	radius of meniscus curvature	m
R	radius of soil particle	m
R_d	dry atmospheric constant	m/K
R_m	mean meniscus curvature	m
R_v	moist atmospheric constant	m/K
R_1, R_2	principal radii of air-water interface	m
r	radius of capillary tube	m
r	equivalent pore radius	m
r_k	Kelvin pore radius	m
r_p	pore radius	m
r_1, r_2	radii of air-water interface defining toroidal water lens	m
S	degree of saturation	%
S	specific surface area	m²/kg
S_e	effective degree of saturation	%
S_r	residual degree of saturation	%
S_s	specific storage	1/m
s	sorptivity	m/s^{1/2}
T	absolute temperature	K
T	dimensionless time for capillary rise	Dimensionless
T_d	dew-point temperature	K
T_s	surface tension	N/m
T_v	virtual temperature	K
T_0	reference temperature	K
t	time	s
t	thickness of water film	m
u	pore water pressure	kPa
u_a	pore air pressure; air pressure	kPa
u_b	air entry (bubbling) pressure	kPa
u_d	dry air pressure	kPa
u_e	transition suction between saturated and unsaturated states	kPa
u_g	gauge pressure	kPa
u_i	partial pressure of gaseous species i	kPa
u_{sat}	saturated vapor pressure	kPa
u_v	partial vapor pressure	kPa
$u_{v,sat}$	saturated vapor pressure	kPa
u_{v0}	saturated vapor pressure	kPa
u_w	pore water pressure; water pressure	kPa
u_x	absolute pressure of phase x	kPa

u_y	absolute pressure of phase y	kPa
$(u_a - u_w)$	matric suction	kPa
$(u_a - u_w)_b$	air-entry suction	kPa
V_i	volume of species i	m³
V_t	total volume	m³
V_v	void volume	m³
v	kinematic viscosity	m²/s
v	discharge velocity	m/s
v_i	partial volume of gaseous species i	L
w	gravimetric water content	%
w	weight of pore water	kg
w_{sat}	gravimetric water content at 100% saturation	%
w_{sc}	gravimetric water content in simple cubic packing	%
w_{th}	gravimetric water content in tetrahedral close packing	%
w_{fp}	filter paper water content	%
X	correction factor for coefficient of air solubility in water	Dimensionless
x, y, z	Cartesian coordinate directions	m
Z	dimensionless distance	Dimensionless
Z_c	cracking depth	m
α	principal stress direction	deg
α	contact angle	deg
α	pore size distribution parameter; SWCC modeling parameter	1/kPa
α_d	drying contact angle	deg
α_s	bulk compressibility of soil	m²/N
α_w	wetting contact angle	deg
β	pore size distribution parameter; SWCC modeling parameter	1/m
β	air diffusion parameter	Dimensionless
β_w	compressibility of water	m²/N
Γ	shape function for van der Waals attraction	Dimensionless
γ	bulk (total) unit weight	kN/m³
γ_w	unit weight of water	kN/m³
δ	increment or change	Dimensionless
ε	partial dielectric constant	Dimensionless
ε	initial phase angle	rad
ε	strain	Dimensionless
ε	porosity	%
$\varepsilon_x, \varepsilon_y, \varepsilon_z$	strain components in x, y, and z directions	Dimensionless

η	SWCC modeling parameter	Dimensionless
η	vapor diffusion enhancement factor	Dimensionless
θ^*	normalized volumetric water content	%
θ	filling angle	deg
θ	volumetric water content	%
θ_r	residual volumetric water content	%
θ_s	saturated volumetric water content	%
θ_w	volumetric water content	%
κ	fitting parameter for effective stress parameter	Dimensionless
λ	latent heat of vaporization	J/kg
λ	standard atmospheric lapse rate	K/km
λ	Boltzmann transformation variable	Dimensionless
λ	SWCC modeling parameter	Dimensionless
μ	dynamic viscosity	kg/m · s
μ	chemical potential	J/kg or J/mol
μ	Poisson's ratio	Dimensionless
μ_c	chemical potential due to interfacial curvature	J/kg or J/mol
μ_{da}	chemical potential of dry air	J/kg
μ_f	chemical potential due to van der Waals attraction	J/kg or J/mol
μ_i	chemical potential of species i	J/kg or J/mol
μ_o	chemical potential due to solute concentration	J/kg or J/mol
μ_t	total chemical potential	J/kg or J/mol
μ_v	chemical potential of water vapor	J/kg
μ_0	chemical potential of reference state	J/kg
ν_{da}	partial molar volume of dry air	m³/mol
ν_v	partial molar volume of water vapor	m³/mol
ν_w	partial molar volume of water	m³/mol
π	osmotic pressure	kPa
ρ	radius of air-water interface	m
ρ_a	density of air	kg/m³
$\rho_{a,moist}$	density of moist air	kg/m³
ρ_{a0}	initial air density	kg/m³
ρ_d	density of dry air	kg/m³
ρ_s	density of soil solids	kg/m³
ρ_v	density of water vapor (absolute humidity)	kg/m³
ρ_w	density of water	kg/m³
σ	total stress	kPa
σ'	effective stress	kPa
σ_c	stress due to capillary cohesion	kPa
σ_h	total horizontal stress	kPa

Symbol	Description	Units
σ_n	total normal stress	kPa
σ'_n	effective normal stress	kPa
σ_v	total vertical stress	kPa
σ_1	major principal stress	kPa
σ_2	intermediate principal stress	kPa
σ_3	minor principal stress	kPa
$\sigma - u_a$	net normal stress	kPa
$(\sigma_f - u_a)_f$	net normal stress on failure plane	kPa
τ	shear stress	kPa
τ	tortuosity factor	Dimensionless
τ	diameter of adsorbed water molecule	m
τ_f	shear stress at failure	kPa
ϕ	angle of dip for capillary barrier	deg
ϕ^b	angle of internal friction with respect to matric suction	deg
ϕ'	effective angle of internal friction	deg
χ	effective stress parameter	Dimensionless
$\chi(u_a - u_w)$	suction stress	kPa
χ_f	effective stress parameter at failure	Dimensionless
ψ	suction pressure	kPa
ψ_{aev}	air-entry pressure	kPa
ψ_b	air-entry (bubbling) pressure	kPa
ψ_m	matric suction	kPa
ψ_o	osmotic suction	kPa
ψ_t	total suction	kPa
ψ_0	matric suction beyond wetting front	kPa
ω_i	molecular mass of species i	kg/mol
ω_a	molecular mass of air	kg/mol
ω_d	molecular mass of dry air	kg/mol
ω_{da}	molecular mass of dry air	kg/mol
ω_w, ω_v	molecular mass of water or water vapor	kg/mol
ω	capillary barrier efficiency	Dimensionless
ω	angular velocity	rad/s
IPM	Instantaneous Profile Method	
LL	liquid limit	%
PL	plastic limit	%
PI	plasticity index	%
Re	Reynolds' number	Dimensionless
RH	relative humidity	%
SWCC	soil-water characteristic curve	
SC	simple cubic packing order	
TH	tetrahedral close packing order	
USCS	Unified Soil Classification System	

INTRODUCTION

CHAPTER 1

STATE OF UNSATURATED SOIL

1.1 UNSATURATED SOIL PHENOMENA

1.1.1 Definition of Unsaturated Soil Mechanics

To provide and agree upon a precise definition of unsaturated soil mechanics is an academic challenge in itself. Perhaps one can draw some areas and boundaries by revisiting the classical definition of soil mechanics posed by Karl Terzaghi some 60 years ago. In his seminal book of 1943, *Theoretical Soil Mechanics,* Terzaghi defined soil mechanics as "the application of the laws of mechanics and hydraulics to engineering problems dealing with sediments and other unconsolidated accumulations of solid particles produced by the mechanical and chemical disintegration of rocks, regardless of whether or not they contain an admixture of organic constituents." In drawing this silhouette of soil mechanics, Terzaghi refers to three basic requirements: (1) earthen materials, (2) the principles of mechanics and hydraulics, and (3) engineering problems.

The emerging appreciation of unsaturated soil in geotechnical engineering practice and education requires refinement of Terzaghi's basic definition. The earthen materials dealt with in problems of unsaturated soil mechanics are arguably the same as in Terzaghi's soil mechanics, referred to as "soils," but under a very specific "unsaturated" condition. The qualifier "unsaturated" bears the same meaning as its alternative "partially saturated" and simply indicates that the degree of pore water saturation is any value less than unity or, more specifically, that a third phase of matter is introduced into the two-phase, saturated soil system. In the modern educational and professional geotechnical engineering environment, where the emphasis has historically been limited to the arena of saturated cohesive materials and completely dry or completely saturated cohesionless materials, the "unsaturated" qualifier is indeed significant.

In dealing with unsaturated soil, one requires not only the principles of mechanics and hydraulics but also of fundamental interfacial physics. Physics in this regard refers primarily to the thermodynamic principles describing equilibrium among gas, solid, and liquid phases, the transition of matter from one phase to another, and the adsorption or desorption of one phase of matter onto or from an adjacent phase of different matter. The forces and energies associated with these multiphase interactions by their very nature separate unsaturated soil behavior from saturated soil behavior. In many practical problems, where the hydrologic and stress-strain behavior of natural or engineered systems comprised of soil is strongly influenced by the presence, absence, or changes in these interfacial interactions, the traditional saturated soil mechanics framework often fails to satisfactorily describe or predict the behavior of the system.

Terzaghi's reference to engineering problems was developed in the wake of a period of great uncertainty in the basic understanding of soil behavior. His formalization of soil mechanics provided a rational basis for tackling many of the pressing engineering problems of the day, most notably bearing capacity, consolidation and settlement, slope stability, lateral earth pressure, and seepage-related problems. In addition to these traditional geotechnical engineering problems, the practical problems of interest today might also include geo-environmental, seismic, land reclamation, and other challenges that have come to light over the past 30 years or so. These emerging problems have created important subdisciplines within the more general field of geotechnical engineering, which often benefit from a thorough understanding of the physical and thermodynamic principles governing unsaturated soil behavior.

Extending Terzaghi's classical definition, therefore, unsaturated soil mechanics might be defined as "the application of the laws of mechanics, hydraulics, and interfacial physics to engineering problems dealing with partially saturated soils." The spirit of this definition and the laws, concepts, and problems that characterize it will be addressed throughout this book. Of course, as new technical discoveries are made, as new and unforeseen types of problems emerge, and as the once distinct boundaries between the traditional engineering and science disciplines continue to blur, there is no doubt that this definition may one day also require refinement.

1.1.2 Interdisciplinary Nature of Unsaturated Soil Mechanics

The history of unsaturated soil mechanics is embedded in the history of hydrology, soil mechanics, and soil physics. Engineering problems involving unsaturated soil span numerous subdisciplines and practices within the general field of civil engineering. Hydrologists, for example, have long recognized that modeling of regional or local surface water and groundwater systems and cycles must consider infiltration, evaporation, and transpiration processes occurring in the near-surface unsaturated soil zone. Quantitative evaluation of moisture flux at the atmosphere-subsurface boundary requires not only knowledge of the relevant soil and pore water properties but also the predominant environmental conditions at the soil-atmosphere interface. Unsaturated soil often comprises cover or barrier materials for landfills and hazardous waste storage facilities of interest to the geo-environmental community. Contaminant transport and leaching processes are often strictly unsaturated fluid transport phenomena, occurring in many cases as multiphase transport problems. As national and international policy with regard to the health of the natural environment is becoming increasingly more regulated, recognition of these types of geo-enviromental issues and development of solutions from an unsaturated soil mechanics framework is becoming more and more common.

Many of the more traditional geotechnical engineering problems also fall wholly or partly into the category of unsaturated soil mechanics problems. Compaction, for example, a classical application involving unsaturated soil, has been routine practice for improving the mechanical and hydraulic properties of soil since far before the formation of civil engineering as a formal discipline in the mid-nineteenth century. Compacted soil comprising the many earthworks constructed all over the world is most appropriately considered from an unsaturated soils framework. It has long been recognized that expansive soils pose a severe threat to civil engineering infrastructure such as roads, housing, and transportation facilities nationally and internationally. Expansive soil formations in the United States alone are responsible for billions of dollars in damage costs each year, an amount exceeding that of all other natural hazards combined, including earthquakes, floods, fires, and tornados (Jones and Holtz, 1973). Expansive soils have been the subject, if not the driving force, of unsaturated soil research since the early stages in the formulation of unsaturated soil mechanics principles. Collapsing soils also pose a significant threat in many areas of the world. These problematic soils, which are typified by the massive loess deposits of the central United States, are marked by a structurally sensitive fabric weakly cemented by a small clay fraction. Upon wetting, usually occurring either as a sudden precipitation event or gradual process associated with urbanization and development, the cementation bonds are weakened and the inititally loose fabric collapses and densifies, often resulting in dramatic and damaging settlement. Any fundamental approach to mitigating collapsing soil hazards requires insight into the role of pore water interactions on the microscopic scale of the solid-liquid-air interface, a hallmark of unsaturated soil mechanics.

Reconsideration of the traditional saturated soil mechanics approach in light of these types of problems began to emerge during the late 1970s and continues today. In the authors' opinion, the soil mechanics community is far from achieving a comprehensive and satisfactory framework for approaching these and other unsaturated soil mechanics problems, but new insights and technical advances are continuously being made.

1.1.3 Classification of Unsaturated Soil Phenomena

While the development of theory and techniques in unsaturated soil mechanics requires principles drawn from mechanics, hydraulics, and interfacial physics, it is convenient to classify the various geotechnical engineering problems involving unsaturated soil into three general phenomena, specifically, flow phenomena, stress phenomena, and deformation phenomena. It should be noted, however, that generalization in this manner is mainly for understanding purposes and for convenience of presenting the principles, not to set up boundaries among different geotechnical problems. The majority of practical engineering problems generally involve all three phenomena concurrently and in coupled fashion. An effective theory describing the deformation behavior of expansive soil, for example, could well require application of the principles of stress, strain, and flow in highly deformable porous media.

Flow Phenomena *Flow phenomena* require mainly the application of hydraulics and interfacial physics principles. One well-known example falling into this class is capillary flow. The search for the driving force for capillary flow had once been the subject of research for many years. As early as the 1900s, Buckingham (1907) systematically studied capillary rise and drainage in laboratory soil columns such as that illustrated in Fig. 1.1. Early data provided evidence of the important effects of soil type, grain size, and pore size properties on capillary rise and pore water retention in unsaturated soil. As part of this early work, the terms *capillary potential* and *capillary conductivity* were introduced as the driving force and controlling material variable, respectively, for capillary fluid flow. Later, others recalled the more

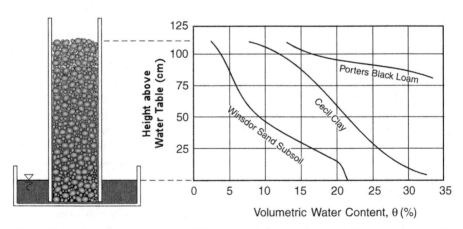

Figure 1.1 Capillary rise and equilibrium moisture content distribution in vertically oriented soil column (data from Buckingham, 1907). The curves shown, which describe the relationship between suction head and moisture content, are commonly called soil-water characteristic curves.

general term *chemical potential,* to include components of the pore water potential resulting from dissolved chemical species, gravity, capillarity, and short-range physicochemical effects occurring at the solid-liquid phase interface (e.g., Gardner and Widstoe, 1921; Richards, 1928; Russell, 1942; Edlefsen and Anderson, 1943). The chemical potential, or free energy, concept for soil pore water has been generalized by Sposito (1981) and others to include the mass of all three phases (gas, solid, and liquid), together with temperature and pressure as independent state variables. As a result, many seepage-related problems in unsaturated soil mechanics may be effectively treated through the application of thermodynamic potential theory with little or no involvement of solid mechanics.

Stress Phenomena Problems requiring consideration of both mechanical and chemical equilibrium are classified as *stress phenomena.* These include traditional geotechnical engineering problems such as lateral earth pressure, bearing capacity, and slope stability analysis. For each of these problems, the strength of the soil at its limit state is the primary concern. Analysis of the stress distribution within the soil mass and the corresponding bulk strength becomes critically important. Limit analysis developed extensively since the 1930s for saturated soil applications formed the basis for solving most of these types of problems. Developing elastoplastic theories for soil became the focus of much of the geomechanics research activity during the 1970s and 1980s. Powerful numerical methods to solve the governing partial differential equations for stress equilibrium under static or dynamic conditions have been developed and applied to many difficult foundation problems in the past 20 years or so.

It has become clear in recent years that improved solutions of many stress-related geotechnical engineering problems require not only sustained activities along the continuum-based solid mechanics approach but also new theories along a microscopic discontinuous approach for describing effective stress under multiphase conditions. Terzaghi's effective stress, which is the cornerstone of soil mechanics under saturated conditions, becomes either ineffective or inappropriate for fully describing the stress distributions or failure conditions in unsaturated soil. It has been recognized that theories for describing the states of stress and failure in unsaturated soil require consideration of the thermodynamic properties of the pore water in terms of soil suction, material variables such as grain size and grain size distribution, state variables such as the degree of saturation, and the consequent interparticle forces such as suction-induced effective stress or suction stress.

Deformation Phenomena Physical processes characterized by large deformations or strains are classified as *deformation phenomena.* In unsaturated soils, these deformations are very often caused or governed by changes in the moisture condition of the soil. Important deformation phenomena include compaction, multiphase consolidation and compressibility, and collapsing soil

8 STATE OF UNSATURATED SOIL

behavior. Arguably, the most notorious unsaturated soil deformation phenomenon is that of swelling or shrinking (i.e., expansive) soil. Figure 1.2, for example, illustrates several important mechanisms commonly occurring in near-surface deposits of expansive soil. Many of these mechanisms, such as heave or subsidence of the ground surface, swelling pressure generation under pavements or foundations, and tension cracking, fall into the general category of deformation phenomena. Others shown in the figure, such as infiltration, evaporation, and the corresponding seasonal fluctuation in the subsurface moisture profile, fall into the general category of unsaturated flow phenomena. The inherent coupling between volume change, pressure generation, and moisture transport in expansive soil demonstrates the importance of the combined roles of deformation, stress, and fluid flow phenomena in this and numerous other types of unsaturated soil mechanics problems.

1.2 SCOPE AND ORGANIZATION OF BOOK

1.2.1 Chapter Structure

Unsaturated Soil Mechanics is organized into four divisible but interrelated parts. The intent of this separation is to provide the reader with a format

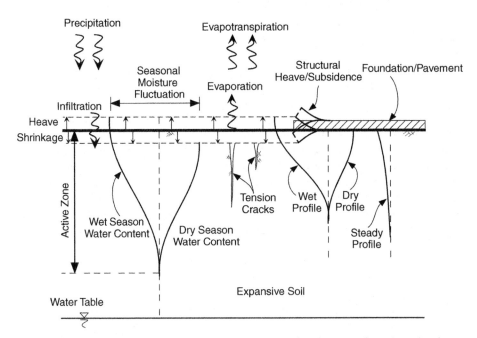

Figure 1.2 Deformation and fluid flow phenomena in a near-surface deposit of unsaturated expansive soil.

where particular concepts, theories, phenomena, or practical applications of interest may be directly accessed in a focused and concise manner. Each chapter concludes with a series of qualitative and/or quantitative problems. The 12 chapters of the book are organized as follows:

Introduction: Chapter 1
Part I: Fundamental Principles, Chapters 2 to 4
Part II: Stress Phenomena, Chapters 5 to 7
Part III: Flow Phenomena, Chapters 8 and 9
Part IV: Material Variable Measurement and Modeling, Chapters 10 to 12

Chapter 1, State of Unsaturated Soil, provides a general introduction to unsaturated soil mechanics. The relevant state variables, material variables, and constitutive laws for describing flow, stress, and deformation phenomena in three-phase unsaturated soil systems are introduced. The important differences between saturated and unsaturated soil systems in terms of subsurface moisture, pore pressure, and stress profiles are described. Common types of practical engineering applications that warrant an unsaturated soil mechanics approach are introduced. The important role of unsaturated soil in terms of naturally occurring phenomena such as the hydrologic cycle, global climatic changes, and soil formation is described. Finally, the important concepts of pore water potential, soil suction, and the constitutive relationship between soil suction and water content, the soil-water characteristic curve, are introduced.

Part I, Fundamental Principles, provides the necessary background for the remainder of the book. Chapter 2, Material Variables, introduces the relevant physical properties of air, water, and water vapor and evaluates their dependency on the state variables that are used to describe multiphase unsaturated soil systems. Relative humidity and surface tension are introduced and described with respect to their roles in the behavior and analysis of unsaturated soil systems. Cavitation phenomena are systematically described. Chapter 3, Interfacial Equilibrium, describes several fundamental concepts within the general realm of interfacial physics. Mechanical and chemical equilibrium for air-water-solid interfaces are described with the introduction of Kelvin's law. Associated interfacial phenomena including vapor pressure lowering, capillary condensation, and the solubility of air in water are described and illustrated through a series of thought experiments and quantitative examples. Finally, the soil-water characteristic curve is introduced from a micromechanical perspective by considering mechanical and chemical equilibrium for idealized systems of unsaturated soil grains. Chapter 4, Capillarity, introduces the Young-Laplace equation for describing equilibrium at an air-water interface, the height and rate of capillary rise, and the estimation of pore size distribution using capillary theory. The concept of suction stress is formulated from a micromechanical perspective to serve as a link between the preceding interfacial equilibrium concepts and the associated interparticle stresses in unsaturated soil systems.

Part II, Stress Phenomena, contains three chapters. Chapter 5, State of Stress, complements the interfacial equilibrium concepts introduced in Chapter 4 by providing a derivation of effective stress among idealized unsaturated soil particles. Mechanisms for hysteresis in the soil-water characteristic curve and suction stress characteristic curve are introduced and evaluated. Tensor notation and graphical representation for the independent stress state variable approach and the effective stress approach to describing the state of stress in unsaturated soil are introduced and explained using example problems. The concept of axis translation for controlling the stress state variables relevant to unsaturated soil is presented. Chapter 6, Shear Strength, describes several alternative theories for interpreting and analyzing shear strength in unsaturated soil. The extended Mohr-Coulomb failure criterion, the shear strength parameters describing it, and its advantages and limitations are introduced. Bishop's effective stress parameter χ and its role in effective stress for unsaturated soil is described. Finally, a unified framework for interpreting and measuring shear strength characteristics in unsaturated soil is suggested. Chapter 7, Suction and Earth Pressure Profiles, includes theoretical development of subsurface suction stress and water content profiles under steady-state infiltration, hydrostatic, and evaporation conditions. Corresponding lateral earth pressure profiles are derived for conditions at rest and under active and passive conditions. These new theories serve as an instructional vehicle to provide insight into the fundamental differences between the states of stress in saturated and unsaturated soil.

Part III, Flow Phenomena, is divided into Chapter 8, Steady Flows, and Chapter 9, Transient Flows. Together, these chapters provide an introduction to the governing principles and solutions for both liquid and gas flow in unsaturated soil systems. Governing flow equations are solved analytically and numerically and illustrated graphically through simple one-dimensional example problems. Capillary barriers for geo-environmental applications are described along with vapor phase transport and diffusion processes. Practical examples involving transient pore airflow by barometric pumping are provided to highlight the important impact of variations in the governing state variables (e.g., temperature and pressure) on pore fluid transport processes in unsaturated soil.

Part IV, Material Variable Measurement and Modeling, provides the practicing and research community with a reference source pertaining to suction and hydraulic conductivity measurement and modeling alternatives. Chapter 10, Suction Measurement, describes the general principles, technical aspects, and performance of many of the more common suction and soil-water characteristic curve measurement techniques. Chapter 11, Hydraulic Conductivity Measurement, describes several common steady-state and transient techniques for measuring the unsaturated hydraulic conductivity function. Finally, chapter 12, Suction and Hydraulic Conductivity Models, describes numerous methodologies by which the soil-water characteristic curve and hydraulic conductivity function may be either modeled, estimated, or predicted from more readily available material properties.

1.2.2 Geomechanics and Geo-environmental Tracks

Unsaturated Soil Mechanics contains sufficient material for a one-semester course tailored to follow either a geomechanics or geo-environmental track. Figure 1.3 illustrates two suggested paths through the chapters of the book corresponding to a geomechanics track, which emphasizes the stress and

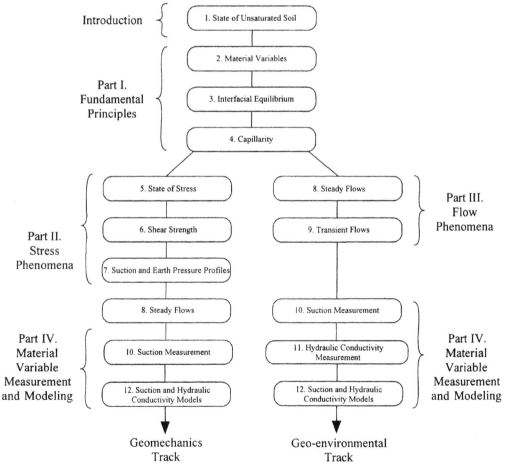

Figure 1.3 Recommended chapter sequences for geomechanics and geo-environmental learning tracks.

strength concepts described in Part II, and a geo-environmental track, which emphasizes the hydrology concepts described in Part III. Content along each track has been included such that each may generally stand on its own.

1.3 UNSATURATED SOIL IN NATURE AND PRACTICE

1.3.1 Unsaturated Soil in Hydrologic Cycle

Figure 1.4 shows a schematic diagram of the unsaturated soil environment and its role in the natural hydrologic cycle. The steady-state position of the water table is controlled by the general topography of the system, the soil properties, and the balance achieved among the natural mechanisms that act to either add or remove water to or from the subsurface. The scale of the corresponding hydrologic cycle could be either local or regional, extending from as small as a local engineering work site to as large as the continental or global scale. Globally, the amount of water in the unsaturated zone located between the water table and the ground surface represents only a small portion of the total water involved in the hydrologic cycle (less than 0.01%). However, because the unsaturated zone forms the necessary transition between the atmosphere and larger groundwater aquifers at depth, the movement of water within this small portion of the cycle is indeed significant.

1.3.2 Global Extent of Climatic Factors

The size and extent of the near-surface unsaturated soil zone are highly sensitive to perturbations in local or regional climate. Precipitation, evaporation,

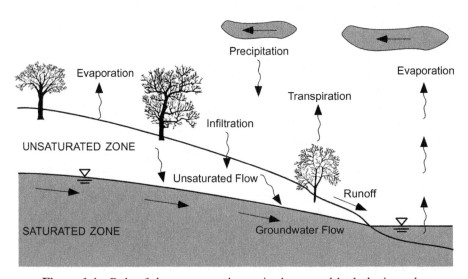

Figure 1.4 Role of the unsaturated zone in the natural hydrologic cycle.

and evapotranspiration are all important natural environmental mechanisms that act to influence the depth and extent of the unsaturated zone.

Figure 1.5 is a map delineating the estimated average annual precipitation on the global scale. The lightest zones on the figure indicate regions receiving less than 25 cm of annual precipitation. The darkest zones indicate regions receiving more than 200 cm annually. Regions receiving average annual precipitation less than 200 cm will generally have a depth to the water table greater than 200 cm. Figure 1.6 shows an estimate of net precipitation (precipitation minus evaporation) as a function of latitude, indicating that regions where evaporation exceeds precipitation are generally concentrated within 40° north and south of the equator.

Figure 1.7 is a global-scale map delineated in terms of the global humidiy index (UNESCO, 1984). The global humidity index is based on the ratio of average annual precipitation and potential evaporation (P/PET), such that hyperarid zones fall into a category where P/PET < 0.05, arid zones indicate 0.05 < P/PET < 0.2, semiarid zones indicate 0.2 < P/PET < 0.5, dry subhumid zones indicate 0.5 < P/PET < 0.65, and humid zones indicate 0.65 < P/PET. Figure 1.8 shows a similar humidity index map for North America. In the semiarid to arid regions of the western United States, the depth of the unsaturated zone may extend to as much as several hundred meters.

1.3.3 Unsaturated Zone and Soil Formation

The unsaturated soil zone plays a critical role in biological, physical, and chemical weathering processes that have occurred throughout the history of Earth. The history of soil formation is the history of the unsaturated zone. As a result of physical and chemical weathering processes largely controlled by environmental factors at the ground surface, parent rock weathers to a residual soil profile of distinct horizons and chemical composition. The evolutionary process from unweathered rock to mature soil is illustrated for a typical profile in Fig. 1.9. A simplified description of nomenclature for each soil horizon follows, based on a complete systematic description by Birkeland (1999).

> *O Horizon* Surface accumulations of mainly organic material are subdivided on the degree of decomposition as measured by the fiber content.
>
> *Oi Horizon* Least decomposed organic materials; the fiber content is greater than 40% by volume.
>
> *Oe Horizon* Intermediate degree of decomposition; the fiber content is between 17 and 40% by volume.
>
> *Oa Horizon* Most decomposed organic materials, the fiber content is less than 17% by volume.
>
> *A Horizon* Accumulations of humidified organic materials mixed with dominant mineral fraction occur at the surface or below an O horizon.

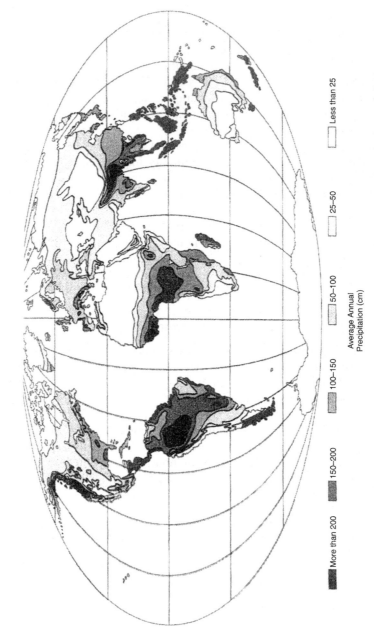

Figure 1.5 Global average annual precipitation. (from Penman, 1970; reproduced with permission).

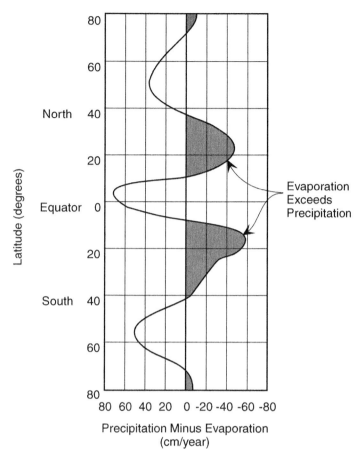

Figure 1.6 Net global precipitation and evaporation as a function of latitide. Positive values indicate latitudes where precipitation exceeds evaporation. Negative values (shaded regions) indicate latitudes where evaporation exceeds precipitation (after Peixoto and Kettani, 1973; illustration by Eric O. Mose).

E Horizon Accumulations usually underlie an O or A horizon and can be used for eluvial horizons within or between parts of B horizon, and are characterized by less organic materials and/or fewer compounds of iron and aluminum (sesquioxides) and/or less clay than the underlying horizon.

B Horizon Soil underlies an O, A, or E horizon, shows little or no evidence of original sediment or rock structure, and is recognized into several subhorizons based on the kinds of materials illuviated into them or residual concentrations of materials.

Bh Horizon Illuvial accumulation of amorphous organic materials-sesquioxide complexes that either coat grains or form sufficient coatings and pore fillings to cement the horizon.

16 STATE OF UNSATURATED SOIL

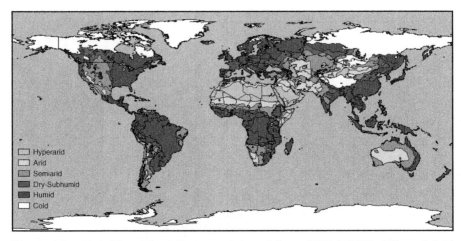

Figure 1.7 Global humidity index map (adapted from GRID/UNEP, Office of Arid Lands Studies, University of Arizona).

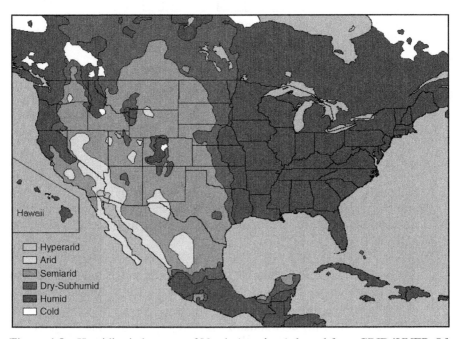

Figure 1.8 Humidity index map of North America (adapted from GRID/UNEP, Office of Arid Lands Studies, University of Arizona).

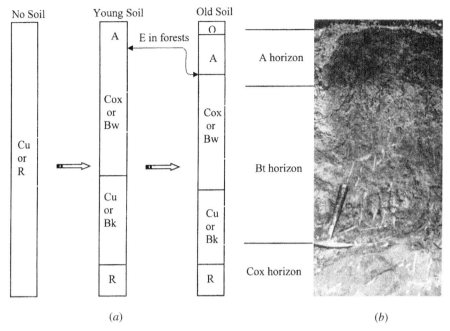

Figure 1.9 Soil profile: (*a*) typical evolution of soil profile and (*b*) soil formed on marine-terrace deposits near San Diego, California [(b) from Birkeland, 1999; adapted by permission of Oxford University Press, Inc.].

Bhs Horizon   Illuvial accumulation of amorphous organic materials-sesquioxide complexes, and sesquioxide component is significant; both color value and chroma are three or less.

Bk Horizon   Illuvial accumulation of alkaline earth carbonates, mainly calcium carbonate; properties do not meet those for the K horizon.

Bl Horizon   Illuvial concentrations primarily of silt.

Bo Horizon   Residual concentration of sesquioxides, the more soluble materials having been removed.

Bq Horizon   Accumulation of secondary silica.

Bs Horizon   Illuvial accumulation of amorphous organic materials-sesquioxide complexes if both color value and chroma are greater than three.

Bt Horizon   Accumulation of silicate clay that has either formed in situ or is illuvial.

Bw Horizon   Development of color (redder hue or higher chroma relative to C horizon) or structure, or both, with little or no apparent illuvial accumulation of material.

By Horizon   Accumulation of secondary gypsum.

Bz Horizon   Accumulation of salts more soluble than gypsum.

K Horizon   Soil is so impregnated with carbonate that its morphology is determined by the carbonate.

C Horizon   Soil lacks properties of A and B horizons, excludes R horizon but includes materials in various stages of weathering.

Cox and Cu Horizons   Oxidized C horizon for Cox and unweathered C horizon for Cu.

Cr Horizon   Weathered rock formed in place.

R Horizon   Consolidated bedrock underlying soil.

The depth of the dynamic weathering zone described above is largely controlled by environmental factors including net precipitation and temperature, as illustrated in Fig. 1.10. If, for example, the weathering process occurs at a location where the net flux of water at the ground surface is downward (i.e., precipitation exceeds evaporation), the weathering front and associated dissolved minerals will extend relatively deep below the ground surface. If, on the other hand, the net flux of water from the ground surface is upward (i.e., evaporation exceeds precipitation), the weathering front will be relatively shallow and dissolved minerals may be deposited in horizons relatively near the ground surface. Near surface deposits of calcium carbonate (caliche) common to the arid regions of the western United States and Australia are an excellent example of the latter phemonenon. The depth and rate of pore water and pore vapor movement are largely controlled by the unsaturated hydrologic characteristics of the deposit, as quantified in Part III.

1.3.4 Unsaturated Soil in Engineering Practice

For many years, unsaturated soils were either ignored in civil engineering design and construction analyses or were approached inappropriately from the traditional framework of saturated soil mechanics. Rapid advancement in our understanding of unsaturated soil behavior over the last 30 to 40 years, however, has led today's civil engineer to realize that there is now an opportunity to approach problems involving unsaturated soil on a much more rational basis. The expanding knowledge base on the fundamental principles of unsaturated soil mechanics is increasingly being incorporated into a diverse array of practical engineering problems. The following lists summarize several of the more common types of engineering problems involving predominantly unsaturated soils.

Flow-Related Problems

1. Water balance at the interface of soil and atmosphere
2. Net recharge rate to the saturated zone or aquifers
3. Design of final covers for underground waste storage and containment
4. Near-surface contaminant transport and remediation
5. Transient and steady seepage in unsaturated embankment dams

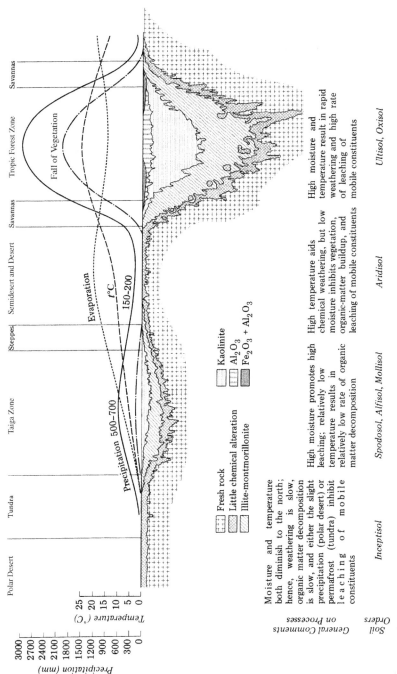

Figure 1.10 Relative depth of weathering and soil formation as related to environmental factors along a transect from the equator to the north polar region (Birkeland, 1999; reproduced by permission of Oxford University Press, Inc.).

Stress-Related Problems

1. Slope stability and land sliding under changing climatic conditions
2. Lateral earth pressure and stability of retaining structures
3. Excavation and bore hole stability
4. Bearing capacity for shallow foundations under moisture loading
5. Stress wave propagation in unsaturated soil

Deformation-Related Problems

1. Swelling and shrinkage of expansive soil
2. Desiccation cracking of clay
3. Collapsing soil
4. Consolidation and settlement of unsaturated soil
5. Soil compaction

1.4 MOISTURE, PORE PRESSURE, AND STRESS PROFILES

1.4.1 Stress in the Unsaturated State

Subsurface moisture, suction, and stress profiles depend on the soil and pore water properties as well as the prevalent environmental or atmospheric conditions. Soil type, particle size distribution, and pore size distribution all act to influence the equilibrium distribution and flow of pore water within the soil profile. Atmospheric conditions, which include relative humidity, temperature, wind speed, and precipitation, all act to influence transient changes in the flow and distribution of the subsurface pore water.

The mechanical stability of any point in the subsurface depends on the strength parameters of the soil and the state of stress at that point. In saturated soil, the state of stress can be described by total stress and pore pressure, unified under the concept of effective stress. Effective stress, which is the difference between total stress and pore pressure, is the stress experienced by soil's solid phase, or skeleton. The state of effective stress controls whether or not a given soil mass is under a state of stability or a state of failure. Soil strength is an intrinsic material property that generally depends on the soil mineralogy, particle morphology, and interparticle arrangement. Macroscopic description of these controlling factors often leads to empirical material parameters, most notably cohesion and internal friction angle. These material parameters, together with the stress state variables, define the boundaries controlling whether soils are in stable or failure conditions.

Total stress can be considered as an external stress and is due to either surcharge load or the soil's self-weight. Pore pressure in saturated soil is generally compressive and isotropic. All pore pressure in saturated soil contributes to total stress according to the effective stress principle. Pore pressure in unsaturated soil, on the other hand, is generally tensile. The contribution of pore pressure to total stress depends on the degree of saturation and pore size distribution. This contribution is not always 100%, making analysis of the state of stress in unsaturated soil far more complicated than the relatively simple case for saturated conditions. A detailed micromechanical analysis of the origin and behavior of this stress follows in Chapter 4. The following sections contain several conceptual examples to illustrate the differences in the states of stress for saturated and unsaturated soil.

1.4.2 Saturated Moisture and Stress Profiles: Conceptual Illustration

Consider a homogeneous soil layer that is initially saturated and free of surcharge loading as shown in Fig. 1.11a. The water table is at the ground surface and the soil layer is bounded by a layer of bedrock below. The total vertical stress profile within the soil layer due to self-weight is a function of depth as follows:

$$\sigma_z = \gamma z \qquad (1.1)$$

where γ is the bulk (total) unit weight of the soil and z is the depth from the ground surface. Horizontal stresses (σ_x and σ_y) may be estimated from the vertical stresses under the at-rest, or K_0, condition:

$$\sigma_x = \frac{\mu}{1-\mu}\sigma_z + \frac{1-2\mu}{1-\mu}u_w \qquad (1.2)$$

where μ is Poisson's ratio and u_w is the pore water pressure. The pore water pressure profile under both the saturated and unsaturated hydrostatic condition is as follows:

$$u_w = \gamma_w z_w \qquad (1.3)$$

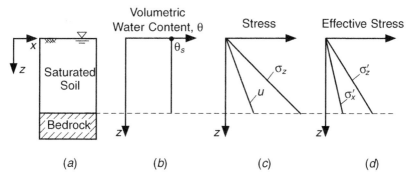

Figure 1.11 Conceptual profiles of water content, stress, and effective stress in a homogeneous, saturated soil layer.

where γ_w is the unit weight of water (9.8 kN/m³) and z_w is the distance from the water table to the point of interest and is positive for points located below the water table. The vertical effective stress under the saturated condition is

$$\sigma'_z = \sigma_z - u_w = \gamma z - \gamma_w z_w \quad (1.4a)$$

and the horizontal effective stress under the K_0 condition is

$$\sigma'_x = \frac{\mu}{1 - \mu} \sigma'_z \quad (1.4b)$$

Conceptual profiles of volumetric water content ($\theta = V_w/V_t$), total vertical stress σ_z, pore pressure u_w, and effective vertical and horizontal stress (σ'_z and σ'_x) for the saturated soil layer shown in Fig. 1.11a are plotted in Figs. 1.11b, 1.11c, and 1.11d. Because the soil is saturated, the saturated volumetric water content is a constant equal to the soil porosity, n (i.e., $\theta_s = V_w/V_t = V_v/V_t = n$). For a quantitative analysis, consider a 10-m-thick, homogeneous, saturated sand layer with Poisson's ratio equal to 0.35, bulk unit weight equal to 18.8 kN/m³, and porosity equal to 30% (Figure 1.12a). The corresponding volumetric water content profile is shown in Fig. 1.12b, the vertical total stress and pore pressure profiles are shown in Fig. 1.12c, and the vertical and horizontal effective stress profiles are shown in Fig. 1.12d. Each of these profiles is a linear function with depth.

1.4.3 Unsaturated Moisture and Stress Profiles: Conceptual Illustration

Unsaturated soils in the field are characterized by a water table located at some depth below the ground surface. If, for example, the water table in the preceding example drops 10 m to the interface of the soil layer and bedrock,

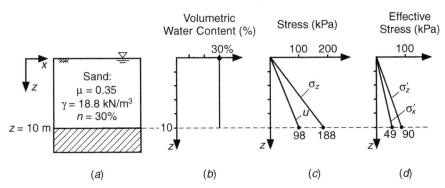

Figure 1.12 Profiles of water content, stress, and effective stress in a sandy soil layer under saturated conditions.

a varying water content profile develops, as conceptualized in Fig. 1.13. At hydrostatic equilibrium, assume that the volumetric water content varies from 5% at the ground surface to 30% at the water table (Fig. 1.13b). The water content of 30% at the water table is the saturated water content, θ_s, equal to the soil porosity. The consequent pore pressure profile is distributed linearly with depth, as shown in Fig. 1.13c. Here, tensile pore water pressure varies from -98 kPa at the ground surface to zero at the water table.

The horizontal total stress may be estimated from the vertical total stress under the K_0 condition formulated in Section 7.3:

$$\sigma_x = \frac{\mu}{1 - \mu}\sigma_z - \frac{1 - 2\mu}{1 - \mu}\chi(u_a - u_w) \tag{1.5a}$$

where χ is Bishop's effective stress parameter, u_a is the pore air pressure, and $u_a - u_w$ is matric suction. The vertical effective stress can be estimated from Bishop's effective stress defined by eq. (1.11):

$$\sigma'_z = (\sigma_z - u_a) + \chi(u_a - u_w) \tag{1.5b}$$

and the horizontal effective stress bears the same relationship as eq. (1.4b).

The vertical total stress profile also changes due to the dewatering process because the self-weight of the material decreases. Assuming the total unit weight is reduced from the original 18.8 kN/m³ to an average of 15.0 kN/m³ at all depths and the average effective stress parameter χ is 0.5, profiles for vertical total stress, pore pressure, and vertical and horizontal effective stress can be determined, as shown in Figs. 1.13c and 1.13d.

1.4.4 Illustrative Stress Analysis

Comparison of the effective stress profiles under saturated and unsaturated conditions for the preceding examples (Figs. 1.12d and 1.13d) reveals that

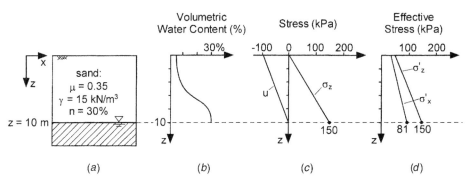

Figure 1.13 Profiles of water content, stress, and effective stress in a sandy soil layer under unsaturated conditions.

both the horizontal and vertical effective stresses increase considerably upon desaturation over the entire depth of the soil layer. The fundamental question for geotechnical engineers is: Which effective stress profile, saturated or unsaturated, is more representative? Considering the traditional Mohr-Coulomb failure criterion provides one answer to this question.

The Mohr-Coulomb failure criterion can be written in terms of cohesion c' and effective internal friction angle ϕ' in the space of shear stress at failure τ_f and normal effective stress σ'_n as follows:

$$\tau_f = c' + \sigma'_n \tan \phi' \tag{1.6}$$

For loose, uncemented sand that is either completely dry or completely saturated, the cohesion term in eq. (1.6) may be considered essentially equal to zero. For wet or moist sand, however, considerable cohesive strength may exist. This "apparent cohesion" unique to unsaturated soil arises from negative pore water pressure and surface tension effects occurring at the interface of the pore water, pore air, and soil solids among the unsaturated soil grains.

Consider a point in the sand layer from the preceding examples located 8 m from the ground surface. Assume a friction angle equal to 35° for both the saturated case (where the water table is at the surface) and the unsaturated case (where the water table is 10 m from the surface). For the saturated case (Fig. 1.14a), where $z = z_w$, and $\gamma = 18.8$ kN/m³, the vertical and horizontal effective stresses at $z = 8$ m are

$$\sigma'_z = (\gamma - \gamma_w)z = (18.8 - 9.8)(8) = 72 \text{ kPa} \tag{1.7a}$$

$$\sigma'_x = \frac{\mu}{1 - \mu} \sigma'_z = \frac{0.35}{1 - 0.35}(72) = 39 \text{ kPa}$$

For the unsaturated case (Fig. 1.14b), where $z = 8$ m, $z_w = -2$ m, and $\gamma = 15.0$ kN/m³, the effective stresses are

$$\sigma'_z = (\sigma_z - u_a) + \chi(u_a - u_w) = [(15)(8) - 0] + (0.5)[0 - (9.8)(-2)]$$

$$= 130 \text{ kPa}$$

$$\sigma'_x = \frac{\mu}{1 - \mu} \sigma'_z = \frac{0.35}{1 - 0.35}(130) = 70 \text{ kPa} \tag{1.7b}$$

If the unsaturated soil at 8 m depth experiences an apparent cohesion of 50 kPa due to capillarity, Mohr's circles for the saturated and unsaturated states of effective stress and the corresponding failure envelopes are shown on Fig. 1.14c. Note that the saturated state of stress falls relatively close to the saturated failure envelope. The unsaturated state of stress, however, becomes far more stable because the envelope shifts upward and the Mohr circle shifts towards the right.

The practical importance of accounting for the unsaturated condition in geotechnical engineering applications can be demonstrated by estimating the

1.4 MOISTURE, PORE PRESSURE, AND STRESS PROFILES   25

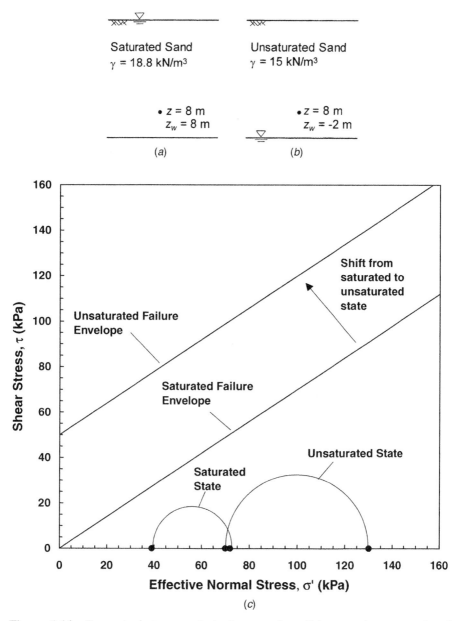

Figure 1.14 Conceptual stress analysis for a sandy soil layer under saturated and unsaturated conditions: (a) saturated soil profile, (b) unsaturated soil profile, and (c) states of stress at $z = 8$ m and Mohr-Coulomb failure envelopes.

ultimate bearing capacity of a strip footing assumed to be constructed on this soil. According to classical theory, the ultimate bearing capacity q_u is as follows:

$$q_u = c'N_c + \gamma D N_q + \tfrac{1}{2} \gamma B N_\gamma \qquad (1.8)$$

where D is the footing embedment depth, B is the footing width, and N_c, N_q, and N_γ are bearing capacity factors. For $\phi = 35°$, $N_c = 46.12$, $N_q = 33.30$, and $N_\gamma = 48.03$. For $D = 1$ m, $B = 1$ m, $\gamma = 18.8$ kN/m³, and $c' = 0$ under the saturated condition, the ultimate bearing capacity is 1078 kPa. For $\gamma = 15.0$ kN/m³ and assuming an apparent cohesion 10 kPa for the unsaturated condition, the ultimate bearing capacity increases to 1321 kPa, an increase of 32%. For an apparent cohesion 50 kPa, the capacity increases to 3166 kPa, an increase of 220%.

Is the notion that unsaturated soil is in a more stable state than saturated soil generally true? Do the changes in degree of saturation and water content that occur in the field under natural and manmade influences such as precipitation, evaporation, irrigation, or water table lowering significantly affect the state of stress and consequent stability of near-surface soil? Does negative pore pressure in unsaturated soil entirely contribute to total stress as positive pore pressure does in saturated soil? How can the differences in the stress conditions for saturated and unsaturated soil and the consequent differences in strength be effectively formulated and quantified? These types of questions have extremely important bearing on stress and deformation problems involving unsaturated soil in geotechnical engineering practice and will be addressed throughout the remainder of this book.

1.5 STATE VARIABLES, MATERIAL VARIABLES, AND CONSTITUTIVE LAWS

1.5.1 Phenomena Prediction

Many physical phenomena are constantly occurring or changing in behavior as functions of both space and time. The requirement for variables to describe these different phenomena results largely from our desire to predict their occurrence or behavior in the future. State variables, material variables, and constitutive laws are commonly used for phenomena prediction.

The number of state and material variables and constitutive laws used to define a given phenomenon depends on the conceptualization. For example, the strength of saturated soil can be represented by the Mohr-Coulomb criterion, which uses normal stress and shear stress as state variables to define the state of stress. To evaluate the stability of the soil under this state of stress, a series of conjugate material properties describing the strength characteristics of the soil must also be introduced. For the Mohr-Coulomb criterion, these properties are friction angle and cohesion. Such properties are usually referred to as *material variables* because they generally vary with the state variables, just as friction angle and cohesion may depend on whether soil is under drained or undrained state conditions.

State variables are used to describe phenomena occurring in nature and engineering practice. By general definition, state variables are those that are required to completely describe the state of the system for the phenomenon at hand (Fung, 1965). For example, to describe today's weather, one may use terminology including temperature, pressure, relative humidity, or wind speed. These quantitative descriptors are the state variables defining the state of the weather conditions. Effective stress is the state variable required to describe strength and deformation in saturated soil. Pressure and temperature are often used as state variables to describe the thermodynamic state of a system.

Following a macroscopic or phenomenological formalism under the framework of continuum mechanics and thermodynamics, state variables are not required to be non–material dependent quantities, but have been done so traditionally. The non–material dependent formalism is most effective for phenomena such as heat transfer, mass transport, wave propagation, and chemical reactions occurring in single-phase media such as liquid, gas, or solid or in equivalent continuum media such as saturated soil. In multiphase systems, however, the relative amount of each phase comprising the system often directly controls physical processes such as flow, stress, and deformation phenomena. In unsaturated soil, for example, a decrease in the relative amount of the pore water phase (i.e., a desaturation process) implies a drier soil or a soil with a lower hydraulic head. Consequently, phenomena involving fluid flow, stress, or deformation are likely to occur. Because the amount of pore water corresponding to a given value of head is highly dependent on the type of soil (clearly a material property), describing the mechanical and hydrological behavior of the multiphase system without involving the characteristic relationship between water content and head is impossible. Thus, for multiphase systems such as unsaturated soil, the commonly used conceptualization defining state variables as independent of material variables may be limiting and ineffective. In such cases, according to continuum mechanics, using material variables in conjunction with state variables to describe the state of a multiphase system is necessary (Fung, 1965).

In soil mechanics, it is convenient to differentiate between stress state variables, deformation state variables, and flow state variables. Common stress state variables are the total stress tensor, pore pressure, the effective stress tensor, Coulombian shear stress, and the first, second, and third stress invariants. Commonly used deformation state variables are the strain tensor, the first, second, and third stress and strain invariants, and void ratio. Widely used flow state variables are the degree of pore water saturation, water content, and total hydraulic head.

Material variables are intrinsic properties that depend on the type of material, usually varying from one material to another material or from one state to another state. Examples of material variables are elastic modulus, permeability, and compressibility. Material variables can also be functions of state variables such as pressure, temperature, and stress. The viscosity of water, for example, a material variable, decreases with increasing temperature, a state variable. Similarly, the hydraulic conductivity of unsaturated soil has been effectively conceptualized as a function of the state variables water content or matric suction. Depending on the type of problem (e.g., flow, stress, or deformation), a number of material variables for describing the physical behavior of unsaturated soil have been widely used, several of which are introduced in Chapter 2.

Constitutive laws or equations describe the interrelationships between or among state variables and material variables. A constitutive equation for a given system allows the prediction of one state variable from others. For example, for an elastic soil, one-dimensional stress can be predicted from one-dimensional strain if the elastic modulus is known. Similarly, discharge velocity in soil can be predicted from the hydraulic gradient using the material variable hydraulic conductivity.

A typical path from physical observation to behavior prediction is shown in Fig. 1.15. Here, physical observation and measurement provides a basis for defining and quantifying the state variables and material variables that describe the phenomenon of interest (e.g., hydraulic conductivity and total head are material and state variables required for predicting steady-state fluid flow behavior). Physical laws provide the connection between the state and material variables governing the physical process (e.g., Darcy's law). Thermodynamic principles (e.g., the first and second laws of thermodynamics) are applied to the physical laws to arrive at governing equations for predicting behavior in space and time (e.g., quantifying a hydraulic potential field). Comparing the predicted behavior with observations of actual behavior provides a basis to refine the state and material variables for improving the prediction.

1.5.2 Head as a State Variable

Total head is often used as a state variable to describe flow phenomena in soil. The total head concept is generally applicable to both saturated and unsaturated conditions. Use of total head for describing fluid flow stems from consideration of thermodynamic law, which assumes that energy flows from a place of higher value to a place of lower value.

Fundamentally, total head is the potential of the water retained in the soil pores. For many geotechnical engineering applications occurring on a relatively macroscopic scale (e.g., larger than the particle scale), the total head h_t responsible for the flow of pore water at a given point can be sufficiently represented by the summation of the elevation head h_e and pressure head h_p at that point as follows:

$$h_t = h_e + h_p \qquad (1.9a)$$

or

1.5 STATE VARIABLES, MATERIAL VARIABLES, AND CONSTITUTIVE LAWS

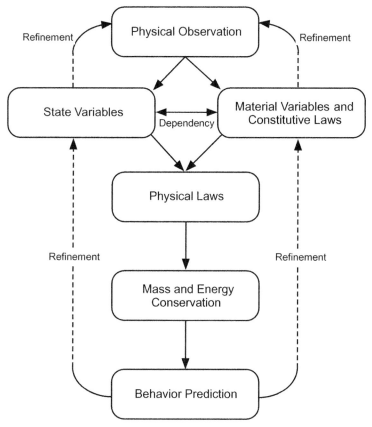

Figure 1.15 Iterative process leading from physical observation and measurement to behavioral prediction.

$$h_t = z + \frac{u_w}{g\rho_w} \tag{1.9b}$$

where z is the vertical coordinate distance from a prescribed datum (m), u_w is the pore water pressure (Pa), g is gravitational acceleration (m/s^2), and ρ_w is the density of water (kg/m^3).

Darcy's law describes discharge velocity v as a function of the gradient of total head (a state variable) and hydraulic conductivity (a material variable) as follows:

$$v = -k \frac{dh_t}{dx} \tag{1.10}$$

where x is the coordinate in the direction of flow (m) and k is hydraulic conductivity (m/s). Darcy's law for fluid flow, therefore, is a direct application

30 STATE OF UNSATURATED SOIL

of thermodynamic law, that is, energy flows from a higher place to a lower place.

The major difference between total head in saturated soil and total head in unsaturated soil is that the pressure head governed by the pore pressure u_w is positive (compressive) in saturated soil and negative (tensile) in unsaturated soil. The pressure head in unsaturated soil is also highly dependent on the degree of saturation or water content and type of soil. Examining pressure and head profiles for a soil layer under the hydrostatic ("no-flow") condition illustrates these major differences. Consider, for example, the homogeneous soil layer shown in Fig. 1.16a. The soil located above the water table can be conceptualized as three regimes: (1) a regime where the soil remains saturated under negative pore water pressure, often referred to as the *capillary fringe,* (2) an unsaturated regime characterized by a continuous water phase, or *funicular* regime, and (3) a residual or *pendular* regime characterized by an isolated, discontinuous water phase. The transitions between each of these regimes, which are by no means well defined in typical field settings, are largely controlled by the pore size and pore size distribution of the soil. As illustrated by the degree of saturation profile shown on Fig. 1.16b, the point where desaturation commences in the soil located above the water table is referred to as the *air-entry* point. The hydraulic head associated with this point is referred to as the *air-entry head.*

According to thermodynamics, the total head in the soil profile under no-flow conditions must be the same everywhere at equilibrium. Accordingly, two piezometers, one placed in the saturated zone at point A (Fig. 1.16a), and the other in the capillary fringe zone at point B, indicate identical total head values. The pressure head at A is positive. The pressure head at B is negative. Assigning an elevation datum to the bottom of the saturated soil layer, the pore pressure profile varies linearly from zero at the water table, to positive values below the water table, and to negative values above the water table. In the unsaturated zone, equilibrium requires the pore water potential to be equal among the three different phases comprising the soil (i.e., pore water, pore gas, and soil solids). This requirement provides the physical basis governing the height of the capillary fringe and the relative humidity of the pore gas in the unsaturated zone. Detailed thermodynamic treatment of these interfacial equilibrium concepts is provided in Chapters 3 and 4.

1.5.3 Effective Stress as a State Variable

Effective stress is considered a fundamental state variable for describing the state of stress in soil. For saturated soil, Terzaghi (1943) defined effective stress as the difference between the total stress and pore pressure. Physically, effective stress describes the stress acting on the soil skeleton. One can fully define the effective stress at any point of interest in saturated soil as long as the total stress and pore pressure are known.

1.5 STATE VARIABLES, MATERIAL VARIABLES, AND CONSTITUTIVE LAWS

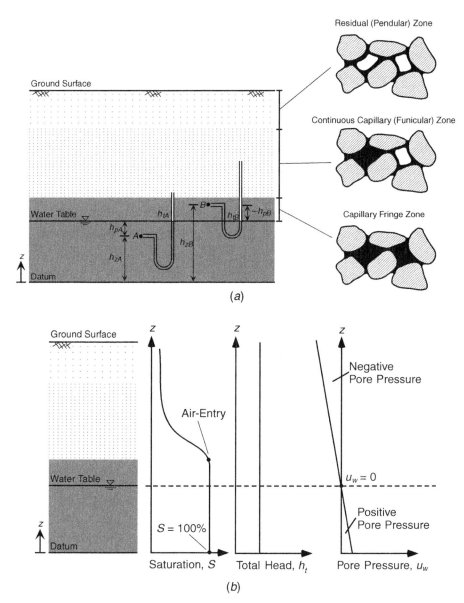

Figure 1.16 Conceptual illustration of the unsaturated soil zone: (*a*) pore water regimes and (*b*) saturation, total head, and pore pressure profiles.

For unsaturated soil, the physical meaning of effective stress remains the same. However, two additional factors must be considered: (1) the stress acting through the air phase (i.e., the pore air pressure, u_a) and (2) the difference between the pore air pressure and the pore water pressure, or matric suction. Bishop's (1959) widely cited effective stress approach for unsaturated soil expands Terzaghi's classic effective stress equation as follows:

$$\sigma' = (\sigma - u_a) + \chi(u_a - u_w) \tag{1.11}$$

The difference $\sigma - u_a$ is referred to as the *net normal stress*, the difference $u_a - u_w$ is *matric suction*, and the *effective stress parameter* χ is a material variable that is generally considered to vary between zero and unity. For $\chi = 0$, corresponding to completely dry soil, and for $\chi = 1$, corresponding to fully saturated soil, eq. (1.11) reduces to Terzaghi's classic effective stress equation for describing the behavior of saturated soil ($\sigma' = \sigma - u_w$).

The material variable χ is captured by its strong dependency on the degree of pore water saturation S:

$$\chi = \chi(S) \tag{1.12}$$

Determination of the effective stress parameter and its dependency on the amount of water in the system is essential in order to evaluate effective stress in unsaturated soil. Figure 1.17 illustrates this dependency on saturation for several types of soil. Measurement of this function is experimentally challenging, particularly near the low saturation range. For relatively high degree of saturation, indirect measurement of χ is possible via shear testing under

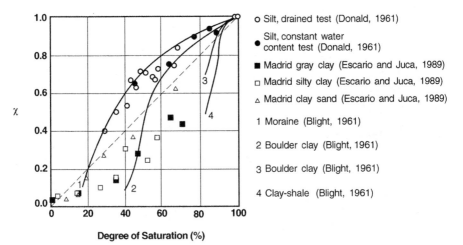

Figure 1.17 Experimental results showing the dependency of Bishop's effective stress parameter χ on degree of saturation.

controlled suction conditions. Theoretical studies have also shown that χ could exceed unity and is highly nonlinear in the transitional regime between conditions of isolated pore water menisci and continuous pore water. The nature of χ, as well as its determination by experimental techniques, are important and wide open subjects in unsaturated soil mechanics. A systematic description of the nature of χ from both microscopic and macroscopic perspectives is provided in Chapters 5 and 6.

1.5.4 Net Normal Stresses as State Variables

Bishop's original treatment of effective stress as a single-valued stress state variable for unsaturated soil [eq. (1.11)] has been challenged from theoretical, experimental, and philosophical perspectives. Jennings and Burland (1962), for example, explored the limitations of using the effective stress concept and suggested that it may not be adequate for describing deformation phenomena such as collapse upon wetting. Khalili et al. (2004) and others contend that this argument has been formulated within the context of a linear elastic framework and that nonrecoverable (plastic) deformations such as collapse can indeed readily be described within an effective stress framework using elastoplastic theories. The effective stress approach for unsaturated soil mechanics continues to be the subject of debate. Of specific issue is the necessity to include the material variable χ, which may or may not be readily determined, in defining effective stress.

In wrestling with these apparent difficulties, Coleman (1962) suggested the use of net normal stress $\sigma - u_a$ and matric suction $u_a - u_w$ as stress variables to describe stress-strain relations for unsaturated soil. Further work by Bishop and Blight (1963) illustrated some advantages of using net normal stress and matric suction as stress state variables. Fredlund and Morgenstern (1977) considered the approach from both experimental and theoretical standpoints and formally proposed the use of net normal stress and matric suction as independent stress state variables for unsaturated soil. Here, it was considered that any two of three stress variables $\sigma - u_a$, $\sigma - u_w$, or $u_a - u_w$ may be used, that is, $\sigma - u_a$ and $u_a - u_w$, $\sigma - u_w$ and $u_a - u_w$, or $\sigma - u_a$ and $\sigma - u_w$. The proposed approach was supported by a series of "null" triaxial tests conducted earlier (Fredlund, 1973) where it was shown that the volume of unsaturated soil specimens remains relatively unchanged if changes in the proposed stress state variables were prohibited.

Studies conducted over the past two decades have demonstrated renewed interest in Bishop's effective stress approach, as well as support for Fredlund and Morgenstern's (1977) independent stress state variable approach. Identifying the most appropriate stress state variables for unsaturated soil remains a highly active area of research. Detailed introductions and analyses of each approach, their major differences, their very different roles in describing the shear strength of unsaturated soil, and reconciliation between these two approaches are provided in Chapters 5 and 6.

1.6 SUCTION AND POTENTIAL OF SOIL WATER

1.6.1 Total Soil Suction

Total soil suction quantifies the thermodynamic potential of soil pore water relative to a reference potential of free water. Free water in this regard is defined as water containing no dissolved solutes, having no interactions with other phases that impart curvature to the air-water interface, and having no external forces other than gravity. The physical and physicochemical mechanisms responsible for total soil suction are those that decrease the potential of the pore water relative to this reference state.

Neglecting temperature, gravity, and inertial effects, the primary mechanisms that decrease the potential of soil pore water include capillary effects, short-range adsorption (particle-pore water interaction) effects, and osmotic effects. The former mechanism is unique to unsaturated soil. The latter two may occur under either saturated or unsaturated conditions.

Capillary effects, which include curvature of the air-water interface and the associated negative pore water pressures in the three-phase unsaturated soil system, are described in detail in Chapters 3 and 4.

Short-range adsorption effects arise primarily from electrical and van der Waals force fields occurring within the vicinity of the solid-liquid (i.e., soil-pore water) interface and are most important for fine-grained soils. Electrical fields emanate from the net negative charge on the surface of clay minerals. van der Waals fields arise from atomic scale interactions between the molecules comprising the surface of the solid phase (i.e., the soil particles) and the molecules comprising the liquid phase (i.e., the pore water) and occur for all types of soil. The effect of each of these fields is most pronounced for water adsorbed by clay particles, which posses both significant net surface charge and relatively large surface area. The strength of electrical and van der Waals fields decays rapidly as the distance from the particle surface increases. Accordingly, short-range adsorption effects are most relevant at relatively low water content or degree of saturation when the adsorbed pore water is primarily in the form of thin films coating the particle surfaces.

Osmotic effects are the result of dissolved solutes in the pore water. Dissolved solutes may arise from two sources: as externally introduced solutes (e.g., through natural leaching processes), or as naturally occurring solutes adsorbed by the soil mineral surfaces (e.g., exchangeable cations adsorbed by clay particles). Hydration and solvation of such dissolved solutes and the associated structural ordering of neighboring water molecules reduces the chemical potential of the pore water to a degree dependent on the dissolved solute concentration.

Suction arising from the combined effects of capillarity and short-range adsorption is usually grouped under the more general term *matric* suction, which may be designated in units of pressure as ψ_m. The term *matric* reflects earlier usage of the term *matrix*, which was intended to describe the component of suction arising from interactions between the pore water and the soil solids, or soil matrix. Suction arising from the presence of dissolved solutes is referred to as *osmotic* suction, or ψ_o. Total soil suction ψ_t is generally considered the algebraic sum of the matric and osmotic components, which may be written as follows:

$$\psi_t = \psi_m + \psi_o \tag{1.13}$$

1.6.2 Pore Water Potential

The thermodynamic potential of soil pore water is most rigorously described in terms of chemical potential. Chemical potential, typically designated μ, has units of energy per unit mass, measured in either joules per mole (J/mol) or joules per kilogram (J/kg). Pore water chemical potential represents the amount of energy stored per unit mass of pore water. Potential in the latter units of joules per kilogram is often referred to as free energy per unit mass, or E. Chemical potential is the primary criterion for equilibrium with respect to the transfer of energy within any given phase of matter (e.g., water) or from one phase of certain matter to another phase of the same matter (e.g., from liquid water to water vapor). Equilibrium requires that energy be transferred from locales or phases of relatively high chemical potential to locales or phases of relatively low chemical potential. In a closed system at equilibrium, the chemical potential of the matter under consideration is the same at every point within each phase and among all of the phases.

Describing the energy state of soil pore water is best accomplished by considering the change from a reference condition for free water. The total change in pore water potential $\Delta\mu_t$ resulting from the various physical and physicochemical suction mechanisms in unsaturated soil can be written as follows:

$$\Delta\mu_t = \Delta\mu_c + \Delta\mu_o + \Delta\mu_e + \Delta\mu_f \tag{1.14}$$

where $\Delta\mu_c$ is the change in potential due to curvature at the air-water interface (i.e., capillary effects), $\Delta\mu_o$ is the change due to dissolved solute effects (i.e., osmotic effects), $\Delta\mu_e$ is the change to due the presence of electrical fields, and $\Delta\mu_f$ is the change due to van der Waals fields. Each term on the right-hand side of eq. (1.14) is a negative value, reflecting a decrease or decrement in chemical potential associated with each mechanism. Soil suction is a positive value because it describes this decrement relative to a reference potential for free water equal to zero.

As described in detail in Chapter 4, curvature at the air-water interface in unsaturated soil decreases the chemical potential (J/mol) of soil pore water an amount described by a form of the Young-Laplace equation as follows:

$$\Delta\mu_c = -T_s v_w \left(\frac{1}{R_1} + \frac{1}{R_2}\right) \tag{1.15}$$

where T_s is the surface tension of the water (mN/m), R_1 and R_2 are principal radii describing the curvature of the air-water interface (m), and v_w is the partial molar volume of water (m³/mol). As the net curvature of the air-water interface located between and among unsaturated soil grains increases (i.e., as the soil desaturates and the pore pressure becomes more negative), the decrement in chemical potential becomes greater.

For ideal and dilute solutions, the decrement in chemical potential (J/mol) due to the presence of dissolved solutes may be approximated by a form of the van't Hoff equation:

$$\Delta\mu_o = -CRTv_w = -\pi v_w \tag{1.16}$$

where C is the molar concentration of the pore solute solution (mol/m³), R is the universal gas constant (J/mol · K), and T is temperature (K). The product CRT in the above equation is commonly referred to as *osmotic pressure*, or π. Under more general, nondilute conditions, osmotic pressure is described by the viral equation (e.g., Shaw, 1992):

$$\pi = CRT(1 + B_2 C^2 + B_3 C^3 + \cdots) \tag{1.17}$$

where $B_2, B_3, \ldots$ are viral coefficients. As the concentration term approaches zero, the viral equation (1.17) approaches the van't Hoff approximation (1.16). As shown in Fig. 1.18, as the concentration of dissolved solutes increases, the osmotic pressure increases. The corresponding chemical potential of the pore water solution decreases.

Because H$_2$O is a polar molecule, the physical consequence of short-range electrical fields emanating from soil particle (clay mineral) surfaces is to attract, align, and impart order into the molecular arrangement of neighboring pore water. Considering an interaction with a single particle surface, the corresponding decrement in chemical potential of the pore water is dependent on the location of the water relative to the particle surface and may be quantified as follows (e.g., Iwata et al., 1995):

$$\Delta\mu_e = \int_0^D \frac{Dv_w}{4\pi}\left(\frac{1}{\varepsilon} - 1\right) dD \tag{1.18}$$

where ε is the partial dielectric constant of the pore water and D is the value of electric displacement at the point where the water exists. The value of D depends on the shape and size of the soil particle, its surface charge density, and the distance from the particle surface to the water under consideration.

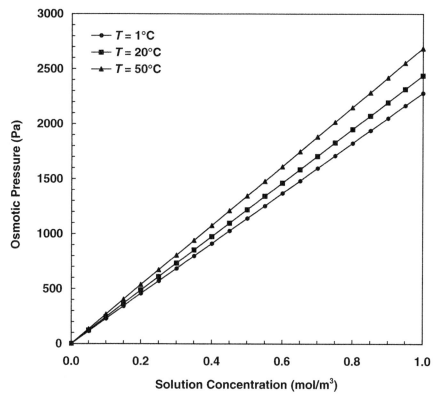

Figure 1.18 Osmotic pressure as a function of dissolved solute concentration as predicted by eq. (1.16).

As water molecules move closer to the surface of charged soil particles, such as in the thin films surrounding clay particles at very low water content, their potential energy is reduced just as if they were falling in a gravity field or being driven along by a pressure gradient.

The decrement in chemical potential due to van der Waals fields is also dependent on the location of pore water molecules relative to the soil particle surface. The magnitude of this decrement generally depends on the assumed shape of the particle surface (e.g., planar, spherical), which may be captured by the following functional dependency:

$$\Delta\mu_f = -Af(z, \Gamma) \qquad (1.19)$$

where A is Hamaker's constant for the soil-water interaction, z is the distance from the pore water molecule to the particle surface, and Γ is a shape function.

Equations (1.18) and (1.19) dictate that the decrements in chemical potential due to electrical and van der Waals fields are much less at locations relatively far from particle surfaces than at locations near the surfaces. However, local thermodynamic equilibrium requires the chemical potential throughout the entire pore water phase to be the same. To satisfy this requirement, positive internal water pressure, which acts to increase the potential, builds up in the water films immediately adjacent to the soil particles. The change in potential at any point due to the internal pressure buildup is equal and opposite in magnitude to the change produced by the electrical field and van der Waals fields. Accordingly, the magnitude of the internal pressure decreases as the distance from the particle surface increases. Macroscopically, the positive pore pressure immediately adjacent to soil particles is manifested as a convexly curved film coating the particles (much like a film of oil coating a ball). Thus, the pore water pressure in unsaturated soil may actually be either positive or negative depending on the location of the pore water relative to the particle surface and the scale of the problem under consideration (e.g., Olson and Langfelder, 1965; Nitao and Bear, 1996). For most practical geotechnical engineering purposes, the problems of interest occur on a relatively macroscopic scale and equivalent negative pore water pressures are considered.

The total reduction in chemical potential associated with each of the above mechanisms defines the total suction of the soil-water system. The magnitude of the potential reduction is dependent on the amount of pore water in the system. The constitutive relationship that describes this dependency is referred to as the soil-water characteristic curve, or SWCC.

1.6.3 Units of Soil Suction

The potential of soil pore water may be expressed as an energy per unit mass, a chemical potential (i.e., J/kg or J/mol), as an energy per unit volume, a pressure potential (i.e., $J/m^3 = N \cdot m/m^3 = N/m^2 = Pa$), or as an energy per unit weight, a head potential (i.e., $J/N = N \cdot m/N = m$). Conversion among units of potential μ, pressure ψ, and head h may be achieved by considering the following equivalency:

$$\mu = \psi v_w = hg\omega_w \tag{1.20}$$

where g is gravitational acceleration (m/s²), ω_w is the molecular mass of water (kg/mol), v_w is the partial molar volume of water (m³/mol), and chemical potential μ is in units of Joules per mole. Equations for direct conversion among these various units are summarized in Table 1.1.

Pore water potential in units of either head h or pressure ψ are preferred for describing flow, stress, and deformation phenomena in unsaturated soil mechanics. Use of the term *suction head* generally refers to pore water potential in units of head. The preferred term in geotechnical engineering practice, *soil suction*, refers to pore water potential in units of pressure. The International System of Units (SI) of suction pressure are pascals (Pa), which are typically described in terms of kilopascals (kPa) for the range of magni-

TABLE 1.1 Conversion Chart for Pore Water Potential Terms

	Potential	Head	Pressure
Potential (J/mol)	—	$\mu = hg\omega_w$	$\mu = \psi v_w$
Head (m)	$h = \dfrac{\mu}{g\omega_w}$	—	$h = \dfrac{\psi v_w}{g\omega_w} = \dfrac{\psi}{\rho_w g}$
Pressure (kPa)	$\psi = \dfrac{\mu}{v_w}$	$\psi = \dfrac{hg\omega_w}{v_w} = hg\rho_w$	—

$\rho_w = \omega_w / v_w$.

tude relevant to most practical unsaturated soil mechanics applications. Alternative units of (pF), which were common in the early literature, are defined as the logarithm of pore water potential in units of head in centimeters of water (Schofield, 1935):

$$\text{pF} = \log(\text{cm}_{\text{H}_2\text{O}}) \qquad (1.21)$$

For convenience, units of pF may be approximated in terms of kilopascals as follows:

$$10^{(\text{pF}-1)} \approx \text{kPa} \qquad (1.22)$$

for example, for pF = 4:

$$10^{(4-1)} \approx 10^3 \text{ kPa} = 1000 \text{ kPa}$$

An additional series of commonly used equivalent units for pore water potential is summarized in Table 1.2.

1.6.4 Suction Regimes and the Soil-Water Characteristic Curve

The relative importance of the individual physical and physicochemical mechanisms responsible for soil suction depends on the water content of the unsaturated soil-water-air system. At relatively low values of water content and correspondingly high values of suction, where pore water is primarily in the form of thin films on the particle surfaces, the dominant mechanisms contributing to suction are the relatively short-range adsorption effects governed by the surface properties of the soil solids. On the other hand, at relatively high values of water content and correspondingly low values of suction, the dominant pore water retention mechanism becomes capillarity, governed primarily by the particle and pore structure and pore size distribution. Osmotic suction is constant over the entire range of water content unless the concentration of dissolved solutes changes.

TABLE 1.2 Equivalent Units for Describing Potential, Head, and Pressure

	Unit	Equivalent Value
Potential Units	J/kg	100
	J/mol	1.8016
Head Units	m H_2O	10.2
	cm H_2O	1020
	inches H_2O	401.5
	feet H_2O	33.42
	mm Hg	750
	inches Hg	29.53
	pF	3.01
Pressure Units	Pa	100000
	kPa	100
	MPa	0.1
	bars	1.0
	millibars	1000
	atmospheres	0.987
	lb/in² (psi)	14.5
	lb/ft² (psf)	2088
	US tons/ft² (tsf)	1.044

The transition between the high suction regime dominated by short-range adsorption mechanisms and the low suction regime dominated by capillary mechanisms is highly dependent on soil type. In fine-grained materials such as clays, for example, a much greater amount of pore water is required to satisfy the relatively large surface hydration energies associated with the high suction regime. In sands, however, very little water is adsorbed under initial surface hydration mechanisms and capillary effects dominate over the majority of the unsaturated water content range. The SWCC describes the corresponding constitutive relationship between soil suction and soil-water content.

The general shape of the SWCC for various soils reflects the dominating influence of material properties including pore size distribution, grain size distribution, density, organic material content, clay content, and mineralogy on the pore water retention behavior. Understanding the general behavior of the SWCC and its relationship to the physical properties of the soil that it describes is a critical component of unsaturated soil mechanics. McQueen and Miller (1974) developed an instructive conceptual model based on empirical evidence for describing the general shape and behavior of the SWCC. As illustrated graphically in Fig. 1.19, it was suggested that any SWCC could be approximated as a composite of three straight-line segments on a semilog plot of suction versus moisture content ranging from zero to saturation. These line segments include one extending from 10^6 to 10^4 kPa designated the *tightly adsorbed segment*, a second extending from 10^4 kPa to approximately 100

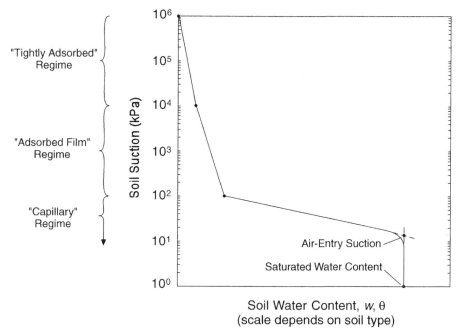

Figure 1.19 Illustration of McQueen and Miller's (1974) conceptual model for general behavior of the soil-water characteristic curve.

kPa designated the *adsorbed film segment,* and a third extending from 100 to 0 kPa (saturation) designated the *capillary segment.* Each segment is characterized by a change in slope at the transition points.

Within the so-called tightly adsorbed regime, pore water is retained by molecular bonding mechanisms, primarily hydrogen bonding with exposed oxygen or hydroxyl on the surfaces of the soil minerals. Within the adsorbed film segment, water is retained in the form of thin films on the particle surfaces under the influences of short-range solid-liquid interaction mechanisms (e.g., electrical field polarization, van der Waals attraction, and exchangeable cation hydration). The amount of water adsorbed within the first two regimes (i.e., the slopes of the line segments) is a function of the surface area of the soil particles, the surface charge density of the soil mineral, and the type and valency of any adsorbed exchangeable cations. When the adsorbed films on the particle surfaces grow thick enough to extend beyond the range of influence of the short-range solid-liquid interaction effects, the characteristic curve enters a regime dominated by capillary pore water retention mechanisms. The amount of water adsorbed here is a function of the particle and pore size properties, terminating at the air-entry pressure where the capillary air-water interfaces begin to disappear as the system approaches saturation.

42 STATE OF UNSATURATED SOIL

Figure 1.20 shows SWCCs representative of sand, silt, and clay that illustrate the general behavior of characteristic curves and their associated dependency on soil type. For sandy soil, the surface adsorption regime in the high suction range is generally very limited because the specific surface and surface charge properties of sand are relatively small. Capillarity is the dominant suction mechanism over the majority of the unsaturated water content range, terminating at a relatively low air-entry pressure controlled by the relatively large pore throats formed between and among the sand particles. The overall slope and shape of the capillary regime is controlled primarily by the pore size distribution of the material. Soils with a relatively narrow pore size distribution are marked by relatively flat characteristic curves in the capillary regime because the majority of pores are drained over a relatively narrow range of suction. Silty soil may adsorb a significantly greater amount of water under short-range adsorption mechanisms becaue the specific surface area of silt is much larger than sand. The air-entry pressure of silt is also larger as controlled by the relatively small pores. Clay has the highest capacity for water adsorption under short-range surface interaction effects because clay particles have charged surfaces and very high specific surface area. As illustrated by the experimental data in Fig. 1.21, highly expansive clays (e.g., smectite) are capable of adsorbing as much as 20% water by mass during the initial surface adsorption regime and may sustain extremely high suction over

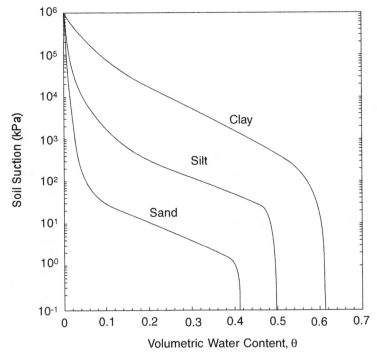

Figure 1.20 Representative soil-water characteristic curves for sand, silt, and clay.

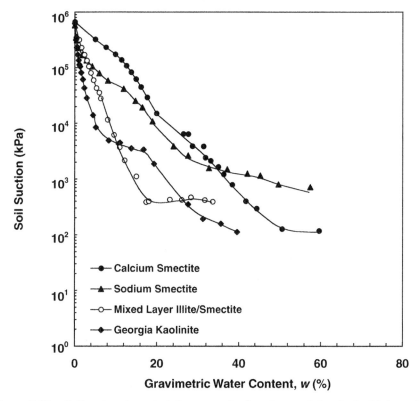

Figure 1.21 Soil-water characteristic curves for four types of clay in the high suction range (Likos, 2000).

a wide range of water content. Nonexpansive clays (e.g., kaolinite), on the other hand, adsorb much less water in the high suction regime. For expansive clays, the SWCC is more physically meaningful in terms of gravimetric water content since the volume is a variable during the sorption process.

PROBLEMS

1.1. Where are the regions in the United States where unsaturated soils are likely encountered to significant depth below the ground surface?

1.2. What kind of climatic conditions tend to lead to the formation of a thick unsaturated zone?

1.3. What is the fundamental difference between saturated soils and unsaturated soils in terms of pore water pressure?

1.4. Describe and illustrate the Mohr-Coulomb failure criterion.

1.5. When the state of stress (i.e., Mohr circle) in a soil reaches the Mohr-Coulomb criterion, what is the state of the stress called?

1.6. Give three examples of unsaturated soil mechanics problems in geotechnical engineering.

1.7. For a given unsaturated soil under either a dry or wet condition, which one has a higher suction?

1.8. What are state variables, material variables, and constitutive laws?

1.9. What are the principal differences between saturated and unsaturated soil profiles of pore water pressure, total stress, and effective stress?

1.10. According to Bishop's effective stress concept, which state, saturated or unsaturated, has a higher effective stress? Why?

1.11. What is the shape of the pore pressure profile under the hydrostatic condition in saturated and unsaturated states, respectively?

1.12. If an unsaturated soil has a water potential of -1000 J/kg, what is the equivalent soil suction value? If the soil at the air dry condition has a matric suction of 100 MPa, what is the soil water potential in joules per kilogram?

1.13. Three soils—clay, silt, and sand—are all equilibrated at the same matric suction, which soil has the highest water content and why?

1.14. Describe the major physical and physicochemical mechanisms responsible for soil suction.

PART I

FUNDAMENTAL PRINCIPLES

CHAPTER 2

MATERIAL VARIABLES

2.1 PHYSICAL PROPERTIES OF AIR AND WATER

2.1.1 Unsaturated Soil as a Multiphase System

Unsaturated soil is a multiphase system comprised of three phases of matter: gas, liquid, and solid. The gas phase is generally bounded by the pore space not occupied by liquid. The matter within this pore space may be any gas, vapor, or combination thereof. The liquid phase is generally bounded by the pore space not occupied by gas. The matter within this pore space may be any liquid or miscible or immiscible combination of two or more liquids (water, oil, non–aqueous phase liquids, etc.) The solid phase consists of the soil grains or particles and may range from relatively fine-grained materials such as silts and clays, to organic material, to relatively coarse-grained materials such as sand or gravel. Throughout the remainder of this book, the gas, liquid, and solid phases of unsaturated soil are assumed to be air, water, and soil solids, respectively. It should be borne in mind, however, that each of these species is by no means pure. It will be shown, for example, that water vapor dissolved in the pore air plays a critical role in the physical behavior and characterization of unsaturated soil. Gases and solids dissolved in the pore water (e.g., dissolved air, dissolved salts) play an important role as well.

Each phase of matter in an unsaturated soil system possesses unique material properties, the values of which are often very different from one phase to another. For example, the density of air is about 1 kg/m^3. The density of water, on the other hand, is about 1000 times higher. Many properties, such as the surface tension of water, are unique not only to the phase that they describe but also play an important role in governing the interactions among the various phases. Surface tension has a clear and profound influence on fluid flow, pore pressure, and the state of stress in unsaturated soil through its role in the state variable matric suction.

Many of the material properties characterizing the phases of an unsaturated soil system are dependent on the state variables governing the system. Important state variables in this regard include temperature, pressure, water content, relative humidity, and stress. The viscosity of air, for example, increases with increasing temperature. The viscosity of water on the other hand decreases with increasing temperature. Material properties are more rigorously referred to as *material variables* in order to capture these types of dependencies.

Material variables can be divided into two general types: (1) physical properties and (2) constitutive functions. In this section, the relevant physical properties of air and water will be described, namely, density and viscosity. Other physical properties pertinent to unsaturated soil mechanics, such as the surface tension of water, the solubility of air in water, and the density of water vapor in air, will be systematically introduced in following sections of the chapter. The important constitutive functions in unsaturated soil mechanics, which include the soil-water characteristic curve, the hydraulic conductivity function, and the suction stress characteristic curve will be described in detail in subsequent chapters.

2.1.2 Density of Dry Air

The density of air is defined as the mass of air per unit volume of air. Air density can vary significantly in shallow unsaturated soil under the influence of varying atmospheric conditions, most notably temperature and pressure. Local or regional gradients in air density provide a driving force for the flow of pore air in unsaturated soil, often becoming the dominant transport mechanism for the vapor phase transport of pore fluids in soil located near the ground surface and a mechanism by which geochemical reactions are catalyzed under changes in pore air chemistry.

Because air is comprised of different gaseous mixtures of oxygen (20.95% by volume), nitrogen (78.09%), and other trace gases, its density varies slightly with composition. For most practical purposes, the density of dry air can be determined by assuming ideal gas behavior. The ideal gas law describes the density of dry air, ρ_a, in terms of the relationship among temperature T, pressure u_a, volume V_a, mass M_a, molecular mass ω_a, and the universal gas constant R as follows:

$$\rho_a = \frac{M_a}{V_a} = \frac{u_a \omega_a}{RT} \tag{2.1}$$

2.1 PHYSICAL PROPERTIES OF AIR AND WATER

For example, if the air pressure is 100 kPa and temperature is 298 K and given that the molecular mass of air is about 29 kg/kmol and the universal gas constant is 8.314 N · m/mol · K, the corresponding air density is

$$\rho_a = \frac{u_a \omega_a}{RT} = \frac{(100 \times 10^3 \text{ N/m}^2)(29 \times 10^{-3} \text{ kg/mol})}{(8.314 \text{ N} \cdot \text{m/mol} \cdot \text{K})298\text{K}} = 1.17 \text{ kg/m}^3$$

The sensitivity of air density to the state variables temperature or pressure may also be investigated using eq. (2.1). For example, the change in air density ($\Delta\rho_a$) relative to some initial value (ρ_{a0}) resulting from a change in either state variable (Δu_a or ΔT) can be obtained by the following:

$$\frac{\Delta\rho_a}{\rho_{a0}} = \frac{\omega_a}{RT\rho_{a0}} \Delta u_a - \frac{\omega_a u_a}{RT^2 \rho_{a0}} \Delta T = \frac{\Delta u_a}{u_a} - \frac{\Delta T}{T} \tag{2.2}$$

The positive term on the right-hand side of eq. (2.2) indicates that an increase in air pressure for a given temperature results in an increased air density, a direct manifestation of the compressibility of air. Conversely, the negative term on the right-hand side indicates that an increase in temperature for a given pressure results in a decrease in air density, a direct manifestation of thermal expansion effects.

Consider the following practical example. If atmospheric air pressure remains constant at 101.3 kPa (1 atm) and the air temperature varies from 263 to 323 K with an average temperature of 300 K during a typical yearly cycle, the relative change in air density due to the temperature variation can be calculated with respect to its average value at 300 K as follows:

$$\frac{\Delta\rho_a}{\rho_{a0}} = -\frac{\Delta T}{T} \tag{2.3}$$

Table 2.1 shows this variation for temperature increments of 10 K.

Similarly, if atmospheric temperature remains constant at 300 K, and the pressure varies from 80 to 110 kPa with an average pressure of 100 kPa during a windy summer night, the relative change in air density is as follows:

$$\frac{\Delta\rho_a}{\rho_{a0}} = \frac{\Delta u_a}{u_a} \tag{2.4}$$

Table 2.2 shows this variation for pressure increments of 5 kPa.

Such variations in temperature and pressure are commonly encountered in near-surface unsaturated soil under natural environmental fluctuations. Because these fluctuations may indeed cause significant changes in the density of the pore air, they often become important mechanisms in governing gas

TABLE 2.1 Percent Change in Density of Air with Respect to Changes in Temperature[a]

Temperature (K)	Relative Change in Air Density (%)
263	12.3
273	9.0
283	5.7
293	2.3
300	0.0
303	−1.0
313	−4.3
323	−7.7

[a] Air density changes are relative to a reference temperature of 300 K.

flow and stress distribution processes in unsaturated soil. Changes in pore air pressure in response to barometric pressure fluctuations, for example, can cause periodic pore air flow into or out of the soil, often leading to significant vapor phase transport, or *barometric pumping,* in soil located near the ground surface. Temperature variations occurring on daily, weekly, or annual cycles can also cause significant changes in air density and thus drive pore air flow. These types of potentially important unsaturated fluid flow phenomena and the practical applications in which they arise are described in detail in Chapters 8 and 9.

2.1.3 Density of Water

Because many material variables pertaining to the water phase in unsaturated soil depend on the density of the pore water (e.g., viscosity and surface tension), changes in pore water density can directly influence the mechanical and hydrological behavior of the soil system. The primary state variables

TABLE 2.2 Percent Change in Density of Air with Respect to Changes in Total Air Pressure[a]

Air Pressure (kPa)	Relative Change in Air Density (%)
80	−20.0
85	−15.0
90	−10.0
95	−5.0
100	0.0
105	5.0
110	10.0

[a] Air density changes are relative to a reference value of 100 kPa.

controlling the density of water are temperature and pressure. Because water is relatively incompressible, however, the pressure dependency is relatively small, typically less than about 0.1% for the range of pressure significant to most geotechnical engineering problems. Variations in the density of water due to temperature changes, on the other hand, can be significant.

Figure 2.1 shows the relationship between the density of water and temperature at one atmosphere of pressure for temperature ranging from -4 to $18°C$. The density reaches a maximum of 1.000 g/cm^3 at approximately $4°C$. Increasing or decreasing the temperature from this point causes the density to decrease. At $50°C$, the density of water is 0.988 g/cm^3, or a reduction of 1.2%.

In many unsaturated soil mechanics problems, it is often necessary to consider the properties of *adsorbed* water, particularly at very low degrees of saturation or *residual* conditions where the majority of the pore water in the system exists as thin films surrounding the soil particle surfaces. The properties of adsorbed water, which is under the influence of short-range physical and physicochemical interactions with the soil surface, are quite different from those of free water. When the surface area and surface charge density of the soil type under consideration are relatively high, such as for expansive clay minerals, the interaction effects that occur at the pore water–soil solid interface are particularly strong and may indeed be of practical significance.

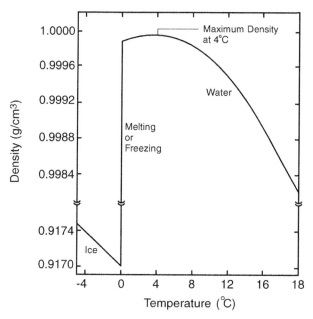

Figure 2.1 Density of water as a function of temperature (after Berner and Berner, 1987).

Properties of pore water affected by adsorption include density, viscosity, dissolved ion mobility, dielectric and magnetic properties, and freezing point temperature. For convenience, the state variable most commonly used to describe the variation in the properties of adsorbed water is the soil water content. Figure 2.2, for example, shows the variation in adsorbed pore water density as a function of water content for water adsorbed by highly expansive sodium montmorillonite. The data may be interpreted to indicate that the pore water molecules within the relatively low water content regime (<0.3 g/g) (i.e., located very close to the particle surface) are solvated about surface ions and the clay particle surfaces in an orderly, relatively dense manner. As water content increases and the thickness of the adsorbed water film grows, the density recovers to that of free water near 1.0 g/m^3. Thermodynamically, the adsorbed water has a lower chemical potential than water located increasingly far from the particle surface or perfectly free water. Physically, the adsorbed water is less mobile than free water. Studies regarding the contribution of this "immobile" pore water fraction to macroscopic fluid flow have long been of interest. Ongoing research using high-resolution X-ray reflectivity (e.g., Cheng et al., 2001), molecular dynamics simulations (e.g., Park and Sposito, 2002), and other emerging technologies is continuing to provide remarkable insight into the structure and properties of water adsorbed by mineral surfaces.

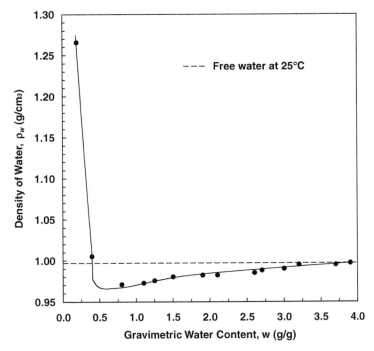

Figure 2.2 Density of water adsorbed by sodium montmorillonite as a function of water content (modified from Martin, 1960).

2.1.4 Viscosity of Air and Water

Viscosity is a material variable describing the ability of a given fluid to resist flow. Dynamic, or "absolute," viscosity, typically designated μ, has units of $N \cdot s/m^2$, $Pa \cdot s$, or $kg/m \cdot s$. Another unit for dynamic viscosity is the poise (P), equal to $1 \text{ dyn} \cdot s/cm^2$, which is more commonly referenced in terms of the centipoise (cP). Kinematic viscosity, typically designated v, is the ratio of dynamic viscosity to fluid density ρ, or

$$v = \frac{\mu}{\rho} \qquad (2.5)$$

The dynamic viscosity of pure water at 20°C is about 1.002 cP (1×10^{-3} $N \cdot s/m^2$). By comparison, the dynamic viscosity of pure air at 20°C is about 0.018 cP (1.8×10^{-5} $N \cdot s/m^2$). As illustrated in Fig. 2.3 and summarized on Tables 2.3 and 2.4, the controlling state variable for the viscosity of air and water is temperature. Note that the viscosity of water decreases as temperature increases. The viscosity of air on the other hand increases as tem-

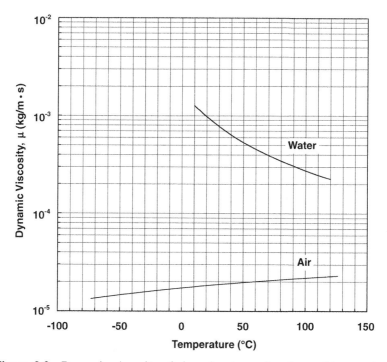

Figure 2.3 Dynamic viscosity of air and water as functions of temperature.

perature increases. Note also that the viscosity of water is far more sensitive to temperature than the viscosity of air.

The dynamic viscosity of water from 0 to 150°C can be quantitatively expressed using the following empirical relationship (Touloukian et al., 1975):

$$\mu_w = 2.5 \times 10^{-5} \times 10^{248/(T+133)} \text{ kg} \cdot \text{m}^{-1} \cdot \text{s}^{-1} \quad (2.6a)$$

where T is in degrees Celsius.

Similarly, the dynamic viscosity of air from -20 to $50°C$ can be expressed as follows (Streeter et al., 1997):

$$\mu_a = 10^{-5}[1.604 + 0.9(1 - e^{-(120+6T)/1000})] \text{ kg} \cdot \text{m}^{-1} \cdot \text{s}^{-1} \quad (2.6b)$$

Viscosity has an important influence on the conductivity and flow behavior of gases and liquids in unsaturated soil. High viscosity leads to relatively low liquid or gas conductivity and generally decreases flow velocity. Low viscosity leads to relatively high conductivity and generally increases flow velocity. The viscosity of the soil pore water also controls the compressibility and rheological behavior of the overall soil system.

TABLE 2.3 Dynamic Viscosity of Air as Function of Temperature

Air Temp. (°C)	Dynamic Visc. [kg/(m · s)]
−73	1.34×10^{-5}
−70	1.35×10^{-5}
−60	1.41×10^{-5}
−50	1.46×10^{-5}
−40	1.52×10^{-5}
−30	1.57×10^{-5}
−20	1.62×10^{-5}
−10	1.67×10^{-5}
0	1.72×10^{-5}
10	1.77×10^{-5}
20	1.82×10^{-5}
30	1.87×10^{-5}
40	1.91×10^{-5}
50	1.96×10^{-5}
60	2.01×10^{-5}
70	2.05×10^{-5}
80	2.09×10^{-5}
90	2.14×10^{-5}
100	2.18×10^{-5}
110	2.22×10^{-5}
120	2.26×10^{-5}
126	2.29×10^{-5}

TABLE 2.4 Dynamic Viscosity of Water as Function of Temperature

Water Temp. (°C)	Dynamic Visc. [kg/(m · s)]
1	1.64×10^{-3}
10	1.27×10^{-3}
20	9.77×10^{-4}
30	7.77×10^{-4}
40	6.35×10^{-4}
50	5.32×10^{-4}
60	4.54×10^{-4}
70	3.94×10^{-4}
80	3.46×10^{-4}
90	3.07×10^{-4}
100	2.75×10^{-4}
110	2.48×10^{-4}
120	2.25×10^{-4}

2.1.5 Flow Regimes

Viscosity has a direct effect on the nature of fluid flow in saturated or unsaturated soil. The Reynolds number has been used as an important criterion to identify the various flow regimes and the applicability of Darcy's law. The Reynolds number Re is defined in dimensionless form by the density of fluid, ρ, flow velocity v, dynamic viscosity μ, and a representative length dimension d by eq. (2.7). For applications in soil, the representative length dimension d is often approximated as the mean grain diameter or the mean pore size:

$$\text{Re} = \frac{\rho v d}{\mu} \qquad (2.7)$$

Small Reynolds numbers indicate that fluid viscosity dominates the flow process and that the flow is laminar. Large Reynolds numbers indicate that kinetic or inertial effects dominate the flow process and that the flow is turbulent. Figure 2.4 approximates the boundaries between laminar and turbulent flow regimes as a function of Reynolds number for flow processes in soil. When the Reynolds number is less than some value between 1 and 10, Darcy's law for laminar flow is generally valid. When the Reynolds number is greater than 100, flow enters the turbulent regime. When the Reynolds number is less than 100 but greater than some value between 1 and 10, the flow remains in the laminar regime but behaves nonlinearly (Bear, 1972). Reynolds numbers for the majority of liquid and gas flow conditions in the field are less than unity, implying that Darcy's law is in most cases valid. In some cases, however, such as fracture flow, pore air flow in relatively coarse-grained unsatu-

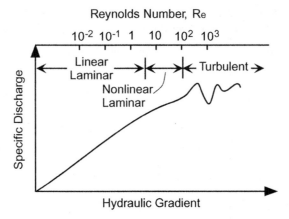

Figure 2.4 Flow regimes in porous media (after Freeze and Cherry, 1979).

rated soils, or flow driven by an extremely large gradient, the flow conditions may approach the nonlinear laminar or turbulent regimes.

Example Problem 2.1 Liquid and gas conductivity testing was conducted using separate constant-head permeameters for sandy soil with a mean pore size of 10^{-3} m. The steady-state velocity of air through the soil was 10^{-2} m/s. The steady-state velocity of water was 10^{-4} m/s. Identify the air and water flow regimes at a temperature of 25°C and determine whether or not Darcy's law may be used to analyze the test results.

Solution The viscosities of air and water at 25°C can be found from Fig. 2.3 as about 1.8×10^{-5} (kg/s · m) and 0.9×10^{-3} (kg/s · m), respectively. The density of air at this temperature is about 1.2 kg/m^3, and the density of water is about 1000 kg/m^3. From eq. (2.7), the Reynolds number for the flow of air is as follows:

$$\text{Re} = \frac{\rho v d}{\mu} = \frac{(1.2 \text{ kg/m}^3)(10^{-2} \text{ m/s})(10^{-3} \text{ m})}{1.8 \times 10^{-5} \text{ kg/s} \cdot \text{m}} = 0.67$$

The Reynolds number for the flow of water is

$$\text{Re} = \frac{\rho v d}{\mu} = \frac{(10^3 \text{ kg/m}^3)(10^{-4} \text{ m/s})(10^{-3} \text{ m})}{0.9 \times 10^{-3} \text{ kg/s} \cdot \text{m}} = 0.1$$

From Fig. 2.4, both the air and water flow are within the linear laminar regime and Darcy's law may generally be assumed valid. Since the discharge velocity and mean pore size in this example approximate the upper bound for most water seepage problems in the field, the Reynolds number is unlikely to exceed unity. On the other hand, air velocity due to blowing wind, topographic relief, or other environmental factors could be greater than 1 m/s, leading to a high Reynolds number exceeding 1 to 10. This can cause the airflow regime in unsaturated soil to reach the nonlinear laminar or even turbulent state.

2.2 PARTIAL PRESSURE AND RELATIVE HUMIDITY

2.2.1 Relative Humidity in Unsaturated Soil Mechanics

Relative humidity (RH) describes the state of thermodynamic equilibrium between air and water. A comprehensive understanding of relative humidity is a prerequisite for understanding and appreciating many of the forthcoming concepts in this book. As described in Chapter 3, for example, the relative humidity of the pore air phase in unsaturated soil is fundamentally linked to the chemical potential and suction of the pore water phase. As introduced in Chapter 10, measuring relative humidity becomes a powerful means to quantify soil suction and determine the corresponding soil-water characteristic curve. Chapters 8 and 9 demonstrate that variations in atmospheric relative humidity from time to time or from place to place become an important driving potential for unsaturated fluid flow and phase change phenomena such as evaporation and condensation.

2.2.2 Composition and Partial Pressure of Air

Although relative humidity has become common terminology in our daily life, its definition and usage embeds three major assumptions that are often easily overlooked. Appreciating these assumptions facilitates our general understanding of relative humidity and its importance to unsaturated soil mechanics. All three assumptions have been shown to be valid for most practical geotechnical applications.

The first assumption states that, excluding the water vapor component, the composition of air at a given point of interest (e.g., in the atmosphere or the pores of unsaturated soil) remains essentially unchanged over time. As shown in Table 2.5 and Fig. 2.5, perfectly "dry" air (i.e., having no water vapor component) at standard temperature and pressure consists by volume of approximately 78.09% nitrogen, 20.95% oxygen, and less than 1% trace gases (primarily argon and carbon dioxide). In "moist" air, however, a water vapor component exists that may vary significantly from time to time or place to place under changing environmental conditions, including changes in pressure, temperature, and the availability of water in the liquid (or solid) phase. In general, the water vapor component may be considered the single most important factor by which the overall chemical composition of atmospheric or soil pore air may change. Accordingly, the prevailing amount of water

TABLE 2.5 Composition and Physical Properties of Dry Air and Its Major Component Gases

Gas	Molecular Weight (g)	Percent by Volume (%)	Partial Pressure (kPa)
Nitrogen	28.01	78.09	79.11
Oxygen	32.00	20.95	21.22
Argon	39.98	0.93	0.94
Carbon dioxide	44.01	0.03	0.03
Air	29.00	100.00	101.3

vapor present in the air may be used to define the changes in the energy level or chemical potential of the multiphase system.

The second assumption regarding relative humidity states that each of the component gases that make up air, as well as the mixture of the component gases as a whole, follows ideal gas behavior. The practical implication of this assumption is that the partial pressure of each gas component may be quantified from the molar fraction of each. Considering the oxygen component, for example, its partial pressure is

$$\frac{u_{O_2}}{u_a} = \frac{n_{O_2}}{\Sigma_i n_i} \quad (2.8a)$$

where u_{O_2} is the partial pressure of oxygen, u_a is the total air pressure, n_{O_2} is the molar quantity of oxygen, and $\Sigma_i n_i$ represents the total molar quantity of all i gaseous species comprising air as the summation of the molar quantities of each.

Because every ideal gas has the same volume (22.4 L/mol), the partial volume v_i for the ith gas is as follows:

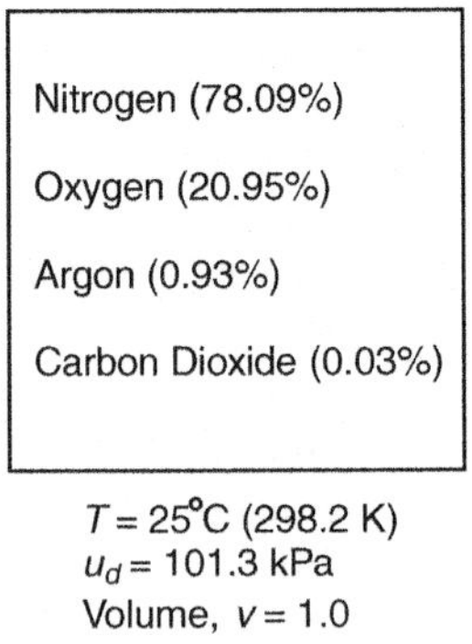

Nitrogen (78.09%)

Oxygen (20.95%)

Argon (0.93%)

Carbon Dioxide (0.03%)

$T = 25°C$ (298.2 K)
$u_d = 101.3$ kPa
Volume, $v = 1.0$

Figure 2.5 Composition of a unit volume of dry air in a closed chamber under standard temperature and pressure conditions.

$$\frac{v_i}{n_i} = 22.4 \text{ L} \tag{2.8b}$$

Substituting eq. (2.8b) into eq. (2.8a) leads to

$$\frac{u_{O_2}}{u_a} = \frac{v_{O_2}/22.4}{\Sigma_i(v_i/22.4)} = \frac{v_{O_2}}{\Sigma_i v_i} \tag{2.9a}$$

or

$$u_{O_2} = \frac{v_{O_2}}{\Sigma_i v_i} u_a \tag{2.9b}$$

which implies more generally that the partial pressure of any specific gas comprising air is equal to the total air pressure multiplied by the volume fraction of that gas. Following Table 2.5, therefore, for perfectly dry air, nitrogen will have a partial pressure of 79.11 kPa, oxygen 21.22 kPa, argon 0.94 kPa, and carbon dioxide 0.03 kPa. Summation of all these partial pressures leads to a total air pressure at standard conditions equal to 101.3 kPa.

The third assumption regarding relative humidity states that all components of air, including the water vapor component, reach local thermodynamic equilibrium. Thermodynamic equilibrium requires that the chemical potentials among all components of all phases in the system are the same. The equilibrium amount of water vapor present in a two-phase system of pure water and air generally depends on three factors: temperature, pressure, and the availability of water (i.e., sufficient water must exist for phase transition processes such as evaporation and condensation to reach equilibrium). As introduced in Section 1.6, in the pore space of unsaturated soil, additional factors such as the pore water salinity, the air-water interface geometry, and the soil surface area and mineralogy also act to affect the chemical potential of the water phase and the corresponding state of equilibrium between the pore water and the pore air. Accordingly, the equilibrium amount of water vapor in the pore air becomes a direct means to quantify the chemical potential or total suction of the pore water.

2.2.3 Equilibrium between Free Water and Air

Figure 2.6a illustrates equilibrium partial gas pressures for a simple two-phase system of air and pure water at a temperature of 25°C and a total air pressure of 101.3 kPa. At equilibrium, there is no further mass exchange between the water in the liquid phase and the water vapor in the gas phase. The corresponding partial pressure of the water vapor, u_v, is equal to 3.17 kPa. The partial pressure of the remaining components of the air (i.e., the "dry" air), u_d, is equal to 98.13 kPa. The total mass of water vapor per unit volume at

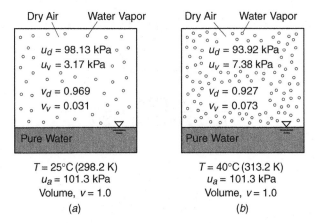

Figure 2.6 Equilibrium condition between pure water and air for (*a*) 25°C and 101.3 kPa and (*b*) 40°C and 101.3 kPa.

equilibrium (ρ_v), also referred to as *vapor density* or *absolute humidity*, can be calculated by applying the ideal gas law to the water vapor component as follows:

$$\rho_v = \frac{\omega_v u_v}{RT} = \frac{(18 \text{ kg/kmol})(3.17 \text{ kPa})}{(8.314 \text{ J/mol} \cdot \text{K})(298.2 \text{ K})} = 22.99 \text{ g/m}^3 \quad (2.10)$$

Thus, at 25°C and 101.3 kPa, the maximum amount of water that can be vaporized in air is 22.99 g/m³. Further vaporization is not physically possible without changing either the temperature or the total air pressure of the system. The vapor pressure corresponding to this equilibrium state is referred to as *saturated vapor pressure,* or $u_{v,\text{sat}}$.

The saturated vapor pressure for a system of pure water and air depends on temperature and total air pressure. The dependency on air pressure is much less pronounced than the dependency on temperature. In most atmospheric and shallow subsurface environments, the pressure dependency can be safely ignored. The temperature dependency, however, necessitates consideration, particularly because temperature may vary quite widely in many practical situations and because the sensitivity to temperature is significant. For example, if the temperature of the air-water system from the previous example (Fig. 2.6*a*) is increased to 40°C while maintaining the total air pressure at 101.3 kPa, the saturated vapor pressure at equilibrium will increase to 7.38 kPa (Fig. 2.6*b*), more than twice the vapor pressure at 25°C. The vapor density increases to 51.00 g/m³. Table 2.6 shows exact values of saturated water vapor pressure and the corresponding vapor density [eq. (2.10)] for temperature ranging from −5 to 45°C.

Rigorous solution for the dependency of saturated vapor pressure on temperature requires integration of the so-called Clausius-Clapeyron equation.

TABLE 2.6 Saturated Vapor Pressure and Absolute Humidity as Functions of Temperature at 101.3 kPa Total Air Pressure

Temperature		Saturated Vapor Pressure $u_{v,\text{sat}}$ (kPa)	Absolute Humidity ρ_v (g/m³)	Temperature		Saturated Vapor Pressure $u_{v,\text{sat}}$ (kPa)	Absolute Humidity ρ_v (g/m³)
T (°C)	T (K)			T (°C)	T (K)		
−5	268.2	0.421	3.398	21	294.2	2.486	18.294
−4	269.2	0.455	3.659	22	295.2	2.643	19.384
−3	270.2	0.490	3.926	23	296.2	2.809	20.532
−2	271.2	0.528	4.215	24	297.2	2.983	21.730
−1	272.2	0.568	4.518	25	298.2	3.167	22.993
0	273.2	0.611	4.842	26	299.2	3.361	24.320
1	274.2	0.657	5.187	27	300.2	3.565	25.710
2	275.2	0.705	5.546	28	301.2	3.780	27.170
3	276.2	0.758	5.942	29	302.2	4.006	28.700
4	277.2	0.813	6.350	30	303.2	4.243	30.297
5	278.2	0.872	6.786	31	304.2	4.493	31.977
6	279.2	0.935	7.250	32	305.2	4.755	33.731
7	280.2	1.001	7.734	33	306.2	5.031	35.572
8	281.2	1.072	8.253	34	307.2	5.320	37.493
9	282.2	1.147	8.800	35	308.2	5.624	39.507
10	283.2	1.227	9.380	36	309.2	5.942	41.606
11	284.2	1.312	9.995	37	310.2	6.276	43.803
12	285.2	1.402	10.643	38	311.2	6.662	46.347
13	286.2	1.497	11.324	39	312.2	6.993	48.494
14	287.2	1.598	12.046	40	313.2	7.378	51.001
15	288.2	1.704	12.801	41	314.2	7.780	53.608
16	289.2	1.817	13.602	42	315.2	8.202	56.337
17	290.2	1.937	14.451	43	316.2	8.642	59.171
18	291.2	2.063	15.338	44	317.2	9.103	62.131
19	292.2	2.196	16.271	45	318.2	9.586	65.222
20	293.2	2.337	17.256				

Vapor pressure data from Monteith and Unsworth, 1990.

The procedure to interpret the mathematical solution, however, is quite cumbersome. In practice, it is often preferred to refer to empirical relationships. Tetens (1930), for example, presented an exponential equation to express the saturated vapor pressure of water (kPa) in terms of temperature (K) as follows:

$$u_{v,\text{sat}} = 0.611 \exp\left(17.27 \frac{T - 273.2}{T - 36}\right) \tag{2.11}$$

Figure 2.7 shows a comparison of saturated vapor pressure calculated using eq. (2.11) with the exact values shown in Table 2.6. The calculated values are within 1 Pa of the exact values for temperatures up to 45°C.

2.2.4 Equilibrium between Pore Water and Air

The vapor pressure of soil pore water at thermodynamic equilibrium depends not only on temperature, pressure, and the availability of water but also on dissolved solute effects controlled by the chemistry of the pore water and solid-liquid interaction effects controlled by the pore structure, water content, and soil mineralogy. The combined effect of these additional factors causes

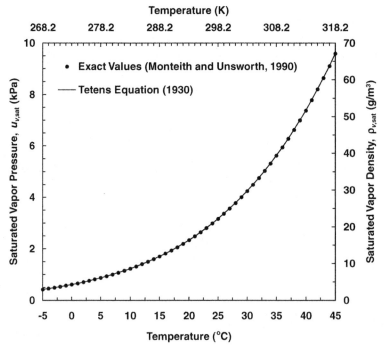

Figure 2.7 Saturated water vapor pressure and water vapor density as functions of temperature.

the pore water in unsaturated soil to be thermodynamically equivalent to a liquid with a lower chemical potential than free water. Because the chemical potential of soil pore water is reduced relative to free water, vapor pressures lower than the saturated vapor pressure are attained at equilibrium.

Consider a simple analogy between soil pore water and a salt solution. Equilibrium requires that the chemical potential of the water vapor in the gas phase be the same as the chemical potential of the solution in the liquid phase. Because the addition of dissolved solutes has reduced the chemical potential of the solution relative to the condition for pure water, a smaller amount of water tends to evaporate to the vapor phase. In other words, the reduced energy of the liquid phase inhibits the transfer of mass and energy to the vapor phase. Figure 2.8, for example, shows the state of thermodynamic equilibrium between air and a saturated LiCl solution. For the same temperature (25°C) and pressure (101.3 kPa) as the previous case for pure water (Fig. 2.6a), only 0.357 kPa of vapor pressure develops. Following eq. (2.10), the corresponding vapor density is reduced to 2.597 g/m^3.

2.2.5 Relative Humidity

Relative humidity is defined as the ratio of the absolute humidity (ρ_v) in equilibrium with any solution to the absolute humidity in equilibrium with free water ($\rho_{v,\text{sat}}$) at the same temperature. Following eqs. (2.8) and (2.9), it can be shown that this ratio is identical to the ratio of vapor pressure in equilibrium with the solution (u_v) and the saturated vapor pressure in equilibrium with free water ($u_{v,\text{sat}}$). The word *relative* refers to the reference with free water. Considering the previous example for a saturated LiCl solution, the relative humidity in the headspace above the solution is as follows:

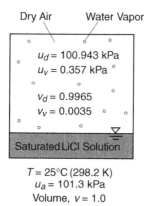

Figure 2.8 Equilibrium condition between saturated LiCl solution and air at 25°C and 101.3 kPa.

$$\text{RH} = \frac{\rho_v}{\rho_{v,\text{sat}}} = \frac{u_v}{u_{v,\text{sat}}} = \frac{2.597 \text{ g/m}^3}{22.99 \text{ g/m}^3} = \frac{0.357 \text{ kPa}}{3.167 \text{ kPa}} \approx 0.113 = 11.3\% \quad (2.12)$$

Relative humidity can be directly used to calculate the free energy per unit mass of the solution, E (J/kg), or chemical potential, μ (J/mol), as follows:

$$E = \frac{\mu}{\omega_w} = -\frac{RT}{\omega_w} \ln \frac{u_v}{u_{v,\text{sat}}} = -\frac{RT}{\omega_w} \ln(\text{RH}) \quad (2.13)$$

Table 2.7 illustrates equilibrium relative humidity, vapor density, and vapor pressure for several common saturated salt solutions at 25°C. For the case of a saturated LiCl solution at 25°C, the free energy per unit mass calculated by eq. (2.13) is about 300,000 J/kg, implying that the process of dissolving LiCl to the saturated solution condition lowers each kilogram of free water by 300,000 J. This also implies that if the pore air of an unsaturated soil has an equilibrium relative humidity of 11.3%, the pore water energy level is 300,000 J/kg lower than free water. Following Table 1.2, this is numerically equivalent to a suction of 300,000 kPa.

2.2.6 Dew Point

Relative humidity in the atmosphere and the pores of unsaturated soil can vary significantly between 0 and 100% between night and day and from season to season. As a result, phase transitions in the form of condensation and evaporation commonly occur. Condensation occurs as either dew or frost formation when the relative humidity of the pore air reaches 100%. Dew formation refers specifically to a phase transition from vapor to liquid. Frost formation refers specifically to a phase transition from vapor to solid.

Figure 2.9 illustrates two ideal scenarios to facilitate the understanding of dew formation. Let the pore air in an unsaturated soil be described at a certain initial condition (point A) by temperature T_0 and vapor pressure u_{vA}. The solid line on the figure defines the saturated vapor pressure condition defined by

TABLE 2.7 Equilibrium Relative Humidity, Vapor Density, and Vapor Pressure for Several Common Saturated Salt Solutions at 25°C

Solution	RH (%)	Vapor Density (g/m³)	Vapor Pressure (kPa)
Distilled water	100.00	22.990	3.167
$(NH_4)H_2PO_4$	92.7	21.340	2.936
KCl	84.2	19.383	2.667
NaCl	75.1	17.288	2.378
$NaNO_2$	64.4	14.825	2.040
$Mg(NO_3)_2 \cdot 6H_2O$	52.8	12.155	1.672
$LiCl \cdot H_2O$	11.28	2.597	0.357

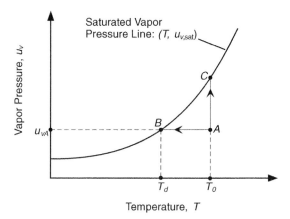

Figure 2.9 Dew formation under constant vapor pressure conditions (path from point A to point B) and constant temperature conditions (path from point A to point C).

eq. (2.11). The relative humidity at point A is defined as the ratio of the vapor pressure to the saturated vapor pressure ($RH_A = u_{vA}/u_{v,\text{sat}}$). Because point A is located below the saturated vapor pressure line, RH_A is some value less than 100%.

If the environment changes in such a way that the vapor pressure remains constant but the temperature drops, a 100% relative humidity condition will occur when the temperature reaches a critical value referred to as the *dew-point* temperature, or T_d. The dew point temperature is reached when the cooling path (shown as the horizontal line from point A to point B) intersects the saturated vapor pressure line and can be calculated from eq. (2.11) as follows, where u_v is in kilopascals:

$$T_d = \frac{36 \ln(u_v) - 4700}{\ln(u_v) - 16.78} \tag{2.14}$$

Dew formation may also occur if the initial temperature at point A remains unchanged but the vapor pressure increases to the saturated vapor pressure $u_{v,\text{sat}}$ (shown as the vertical path from point A to point C). The pressure at which this occurs can be assessed quantitatively from eq. (2.11).

2.3 DENSITY OF MOIST AIR

2.3.1 Effect of Water Vapor on Density of Air

Excluding the water vapor component, the composition of air is relatively constant for a particular location over most time spans important to geotechnical engineering problems. The relative amount of water vapor, however, can vary dramatically. Consider the Denver, Colorado, area, which typically averages about 10% relative humidity in the winter season and as much as 100% in the early summer. When coupled with naturally occurring changes in temperature and pressure, the prevailing atmospheric vapor pressure may vary anywhere between approximately 0.04 and 7.4 kPa. Consequently, the overall air density can vary up to as much as 7%.

Moist air is less dense than dry air because the molecular mass of water vapor [18.016 kg/kmol at standard temperature (0°C, or 273.14 K) and pressure (1 atm, or 101.325 kPa)] is considerably less than the molecular mass of dry air (28.966 kg/kmol). Figure 2.10 shows a comparison between the density of dry air (calculated from the ideal gas law) and the density of moist air at 100% relative humidity (measured experimentally) as functions of temperature and pressure. Note that the highest difference between the dry and moist air density occurs when the temperature is high and the absolute air pressure is low, possibly corresponding to early summer weather conditions.

2.3.2 Formulation for Moist Air Density

A quantitative expression for moist-air density can be derived by assuming that dry air and water vapor both follow ideal gas behavior. Thus, dry-air density can be written as:

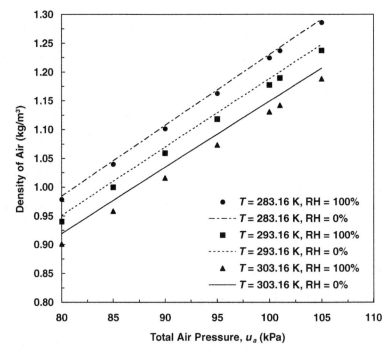

Figure 2.10 Density of dry air and moist air (100% RH) as functions of temperature and total air pressure (100% RH data from Kaye and Laby, 1973).

2.3 DENSITY OF MOIST AIR

$$\rho_d = \frac{u_d \omega_d}{RT} \qquad (2.15)$$

where ρ_d is the dry-air density at temperature T, u_d is the dry-air pressure, ω_d is the molecular mass of dry air, and R is the universal gas constant.

For water vapor density, the ideal gas law may be written in terms of the molecular mass of water vapor, ω_v, saturated vapor pressure $u_{v,\text{sat}}$, or saturated vapor density $\rho_{v,\text{sat}}$ and relative humidity RH as follows:

$$\rho_v = \frac{u_v \omega_v}{RT} = \frac{u_{v,\text{sat}} \omega_v}{RT} \text{RH} = \rho_{v,\text{sat}} \text{RH} \qquad (2.16)$$

Because moist air is comprised of both dry air and a vapor component, the dry-air pressure u_d is equal to $u_a - u_v$, where u_a is the total or absolute air pressure. Accordingly, the density of moist air ($\rho_{a,\text{moist}}$) may be written as the sum of the density of dry air and water vapor as follows:

$$\rho_{a,\text{moist}} = \frac{u_v \omega_v + (u_a - u_v) \omega_d}{RT}$$

$$= \frac{u_a \omega_d}{RT} - \frac{(\omega_d - \omega_v) u_v}{RT}$$

$$= \frac{u_a \omega_d}{RT} - \frac{(\omega_d - \omega_v) \omega_v u_v}{RT \omega_v} \qquad (2.17a)$$

which, in light of eq. (2.16), may be rewritten as

$$\rho_{a,\text{moist}} = \rho_d - \left(\frac{\omega_d}{\omega_v} - 1\right) \rho_v \qquad (2.17b)$$

where ρ_d is the dry-air density at the same temperature and pressure as the moist air.

Equation (2.17b) states that moist-air density can be calculated for any given temperature and pressure condition provided that both dry-air density and water vapor density are known. Substituting eqs. (2.11) and (2.16) into eq. (2.17), therefore, moist-air density can be rewritten as follows:

$$\rho_{a,\text{moist}} = \rho_d - \left(\frac{\omega_d}{\omega_v} - 1\right) \rho_{v,\text{sat}} \text{RH}$$

$$= \frac{u_d \omega_d}{RT} - 0.611 \left(\frac{\omega_d}{\omega_v} - 1\right) \exp\left(17.27 \frac{T - 273.2}{T - 36}\right) \frac{\omega_v \text{RH}}{RT} \qquad (2.18)$$

TABLE 2.8 Dry-Air Density (kg/m³) as Function of Temperature and Pressure

						Temperature (K)					
kPa	268.2 −5°C	273.2 0°C	278.2 5°C	283.2 10°C	288.2 15°C	293.2 20°C	298.2 25°C	303.2 30°C	308.2 35°C	313.2 40°C	318.2 45°C
80	1.039	1.020	1.002	0.984	0.967	0.951	0.935	0.919	0.904	0.890	0.876
81	1.052	1.033	1.014	0.996	0.979	0.962	0.946	0.931	0.916	0.901	0.887
82	1.065	1.046	1.027	1.009	0.991	0.974	0.958	0.942	0.927	0.912	0.898
83	1.078	1.058	1.039	1.021	1.003	0.986	0.970	0.954	0.938	0.923	0.909
84	1.091	1.071	1.052	1.033	1.015	0.998	0.981	0.965	0.950	0.934	0.920
85	1.104	1.084	1.064	1.046	1.028	1.010	0.993	0.977	0.961	0.946	0.931
86	1.117	1.097	1.077	1.058	1.040	1.022	1.005	0.988	0.972	0.957	0.942
87	1.130	1.109	1.090	1.070	1.052	1.034	1.016	1.000	0.983	0.968	0.953
88	1.143	1.122	1.102	1.083	1.064	1.046	1.028	1.011	0.995	0.979	0.963
89	1.156	1.135	1.115	1.095	1.076	1.058	1.040	1.023	1.006	0.990	0.974
90	1.169	1.148	1.127	1.107	1.088	1.069	1.051	1.034	1.017	1.001	0.985
91	1.182	1.160	1.140	1.119	1.100	1.081	1.063	1.046	1.029	1.012	0.996
92	1.195	1.173	1.152	1.132	1.112	1.093	1.075	1.057	1.040	1.023	1.007
93	1.208	1.186	1.165	1.144	1.124	1.105	1.087	1.069	1.051	1.034	1.018
94	1.221	1.199	1.177	1.156	1.136	1.117	1.098	1.080	1.063	1.046	1.029

95	1.234	1.211	1.190	1.169	1.148	1.129	1.110	1.092	1.074	1.057	1.040
96	1.247	1.224	1.202	1.181	1.161	1.141	1.122	1.103	1.085	1.068	1.051
97	1.260	1.237	1.215	1.193	1.173	1.153	1.133	1.115	1.096	1.079	1.062
98	1.273	1.250	1.227	1.206	1.185	1.164	1.145	1.126	1.108	1.090	1.073
99	1.286	1.262	1.240	1.218	1.197	1.176	1.157	1.138	1.119	1.101	1.084
100	1.299	1.275	1.252	1.230	1.209	1.188	1.168	1.149	1.130	1.112	1.095
101	1.312	1.288	1.265	1.242	1.221	1.200	1.180	1.161	1.142	1.123	1.106
102	1.325	1.301	1.277	1.255	1.233	1.212	1.192	1.172	1.153	1.135	1.117
103	1.338	1.313	1.290	1.267	1.245	1.224	1.203	1.184	1.164	1.146	1.128
104	1.351	1.326	1.302	1.279	1.257	1.236	1.215	1.195	1.176	1.157	1.139
105	1.364	1.339	1.315	1.292	1.269	1.248	1.227	1.207	1.187	1.168	1.150
106	1.377	1.352	1.327	1.304	1.281	1.260	1.238	1.218	1.198	1.179	1.161
107	1.390	1.364	1.340	1.316	1.293	1.271	1.250	1.229	1.210	1.190	1.172
108	1.403	1.377	1.352	1.329	1.306	1.283	1.262	1.241	1.221	1.201	1.182
109	1.416	1.390	1.365	1.341	1.318	1.295	1.273	1.252	1.232	1.212	1.193
110	1.429	1.403	1.378	1.353	1.330	1.307	1.285	1.264	1.243	1.224	1.204

which completely defines moist-air density in terms of temperature, total air pressure, and relative humidity. The dry-air density [the first term of eq. (2.18)] is tabulated in Table 2.8, and the corresponding correction factor [the second term in eq. (2.18)] is tabulated in Table 2.9. Given Tables 2.8 and 2.9, one can estimate moist-air density for any given set of temperature, pressure, and relative humidity conditions.

Example Problem 2.2 Calculate the density of air at RH = 0, 50, 100% when the air temperature is 45°C and the total air pressure is 101.3 kPa. Calculate the density of air at RH = 0, 50, 100% when the air temperature is 0°C and the total air pressure is 80 kPa. At RH = 50% and an air pressure of 100 kPa, which temperature leads to a higher air density, 0 or 45°C?

Solution For temperature of 45°C (318.2 K) and pressure 101.3 kPa:
a. RH = 0%:

$$\rho_d = \frac{u_d \omega_d}{RT} = \frac{(101.3 \times 10^3 \text{ N/m}^2)(28.966 \times 10^{-3} \text{ kg/mol})}{(8.314 \text{ N} \cdot \text{m/mol} \cdot \text{K})(318.2 \text{ K})}$$

$$= 1.109 \text{ kg/m}^3$$

b. RH = 50%:

$$\rho_{a,\text{moist}}$$
$$= 1.109 - 0.611 \left(\frac{\omega_a}{\omega_v} - 1\right) \exp\left(17.27 \frac{T - 273.2}{T - 36}\right) \frac{\omega_v(\text{RH})}{RT}$$

$$= 1.109 - 0.611 \left(\frac{28.966}{18.016} - 1\right) \exp\left(17.27 \frac{318.2 - 273.2}{318.2 - 36}\right) \frac{(18.016)(0.5)}{(8.314)(318.2)}$$

$$= 1.109 - (0.0397)(0.5)$$

$$= 1.089 \text{ kg/m}^3$$

c. RH = 100%:

$$\rho_{a,\text{moist}} = 1.109 - 0.611 \left(\frac{28.966}{18.016} - 1\right) \exp\left(17.27 \frac{318.2 - 273.2}{318.2 - 36}\right)$$

$$\frac{(18.016)(1.0)}{(8.314)(318.2)}$$

$$= 1.109 - (0.0397)(1.0)$$

$$= 1.069 \text{ kg/m}^3$$

For temperature of 0°C (273.2 K) and air pressure 80 kPa:

TABLE 2.9 Correction Factors for Moist-Air Density as Function of Temperature and Relative Humidity

RH (%)	268.2 −5°C	273.2 0°C	278.2 5°C	283.2 10°C	288.2 15°C	293.2 20°C	298.2 25°C	303.2 30°C	308.2 35°C	313.2 40°C	318.2 45°C
0	0.000	0.000	0.000	0.000	0.000	0.000	0.000	0.000	0.000	0.000	0.000
5	0.000	0.000	0.000	0.000	0.000	−0.001	−0.001	−0.001	−0.001	−0.002	−0.002
10	0.000	0.000	0.000	−0.001	−0.001	−0.001	−0.001	−0.002	−0.002	−0.003	−0.004
15	0.000	0.000	−0.001	−0.001	−0.001	−0.002	−0.002	−0.003	−0.004	−0.005	−0.006
20	0.000	−0.001	−0.001	−0.001	−0.002	−0.002	−0.003	−0.004	−0.005	−0.006	−0.008
25	−0.001	−0.001	−0.001	−0.002	−0.002	−0.003	−0.004	−0.005	−0.006	−0.008	−0.010
30	−0.001	−0.001	−0.001	−0.002	−0.002	−0.003	−0.004	−0.006	−0.007	−0.009	−0.012
35	−0.001	−0.001	−0.001	−0.002	−0.003	−0.004	−0.005	−0.006	−0.008	−0.011	−0.014
40	−0.001	−0.001	−0.002	−0.002	−0.003	−0.004	−0.006	−0.007	−0.010	−0.012	−0.016
45	−0.001	−0.001	−0.002	−0.003	−0.004	−0.005	−0.006	−0.008	−0.011	−0.014	−0.018
50	−0.001	−0.001	−0.002	−0.003	−0.004	−0.005	−0.007	−0.009	−0.012	−0.016	−0.020
55	−0.001	−0.002	−0.002	−0.003	−0.004	−0.006	−0.008	−0.010	−0.013	−0.017	−0.022
60	−0.001	−0.002	−0.002	−0.003	−0.005	−0.006	−0.008	−0.011	−0.014	−0.019	−0.024
65	−0.001	−0.002	−0.003	−0.004	−0.005	−0.007	−0.009	−0.012	−0.016	−0.020	−0.026
70	−0.001	−0.002	−0.003	−0.004	−0.005	−0.007	−0.010	−0.013	−0.017	−0.022	−0.028
75	−0.002	−0.002	−0.003	−0.004	−0.006	−0.008	−0.011	−0.014	−0.018	−0.023	−0.030
80	−0.002	−0.002	−0.003	−0.005	−0.006	−0.008	−0.011	−0.015	−0.019	−0.025	−0.032
85	−0.002	−0.003	−0.004	−0.005	−0.007	−0.009	−0.012	−0.016	−0.020	−0.026	−0.034
90	−0.002	−0.003	−0.004	−0.005	−0.007	−0.009	−0.013	−0.017	−0.022	−0.028	−0.036
95	−0.002	−0.003	−0.004	−0.005	−0.007	−0.010	−0.013	−0.018	−0.023	−0.030	−0.038
100	−0.002	−0.003	−0.004	−0.006	−0.008	−0.011	−0.014	−0.018	−0.024	−0.031	−0.040

Temperature (K)

MATERIAL VARIABLES

a. RH = 0%:

$$\rho_d = \frac{u_d \omega_d}{RT} = \frac{(80 \times 10^3 \text{ N/m}^2)(28.966 \times 10^{-3} \text{ kg/mol})}{(8.314 \text{ N} \cdot \text{m/mol} \cdot \text{K})(273.2 \text{ K})} = 1.020 \text{ kg/m}^3$$

b. RH = 50%:

$$\rho_{a,\text{moist}} = 1.020 - 0.611 \left(\frac{28.966}{18.016} - 1\right) \exp\left(17.27 \frac{273.2 - 273.2}{273.2 - 36}\right)$$

$$\frac{(18.016)(0.5)}{(8.314)(273.2)}$$

$$= 1.020 - (0.0030)(0.5)$$

$$= 1.019 \text{ kg/m}^3$$

c. RH = 100%:

$$\rho_{a,\text{moist}} = 1.020 - 0.611 \left(\frac{28.966}{18.016} - 1\right) \exp\left(17.27 \frac{273.2 - 273.2}{273.2 - 36}\right)$$

$$\frac{(18.016)(1.0)}{(8.314)(273.2)}$$

$$= 1.020 - (0.0030)(1.0)$$

$$= 1.017 \text{ kg/m}^3$$

For RH = 50% and air pressure 100 kPa:
a. $T = 45°C$ (318.2 K):

$$\rho_d = \frac{u_d \omega_d}{RT} = \frac{(100 \times 10^3 \text{ N/m}^2)(28.966 \times 10^{-3} \text{ kg/mol})}{(8.314 \text{ N} \cdot \text{m/mol} \cdot \text{K})(318.2 \text{ K})} = 1.095 \text{ kg/m}^3$$

$$\rho_{a,\text{moist}} = 1.095 - 0.611 \left(\frac{28.966}{18.016} - 1\right)$$

$$\exp\left(17.27 \frac{318.2 - 273.2}{318.2 - 36}\right) \frac{(18.016)(0.5)}{(8.314)(318.2)}$$

$$= 1.075 \text{ kg/m}^3$$

b. $T = 0°C$ (273.2 K):

$$\rho_d = \frac{u_d \omega_d}{RT} = \frac{(100 \times 10^3 \text{ N/m}^2)(28.966 \times 10^{-3} \text{ kg/mol})}{(8.314 \text{ N} \cdot \text{m/mol} \cdot \text{K})(273.2 \text{ K})} = 1.275 \text{ kg/m}^3$$

$$\rho_{a,\text{moist}} = 1.275 - 0.611 \left(\frac{28.966}{18.016} - 1\right)$$

$$\exp\left(17.27 \frac{273.2 - 273.2}{273.2 - 36}\right) \frac{(18.016)(0.5)}{(8.314)(273.2)} = 1.274 \text{ kg/m}^3$$

Thus the higher the temperature, the lower the air density. Several relationships also become clear from this example: (1) the higher the air pressure, the higher the air density, (2) the higher the relative humidity, the lower the air density, and (3) air density can vary as much as 26% for the given RH and temperature changes.

2.4 SURFACE TENSION

2.4.1 Origin of Surface Tension

The geometry of the interface between any two fluids is governed by the balance of forces existing on both sides of the interface. In a liquid-liquid system, such as a drop of oil in water, these forces include the pressure in each liquid and an *interfacial tension* that acts between the two. In a gas-liquid system, such as the air-water interface in unsaturated soil, the surface tension of the air phase can be practically ignored, leading to only three components necessary for mechanical equilibrium: air pressure, water pressure, and the *surface tension* of the water phase.

Surface tension is often defined as the maximum energy level a fluid can store without breaking apart. In more specific terms, surface tension may be defined as the energy required to either open or close a unit area at a phase interface. Accordingly, surface tension has units of joules per square meter (J/m^2), which is equivalent to newtons per meter (N/m). The latter dimension of force per unit length may be conceptualized as stress on a thin skin and is a reflection of the amount of force applied to a given length of the interface. Drawing an analogy with a rubber string under a tensile force, the energy stored in the string can be described by force per unit length. Other common units for surface tension include dynes per centimeter and ergs per square centimeter. One dyne per centimeter is equal to one erg per square centimeter and one millinewton per meter.

In a gas-liquid interface, surface tension arises from imbalanced intermolecular forces acting on molecules comprising the liquid phase. At an air-water interface, for example, the water molecules located some finite distance away from the interface do not experience equal cohesive force in all directions. Consequently, the near-surface molecules cohere more strongly to those directly associated with them on the surface, creating an unbalanced force toward the interior of the water phase. For the system to remain in mechanical equilibrium, a resultant force, surface tension, develops along the interface. For liquids comprised of polar molecules such as water, the intermolecular cohesive forces at the surface and the resulting surface tension are relatively high.

For mathematical convenience, the surface tension of a liquid-gas interface is often regarded as a concentrated force acting only along the surface boundary. In reality, however, this is not the case. Rather, surface tension is the resultant of a distributed stress that acts not only at the interface, but also to some depth within the liquid phase. This stress distribution for an air-water interface is illustrated conceptually in Fig. 2.11. If there were no pressure difference across the interface (i.e., $u_a = u_w$), a perfectly flat interfacial surface would be expected. In this case, surface stress may not be required to exist because the system can be under mechanical equilibrium with or without the surface stress anomaly. However, when there is a pressure difference between the two phases, an additional force, surface tension, is required for equilibrium. According to Saint Venant's principle, the stress anomaly can only occur in the vicinity where the stress difference occurs. The thickness of the boundary layer (d) in an air-water system is about 10 layers of water molecules (about 3×10^{-9} m or 3 nm). The resultant of the stress increase within the boundary layer, conveniently called surface tension T_s, may therefore be defined mathematically as

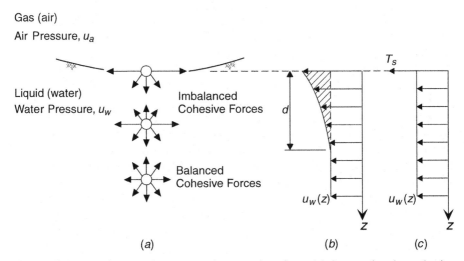

Figure 2.11 Surface tension at an air-water interface: (*a*) intermolecular cohesive forces among water molecules near the interface, (*b*) conceptual pressure distribution with depth from the interface, and (*c*) surface tension model showing T_s as the resultant of imbalanced intermolecular forces acting along interface.

$$T_s = \int_0^d (\sigma - u_w)\, \delta z \qquad (2.19)$$

where σ is the total stress in the water phase and d is the thickness of the boundary layer where the stress increase occurs.

Methods for directly measuring surface tension typically involve actual force measurements at the interface using various types of mechanical probes (e.g., DuNouy rings, Wilhelmy plates, and needle probes), capillary tube analysis, or the analysis of the shape and size of a hanging drop of liquid (Adamson, 1976). Water at 20°C has a surface tension of 72.75 mN/m. This is to say that it would take a force greater than 72.75 mN to "break" a 1-m-long surface film of water at 20°C. By comparison, ethyl alcohol (a nonpolar liquid) has a relatively low surface tension of 22.3 mN/m. Mercury, which has a propensity to develop relatively large cohesive forces, has a surface tension of about 465 to 480 mN/m.

The surface tension of water is dependent on temperature, generally decreasing as temperature increases. Figure 2.12 illustrates this dependence for temperature ranging between −8 and 100°C. Exact values are summarized in Table 2.10. The variation between these extremes is roughly linear. The de-

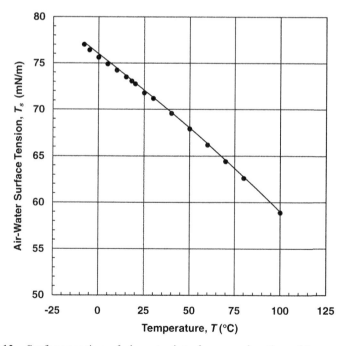

Figure 2.12 Surface tension of air-water interface as a function of temperature (data from Weast et al., 1981).

MATERIAL VARIABLES

TABLE 2.10 Air-Water Surface Tension as Function of Temperature

Temp. (°C)	T_s (mN/m)
−8	77
−5	76.4
0	75.6
5	74.9
10	74.22
15	73.49
18	73.05
20	72.75
25	71.79
30	71.18
40	69.56
50	67.91
60	66.18
70	64.4
80	62.6
100	58.9

crease in surface tension with increasing temperature is partly responsible for the notion that hot water is a better cleaning agent than cold water; the lower surface tension makes hot water a more efficient *wetting agent* to get into minute pores and fissures rather than bridging them with surface tension. Soaps and detergents or other such *surfactants* further lower the surface tension to enhance the cleansing process.

2.4.2 Pressure Drop across an Air-Water Interface

The existence of a curved air-water interface is a direct indication of a pressure difference existing between the air and water phases. In light of the nature and origin of surface tension, however, it should be emphasized that it is not the surface tension that results in the pressure drop across the interface. Rather, it is the pressure drop that causes the surface to change its geometry and to induce the surface tension. In all cases, the phase with the smaller pressure tends to expand, resulting in the interface surface oriented concave to the high-pressure side. The smaller pressure side could be either in the air phase or the water phase. Examples where lower pressure exists in the water phase include an air bubble in water (Fig. 2.13a) and a meniscus in a capillary tube. A case indicative of higher pressure in the water phase is a raindrop in air (Fig. 2.13b). In a three-phase unsaturated soil system, whether the concave side of the interface corresponds to the water or air phase depends on the properties of the soil solid, the air pressure, and the location

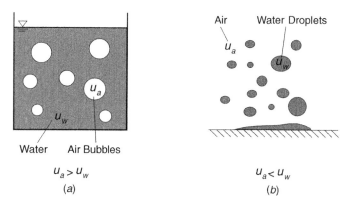

Figure 2.13 Pressure differences across air-water interfaces: (*a*) air bubbles in water; the air pressure is higher than the water pressure ($u_a > u_w$) and (*b*) water droplets in air; the water pressure is higher than the air pressure ($u_a < u_w$).

of the pore water in the system. Under most circumstances of practical interest, the soil solid is hydrophilic, the air pressure is atmospheric, and the pore water exists in the form of menisci located relatively far from the particle surface. Accordingly, the concave side is typically associated with the air phase and the water pressure is lower than the air pressure.

The pressure drop acting across a spherical air-water interface can be evaluated by investigating the requirement for mechanical equilibrium. In the analysis that follows, no assumption is made regarding which phase has a higher pressure. The equations presented can be applied to any general air-water interface.

Figure 2.14*a* shows a free-body diagram for a two-dimensional curved interface. The interface is analogous to a meniscus in a capillary tube, where

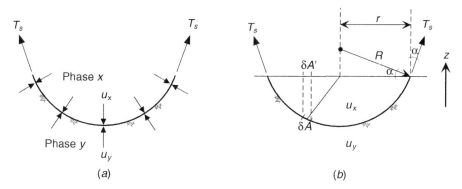

Figure 2.14 Free-body diagrams for pressure and surface tension across a spherical phase interface.

phase y would represent water and phase x would represent the overlying air, or the surface of a drop of mercury, where phase x represents the mercury and phase y represents the surrounding air. The free-body diagram includes the pressure on both sides of the interface and surface tension, which is represented as a tensile force acting perpendicular to each plane where the interface has been cut. In this case, the curvature of the interface dictates that the pressure u_x in phase x is greater than the pressure u_y in phase y. If u_x increases, the interface expands.

Referring to Fig. 2.14b, the projection of incremental force due to pressure on both sides of the interface over an area δA in the vertical direction is as follows:

$$\delta F_v \downarrow = -(u_x - u_y)\,\delta A \cos \alpha = -(u_x - u_y)\,\delta A' \qquad (2.20)$$

where $\delta A'$ is the projection of δA in the horizontal axis. The total vertical force due to the pressure difference acts over the area of the interface as follows:

$$F_v \downarrow = -(u_x - u_y)\pi r^2 \qquad (2.21)$$

The projection of surface tension around the circumference of the cut in the vertical direction is

$$F_v \uparrow = 2\pi r T_s \cos \alpha \qquad (2.22)$$

Applying force equilibrium leads to

$$2\pi r T_s \cos \alpha - (u_x - u_y)\pi r^2 = 0 \qquad (2.23)$$

or

$$u_x - u_y = \frac{2T_s}{r/\cos \alpha} = \frac{2T_s}{R} \qquad (2.24a)$$

This simple equation describes the interrelation among surface tension, pressure change, and surface curvature. When $R \to \infty$, eq. (2.24a) leads to $u_x = u_y$, indicating a null pressure difference and a flat interface. When $u_x > u_y$, $R > 0$, whereas when $u_x < u_y$, $R < 0$.

With respect to water rising in a capillary tube, eq. (2.24a) can be written more specifically as

$$u_a - u_w = \frac{2T_s}{R} \tag{2.24b}$$

where u_a is a positive or zero air pressure, u_w is a negative water pressure, and R is the radius of curvature of the capillary meniscus. With respect to unsaturated soil, the difference $u_a - u_w$ is referred to as matric suction.

Example Problem 2.3 Typical water drops from a sprinkler nozzle are spherical and on the order of 0.2–0.02 mm in radius. If the surface tension of water is 72 mN/m at 25°C and the ambient air pressure is 100 kPa, what is the highest pressure inside the water drops?

Solution The highest pressure occurs in the smallest water drops. Assigning phase x as air and phase y as water, $R < 0$ because the curvature of a water drop dictates that the water pressure be higher than the air pressure. A radius $R = -2 \times 10^{-5}$ m is used to represent the smallest water drop. Rearranging eq. (2.24) and solving for u_w leads to

$$u_w = u_a - \frac{2T_s}{R} = 100 \text{ kPa} - \frac{2 \times 72 \times 10^{-3} \text{ N/m}}{-2 \times 10^{-5} \text{ m}} \frac{1 \text{ kPa}}{1000 \text{ Pa}}$$

$$= 100 + 7.2 = 107.2 \text{ kPa}$$

Example Problem 2.4 Assume that the air-water interface for a certain degree of saturation among particles in fine sand is spherical and 0.1 to 0.01 mm in diameter. If the surface tension of water is 72 mN/m at 25°C and the pore air pressure is 100 kPa, what is the range of pore pressure in the unsaturated sand?

Solution Assigning phase x as air and phase y as water, $R > 0$ because the curvature of the meniscus dictates that the water pressure be lower than the air pressure. Accordingly, the range of R is from 5×10^{-5} to 5×10^{-6} m. Rearranging eq. (2.24) and solving for u_w leads to

$$u_w|_{max} = u_a - \frac{2T_s}{R_{max}} = 100 \text{ kPa} - \frac{2 \times 72 \times 10^{-3} \text{ N/m}}{5 \times 10^{-5} \text{ m}} \frac{1 \text{ kPa}}{1000 \text{ Pa}}$$

$$= 100 - 2.88 = 97.12 \text{ kPa}$$

$$u_w|_{min} = u_a - \frac{2T_s}{R_{min}} = 100 \text{ kPa} - \frac{2 \times 72 \times 10^{-3} \text{ N/m}}{5 \times 10^{-6} \text{ m}} \frac{1 \text{ kPa}}{1000 \text{ Pa}}$$

$$= 100 - 28.8 = 71.2 \text{ kPa}$$

Note that if a tensiometer were used to measure the magnitude of the pore water pressure (Section 10.2) atmospheric pressure would have been used as a reference value. Accordingly, the tensiometer would have recorded readings of negative water pressure as follows:

$$u_w|_{max} = u_a - \frac{2T_s}{R_{max}} - u_a = 100 \text{ kPa} - \frac{2 \times 72 \times 10^{-3} \text{ N/m}}{5 \times 10^{-5} \text{ m}} \frac{1 \text{ kPa}}{1000 \text{ Pa}}$$

$$- 100 \text{ kPa} = -2.88 \text{ kPa}$$

$$u_w|_{min} = u_a - \frac{2T_s}{R_{min}} - u_a = 100 \text{ kPa} - \frac{2 \times 72 \times 10^{-3} \text{ N/m}}{5 \times 10^{-6} \text{ m}} \frac{1 \text{ kPa}}{1000 \text{ Pa}}$$

$$- 100 \text{ kPa} = -28.8 \text{ kPa}$$

Thus, the reading from a tensiometer is a value indicating the pressure deficit with respect to the prevailing atmospheric pressure. The absolute value is equal to matric suction.

2.5 CAVITATION OF WATER

2.5.1 Cavitation and Boiling

The terms cavitation and boiling refer to the same phase transformation process yet under fundamentally different conditions. Cavitation and boiling are identical in that they each describe the same physical result, specifically, the formation or *nucleation* of vapor bubbles in liquid. Each occurs when the liquid vapor pressure is higher than the absolute liquid pressure. The paths describing the change in the thermodynamic state variables that precipitate vapor bubble formation in each case, however, are quite different. A useful way to distinguish these two processes is to define cavitation as the process of vapor nucleation in a liquid when the absolute pressure falls below the vapor pressure. Boiling, on the other hand, may be defined as the process of vapor nucleation in a liquid when the temperature is raised above the saturated vapor/liquid temperature (see, e.g., Brennen, 1995).

The phase transformation associated with cavitation or boiling may be better understood by considering a thermodynamic phase diagram for water (Fig. 2.15). Water can, of course, be found in one of three phases: the solid phase as ice, the liquid phase as water, and the gaseous phase as water vapor. Within the space of the state variables pressure and temperature, boundaries can be drawn to define the state of water at any given pressure and temperature condition. The boundary between the liquid and vapor state is referred to as the *vaporization curve* (designated by points *a* and *b*), the boundary between the solid and liquid state (*ac*) as the *fusion curve,* and the boundary between the solid and vapor state (*ad*) as the *sublimation curve.* The triple point (*a*) is the point representing conditions at which solid, liquid, and vapor

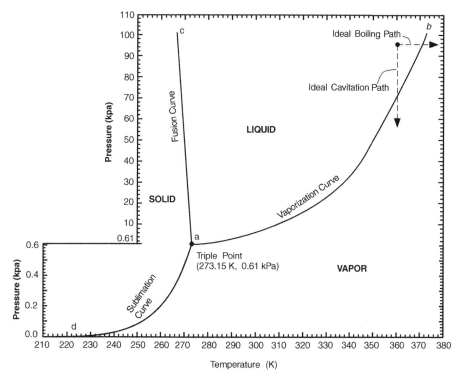

Figure 2.15 Thermodynamic phase diagram for pure water.

states may coexist. The triple point of water occurs at a temperature of 0°C (273.2 K) and a pressure of 0.61 kPa.

The vaporization curve (*ab*) describes the combination of pressure and temperature conditions for which the liquid and vapor states of water can exist in equilibrium. At all points along this line, evaporation (i.e., phase transformation from liquid to vapor) and condensation (i.e., phase transformation from vapor to liquid) occur simultaneously and at the same statistical frequency. The chemical potentials of the two coexisting phases are equal and the vapor pressure is equal to the saturated vapor pressure.

Cavitation and boiling each describe the process of translation across the vaporization curve from the liquid state to the vapor state. In cavitation, the vaporization curve is crossed along a path of decreasing pressure. Along an ideal cavitation path at constant temperature, such as that hypothesized in Fig. 2.15, the vapor pressure in both the liquid and gas phase remains relatively unchanged. In boiling, the vaporization curve is crossed along a path of increasing temperature. Along an ideal boiling path at constant pressure (also shown), the vapor pressure is highly dependent on temperature. The nucleation of vapor bubbles along either a cavitation or boiling path does not necessarily occur as soon as the vaporization curve is "crossed." Rather, an intermediate, or *metastable,* liquid phase zone is entered within which the liquid is under tension. Phase transformation from liquid to vapor occurs if and when the tensile strength of the water is exceeded.

2.5.2 Hydrostatic Atmospheric Pressure

To accurately assess whether the absolute pressure of water reaches the local vapor pressure along a path toward cavitation, knowledge of the vapor pressure, atmospheric air pressure, temperature, and relative humidity is required at a particular elevation of interest. The dependency of liquid vapor pressure on temperature in free water can be quantified by eq. (2.11) or Fig. 2.7 if the local temperature is known. Liquid vapor pressure is relatively insensitive to change in total pressure.

The mean absolute atmospheric pressure u_a at an elevation z can be obtained by the following expression (Ross et al., 1992):

$$u_a = u_0 \left(1 - \frac{\lambda z}{T_0}\right)^{1/R_v \lambda} \tag{2.25}$$

where T_0 is a standard reference temperature ($T_0 = 288.15$ K), u_0 is a standard atmospheric pressure at mean sea level ($u_0 = 1013.25$ mbar), z is the elevation above mean sea level (m), λ is the *standard atmospheric lapse rate* in the tropopause zone (i.e., $z < 11000$ m), and R_v is the *moist atmospheric constant.* The atmospheric lapse rate λ describes the decrease in temperature with increasing elevation. The standard lapse rate is $\lambda = -dT/dz = 6.5$ K · km^{-1}.

When the vapor pressure is much smaller than the total air pressure, the moist atmospheric constant R_v may be defined as

$$R_v = R_d \left(1 + 0.38 \text{RH} \frac{u_{v0}}{u_0}\right) \tag{2.26}$$

where u_{v0} is the saturated vapor pressure at the reference temperature T_0, RH is relative humidity, and R_d is an atmospheric constant for dry air ($R_d = 29.271$ m · K^{-1}).

For example, if the elevation of interest is 1500 m above mean sea level (roughly the elevation of Denver, Colorado) and the local mean relative humidity is 100%, then the mean atmospheric pressure at the ground surface according to eqs. (2.25) and (2.26) is 84.653 kPa. By comparison, the mean atmospheric pressure is 84.817 kPa if the local mean relative humidity is zero.

Equation (2.25) is a generalization of several well-known laws describing the atmospheric pressure profile. Ross et al. (1992), for example, showed that eq. (2.25) could be reduced to the well-known pressure law for dry (RH = 0%), isothermal conditions (e.g., Iribarne and Godson, 1981):

$$u_a = u_0 e^{-z/R_d T_0} \tag{2.27}$$

If it is assumed that the atmosphere is described by a constant lapse rate (i.e., temperature decreases linearly with elevation), it has been shown that the atmospheric pressure profile follows a law similar in form to eq. (2.25) (e.g., Iribarne and Godson, 1981):

$$u_a = u_0 \left(1 - \frac{\lambda z}{T_{v0}}\right)^{1/R_d \lambda} \tag{2.28}$$

where T_{v0} is the *virtual temperature* at the true temperature T_0. Virtual temperature is a widely used term in atmospheric science and is defined as the temperature required for dry air to have the same density as moist air at the true temperature. For the same volume and pressure, dry air requires a higher temperature than moist air to achieve the same density.

Depending on relative humidity, the virtual temperature T_{v0} is generally equal to or higher than the true temperature T. The relationship between virtual temperature and true temperature when the vapor pressure is much smaller than the total air pressure is

$$T_{v0} = \frac{T_0}{1 - \text{RH}\,\dfrac{\omega_a - \omega_v}{\omega_a}\,\dfrac{u_{v0}}{u_0}} \approx \left(1 + 0.38\,\text{RH}\,\frac{u_{v0}}{u_0}\right) T_0 \tag{2.29}$$

which is illustrated on Fig. 2.16 for RH = 0, 50, 100%. As relative humidity and temperature increase, the difference between the virtual temperature and true temperature increases.

Table 2.11 shows a comparison of the predicted relationships between mean atmospheric pressure and elevation using eqs. (2.25), (2.27), and (2.28). Equation (2.28) is identical to eq. (2.25) when the relative humidity is zero. It can be observed that the predicted mean atmospheric pressures are higher for the moist atmosphere than that of the dry atmosphere at low elevation, but the predicted mean atmospheric pressures for the moist atmosphere are lower than that of the dry atmosphere at high elevation. The equal pressure elevation for the moist atmosphere predicted by eq. (2.25) is about 500 m, and the equal pressure elevation for the moist atmosphere predicted by eq. (2.28) is about 1100 m. The reason for a higher atmospheric pressure for the moist atmosphere at low elevation is that both the moist atmospheric models [eqs. (2.25) and (2.28)] consider the linear temperature drop so that the density of air at lower elevation is higher than the dry air. On the other hand, as the elevation increases, density reduction due to water vapor becomes more and more pronounced. Consequently, the pressure of moist air becomes smaller than that of dry air. The maximum difference in the predicted atmos-

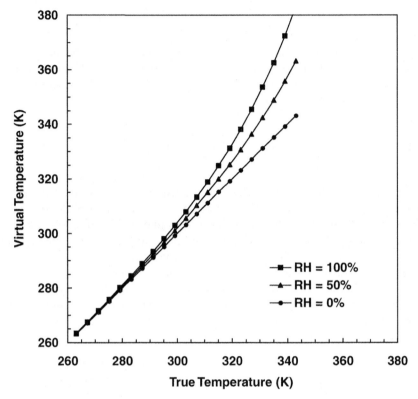

Figure 2.16 Virtual temperature as a function of relative humidity and true temperature.

pheric pressure among these different laws could be as high as 0.317 kPa at an elevation of 2000 m.

2.5.3 Cavitation Pressure

The concern with the cavitation of water in the context of unsaturated soil mechanics is primarily a practical one. Tensiometers, for example, are commonly used for direct measurements of negative pore water pressure in unsaturated soil. If cavitation occurs under increasingly negative water pressure, continuity in the liquid phase between the measurement system and the soil pore water is lost and the measurement becomes unreliable.

Cavitation can occur in free water, pore water, porous stones, or capillary tubes when the liquid phase pressure u_w approaches its vapor pressure u_v. In measurement devices such as tensiometers, the liquid phase pressure is recorded as a deficit with respect to the local atmospheric pressure, generally referred to as a *negative gauge pressure measurement*. Cavitation pressure for

TABLE 2.11 Relationship between Elevation above Sea Level and Mean Atmospheric Pressure Predicted by Several Laws

Elevation (z) (m)	Moist Atm. (RH = 100%) eq. (2.25) (kPa)	Dry Atm. (RH = 0%) eq. (2.27) (kPa)	Moist Atm. (RH = 100%) eq. (2.28) (kPa)
0	101.325	101.325	101.325
100	100.137	100.131	100.145
200	98.960	98.951	98.975
300	97.795	97.784	97.817
400	96.640	96.632	96.670
500	95.497	95.493	95.533
600	94.365	94.367	94.408
700	93.243	93.255	93.293
800	92.132	92.156	92.189
900	91.033	91.070	91.095
1000	89.943	89.997	90.012
1100	88.864	88.936	88.939
1200	87.796	87.888	87.877
1300	86.738	86.852	86.825
1400	85.691	85.828	85.783
1500	84.653	84.817	84.752
1600	83.626	83.817	83.730
1700	82.609	82.829	82.719
1800	81.602	81.853	81.717
1900	80.605	80.888	80.725
2000	79.618	79.935	79.743

a negative gauge instrument u_g is the difference between the local atmospheric pressure and the liquid vapor pressure:

$$u_g = u_a - u_v = u_a - \text{RH}\, u_{v0} \qquad (2.30)$$

where u_a can be evaluated from eq. (2.25), (2.27), or (2.28), the relative humidity RH can be measured from a relative humidity probe, and the saturated vapor pressure can be obtained from eq. (2.11). Because atmospheric pressure is a function of elevation, the cavitation pressure for a negative gauge instrument may be calculated as a function of elevation. A conceptual illustration of gauge cavitation pressure u_g in pure water as a function of elevation z is illustrated in Fig. 2.17. Temperature variations and impurities in the water under consideration (e.g., dissolved gases, minute air bubbles, minute solids) may further lower cavitation pressure by creating sites for nucleation to occur.

Example Problem 2.5 A tensiometer measurement is made in Denver, Colorado, where the elevation is about 1500 m, the relative humidity in early

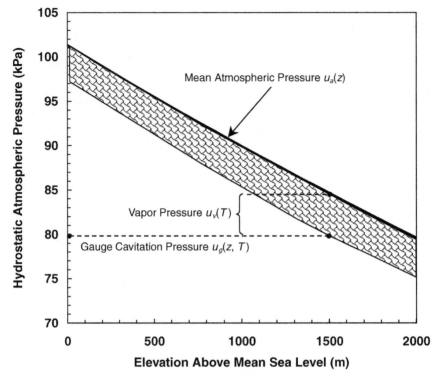

Figure 2.17 Gauge cavitation pressure for free water as function of elevation above sea level and hydrostatic atmospheric pressure.

summer can reach 90%, and temperature can reach 35°C (308.18 K). Estimate the maximum matric suction that may be measured. Assume that this limit is controlled by the cavitation pressure of free water.

Solution According to eqs. (2.25), (2.27), and (2.28), the predicted mean atmospheric pressures are 84.65, 84.82, and 84.75 kPa, respectively. The vapor pressure, from eqs. (2.11) and (2.12), is (5.624)(0.90) = 5.06 kPa. Therefore, cavitation may occur when the reading of the tensiometer reaches 84.65 − 5.06 = 79.59 kPa. By the same token, if the same tensiometer is used at sea level under the same climatic conditions, cavitation may not occur until the reading reaches 96.24 kPa.

PROBLEMS

2.1. What are the state variables that control the density of air? What is the average air density at your location?

2.2. What is the physical meaning of relative humidity?

2.3. At 25°C and 101.3 kPa (1 atm), what is the ratio of the viscosity of water to the viscosity of air? The viscosity of which phase, air or water, is more sensitive to temperature changes between 0 and 100°C?

2.4. Temperature varies between 15°C in the night and 30°C in the afternoon at a certain location. If the ambient vapor pressure remains constant at 1.6 kPa, what is the range of the relative humidity variation? If the vapor pressure remains unchanged, at what temperature will dew formation occur?

2.5. If a saturated swelling soil has a specific gravity of 2.7 and gravimetric water content of 300%, what is the volumetric water content?

2.6. A closed room is filled with humid air. If the temperature rises significantly, does the relative humidity increase or decrease?

2.7. Can the vapor pressure of soil gas be greater than the saturation pressure at the same temperature and pressure? Why or why not?

2.8. Can volumetric water content be greater than 100% in unsaturated soil?

2.9. Is degree of saturation a mass-based or volume-based quantity?

2.10. When the temperature of unsaturated soil increases, does the surface tension at the air-water interface increase or decrease?

2.11. What is the density of dry air if the prevailing temperature and pressure are 25°C and 95 kPa, respectively? What is the relative change in dry-air density if the temperature rises to 40°C and the air pressure remains unchanged? If the temperature is kept at a constant value of 25°C, how much pressure change is required to cause the dry-air density to decrease by 15% compared to 95 kPa?

2.12. Estimate the viscosity of air and water at a temperature of 50°C. Given a mean pore size for a sandy soil as 10^{-3} m, and a specific discharge for both air and water as 10^{-2} m/s, identify the flow regimes for the air and water, respectively.

2.13. The relative humidity at equilibrium in an unsaturated soil is measured to be 80% at 22°C. (a) What is the vapor pressure in the soil? (b) What is the vapor density in the soil? (c) What is the dew-point temperature if the vapor density is maintained constant but the temperature drops during the night? (d) What is the absolute humidity if temperature in the soil is maintained constant but vaporization is allowed to occur? (e) What is the free energy per unit mass of the pore water?

2.14. At a prevailing temperature of 25°C and pressure of 95 kPa, how much does the density of air change from a completely dry state to a 100% relative humidity state?

2.15. If the ambient air pressure is 101.3 kPa and the temperature is 20°C, what is the pressure inside the water meniscus for a capillary tube with a diameter of 0.001 mm? If the temperature increases to 50°C, what is the pressure inside the meniscus?

2.16. If a tensiometer were used to measure matric suction of unsaturated soil at an elevation of 500 m above sea level, what would be the approximate maximum possible reading of the tensiometer?

CHAPTER 3

INTERFACIAL EQUILIBRIUM

3.1 SOLUBILITY OF AIR IN WATER

3.1.1 Henry's Law

Air and water mainly exist as separate phases in unsaturated soil. However, at mechanical and chemical equilibrium, a portion of water may exist in the air phase as vapor and a portion of air may exist in the water phase as solute. For any given set of temperature and pressure conditions, there are two fundamental questions associated with many geotechnical engineering problems: (1) How much water can be vaporized in air? (2) How much air can be dissolved in water? The principles relevant to answering the former question were addressed in the previous chapter by introducing the concepts of saturated vapor pressure, the ideal gas law, and relative humidity. This section describes the principles regarding the latter question.

Figure 3.1 illustrates two important principles regarding the dissolution of gases into liquids, in this case, air into water. First, the relative amount of a component gas that dissolves into liquid is proportional to the relative concentration of that species present in the gas phase. This relative concentration can be described in terms of partial pressure. For example, air, which is composed primarily of N_2 (~78%) and O_2 (~21%), will dissolve a relatively higher amount of N_2 into water (Fig. 3.1a). Second, the total amount of dissolved species in the liquid phase is proportional to the total pressure of the gas phase. This is illustrated by Fig. 3.1b, which represents an increase in pressure from the initial condition of Fig. 3.1a by a larger piston force F_2. Consequently, more N_2 and O_2 are dissolved into the liquid phase.

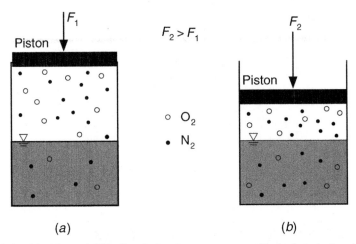

Figure 3.1 Air (O_2 and N_2) dissolution in water at mechanical and chemical equilibrium. As chamber pressure increases from (a) to (b), greater mass of air is dissolved. Total volume of dissolved air, however, remains relatively unchanged.

Henry's law states that the molar mass of a dissolved gas in a given volume of liquid is proportional to the partial pressure of the gas in the gas phase at equilibrium, that is,

$$\frac{M_i/\omega_i}{V_l} = K_{H_i} u_i \qquad (3.1)$$

where M_i is the mass of the gas i (kg), ω_i is the molecular mass of gas i (kg/mol), V_l is the volume of liquid (L), u_i is the partial pressure of gas i (bar) and K_{H_i} is the Henry's law constant for the dissolution of gas i in that particular liquid. Henry's law constant is commonly expressed in units of mass concentration per unit pressure, typically M/bar (mol · L^{-1} · bar^{-1}, where 1 bar ≈ 100 kPa). The larger the value of K_H, the more soluble the gas and vice versa.

The equilibrium amount of a multicomponent gas dissolved in a unit volume is expressed by the sum of the dissolved species. For example, the dissolution of air in water can be expressed as

$$\frac{M_{O_2}/\omega_{O_2} + M_{N_2}/\omega_{N_2} + \cdots}{V_l} = K_{H_{O_2}} u_{O_2} + K_{H_{N_2}} u_{N_2} + \cdots = K_{H_a} u_a \qquad (3.2)$$

where K_{H_a} is the Henry's law constant for air and u_a is the sum of the partial pressures of the individual gases comprising air. According to Dalton's laws of partial pressures

$$u_a = u_{O_2} + u_{N_2} + \cdots \tag{3.3}$$

Table 3.1 shows the partial pressure, Henry's law constant, and mass (molar) concentration for the major components of air at 25°C and 1 bar total air pressure.

Example Problem 3.1 Given the information in Table 3.1, find the solubility of atmospheric nitrogen gas (N_2) in liquid water at 25°C.

Solution The solubility of nitrogen, M_{N_2}/V_l, can be calculated from Henry's law as follows:

$$\frac{M_{N_2}}{V_l} = \omega_{N_2} K_{N_2} u_{N_2} = (28.016 \text{ g/mol})(6.40 \times 10^{-4} \text{ mol/L} \cdot \text{bar})(0.7808 \text{ bar})$$

$$= 0.014 \text{ g/L} = 14 \text{ mg/L}$$

which is to say that in each liter of water there will be 14 mg of dissolved nitrogen gas at a temperature of 25°C and 1 bar of total air pressure.

3.1.2 Temperature Dependence

Henry's law constant is dependent on temperature. This dependence is shown in Table 3.2 for the major constituents of air over a temperature range relevant to geotechnical engineering practice. Note that Henry's law constant may vary over an order of magnitude as temperature varies from 0 to 50°C. The constants tend to decrease with increasing temperature, indicating that a lesser amount of a particular gas may be dissolved as temperature increases. Any experienced trout fisherman knows to concentrate in areas of relatively cold water where more oxygen is dissolved for his or her prey.

TABLE 3.1 Partial Pressure, Henry's Law Constant, and Molar Concentration of Major Air Components in Water at 25°C and 1 bar Total Pressure

Gas	Partial Pressure u_i (bar)	Henry's Law Constant K_H (M/bar)	Molar Concentration (M)
O_2	0.2095	1.26×10^{-3}	0.2646×10^{-3}
N_2	0.7808	6.40×10^{-4}	4.9920×10^{-4}
CO_2	0.0003	3.39×10^{-2}	0.0011×10^{-2}

TABLE 3.2 Henry's Law Constant (K_H) for Major Components of Air as Function of Temperature at 1 bar Total Pressure

Temperature (°C)	Henry's Law Constant K_H (M/bar)		
	N_2	O_2	CO_2
0	1.05×10^{-3}	2.18×10^{-3}	7.64×10^{-2}
5	9.31×10^{-4}	1.91×10^{-3}	6.35×10^{-2}
10	8.30×10^{-4}	1.70×10^{-3}	5.33×10^{-2}
15	7.52×10^{-4}	1.52×10^{-3}	4.55×10^{-2}
20	6.89×10^{-4}	1.38×10^{-3}	3.92×10^{-2}
25	6.40×10^{-4}	1.26×10^{-3}	3.39×10^{-2}
30	5.99×10^{-4}	1.16×10^{-3}	2.97×10^{-2}
35	5.60×10^{-4}	1.09×10^{-3}	2.64×10^{-2}
40	5.28×10^{-4}	1.03×10^{-3}	2.36×10^{-2}
50	4.85×10^{-4}	9.32×10^{-4}	1.95×10^{-2}

Source: From Pagenkopf (1978).

3.1.3 Volumetric Coefficient of Solubility

Another useful way to describe the amount of gas dissolved in liquid is in terms of volumetric concentration (L/L). Motivation for using volumetric solubility stems from the fact that unlike dissolved mass, the volume of dissolved gas in a unit volume of liquid is relatively insensitive to the gas or liquid pressure.

According to Henry's law, the volumetric concentration for any gaseous component of air dissolved in water may be written as follows:

$$\frac{V_i}{V_l} = h_i \frac{u_i}{u_a} \tag{3.4}$$

where V_i and V_l are the volume of gaseous species i and water, respectively (L), and h_i is the volumetric coefficient of solubility (L/L) for the gas i.

The total volumetric concentration of a multicomponent gas is the sum of the product of the partial pressures and volumetric concentrations of each component, which for air is

$$\frac{V_{O_2} + V_{N_2} + \cdots}{V_l} = h_{O_2} \frac{u_{O_2}}{u_a} + h_{N_2} \frac{u_{N_2}}{u_a} + \cdots = h_a \tag{3.5}$$

where h_a is the volumetric coefficient of solubility for air.

Table 3.3 shows the volumetric coefficient of solubility for the major components of air in water as a function of temperature at 1 bar total pressure and 100% relative humidity.

TABLE 3.3 Volumetric Coefficient of Solubility (h) for Major Components of Air in Water at 1 bar Air Pressure and 100% Relative Humidity

Temperature (°C)	N_2	O_2	CO_2	Air
0	0.0235	0.0489	1.713	0.0292
10	0.0186	0.0380	1.193	0.0228
20	0.0154	0.0310	0.878	0.0187
30	0.0134	0.0261	0.665	0.0156
40	0.0118	0.0231	0.530	0.0141

If nitrogen and oxygen are considered the primary components of air and the trace constituents are neglected, a fairly good estimate for the volumetric solubility of air can be obtained using eq. (3.5) from the component solubility (Table 3.3) and the partial pressure (Table 3.1) of nitrogen and oxygen. For example, the volumetric coefficient of solubility of air in water at 30°C and 1 bar pressure may be approximated as

$$h_a \approx h_{O_2}\frac{u_{O_2}}{u_a} + h_{N_2}\frac{u_{N_2}}{u_a} = (0.0261)(0.21) + (0.0134)(0.78) = 0.0159 \text{ L/L}$$

which is quite close to the more accurate value for all the components of air shown in Table 3.3 ($h_a = 0.0156$).

3.1.4 Henry's Law Constant and Volumetric Coefficient of Solubility

A relationship between Henry's law constant and the volumetric coefficient of solubility can be established by applying the ideal gas law introduced in Chapter 2. For any gaseous species i in air, the ideal gas law states:

$$V_{ai} = \frac{M_{ai}}{u_{ai}} \frac{RT}{\omega_{ai}} \tag{3.6}$$

For two different pressure conditions (designated 1 and 2), Henry's law leads to the following relationship:

$$\frac{M_{ai}^1}{u_{ai}^1} = \frac{M_{ai}^2}{u_{ai}^2} = K_{H_i} V_l \omega_{ai} = \text{const} \tag{3.7}$$

Substituting eq. (3.7) into eq. (3.6) leads to

$$V_{ai} = \frac{M_{ai}}{u_{ai}} \frac{RT}{\omega_{ai}} = K_{H_i} V_l \omega_{ai} \frac{RT}{\omega_{ai}} = K_{Hi} V_l RT \tag{3.8a}$$

or

$$\frac{V_{ai}}{V_l} = K_{H_i} RT = h_{ai} \frac{u_{ai}}{u_a} \tag{3.8b}$$

which can be written for all component gases as

$$\frac{\sum V_{ai}}{V_l} = \sum K_{H_i} RT = \sum h_{ai} \frac{u_{ai}}{u_a} \tag{3.9a}$$

or

$$\frac{V_a}{V_l} = K_{H_a} RT = h_a \tag{3.9b}$$

Equation (3.8) states that for a given volume of liquid water V_l, the volume of dissolved air is a constant for a given temperature. Equation (3.9) connects Henry's law constant to the volumetric coefficient of solubility. Rearranging eq. (3.8a) leads to

$$M_{ai} = u_{ai} K_{H_i} V_l \omega_{ai} = \frac{u_{ai} \omega_{ai}}{RT} RT K_{H_i} V_l = \rho_{ai} RT K_{H_i} V_l \tag{3.10}$$

Since dissolved air may be considered an ideal gas, and thus is compressible, the mass of dissolved air and the corresponding density increase as the partial gas pressure increases. As illustrated previously by Fig. 3.1, this situation occurs if a series of mixed gases is placed in a closed container followed by an increase in total pressure.

3.1.5 Vapor Component Correction

Henry's law constants typically correspond to dry gas conditions (i.e., no water vapor; RH = 0%) at 1 bar total pressure. Volumetric coefficients of solubility, on the other hand, are typically reported under a 100% relative humidity and 1-bar total pressure condition. Because finite values of vapor pressure exist under typical circumstances in unsaturated soil, a correction is necessary when estimating volumetric coefficients of solubility from Henry's law constants using eq. (3.9).

The volumetric coefficient of solubility for a gas i under consideration of vapor pressure and total pressure may be written as

$$h_i = K_{H_i} RTX \tag{3.11a}$$

where

$$X = \frac{u_i}{u_a} \frac{u_a - u_v}{u_i} = \frac{u_a - u_v}{u_a} \tag{3.11b}$$

Example Problem 3.2 At 25°C and 1 bar (101.3 kPa), the Henry's law constant for N_2 is 6.4×10^{-4} M/bar. Estimate the volumetric coefficient of solubility from the Henry's law constant for a condition of 100% relative humidity.

Solution The vapor pressure at 100% RH is equal to the saturated water vapor pressure. From Section 2.2, saturated vapor pressure at 25°C is 3.167 kPa (Table 2.6). The volumetric coefficient of solubility of N_2 at 100% RH, therefore, is

$$h_{N_2} = K_{H_{N_2}} RT \frac{u_a - u_v}{u_a}$$

$$= (6.4 \times 10^{-4} \text{ mol/L} \cdot \text{bar})(8.31432 \text{ N} \cdot \text{m/mol} \cdot \text{K})$$

$$(298.2 \text{ K}) \frac{101.3 - 3.167}{101.3}$$

$$= (6.4 \times 10^{-4} \text{ mol/L} \cdot 10^5 \text{ N/m}^2)(8.31432 \text{ N} \cdot \text{m/mol} \cdot \text{K})$$

$$(298.2 \text{ K})(0.9687)$$

$$= (6.4 \times 10^{-4} \text{ m}^3/\text{L} \cdot 10^5)(8.31432)(298.2)(0.9867)$$

$$= 0.01537 \text{ L/L}$$

which is about 3.2% less than the uncorrected value ($h_{N_2} = 0.01586$) and is quite close to the actual value of 0.0144 L/L interpolated from Table 3.3.

3.1.6 Mass Coefficient of Solubility

For problems where the total air pressure does not significantly vary, a mass coefficient of solubility (mol/mol or kg/kg) is often used. This quantity is often convenient because it gives the mass of dissolved air per unit mass of water. Considering a definition consistent with the volumetric coefficient of solubility, the mass coefficient of solubility is

$$H_{ai} = \frac{M_{ai}/M_l}{u_{ai}/u_a} \tag{3.12}$$

Combining eqs. (3.9) and (3.12), the mass coefficient of solubility can be related to the volumetric coefficient of solubility as

$$\frac{M_{ai}/V_{ai}}{M_l/V_l} = \frac{H_{ai}}{h_{ai}} = \frac{\rho_{ai}}{\rho_w} \qquad H_{ai} = \frac{\rho_{ai}}{\rho_w} h_{ai} \tag{3.13}$$

which can also be written for the sum of the individual gas components comprising the air phase:

$$\frac{\sum M_{ai}}{M_l} = \sum H_{ai} \frac{u_{ai}}{u_a}$$
$$\frac{M_a}{M_l} = H_a \qquad H_a = \frac{\rho_a}{\rho_w} h_a \tag{3.14}$$

3.2 AIR-WATER-SOLID INTERFACE

3.2.1 Equilibrium between Two Water Drops

The interaction between air and water often creates some interesting and counterintuitive phenomena. Consider, for example, the scenario illustrated in Fig. 3.2a. Two water drops with initially different diameters are connected through a water-filled thin pipe. The radii of the smaller and larger drops are R_1 and R_2, respectively, and each has an internal water pressure u_{w1} and u_{w2}. What happens when the valve in the middle of the pipe is opened? Will water

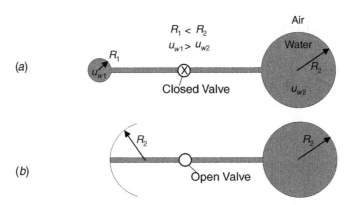

Figure 3.2 Equilibrium between two interconnected drops of water; the rich get richer and the poor get poorer.

flow from the smaller drop to the larger one? Will it flow from the larger drop to the smaller? Will nothing at all happen?

According to the concept of surface tension and the spherical interface equation described in Section 2.4, the water pressure in the smaller drop is higher than the pressure in the larger drop because the radius of curvature R_1 is smaller than R_2. Therefore, upon opening the valve, the larger drop becomes larger and the smaller drop becomes smaller! Much like an unbalanced economy, the rich get richer and the poor get poorer. Fluid flow stops when the smaller water drop enters the pipe and forms a convex meniscus with a radius equal to R_2 (Fig. 3.2b).

3.2.2 Equilibrium at an Air-Water-Solid Interface

In the case of air-water interaction, the interface geometry is often controlled by the phase with a smaller volume. As introduced in Section 2.4, this may be in the form of small water drops in a surrounding air phase or small air bubbles in a surrounding water phase. The phase with the smaller volume generally assumes a spherical shape. The diameter of the sphere and the surface tension of the denser phase control the pressure change across the two-phase interface.

In the case of a three-phase system (e.g., gas, liquid, and solid), the geometry of the solid and the liquid-solid contact angle provide two additional factors controlling the forces and pressures among the phases. Consider, for example, a solid sphere in the vicinity of an air-water interface. Figure 3.3 shows two possibilities for this three-phase interaction. If the solid is a wetting, or *hydrophilic*, material (Fig. 3.3a), surface tension at the air-water interface may provide a downward pulling force such that a solid with a density lighter than water ($\rho_s < \rho_w$) can be submerged. On the other hand, if the solid is a water repellent, or *hydrophobic*, material, an upward pulling force provided by the air-water interface may counteract the gravity force such that a solid denser than water ($\rho_s > \rho_w$) can float (Fig. 3.3b).

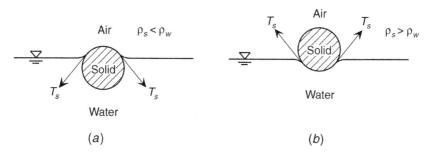

Figure 3.3 Manifestation of surface tension forces at air-water-solid interface showing (a) how lighter solid can be submerged in water and (b) how heavier solid can float in water.

Figure 3.4a shows the interaction between air, water, and solid in a wide, closed container. Point 1 is located just below the air-water interface toward the middle of the container. Point 2 is located within the meniscus formed at the air-water-solid interface near the container wall. The water pressure at each point is designated u_{w1} and u_{w2}, respectively. At equilibrium, the water pressure at point 1 is equal to the air pressure u_a. This is because the air-water interface located in the interior of the container is flat. At point 2, however, the water pressure is less than the air pressure, as reflected by the curvature of the air-water interface. The pressure difference can be inferred by either the meniscus geometry as $u_a - u_{w2} = 2T_s/R$, where T_s is the surface tension of the air-water interface and R is the radii of curvature.

A flat air-water interface is not likely to occur in unsaturated soil. Here, a capillary tube is a slightly more realistic model. As shown for a capillary tube in Fig. 3.4b, the same water pressure at points 1 and 2 are observed because both are under the same spherical meniscus having radius R. The pressure drop across the interface can be readily calculated by the same equation as that from the previous example for point 2 under the meniscus at the edge of the container. In this case, however, the magnitude of R may now be determined by the geometrical constraint imposed by the solid phase (radius r) and the liquid-solid contact angle α as

$$R = \frac{r}{\cos \alpha} \tag{3.15}$$

Accordingly, the pressure drop across the air-water interface is

$$u_a - u_w = \frac{2T_s \cos \alpha}{r} \tag{3.16}$$

Now consider Fig. 3.5, where the wide container and capillary tube are connected by a water-filled pipe and valve. The air phase in both containers

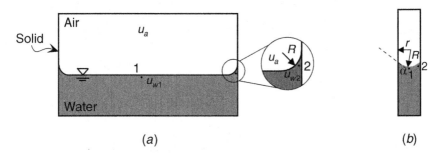

Figure 3.4 Air-water-solid interface in (a) a wide container and (b) a capillary tube.

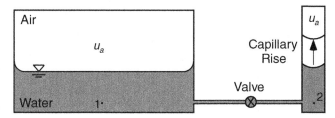

Figure 3.5 Equilibrium between water in a wide container and a capillary tube.

is at the same pressure. If the valve is closed, the previous two analyses dictate that the pressures at points 1 and 2 will be different, with the pressure at point 2 being smaller than that at point 1. If the valve is opened, the total head in both water phases will tend toward equilibrium. Reflecting the lower pressure in the capillary tube, its water level will rise to the height of $2T_s \cos \alpha / r\gamma_w$ above the flat interface in the wide container, where γ_w is the unit weight of water.

3.2.3 Contact Angle

Contact angle α is an intrinsic property of any two contacting phases in a solid-liquid-gas system. For unsaturated soil systems, contact angle may be defined as the angle between a line tangent to the air-water interface and a line defined by the water-solid interface. The solid is either wetted by the liquid, which is the case for most soil solids and water, or not wetted by the liquid, which is the case for most solids and a liquid such as mercury.

Figure 3.6 shows examples of typical air-water-solid interactions and the corresponding location of the solid-liquid contact angle. In the case of a wetting interaction (Fig. 3.6a), the contact angle varies between 0° and 90°. Nonwetting interactions (Fig. 3.6b) exhibit contact angles between 90° and 180°. Neutral interactions have a contact angle equal to 90°.

Contact angle has an important influence on the geometry of solid-liquid-gas interfaces and the consequent physical behavior of the system. In a cap-

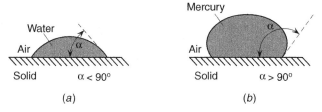

Figure 3.6 Air-water-solid interaction showing location of solid-liquid contact angle for drop of liquid on solid surface: (a) wetting interaction and (b) repellent interaction.

illary tube filled with water, for example, a wetting contact angle will lead to capillary rise (Fig. 3.7a). On the other hand, the repellent contact angle in a capillary tube filled with mercury will lead to capillary depression (Fig. 3.7b). The pressure drop across the interface in a capillary tube for any contact angle between 0° and 180° can be estimated by eq. (3.16).

All five possibilities for the magnitude of contact angle in a capillary tube are summarized below:

1. $\alpha = 0°$; a perfectly wetting surface; the air-water interface curvature R is identical to the radius of the capillary tube r. Most soil exhibits a near-zero contact angle during drying processes with water.
2. $0° < \alpha < 90°$; a partially wetting surface; the air-water interface curvature R is equal to the radius of the capillary tube r divided by $\cos \alpha$. Most soil exhibits considerable contact angle during wetting processes with water. Contact angles as high as 65° are commonly reported in the literature (e.g., Letey et al., 1962; Kumar and Malik, 1990).
3. $\alpha = 90°$; a neutral surface; the air-water interface curvature R is infinite. There is no capillary rise or depression and no pressure change across the air-water interface.
4. $90° < \alpha < 180°$; a partially repellent surface. The air-water interface radius R is equal to the radius of the capillary tube r divided by $\cos \alpha$ and is a negative value. Capillary depression occurs. Equation (3.16) dictates that the water pressure is higher than the air pressure. This may occur for soil subjected to extremely high temperatures (e.g., following forest fires) or for certain organic pore liquids or organics-rich soils (e.g., DeBano, 2000).

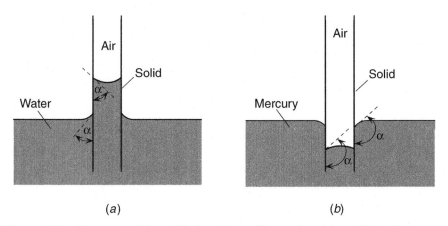

Figure 3.7 Air-water-solid equilibrium in capillary tube: (a) capillary rise corresponding to wetting contact angle ($\alpha < 90°$) and (b) capillary depression corresponding to repellent contact angle ($\alpha > 90°$).

5. $\alpha = 180°$; a perfectly repellent surface. The air-water interface curvature R is equal to the negative radius of the capillary tube $(-r)$. Equation (3.16) dictates that the water pressure is higher than the air pressure. This is an unlikely case for soil under unsaturated conditions.

3.2.4 Air-Water-Solid Interface in Unsaturated Soil

The spherical interface model developed for capillary tubes provides a conceptual model to describe the pressure change across an air-water interface and a physical explanation for an important component of suction in unsaturated soil. In real soil, however, a spherical interface is rarely the case. Rather, the existence of particles with various shapes and sizes and the complex pore fabric formed among adjacent particles also control the interface geometry. Assumptions must be made about this complex pore geometry in order to extend the simple capillary tube model to analyses of unsaturated soil. The following analysis assumes two identical spherical sand particles and an air-water interface described by the so-called *toroidal approximation*.

An idealized geometry of the air-water interface between two spherical soil grains can be characterized by two radii of curvature r_1 and r_2, as shown in Fig. 3.8a. The fundamental question here is whether the interface described by r_1 and r_2 causes a pressure increase or a pressure decrease in the water meniscus formed between the particles.

The answer to this question is less straightforward than the previous discussion for a spherical interface in a capillary tube. According to the spherical model, the curvature described by r_1 rotated about an axis perpendicular to the plane of the figure should cause a pressure drop in the pore water since the spherical interface is toward the water phase (i.e., r_1 describes a curvature concave toward the water phase). On the other hand, the radius r_2 rotated about axis $A - A'$ describes a liquid "cylinder" that leads to a pressure increase in the water (i.e., r_2 describes a curvature convex from the water phase). In three dimensions, the toriodal meniscus described by r_1 and r_2 resembles a horse's saddle, rotated 90° in Fig. 3.8. The pressure changes governed by these curvatures act to oppose each other.

Consider force balance in the horizontal direction and the free-body diagram shown in Fig. 3.8b. There are three force contributions in the free-body diagram: surface tension along the interface described by r_1 that results in the positive direction horizontally, surface tension along the interface described by r_2 that results in the negative direction horizontally, and air and water pressure applied on either side of the interface. The projection of surface tension in the positive horizontal direction is

$$F_1 = (T_s \sin \alpha)(2r_3)(2) = 4r_3 T_s \sin \alpha \tag{3.17}$$

The projection of the surface tension in the negative horizontal direction is

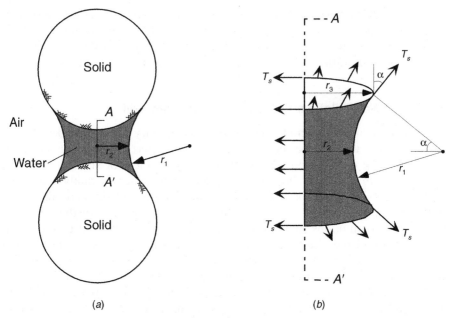

Figure 3.8 Idealized air-water interface geometry in unsaturated soil: (a) water meniscus between two spherical soil particles and (b) free-body diagram for water meniscus.

$$F_2 = -(T_s)(r_1 \sin \alpha)(2)(2) = -4r_1 T_s \sin \alpha \qquad (3.18)$$

and the projection of air and water pressure u_a and u_w in the horizontal direction (assuming $r_2 = r_3$) is

$$F_3 = (u_a - u_w)(2r_1 \sin \alpha)(2r_2) = 4r_1 r_2 (u_a - u_w) \sin \alpha \qquad (3.19)$$

Balancing all three forces leads to

$$T_s(r_2 - r_1) = (u_a - u_w) r_1 r_2 \qquad (3.20a)$$

or

$$u_a - u_w = T_s \left(\frac{1}{r_1} - \frac{1}{r_2} \right) \qquad (3.20b)$$

The above equation provides a simple mathematical expression describing the pressure change across an air-water-solid interface between two idealized soil grains. The quantity $(u_a - u_w)$ is the matric suction and, depending on the relative magnitudes of r_1 and r_2, could be positive, zero, or negative. Most likely, the value of matric suction is positive due to the fact that r_1 is mostly less than r_2 under unsaturated conditions. Note that for a given set of spherical radii r_1 and r_2, the magnitude of matric suction is independent of the contact angle α. In principle, there are three possible regimes for the magnitude of the pressure difference $(u_a - u_w)$ depending on the values of r_1 and r_2.

1. $r_1 < r_2$: $u_a > u_w$, a pressure decrease in the soil water. Examples are when water content is low between two sand grains, as shown in Fig. 3.9a, and water sandwiched between two platy clay particles as shown in Fig. 3.9b.
2. $r_1 = r_2$: $u_a = u_w$, no pressure change across the air-water interface. An example is when sandy soil is nearly saturated.
3. $r_1 > r_2$: $u_a < u_w$, a pressure increase in the pore water. This case is likely to occur in real soil-water systems when the soil is nearly saturated or when there is a large void but narrow distance between platy particles, shown as the dashed curves in Fig. 3.9.

Example Problem 3.3 The pore water meniscus between two sand grains at a certain degree of saturation can be characterized by two spherical radii $r_1 = 10^{-6}$ m and $r_2 = 10^{-3}$ m. Assume the contact angle is zero, the temperature is 25°C, and the pore air pressure is a reference value of zero. What

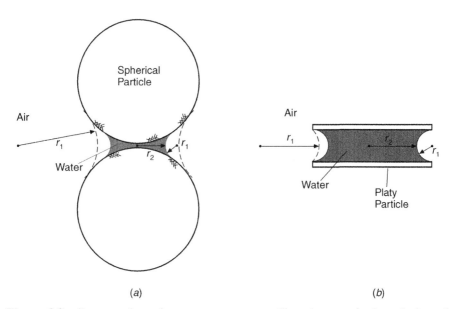

Figure 3.9 Conceptual meniscus geometry controlling the magnitude and sign of pressure drop $(u_a - u_w)$ in unsaturated soil: (a) between two spherical particles and (b) between two platy particles.

is the corresponding matric suction? What is the corresponding pore water pressure? Repeat the analysis for the system at a lower degree of saturation where r_1 decreases to 10^{-8} m and r_2 decreases to 10^{-4} m.

Solution Because $r_2 > r_1$, the pore water pressure is decreased relative to the pore air pressure. According to Table 2.10, the surface tension at 25°C is 71.79 mN/m. Thus, according to eq. (3.20b), the matric suction may be calculated as

$$u_a - u_w = (71.79 \text{ mN/m}) \left(\frac{1}{10^{-6} \text{ m}} - \frac{1}{10^{-3} \text{ m}} \right) = 71.72$$
$$\times 10^6 \text{ mN/m}^2 = 71.72 \text{ kPa}$$

For $u_a = 0$, the pore water pressure is $u_w = -71.72$ kPa. At the lower degree of saturation, the matric suction increases to 7178 kPa and the corresponding pore water pressure is -7178 kPa.

3.3 VAPOR PRESSURE LOWERING

3.3.1 Implications of Kelvin's Equation

The total pressure change $u_a - u_w$ across a curved air-water interface in a capillary tube or idealized soil pore with radius r and contact angle α was derived earlier as

$$u_a - u_w = \frac{2T_s \cos \alpha}{r} \tag{3.21}$$

The quantity on the left-hand side of the above equation, matric suction $u_a - u_w$, is of fundamental importance for understanding stress, strain, and flow phenomena in unsaturated soil. In this section, it is shown that matric suction is a determining state variable for soil-water equilibrium. For a given soil-water-air system, the water content at mechanical and chemical equilibrium is determined by the state of the matric suction.

In 1871, William Thomson (also known as Lord Kelvin) derived a remarkably simple equation that connects the pressure change across a curved air-water interface to the vapor pressure above the interface. In the case of a capillary tube, Kelvin's equation can be expressed as

$$\mu_1 - \mu_0 = -RT \ln \frac{u_{v1}}{u_{v0}} = \frac{2T_s v_w \cos \alpha}{r} \qquad (3.22)$$

where $\mu_1 - \mu_0$ is the change in chemical potential of the water vapor (J/mol) in the capillary tube due to the curvature of the air-water interface (μ_0 is the chemical potential of free water used as a reference point and μ_1 is the chemical potential at the prevailing state), R is the universal gas constant (J/mol · K), T is temperature (K), u_{v0} is the saturated vapor pressure at T in equilibrium with free water, u_{v1} is the prevailing vapor pressure in the capillary tube, T_s is surface tension (J/m^{-2}), and v_w is the partial molar volume of water vapor (m^3/mol). Substituting eq. (3.22) into eq. (3.21) leads to another form of Kelvin's equation:

$$u_a - u_w = -\frac{RT}{v_w} \ln \frac{u_{v1}}{u_{v0}} = -\frac{RT}{v_w} \ln(\text{RH}) = \frac{2T_s \cos \alpha}{r} \qquad (3.23)$$

Kelvin's equation has monumental theoretical and practical implications to unsaturated soil mechanics. Foremost, it implies the following:

1. The vapor pressure in a soil pore in equilibrium with the pore water (i.e., u_{v1}) could be lower or higher than the saturated vapor pressure of the soil pore water in equilibrium with free water (i.e., u_{v0}). Because most soil has a contact angle with water less than 90°, the magnitude of vapor pressure in a soil pore at equilibrium is predominantly lower than the saturated vapor pressure in equilibrium with free water. The phenomenon of *vapor pressure lowering,* therefore, is widely applicable to unsaturated soil mechanics.
2. The magnitude of the equilibrium vapor pressure in a soil pore depends on the soil pore structure. Without losing the general physics of the system, a soil pore can be idealized as a capillary tube with radius r, a contact angle α between the soil solids and pore water, and the surface tension of the liquid phase, T_s.
3. The task of measuring soil pore water pressure can be accomplished by measuring the equilibrium soil vapor pressure at a given temperature if the air pressure is known or assumed and dissolved solute effects are negligible. Total soil suction may be determined directly from measurements of pore water vapor pressure.

Example Problem 3.4 For the same two unsaturated sand grains and degrees of saturation from Example Problem 3.3, what is the relative humidity of the pore water vapor? What is the corresponding vapor pressure?

Solution Equation (3.23) can be rearranged to calculate relative humidity from the known value of matric suction as follows:

$$RH = \exp\left(-\frac{(u_a - u_w)v_w}{RT}\right)$$

$$= \exp(-(71.72 \text{ kN/m}^2)(1.8 \times 10^{-5} \text{ m}^3/\text{mol})/$$

$$(8.314 \text{ J/mol} \cdot \text{K})(298.16 \text{ K}) \, 1000 \text{ N/kN}$$

$$= 0.99948 = 99.948\%$$

The corresponding vapor pressure may be calculated from the saturated vapor pressure at 25°C shown in Table 2.6 (u_{v0} = 3.167 kPa) and eq. (2.12) as follows:

$$u_v = u_{v0}(RH) = (3.167 \text{ kPa})(0.99948) = 3.165 \text{ kPa}$$

At the lower degree of saturation and corresponding matric suction of 7178 kPa, the relative humidity decreases to 94.9% and the vapor pressure decreases to 3.005 kPa.

3.3.2 Derivation of Kelvin's Equation

Kelvin's equation can be better understood through the following thought experiment. Consider a simple three-phase system comprised of air, water, and solid at a state of equilibrium in a closed container (Fig. 3.10). The air phase consists of two components: dry air and water vapor, each having a partial pressure component u_{da} and u_v, respectively. The total air pressure u_a is equal to the sum of the partial pressures of dry air and water vapor ($u_a = u_{da} + u_v$). The composition and amount of dry air will not vary in the container, but the amount of water vapor may indeed vary under concurrent condensation and evaporation processes. Assume that the water phase is free

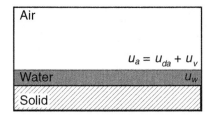

Figure 3.10 Air-water-solid system at mechanical and chemical equilibrium to facilitate derivation of Kelvin's equation between pure water and air with flat interface.

(i.e., free of influence by the solid, the solid container, and dissolved solutes) and that the air-water interface is perfectly flat.

For relatively incompressible materials such as solids, mechanical force considerations are usually the only criteria necessary to arrive at an equilibrium relationship. However, for highly deformable materials such as dry air, water vapor, or liquids, it is necessary to also consider chemical equilibrium. For this thought experiment, mechanical and chemical equilibrium between the air and water phases are considered. Because the air-water interface is flat, mechanical equilibrium requires that the total air pressure u_a be equal to the total water pressure u_w. Chemical equilibrium requires that the total chemical potential, or more conveniently, the change in the total chemical potential, be the same in each coexisting phase (i.e., air and water). These requirements are written for mechanical equilibrium as

$$u_a = u_w \tag{3.24a}$$

or

$$u_{da} + u_v = u_w \tag{3.24b}$$

and for chemical equilibrium

$$\mu_a = \mu_{da} + \mu_v = \mu_w = RT = u_w v_w = u_a v_w = u_{da} v_{da} + u_v v_v \tag{3.25}$$

where v_{da} is the partial molar volume of dry air, v_w is the partial molar volume of liquid water, and v_v is the partial molar volume of water vapor. Assuming ideal gas behavior for the dry air and water vapor, the dry air pressure and vapor pressure can be expressed as

$$u_{da} = \frac{M_{da}}{V_{da}\omega_{da}} RT = \frac{RT}{v_{da}} \tag{3.26a}$$

$$u_v = \frac{M_v}{V_v \omega_v} RT = \frac{RT}{v_v} \tag{3.26b}$$

where M_{da} is the mass of dry air, V_{da} is the volume of dry air, and ω_{da} is the molecular mass of dry air. Similarly, M_v is the mass of water vapor, V_v is the volume of water vapor, and ω_v is the molecular mass of water vapor. Since any ideal gas has 22.4 L/mol, the partial molar volume of either water vapor or dry air can be calculated from the molecular weight and the volume fraction of each respective gas, as shown on the far right-hand side of eqs. (3.26a) and (3.26b). It follows that the chemical equilibrium condition [eq. (3.25)] can be rewritten as

$$\mu_w = \nu_{da} u_{da} + \nu_v u_v \qquad (3.27)$$

At mechanical and chemical equilibrium, the vapor pressure of pure water reaches its saturated value u_{v0} under the prevailing temperature and pressure condition. In other words, a state of 100% relative humidity is reached. This state is defined as a reference state in this and subsequent cases.

Now suppose that all of the water in the container is in the form of spherical droplets having uniform radii r (Fig. 3.11). The solid container consists of a perfectly water repellent material such that the contact angle is 180°, implying that no water potential change can occur due to surface wetting. Here, a new state of pressure and potential for the air and water phases must be established. If water vapor follows the ideal gas law and the change in chemical potential of dry air is negligible compared to the change in chemical potential of the water vapor, the change in chemical potential for the total air phase with respect to the previous case for the flat air-water interface becomes

$$\Delta \mu_a = \Delta \mu_{da} + \Delta \mu_v = \Delta \mu_v = -RT \ln \frac{u_v}{u_{v0}} \qquad (3.28)$$

The latter assumption in the development of eq. (3.28) is based on the fact that the total pressure change is small and that the partial molar volume of dry air remains unchanged in the closed container. Rigorous derivation of eq. (3.28) requires application of Gibbs-Duhem equilibrium criteria and can be found in Defay et al. (1966).

The chemical potential change in the liquid phase is

$$\Delta \mu_w = v_w \, \Delta u_w \qquad (3.29)$$

As shown earlier, the pressure change across the air-water interface is controlled by the geometry of the water droplets:

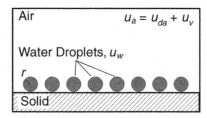

Figure 3.11 Air-water-solid system at mechanical and chemical equilibrium to facilitate derivation of Kelvin's equation between uniformly sized water droplets and air.

3.3 VAPOR PRESSURE LOWERING

$$\Delta u_w = u_a - u_w = \frac{2T_s}{r} \tag{3.30}$$

Substituting eq. (3.30) into eq. (3.29) leads to

$$\Delta \mu_w = v_w(u_a - u_w) = \frac{2T_s v_w}{r} \tag{3.31}$$

If equilibrium between the air and the water are reached, the change in chemical potential of the air phase should be equal to change in potential of the liquid phase:

$$\Delta \mu_w = \Delta \mu_a = -RT \ln \frac{u_v}{u_{v0}} = v_w(u_a - u_w) = \frac{2T_s v_w}{r} \tag{3.32}$$

or

$$-\frac{RT}{v_w} \ln \frac{u_v}{u_{v0}} = \frac{2T_s}{r} = u_a - u_w \tag{3.33}$$

which is Kelvin's equation applied to equilibrium between a water drop and its vapor pressure, implying that vapor pressure increases as the droplet radius decreases and that the pressure increase can be evaluated by measuring the relative humidity at the equilibrium state.

Now continue the thought experiment with an environment more applicable to unsaturated soil by considering the idealized system of capillary tubes shown in Fig. 3.12. The closed container contains air and a series of capillary tubes partially filled with water, each having a radius r and a solid-liquid contact angle α. Following the derivation described previously and noting that eq. (3.28) should be now substituted by eq. (3.21) for a curved air-water interface in a capillary tube, the following form of Kelvin's equation for describing the equilibrium condition is attained:

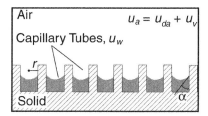

Figure 3.12 Air-water-solid system at mechanical and chemical equilibrium to facilitate derivation of Kelvin's equation among capillary tubes, water, and air.

$$u_a - u_w = \frac{2T_s \cos \alpha}{r} = -\frac{RT}{v_w} \ln \frac{u_v}{u_{v0}} \qquad (3.34)$$

Equation (3.34) predicts that a vapor pressure u_v lower than saturated vapor pressure u_{v0} will occur in the area above the curved air-water interface. In other words, thermodynamic equilibrium in a capillary tube with a wetting solid surface results in positive matric suction. The magnitude of the equilibrium vapor pressure decreases as the capillary tube radius becomes smaller, a phenomenon known as vapor pressure lowering. An important implication of eq. (3.34) is that liquid can exist in a porous medium in equilibrium with undersaturated water vapor.

The influence of capillary radius and droplet radius on relative humidity and the pressure difference between the air and water phases at the equilibrium state (i.e., matric suction) is shown in Table 3.4. The following constants have been assumed: $T = 298.16$ K, $v_w = 0.018$ m^3/kmol, $R = 8.31432$ J/mol · K, $T_s = 72$ mN/m, and $\alpha = 0°$. As predicted by Kelvin's equation, the equilibrium vapor pressure for a system of droplets is higher than the saturated vapor pressure. In other words, to maintain a water drop with a constant radius r, relative humidity greater than unity (RH = 100%) is required. An example of such an environment is a steamy bathroom. By the same token, the vapor pressure in a capillary tube or unsaturated soil with a wetting solid surface is universally lower than the saturated vapor pressure of free water.

Example Problem 3.5 As shown in Figure 3.13, water droplets with initially different sizes ranging from $r = 10^{-5}$ m to 10^{-8} m are in a container that actively sustains an ambient relative humidity of 101%. What will be the final size of the droplets at equilibrium?

Solution From eq. (3.33) the 101% relative humidity condition corresponds to a droplet size of approximately $r = 10^{-7}$ m at equilibrium (see Table 3.4).

TABLE 3.4 Relative Humidity and Pressure Change for Various Sized Droplets and Capillary Menisci

Radius (m)	Droplet RH = u_v/u_{v0}	Capillary Tube RH = u_v/u_{v0}	Pressure Change[a] (kPa) $u_a - u_w$
∞	1.000000	1.000000	0
10^{-4}	1.000011	0.999990	(−)1.4
10^{-5}	1.000105	0.999895	(−)14.4
10^{-6}	1.001046	0.998955	(−)144
10^{-7}	1.010511	0.989599	(−)1,440
10^{-8}	1.110220	0.900722	(−)14,400
10^{-9} (10 Å)	2.845059	0.351487	(−)144,000

[a]The pressure change is negative (−) for the droplets and positive (+) for the capillary menisci.

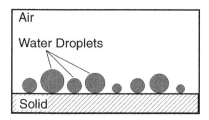

Figure 3.13 Water droplet system for Example Problem 3.5.

Accordingly, the smaller water droplets with radius less than 10^{-7} m will experience condensation and the larger water droplets with radius greater than 10^{-7} m will experience evaporation, leading to uniform droplets with radius of about 10^{-7} m at equilibrium.

3.3.3 Capillary Condensation

Kelvin's equation provides a theoretical basis for explaining many interface phenomena occurring in nature. One of these phenomena important in unsaturated soil behavior is capillary condensation. This section describes a series of thought experiments designed to illustrate and clarify the concept of capillary condensation and its consequent relationship to the soil-water characteristic curve.

Consider the idealized pore system shown in Fig. 3.14a. The pore system is comprised of uniform capillary tubes of radius r and contact angle α in a humidity-controlled chamber. Relative humidity can be continuously increased or decreased to any value between 0 and 100%. The capillary tubes are initially dry.

As menisci begin to form under increasing relative humidity, Kelvin's equation dictates that the vapor pressure in the capillary tubes is lower than the saturated vapor pressure of free water as a result of the curved air-water interfaces. The magnitude of this vapor pressure can be estimated as

$$u_v = \exp\left(-\frac{2T_s v_w \cos \alpha}{rRT}\right) u_{v0} \qquad (3.35a)$$

which can be written in terms of the relative humidity of the capillary water vapor as

$$\text{RH} = \frac{u_v}{u_{v0}} = \exp\left(-\frac{2T_s v_w \cos \alpha}{rRT}\right) \qquad (3.35b)$$

The dependency of relative humidity on the radius of capillary tubes is plotted for three different contact angles on Fig. 3.15.

112 INTERFACIAL EQUILIBRIUM

(a)

(b)

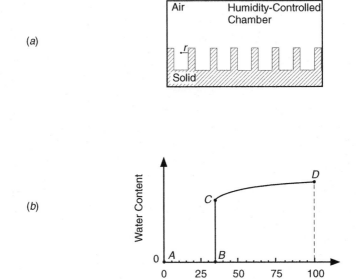

Figure 3.14 Capillary condensation model for idealized pore system comprised of uniform capillary tubes in humidity-controlled chamber: (a) capillary tube system and (b) characteristic curve for capillary tube system.

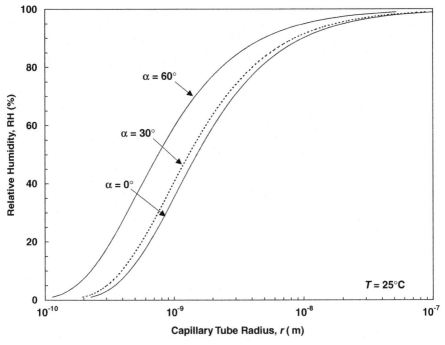

Figure 3.15 Relationship between capillary tube radius and relative humidity for three different contact angles ($T = 25°C$).

The difference between the total air pressure and the water pressure in the tubes is

$$u_a - u_w = -\frac{RT}{v_w} \ln \frac{u_v}{u_{v0}} = \frac{2T_s \cos \alpha}{r} \qquad (3.36)$$

Equations (3.35) and (3.36) show that when the radius of capillary tubes r tends to infinity, a flat air-water interface is observed and the vapor pressure above the tube is equal to the saturated vapor pressure u_{v0}. Correspondingly, there is no pressure drop across the air-water interface. Conversely, both the vapor pressure and the water pressure decrease as the radii of capillary tubes decreases.

Example Problem 3.6 Assume the capillary tube system introduced above is described by the following state and material variables: $r = 10^{-9}$ m (10 Å), $T = 298.16$ K, $\alpha = 0°$, $v_w = 0.018$ m³/kmol, $R = 8.31432$ J/mol · K, and $T_s = 72$ mN/m. Calculate the equilibrium vapor pressure, matric suction, and relative humidity.

Solution According to eqs. (3.35) and (3.36), the vapor pressure, matric suction, and relative humidity are

$$u_v = 1.11 \text{ kPa} \qquad u_a - u_w = 144{,}000 \text{ kPa} \qquad RH = 35\%$$

In other words, 35% RH describes the equilibrium condition for capillary tubes with radii of 10^{-9} m. Given the size of typical soil pores, this value approximates the lower bound for capillary condensation in soil since water molecules (2.7 Å) cannot easily move into pores smaller than 10 Å.

If the ambient vapor pressure in the same system is slowly increased from a perfectly dry condition (RH = 0%) to its saturated vapor pressure (RH = 100%), the amount of capillary condensation that occurs may be quantified in terms of the "water content" of the tube system as a function of relative humidity. As illustrated in Fig. 3.14*b* by the line from point *A* to point *B*, there will be no capillary condensation until the ambient vapor pressure reaches the threshold vapor pressure described by eq. (3.35a), which, based on the above analysis, is assumed to be about 35% RH (point *B*). As the ambient vapor pressure reaches the threshold value, capillary condensation begins to occur and continues spontaneously until all the capillary tubes are nearly filled. Line *BC* on Fig. 3.14*b* shows the increase in water content during this regime. Without further increase in the relative humidity, condensation will cease near point *C*, where the radius of the air-water interface in the tubes has increased to a value that satisfies equilibrium with the ambient vapor pressure. This point is shown for one capillary tube in Fig. 3.16. Further

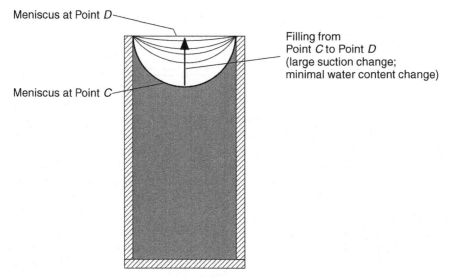

Figure 3.16 Capillary condensation near the mouth of capillary tube. Radius of air-water interface increases dramatically as ambient relative humidity increases from C to D. However, total water content of tube changes only slightly.

increases in the ambient vapor pressure from this point will promote additional capillary condensation; however, the increase in water content during this regime is much less significant (from point C to D in Fig. 3.14b). When the ambient vapor pressure reaches the saturated vapor pressure (i.e., RH = 100%), all of the capillary tubes will be filled up to a point forming a flat air-water interface.

3.4 SOIL-WATER CHARACTERISTIC CURVE

3.4.1 Soil Suction and Soil Water

The soil-water characteristic curve is a fundamental constitutive relationship in unsaturated soil mechanics. In general terms, the soil-water characteristic curve describes the relationship between soil suction and soil water content. More specifically, the soil-water characteristic curve describes the thermodynamic potential of the soil pore water relative to that of free water as a function of the amount of water adsorbed by the soil system. At relatively low water content, the pore water potential is relatively low compared with free water and the corresponding soil suction is high. At relatively high water content, the difference between the pore water potential and the potential of free water decreases and the corresponding soil suction is relatively low. When the potential of the pore water is equal to the potential of free water, the soil suction is equal to zero. For soil with negligible amount of dissolved solutes, suction approaches zero as the degree of saturation approaches unity.

The soil-water characteristic curve can describe either an adsorption (i.e., wetting) process or a desorption (i.e., drying) process. Differentiation between wetting characteristic curves and drying characteristic curves is typically required to account for the significant hysteresis that can occur between the two branches of behavior. More water is generally retained by soil during a drying process than is adsorbed by the soil for the same value of suction during a wetting process. Specific hysteresis mechanisms are described in Section 5.2.

3.4.2 Capillary Tube Model

The capillary condensation model described in the previous section provides a useful framework for understanding the soil-water characteristic curve. In soil, however, the capillary tube system would more realistically have a variable distribution of tube sizes and tube lengths, corresponding to a variable distribution of soil pore sizes.

Consider, for example, the system shown in Fig. 3.17a. If the relative humidity were continuously increased from 0 to 100% in a chamber contain-

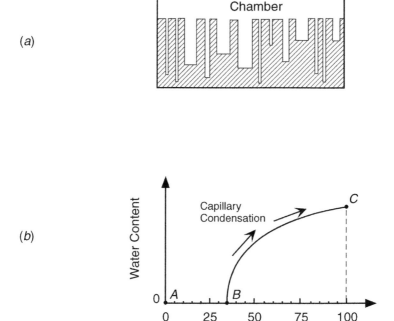

Figure 3.17 Capillary condensation model for idealized pore system comprised of various-sized (distributed radii) capillary tubes in humidity-controlled chamber: (a) capillary tube system, and (b) characteristic curve for capillary tube system.

ing different sized tubes, the effect of the tube size distribution would be evidenced by a spreading out of the capillary condensation regime over a wider range of relative humidity. The "characteristic curve" for the capillary tube system might, for example, follow the trends shown as Fig. 3.17b. Here, an initial increase in the ambient vapor pressure past some threshold value (B) results in a filling up of the smallest tubes. As before, filling proceeds until the water reaches the pore mouths. Condensation in the larger pores initiates when the threshold vapor pressure corresponding to that particular pore size is reached at higher relative humidity. A wide distribution of pore sizes results in a well-distributed capillary condensation regime and the corresponding curvature of the characteristic curve shown from B to C.

It is important to recognize that water retention in unsaturated soil involves far more than just capillarity. All types of soil adsorb water at ambient relative humidity far less than the threshold value for capillary condensation. As introduced in Section 1.6, the primary mechanisms associated with adsorption in this regime include short-range particle surface hydration and, in the case of clays, exchangeable cation hydration. The amount of water associated with the initial hydration regime is dependent primarily on the surface area and surface charge properties of the particles, ranging in terms of gravimetric water content from as little as 2 to 5% for sands to as much as 25% for highly expansive clays. The water adsorbed during hydration occurs in the form of thin films surrounding the particles and, for expansive clays, as monolayers in the interlayer pore space. As described in Chapter 2, the physical properties of adsorbed water may be radically different from those of free water.

If short-range hydration effects were incorporated into the previous capillary tube model for the soil-water characteristic curve, the behavior of the characteristic curve for the system might look like that shown in Fig. 3.18. In this case, thin films of water form on the surfaces of the capillary tubes at relative humidity values less than the threshold value for capillary condensation. At point B, the hydration energies are consumed (i.e., the water films grow to a thickness such that the water is no longer influenced by solid interaction effects) and capillary condensation becomes the dominant adsorption mechanism. Capillary condensation progresses steadily if the soil consists of a wide spectrum of pore sizes until the largest capillary pores are completely filled up (from point B to C). At this stage, soil suction approaches zero and capillary condensation ceases. If relative humidity were then continuously decreased from 100%, hysteresis in the relationship between water content and relative humidity would be observed between the wetting and drying processes (Fig. 3.19).

Figure 3.20 shows the general behavior of soil-water characteristic curves for different types of soil and illustrates the relative importance of the hydration regime as a function of particle size and surface activity. As shown, relatively coarse-grained soil (e.g., sand) adsorbs very little water by hydration. Fine-grained soil, on the other hand, adsorbs a relatively large amount of water by hydration. The commencement of capillary condensation (i.e., the

3.4 SOIL-WATER CHARACTERISTIC CURVE   117

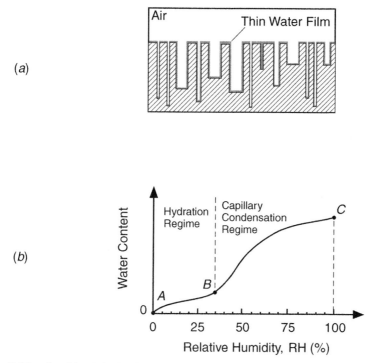

Figure 3.18 Combined hydration and capillary condensation model for idealized pore system comprised of various-sized capillary tubes in humidity-controlled chamber: (*a*) capillary tube system and (*b*) characteristic curve for capillary tube system.

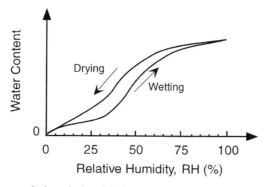

Figure 3.19 Hysteresis in relationship between water content and relative humidity. In general, more water is retained by system during drying than is adsorbed by system at same relative humidity during wetting.

118 INTERFACIAL EQUILIBRIUM

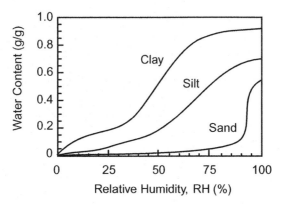

Figure 3.20 Patterns of typical soil-water characteristic curves for different soils.

relative humidity where capillary condensation starts) and the breadth of the capillary condensation regime (i.e., the range of humidity over which capillary condensation occurs) are dependent on the minimum pore size and the pore size distribution, respectively.

3.4.3 Contacting Sphere Model

Significant insight into the quantitative behavior of the soil-water characteristic curve can be gained by isolating the role of capillarity on pore water adsorption behavior. This can be accomplished by considering systems of spherical soil grains arranged in various idealized packing geometries. A relationship between matric suction $u_a - u_w$, surface tension T_s, and two radii r_1 and r_2 describing the geometry of the water meniscus between two spherical particles of identical radius R was established in Section 3.2 as

$$u_a - u_w = T_s \left(\frac{1}{r_1} - \frac{1}{r_2} \right) \tag{3.37}$$

As shown in Fig. 3.21a, a "filling angle" θ can be introduced to describe changes in the size, geometry, and volume of the water lens. Specifically, the filling angle describes the angle between vectors from the axes of rotation of R and r_2 and R and r_1. The volume of the water lens between the particles is equal to zero when θ is equal to zero and increases as θ increases. Dallavalle (1943) presented the following approximations among r_1, r_2, R, and θ for a contact angle equal to zero:

$$r_1 = R\left(\frac{1}{\cos \theta} - 1\right) \quad r_2 = R \tan \theta - r_1 \quad 0 \leq \theta \leq 85° \tag{3.38}$$

Substituting eq. (3.38) into eq. (3.37) leads to a description of matric suction as a function of the filling angle θ (radians):

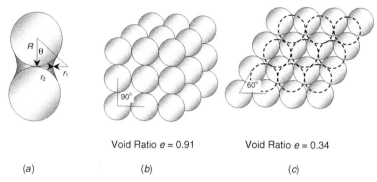

Void Ratio $e = 0.91$ Void Ratio $e = 0.34$

(a)   (b)   (c)

Figure 3.21 Geometrical illustration for three-dimensional meniscus between spherical particles: (*a*) water lens between two particles, (*b*) simple cubic packing representing the loosest packing order, and (*c*) tetrahedral packing representing densest packing order.

$$u_a - u_w = \frac{T_s}{R} \frac{\cos\theta(\sin\theta + 2\cos\theta - 2)}{(1 - \cos\theta)(\sin\theta + \cos\theta - 1)} \quad (3.39)$$

Dallavalle (1943) also showed that the volume of the water lens, V_l, in one orthogonal plane for spheres coordinated in simple cubic (SC) packing order (Fig. 3.21*b*) could be approximated as

$$V_l = 2\pi R^3 \left(\frac{1}{\cos\theta} - 1\right)^2 \left[1 - \left(\frac{\pi}{2} - \theta\right)\tan\theta\right] \quad (3.40)$$

which may be normalized with respect to the volume of one sphere, V_s, as

$$\frac{V_l}{V_s} = \frac{3}{2}\left(\frac{1}{\cos\theta} - 1\right)^2 \left[1 - \left(\frac{\pi}{2} - \theta\right)\tan\theta\right] \quad (3.41)$$

Figure 3.22 illustrates the dependency of the normalized lens volume on the filling angle θ. At a filling angle of 45°, the volume of the water lens is about 5.5% of the particle volume. At a filling angle of 85°, the volume of the water lens is about 42% of the particle volume.

The number of water lenses among one spherical particle and all adjacent particles in a cubical unit volume with dimensions $2R \times 2R \times 2R$ ($8R^3$) in simple cubic packing order is three (six half lenses as shown in Fig. 3.21*b*). Therefore, the gravimetric water content w for a unit volume in three orthogonal planes can be determined from eq. (3.40) by considering the specific gravity of the soil solids G_s:

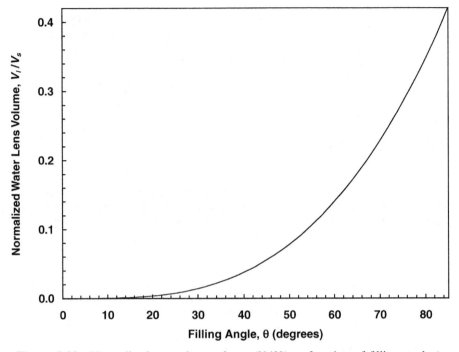

Figure 3.22 Normalized water lens volume (V_l/V_s) as function of filling angle θ.

$$w = \frac{3V_l\rho_w}{V_s\rho_s} = \frac{3V_l}{V_s G_s} = \frac{3V_l}{(4/3)\pi R^3 G_s}$$

$$= \frac{9}{2G_s}\left(\frac{1}{\cos\theta} - 1\right)^2 \left[1 - \left(\frac{\pi}{2} - \theta\right)\tan\theta\right] \quad (3.42)$$

Recognizing the fact that the void ratio e is 0.91 for simple cubic packing order, the saturated water content w_s for solids having a G_s of 2.65 is as follows:

$$w_{s(SC)} = \frac{e}{G_s} = \frac{0.91}{2.65} = 0.343 \quad (3.43)$$

Thus, for the simple cubic packing arrangement of idealized nondeformable sand, 34.3% represents the upper limit of gravimetric water content at full saturation.

The closest packing order for uniform spherical particles is tetrahedral (TH) packing (Fig. 3.21c). Here, each particle has 12 contacts with the surrounding particles and there are 6 full water lenses in each unit volume of $5.66R^3$ (versus $8R^3$ in simple cubic packing). Each sphere is surrounded by 6 spheres in the same plane, 3 on the top, and 3 on the bottom. The water content for a unit volume of $5.66R^3$, therefore, is

$$w = \frac{6V_l\rho_w}{V_s\rho_s} = \frac{6V_l}{V_sG_s} = \frac{6V_l}{(4/3)\pi R^3 G_s}$$

$$= \frac{9}{G_s}\left(\frac{1}{\cos\theta} - 1\right)^2 \left[1 - \left(\frac{\pi}{2} - \theta\right)\tan\theta\right] \quad (3.44)$$

Physically, the closest packing doubles the water content for simple cubic packing. Recognizing the fact that the void ratio is 0.34 for closest packing, the saturated water content for particles with G_s of 2.65 is

$$w_{s(TH)} = \frac{e}{G_s} = \frac{0.34}{2.65} = 0.128 \quad (3.45)$$

Thus, for the closest packing arrangement of idealized nondeformable sand, 12.8% represents the upper limit of gravimetric water content at full saturation.

Equations (3.39) and (3.42) establish a theoretical relationship describing the role of capillarity in the soil-water characteristic curve for uniform spherical particles and simple cubic packing. Equations (3.39) and (3.44) describe the capillary characteristic curve for the case of tetrahedral close packing. Figure 3.23 illustrates soil-water characteristic curves calculated using the contacting sphere model for various particle radii in simple cubic packing order. The upper limit of 6.3% water content corresponds to a filling angle of 45°. This corresponds to the condition where adjacent water lenses begin to overlap to each other and the assumed geometry of the water lens is no longer valid. Filling angle equal to 45° is the upper limit for eq. (3.42). Figure 3.24 illustrates soil-water characteristic curves for various particle radii in closest packing order. Here, the upper limit of 3.2% water content corresponds to a filling angle of 30°. Filling angle equal to 30° is the upper limit for eq. (3.44).

It can be readily observed that the magnitude of matric suction for millimeter size particles is mostly less than 100 kPa. The magnitude of matric suction for micron-size particles can reach 10,000 to 100,000 kPa. For sandy soil, where most of the particle sizes are on the order of millimeters, matric suction values within several hundred kilopascals are important for pore water retention. For clayey soil, where most of the particle sizes are on the order of microns, higher matric suction values are important for pore water retention. The drastic increase in matric suction for all types of soil as the water content approaches zero reflects the drastic reduction in the radii of the air-water interface.

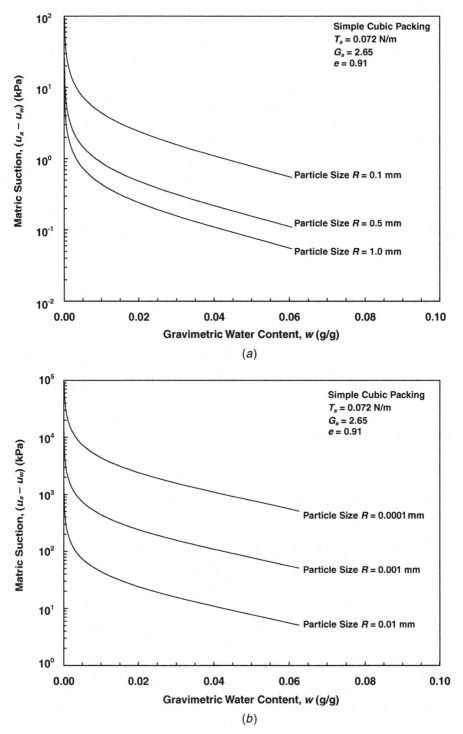

Figure 3.23 Theoretical soil-water characteristic curves for uniform spherical particles in simple cubic packing order: (*a*) sand size particles and (*b*) silt and clay size particles.

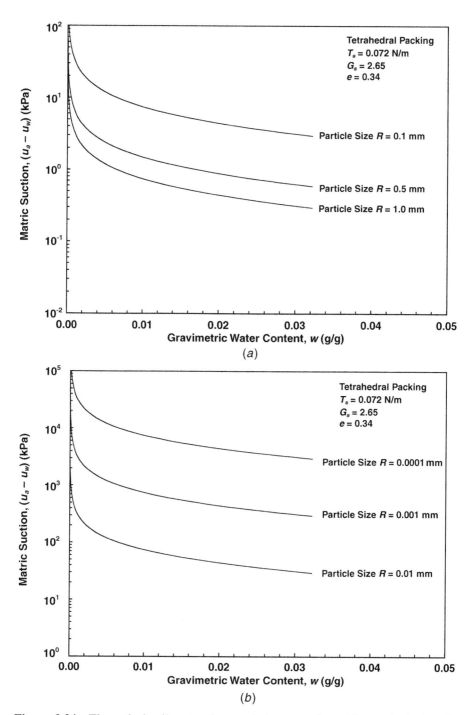

Figure 3.24 Theoretical soil-water characteristic curves for uniform spherical particles in tetrahedral packing order: (*a*) sand size particles and (*b*) silt and clay size particles.

3.4.4 Concluding Remarks

The simplified geometry of toroidal water lenses (characterized by r_1 and r_2) among uniform spherical soil particles described in this chapter provides insight into the micromechanical physics of the soil-water characteristic curve. Real soil particles, however, are often nonspherical and nonuniform in size. Even for perfectly spherical particles, the water lenses among them may not be perfectly toroidal shaped.

In general, the energy reduction or negative pore water pressure in the water lenses between and among particles results from three principal mechanisms: surface tension at the air-water interface, solid-liquid interaction (e.g., hydration effects and van der Waals attraction) at the solid-water interface, and the free energy resulting from adsorbing a given volume of the thermodynamically unfavorable liquid phase (e.g., Orr et al., 1975). A more realistic geometry for the water lens formed under these combined mechanisms has a curvature that varies from one point to another along the air-water interface (i.e., the air-water interface is nonspherical) and a total volume that could be quite different from that captured by an idealized toriodal shape. Orr et al. (1975) showed that the geometry of the meniscus existing between particles is nonunique for a given volume of water and that instability can occur all at minimum energy states. Solutions regarding nontoriodal menisci geometries based on this type of free energy formulation are mathematically and computationally complicated, but have been recently explored by numerous investigators (e.g., Dobbs and Yeomans, 1992, Lian et al., 1993; Molenkemp and Nazemi, 2003). Assessing the impact of nonidealized geometries on the soil-water characteristic curve and the corresponding magnitude of interparticle stress remains an emerging area of research. Findings from the recent studies have indicated that, for a two-particle system, the free energy formulation results in similar principal radii for the menisci compared with the toroidal approximation, but quite different volumes for the water lens. Integrated approaches using free energy and statistical mechanics leading to physics-based soil-water characteristic curve models for rough, nonuniform, and multiparticle systems are on the horizon.

PROBLEMS

3.1. A liter of water at 25°C can dissolve 0.0283 L of oxygen when the pressure of oxygen in equilibrium with the solution is 1 atm. Derive the Henry's law constant for oxygen in water from this information.

3.2. What is the mass coefficient of solubility of air at a temperature of 20°C and a total air pressure of 1 atm if the volumetric coefficient of solubility of air h_{air} is 0.01708?

3.3. If the air pressure changes to 10 bars in the previous problem, what is the mass coefficient of solubility?

3.4. Two different sizes of capillary tubes are in the divided container shown in Fig. 3.25 ($r_1 = 10^{-6}$ m and $r_2 = 10^{-4}$ m). Each side has reached equilibrium between the air in the container and the pore water. Assume the total mass of water vapor in each side is much less than the amount of water in the capillary tubes. Also assume that the initial water levels in the tubes are very low compared to the overall lengths of the tubes. Describe the equilibrium position(s) of the water level in the tubes when the valve is opened.

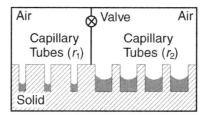

Figure 3.25 Capillary tube system for Problem 3.4.

3.5. Calculate the hydrostatic pressure of water at 28°C in spherical raindrops with (a) 5 mm diameter and (b) 0.2 mm diameter.

3.6. For a bundle of capillary tubes of various sizes ranging between 10^{-7} and 10^{-4} m in radii, assume the contact angle is zero, $T = 25°C$, and answer the following:
   a. What is the range of matric suction?
   b. What is the range of pore water pressure?
   c. What is the range of vapor pressure?
   d. What is the range of relative humidity?

3.7. Table 3.5 shows data comprising the soil-water characteristic curve during a drying process for an unsaturated soil. Assuming the drying process has contact angle of zero, and the wetting process has contact angle of 30°, calculate and plot the soil-water characteristic curve for the wetting process.

TABLE 3.5 Soil-Water Characteristic Curve Data for Problem 3.7

Gravimetric Water Content (g/g)	RH (%)
0.300	100
0.295	90
0.280	85
0.200	75
0.150	65
0.100	55
0.050	45
0.000	40

3.8. a. Plot the relationship between matric suction (kPa, log scale) and relative humidity (%, linear scale) for temperatures of 20, 40, and 60°C.
  b. Plot the relationship between relative humidity (%) and capillary tube radius (m) for a temperature of 20°C and contact angle of 0°, 30°, and 60°.
  c. Discuss the general characteristics of each plot.

3.9. For unsaturated sand undergoing a drying process at 20°C, where the radius of the air-water menisci varies between 10^{-6} and 10^{-5} m, the contact angle is zero, and the air pressure is zero, answer the following:
  a. What is the range of vapor pressure in the soil pores?
  b. What is the range of relative humidity in the soil pores?
  c. What is the range of pore water pressure in the soil pores?

3.10. If the negative pore pressure in the sand from the previous example acts to draw the soil pore water above the water table in the field, what is the corresponding range of the height above the water table?

3.11. Soil-water characteristic curves are shown Fig. 3.26. Figure 3.26a shows curves for the same soil during wetting and drying processes. Figure 3.26b shows the characteristic curve for two different soils. Complete or answer the following:
  a. Label the wetting and drying branches of the characteristic curve for the soil on Fig. 3.26a.
  b. What is the saturated water content during drying for the soil on Fig. 3.26a?
  c. Estimate the air-entry pressure during drying for the soil on Fig. 3.26a.
  d. Which soil on Fig. 3.26b is the more fine-grained?
  e. Estimate the residual water content for soil B.

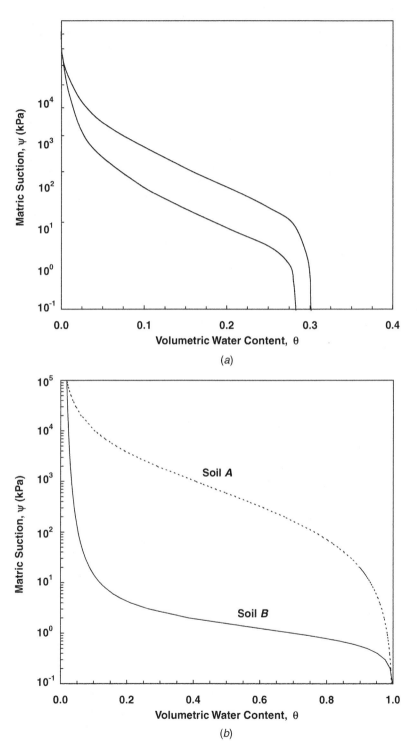

Figure 3.26 Soil-water characteristic curves for Problem 3.11.

CHAPTER 4

CAPILLARITY

4.1 YOUNG-LAPLACE EQUATION

4.1.1 Three-Dimensional Meniscus

The capillary tube models discussed in the previous chapter provide a useful physical interpretation to facilitate understanding of the relationships among fluid pressure, relative humidity, and vapor pressure at an air-water-solid interface. In soil pores, however, the geometry of the pores and fluid menisci are far more complicated, particularly at a scale greater than the largest pore dimension. At a scale less than the largest pore dimension, the air-water-solid interface may be approximated by using simple geometric configurations, including parallel plates, cylinders, ellipsoids, or spheres.

A double-curvature model may be developed on the basis of analytical geometry and mechanical equilibrium to represent the complicated geometry of the air-water-solid interface. The Young-Laplace equation employs this double-curvature concept, providing a general relationship between matric suction and the interface geometry. The Young-Laplace equation may be written as

$$u_a - u_w = T_s \left(\frac{1}{R_1} + \frac{1}{R_2} \right) \qquad (4.1)$$

where u_a and u_w are the air and water phase pressures, respectively, the difference $u_a - u_w$ is the matric suction, T_s is the surface tension of the water phase, and R_1 and R_2 are the two principal radii of curvature of the interface near the area of interest.

P.S. Laplace first derived eq. (4.1) in 1806 on the basis of potential theory, not surface tension. Interestingly, T. Young introduced the concept of macroscopic surface tension in 1805, which was employed by others to prove Laplace's equation on the basis of mechanical equilibrium. The surface tension approach provides an extremely useful means to interpret many interface phenomena. A derivation of eq. (4.1) follows.

Consider mechanical equilibrium near a point O on any arbitrary air-water interface (see Fig. 4.1). Cut an infinitesimal circular element having radius ρ with an axis at point O. The segments AA' and BB' are pairs of any orthogonal lines on the element that pass through point O. The small segments ds at points A, A', B, and B' are subjected to a force arising from surface tension equal to $T_s\, ds$ with projections along the vertical direction (z) equal to $2T_s\, ds \sin \phi$ at points A and A' and $2T_s\, ds \sin \beta$ at points B and B'. Since ρ is small, ϕ is also small, which leads to the following:

$$2T_s\, ds \sin \phi = 2T_s\, ds\, \phi = 2T_s\, ds\, \frac{\rho}{r_1} \qquad (4.2)$$

Similarly, the total vertical force on the segments ds at points B and B' is

$$2T_s\, ds \sin \beta = 2T_s\, ds\, \beta = 2T_s\, ds\, \frac{\rho}{r_2} \qquad (4.3)$$

and the total vertical force on segments along A, A', B, and B' is

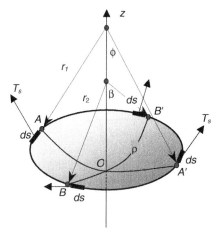

Figure 4.1 Mechanical equilibrium of a three-dimensional double-curvature air-water interface.

$$2T_s \rho \, ds \left(\frac{1}{r_1} + \frac{1}{r_2} \right) \tag{4.4}$$

The values of r_1 and r_2 generally vary from any one pair of lines AA' and BB' to any other pair but can be uniquely linked to the principal radii of curvature R_1 and R_2 by a theorem of Euler as

$$\frac{1}{R_1} + \frac{1}{R_2} = \frac{1}{r_1} + \frac{1}{r_2} \tag{4.5}$$

Thus, the total vertical force on the segments along A, A', B, and B' becomes

$$2T_s \rho \, ds \left(\frac{1}{R_1} + \frac{1}{R_2} \right) = 2T_s \rho \, ds \left(\frac{1}{r_1} + \frac{1}{r_2} \right) \tag{4.6}$$

Since the choice of A, A', B, and B' is completely arbitrary, eq. (4.6) can be integrated along the entire circumference of the meniscus to obtain the total vertical force due to surface tension. Because eq. (4.6) represents the force on four segments on the circumference, the integration requires only a quarter rotation along the circumference, leading to

$$F_z = \pi \rho^2 T_s \left(\frac{1}{R_1} + \frac{1}{R_2} \right) \tag{4.7}$$

At mechanical equilibrium, a force provided by matric suction acting over the projected area of the interface will balance the vertical force F_z:

$$\pi \rho^2 (u_a - u_w) = \pi \rho^2 T_s \left(\frac{1}{R_1} + \frac{1}{R_2} \right) \tag{4.8a}$$

or

$$u_a - u_w = T_s \left(\frac{1}{R_1} + \frac{1}{R_2} \right) \tag{4.8b}$$

which is the familiar form of the Young-Laplace equation.

By introducing the "mean" meniscus curvature R_m, the Young-Laplace equation can be considered a generalized form of the mechanical equilibrium equation for a capillary tube containing a perfectly wetting material. For a three-dimensional meniscus, the mean curvature is

$$\frac{1}{R_m} = \frac{1}{2}\left(\frac{1}{R_1} + \frac{1}{R_2}\right) \tag{4.9}$$

which allows eq. (4.8) to be simplified to the form introduced in previous chapters:

$$u_a - u_w = \frac{2T_s}{R_m} \tag{4.10}$$

Thus, if the geometry of the air-water-solid interface in an unsaturated soil-water system can be represented by an ellipsoidal shape with principal radii ρ_1 and ρ_2, it can be shown (as illustrated in Fig. 4.2) as

$$r_1 \cos \alpha = \rho_1 \qquad r_2 \cos \alpha = \rho_2 \tag{4.11}$$

Substituting eqs. (4.5) and (4.11) into eq. (4.8) results in

$$u_a - u_w = T_s \cos \alpha \left(\frac{1}{r_1} + \frac{1}{r_2}\right) \tag{4.12}$$

4.1.2 Hydrostatic Equilibrium in a Capillary Tube

The negative pore water pressure resulting from interfacial surface tension leads to the redistribution of water in a capillary tube or unsaturated soil. Figure 4.3 demonstrates this *capillary rise* phenomenon for a series of dif-

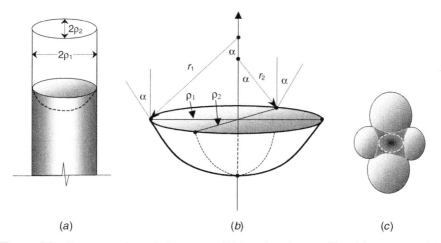

Figure 4.2 Representation of air-water-solid interface by an ellipsoid geometry: (*a*) in a cylindrical tube, (*b*) finite ellipsoid interface, and (*c*) an example in soil pores.

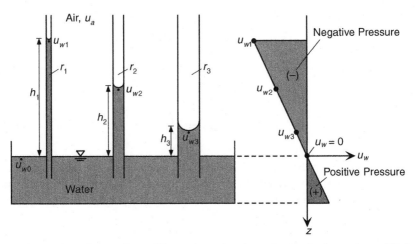

Figure 4.3 Rise of water in capillary tubes of various sizes at hydrostatic equilibrium.

ferent sized capillary tubes at hydrostatic equilibrium. Because the air-water interface in the large tank containing the tubes is flat, the radius of curvature tends to infinity and the matric suction in the bulk fluid tends to zero:

$$u_a - u_{w0} = \frac{2T_s \cos \alpha}{\infty} = 0 \quad (4.13a)$$

or

$$u_a = u_{w0} \quad (4.13b)$$

On the other hand, mechanical equilibrium near the air-water interface in the capillary tubes requires

$$u_a - u_{wi} = \frac{2T_s \cos \alpha}{r_i} \quad (4.14a)$$

or

$$u_{wi} = u_a - \frac{2T_s \cos \alpha}{r_i} = u_{w0} - \frac{2T_s \cos \alpha}{r_i} \quad (4.14b)$$

where the subscript i runs from 1 to 3, corresponding to the three capillary tubes in the figure. As shown in the pressure profile on the right-hand side of Figure 4.3, the water pressure is equal to zero at the water table, increases hydrostatically below the water table, and decreases hydrostatically above the water table.

At mechanical equilibrium, the pore water pressure at the air-water interface u_{wi} is equal to the unit weight of water γ_w multiplied by the height of the capillary rise h_i:

$$u_{wi} = u_a - \frac{2T_s \cos \alpha}{r_i} = u_{w0} - \frac{2T_s \cos \alpha}{r_i} = -h_i \gamma_w \qquad (4.15)$$

or

$$h_i = \frac{2T_s \cos \alpha}{r_i \gamma_w} \qquad (4.16)$$

The above equation states that the height of capillary rise in a capillary tube is directly proportional to surface tension and contact angle, but inversely proportional to the tube radius. In unsaturated soil, the hydrostatic equilibrium position can be inferred from eq. (4.16) if the principal radii of curvature are estimated.

4.2 HEIGHT OF CAPILLARY RISE

4.2.1 Capillary Rise in a Tube

Capillary rise in soil describes the upward movement of water above the water table resulting from the gradient in water potential across the air-water interface at the wetting front. Simple capillary tube models for predicting the ultimate height and rate of capillary rise in soil have been developed based on assumptions of ideal pore geometries and permeability. These models provide excellent insight into the physics of capillary rise and in some cases provide reasonable semiquantitative predictions.

Perhaps the best-known analytical model to quantify the pressure drop across an air-water-solid interface for a nonzero contact angle is the Young-Laplace equation, which was derived in the previous section as

$$u_a - u_w = T_s \cos \alpha \left(\frac{1}{r_1} + \frac{1}{r_2} \right) \qquad (4.17)$$

In an ideal cylindrical capillary tube with a diameter d, $r_1 = r_2 = d/2$ and eq. (4.17) becomes

$$u_a - u_w = \frac{4T_s \cos \alpha}{d} \qquad (4.18)$$

As described in Chapter 3, the contact angle α reflects the ability of water to wet the solid surface at the air-water-solid interface. A contact angle equal to zero describes a perfectly wetting material; 90° describes neutral wetting ability; and an angle greater than 90° describes the interaction between water and a water repellent material. For soil such as sands under drying conditions, the contact angle is often assumed to be equal to 0°.

A simple analysis of mechanical equilibrium can confirm eq. (4.18). Consider the free-body diagram in the area of the small dashed circle shown in Fig. 4.4. Vertical force equilibrium considering $u_a - u_w$ acting over the area of meniscus and the vertical projection of T_s acting over the circumference of the meniscus leads to

$$(u_a - u_w) \frac{\pi}{4} d^2 = T_s \pi d \cos \alpha \qquad (4.19)$$

which can be directly reduced to eq. (4.18).

If the air pressure is set to a reference value of zero, water pressure u_w has a negative value, representing a positive matric suction. The smaller the diameter of the capillary tube d, the greater the matric suction. The greater the wetting ability of the solid surface (i.e., very small contact angle α), the greater the matric suction.

The ultimate height of capillary rise, h_c, can be evaluated by considering mechanical equilibrium in the area of the large dashed circle in Fig. 4.4. Here, the total weight of the water column under the influence of gravity is balanced by surface tension along the water-solid interface as

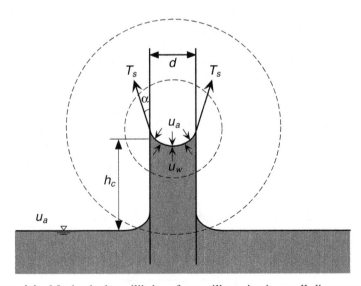

Figure 4.4 Mechanical equilibrium for capillary rise in small-diameter tube.

$$h_c \rho_w g \frac{\pi}{4} d^2 = T_s \pi d \cos \alpha \qquad (4.20)$$

or simply

$$h_c = \frac{4 T_s \cos \alpha}{d \rho_w g} \qquad (4.21)$$

Imposing values of water density ρ_w as 1 g/cm³, gravitational acceleration g = 980 cm/s², T_s = 72 mN/m at 25°C, and a zero contact angle, a simple relationship between capillary rise and capillary tube diameter can be written as

$$h_c \text{ (cm)} = \frac{0.3}{d \text{ (cm)}} \qquad (4.22)$$

The solid curve in Fig. 4.5 shows a plot of eq. (4.22) in terms of the maximum capillary rise versus tube diameter.

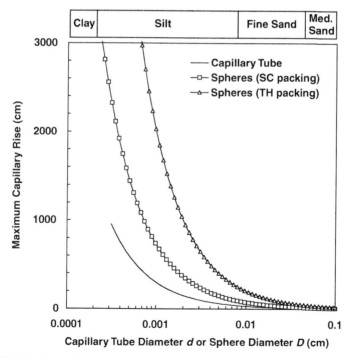

Figure 4.5 Maximum height of capillary rise in capillary tube or idealized soil comprised of uniform spherical particles. Particle diameter is delineated in terms of soil type for comparison.

The upper limit of eq. (4.21) or (4.22) in a glass capillary tube is about 10 m, corresponding to a negative water pressure of about -100 kPa or -1 atm at sea level. As discussed previously in Section 2.5, free water tends to cavitate below this pressure. In soil, however, where the pore water may be under the influence of short-range physicochemical interaction effects at the water-solid interface that lower its chemical potential and alter its physical properties, cavitation may not occur at the same pressure as that for free water. As described in Section 1.6, the intensity of these liquid-solid interaction effects in an unsaturated soil system is a function of the specific surface and surface charge properties of the soil mineral. In clayey soil, for example, which possesses both a very large surface area and a highly "active" surface, capillary rise may be as high as several tens of meters.

4.2.2 Capillary Finger Model

The uniform capillary tube model is often used to describe capillary rise in unsaturated soil and the associated pore water retention characteristics. Although the concept of perfectly uniform tubes in soil is unrealistic, the continuous water fringes or "fingers" that develop above the water table can be conceptualized as *bundled tubes* of various diameters. This conceptualization is illustrated in Fig. 4.6a, where the rising fingers of water are shown with different average diameters and heights at the equilibrium condition. The associated pore water retention curve is shown as Fig. 4.6b in terms of volu-

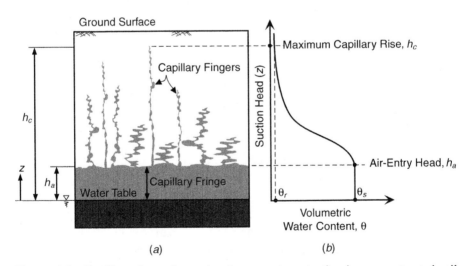

Figure 4.6 Capillary rise and associated pore water retention in an unsaturated soil profile: (*a*) conceptual illustration and (*b*) corresponding soil-water characteristic curve.

metric water content versus suction head (i.e., the height above the water table).

As illustrated in the figure, pore water rises above the water table under capillary suction. The soil remains essentially saturated, described by the *saturated water content* θ_s, until the suction head reaches the *air-entry head,* designated h_a. The air-entry head may be defined as the suction head at which air initially begins to displace water from the soil pores. The saturated zone extending from the water table up to the air-entry head is commonly referred to as the *capillary fringe.* Above the air-entry head, the water content decreases with increasing height, reflecting the fact that fewer and smaller capillary fingers are present for a given cross section of the soil column with increasing elevation. Following the principles developed in the previous section, the narrowest capillary fingers rise to a maximum height h_c, whereas the largest fingers are restrained to relatively low elevations. At relatively large values of suction head, therefore, very little water is retained by the soil. Pore water within this regime is primarily in the form of thin films surrounding the particle surfaces or disconnected "pendular" water menisci. The water content within this regime is commonly referred to as the *residual water content,* or θ_r.

If the soil column above the water table is initially dry, a head gradient exists between the continuous capillary fingers and the overlying soil, which is approximately equal to $(h_c - z)/z$, where z is the height of the advancing wetting front and h_c is the driving head described by eq. (4.21). As the fingers move into higher elevations, the gradient decreases. Eventually, the wetting front reaches a point where $h_c = z$, thus satisfying the requirement for mechanical equilibrium, and the capillary rise ceases. When coupled with an appropriate description for hydraulic conductivity, consideration of the changing driving head as the wetting front advances allows the rate of capillary rise to be evaluated. Two such developments are presented in Section 4.3.

4.2.3 Capillary Rise in Idealized Soil

Equations (4.21) and (4.22) provide a means to estimate the height of capillary rise in a uniform capillary tube. Given this theoretical basis, the upper and lower bounds of capillary rise in idealized soil comprised of uniform spherical particles may be evaluated by considering simple cubic (SC) packing (i.e., loosest possible packing) and tetrahedral (TH) packing (i.e., densest possible packing) as two limiting cases.

Figures 4.7a and 4.7b show simple cubic and tetrahedral packing geometries in plan view for uniform spheres of diameter D, respectively. Minimum pore diameters across these sections corresponding to SC and TH packing are denoted d_{sc} and d_{th}. The relationship between particle size and minimum pore diameter for the case of SC packing is described by:

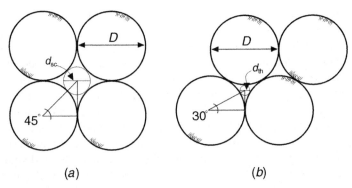

Figure 4.7 Plan view illustration of (*a*) simple (SC) cubic and (*b*) tetrahedral (TH) packing for uniform spherical particles.

$$\cos 45° = \frac{D/2}{D/2 + d_{sc}/2} \qquad (4.23)$$

which leads to the simple relationship

$$d_{sc} = 0.41D \qquad (4.24)$$

Similarly, the relationship between particle size and minimum pore diameter for the case of TH packing is

$$\cos 30° = \frac{D/2}{D/2 + d_{th}/2} \qquad (4.25)$$

or

$$d_{th} = 0.15D \qquad (4.26)$$

A more realistic system of spherical particles would likely have a packing geometry that falls somewhere between these two limiting cases. Substituting eqs. (4.24) and (4.26) into eq. (4.21), therefore, the following bounds for the ultimate height of capillary rise in such a system may be defined:

$$\frac{9.76 T_s \cos \alpha}{D \text{ (cm)} \rho_w g} \leq h_c \text{ (cm)} \leq \frac{26.67 T_s \cos \alpha}{D \text{ (cm)} \rho_w g} \qquad (4.27a)$$

Assuming values for $\rho_w = 1$ g/cm^3, $g = 980$ cm/s^2, $T_s = 72$ mN/m, and zero contact angle leads to

$$\frac{0.73}{D \text{ (cm)}} \leq h_c \text{ (cm)} \leq \frac{2}{D \text{ (cm)}} \tag{4.27b}$$

Equation (4.27b) implies that for the same soil, packing can affect the capillary height by a factor of about 2.75. For example, a sand column prepared in the lab using uniform Ottawa sand with particle diameter of 0.1 cm might have capillary rise ranging anywhere between 7.3 and 20 cm. The maximum capillary rise corresponding to the bounds described by eq. (4.27b) for a wide range of particle diameter is included in Fig. 4.5.

4.2.4 Capillary Rise in Soil

Because real soil is comprised of a range of different particle sizes falling within some size distribution and complex packing geometry, analytical evaluation of the height of capillary rise is extremely difficult. To overcome this difficulty, empirical equations have been developed to relate the height of capillary rise to more easily measured soil properties. These properties most commonly include particle or pore size distribution parameters, void ratio, and air-entry head. In general, hysteresis effects are not considered in the empirical relationships. Most of the empirical equations assume an initially dry soil undergoing a wetting process from a stationary water table.

Peck et al. (1974), for example, describe an empirical equation expressing the height of capillary rise as an inverse function of the product of void ratio, e, and the 10% finer particle size, D_{10}, as

$$h_c = \frac{C}{eD_{10}} \tag{4.28}$$

where h_c and D_{10} are in units of millimeters and C is a constant varying between 10 and 50 mm² depending on surface impurities and grain shape. Because an increase in either void ratio or D_{10} reflects an increase in the average pore diameter of the soil, the corresponding maximum height of capillary rise decreases.

Analysis of capillary rise experiments conducted by Lane and Washburn (1946) for eight different soils indicates that the maximum height of capillary rise may be described by a linear function of D_{10} as

$$h_c = -990(\ln D_{10}) - 1540 \tag{4.29}$$

where both D_{10} and h_c are in units of millimeters and D_{10} ranges from 0.006 to 0.2 mm. Equations (4.28) and (4.29) indicate that the 10% finer particle fraction may adequately describe the effective diameter of the smallest continuous capillary fingers in soil.

Perhaps the most reliable method to determine the height of capillary rise is by direct measurement through open-tube capillary rise tests conducted in the laboratory. Numerous experimental programs in this regard have been described in the literature (e.g., Lane and Washburn, 1946; Malik et al., 1989; Kumar and Malik, 1990). Table 4.1, for example, shows a summary of results from laboratory capillary rise experiments for several different types of soil. Maximum capillary rise h_c was determined in each case by observing the equilibrium wetting front manually. Air-entry head h_a was determined from the soil-water characteristic curve, measured either using representative specimens or by measuring the final equilibrium water content of the soil column as a function of height from the water table. The final column on Table 4.1 shows the dimensionless ratio of maximum capillary rise to air-entry head, h_c/h_a.

The data in Table 4.1 supports the notion of an empirical relationship between air-entry head and the maximum height of capillary rise. For the wide range of soil tested, the ratio h_c/h_a varies from 2 to 5 with only a few exceptions. Thus, if the air-entry head is estimated from independent measurements of grain size distribution or the soil-water characteristic curve, it appears that the upper and lower limits for maximum height of capillary rise may be reasonably estimated.

Kumar and Malik (1990) also found that the difference between the height of capillary rise and the height of capillary fringe is a decreasing function of the square root of an equivalent pore radius r. One such relationship was suggested in the form

$$h_c = h_a + 134.84 - 5.16 \sqrt{r} \tag{4.30}$$

where h_c and h_a are in centimeters and r is in micrometers.

4.3 RATE OF CAPILLARY RISE

4.3.1 Saturated Hydraulic Conductivity Formulation

As early as 1943, Terzaghi formulated a simple theory to predict the rate of capillary rise in a one-dimensional column of soil. To quantify the rate of capillary rise, Terzaghi made two major assumptions: (1) Darcy's law for saturated flow is applicable to unsaturated flow, and (2) the upward hydraulic gradient i responsible for capillary rise at the wetting front can be approximated as

$$i = \frac{h_c - z}{z} \tag{4.31}$$

where z is a distance measured positive upward from the elevation of the water table (see Fig. 4.8). Physically, the maximum capillary height h_c rep-

TABLE 4.1 Experimental Capillary Rise Parameters for Several Different Soils

Test No.[a]	Soil	Gravel (%)	Sand (%)	Silt/Clay (%)	Clay (%)	Void ratio	h_a (cm)	h_c (cm)	h_c/h_a
1	Class 5	25.0	68.0	7.0	—	0.27	41.0	82.0	2.0
2	Class 6	0.0	47.0	53.0	—	0.66	175.0	239.6	1.4
3	Class 7	20.0	60.0	20.0	—	0.36	39.0	165.5	4.1
4	Class 8	0.0	5.0	95.0	—	0.93	140.0	359.2	2.6
5	Ludas sand	—	—	—	—	—	29.1	72.1	2.5
6	Rawalwas sand	—	—	—	—	—	29.6	77.5	2.6
7	Rewari sand	—	—	—	—	—	29.4	60.9	2.1
8	Bhiwani sand	—	—	—	—	—	27.6	65.6	2.4
9	Tohana loamy sand 1	—	—	—	—	—	37.4	117.0	3.1
10	Hisar loamy sand 1	—	—	—	—	—	37.5	149.4	4.0
11	Barwala sandy loam 1	—	—	—	—	—	41.2	158.4	3.8
12	Rohtak sandy loam 1	—	—	—	—	—	48.7	155.7	3.2
13	Hisar sandy loam 1	—	—	—	—	—	47.7	173.5	3.7
14	Pehwa sandy clay loam	—	—	—	—	—	44.5	154.6	3.5
15	Hansi clayey loam 1	—	—	—	—	—	29.6	127.5	4.3
16	Ambala silty clay loam 1	—	—	—	—	—	15.0	141.5	9.4
17	Tohana loamy sand 2	—	89.0	6.0	6.0	0.92	66.7	117.0	1.8
18	Hissar loamy sand 2	—	82.5	11.5	6.0	0.90	72.9	149.4	2.0
19	Barwala sandy loam 2	—	75.0	13.5	11.5	0.94	47.3	158.4	3.3
20	Rohtak sandy loam 2	—	63.0	23.0	14.0	1.01	44.0	155.7	3.5
21	Hissar sandy loam 2	—	63.0	24.0	13.0	0.99	66.0	174.5	2.6
22	Pehowa sandy clay loam	—	55.0	27.0	18.0	1.06	59.6	154.6	2.6
23	Hansi clayey loam 2	—	30.2	26.5	43.3	1.27	16.3	127.5	7.8
24	Ambala silty clay loam 2	—	15.0	49.0	36.0	1.49	16.9	141.5	8.4

[a] 1–4 (Lane and Washburn, 1946), 5–16 (Malik et al., 1989), and 17–24 (Kumar and Malik, 1990).

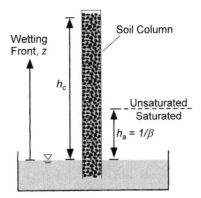

Figure 4.8 System geometry for analytical prediction of rate of capillary rise.

resents the drop in pressure head across the air-water interfaces in the soil pores.

Terzaghi's other assumption, Darcy's law, can be expressed in familiar form as

$$q = k_s i = n \frac{dz}{dt} \tag{4.32}$$

where q is the discharge velocity, k_s is the saturated hydraulic conductivity of the soil column, and n is the porosity.

Solving eqs. (4.31) and (4.32) and imposing an initial condition of a zero capillary rise at zero time, Terzaghi arrived at a solution describing the location of the capillary wetting front z as an implicit function of time t:

$$t = \frac{nh_c}{k_s}\left(\ln\frac{h_c}{h_c - z} - \frac{z}{h_c}\right) \tag{4.33a}$$

which can be rearranged in a compact form by introducing dimensionless time $T = k_s t/nh_c$ and dimensionless distance $Z = z/h_c$ as

$$T = \ln\frac{1}{1 - Z} - Z \tag{4.33b}$$

4.3.2 Unsaturated Hydraulic Conductivity Formulation

Subsequent experimental investigations of capillary rise (e.g., Lane and Washburn, 1946; Krynine, 1948) have shown that Terzaghi's original analytical solution (4.33) significantly overpredicts the rate of rise. The assumption of constant (saturated) hydraulic conductivity had been identified by Terzaghi

(1943) and Krynine (1948) as the cause for the discrepancies. In some cases, a reduction of the saturated hydraulic conductivity by more than 2 orders of magnitude is required to yield a reasonable match between the theory and experimental data.

In reality, capillary rise above the air-entry head is no longer governed by the saturated hydraulic conductivity. As described in Chapter 8, the hydraulic conductivity of soil decreases dramatically with decreases in the degree of saturation, following what is commonly referred to as the *unsaturated hydraulic conductivity function*. By the time the wetting front approaches the maximum height of capillary rise, the degree of saturation may be as low as a few percent, and the hydraulic conductivity may be reduced by 5 to 7 orders of magnitude from its value at saturation. This significant reduction in conductivity, together with the reduction in the available driving head $(h_c - z)/z$ as the wetting front moves upward, leads to a significant decrease in the rate of rise. Consequently, the discrepancies between Terzaghi's theoretical equation and the actual height of capillary rise propagate as time elapses.

The characteristic dependence of hydraulic conductivity with respect to suction, water content, or degree of saturation has been a focus of intensive research since Terzaghi's original work. Numerous models for describing the unsaturated conductivity function have been developed, with the majority accounting for the drastic reduction in conductivity using either exponential, power, or series functions. Several of these models are described in detail in Chapter 12.

Lu and Likos (2004) developed an alternative solution for the rate of capillary rise by incorporating the Gardner (1958) one-parameter model to estimate the unsaturated hydraulic conductivity function. As described in Chapter 12, Gardner's model may be expressed as an exponential function in terms of the saturated hydraulic conductivity and suction head as

$$k(h_m) = k_s \exp(\beta h_m) \tag{4.34}$$

where k is the unsaturated hydraulic conductivity at suction head h_m (cm) and β is a pore size distribution parameter (cm^{-1}) representing the rate of decrease in hydraulic conductivity with increasing suction head. As illustrated on Fig. 4.8, the inverse of β can be interpreted as the air-entry head, or equivalently, as the height of the saturated portion of capillary rise, that is, the capillary fringe.

The general behavior and performance of eq. (4.34) is demonstrated in Fig. 4.9*a* by comparison with experimental data for sand (Richards, 1952) and clay (Moore, 1939). Figure 4.9*b* provides a more general illustration of the model's behavior for parameters (k_s and β) representative of three different soil types.

Incorporating the Gardner model to represent hydraulic conductivity at the wetting front, a governing equation for the rate of capillary rise can be written as

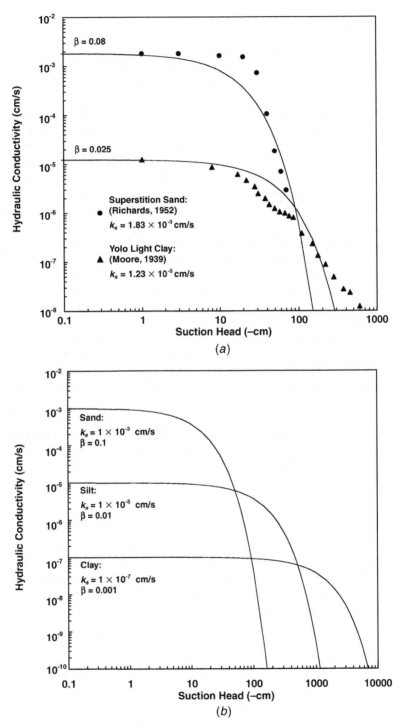

Figure 4.9 Unsaturated hydraulic conductivity function according to the Gardner (1958) one-parameter model: (*a*) comparison with experimental data and (*b*) general pattern for three representative soil types.

$$\frac{dz}{dt} = \frac{k_s}{n}\exp(-\beta z)\frac{h_c - z}{z} \tag{4.35}$$

Analytical solution of eq. (4.35) can be written in series form:

$$t = \frac{n}{k_s}\sum_{j=0}^{m=\infty}\frac{\beta^j}{j!}\left(h_c^{j+1}\ln\frac{h_c}{h_c - z} - \sum_{s=0}^{j}\frac{h_c^s z^{j+1-s}}{j + 1 - s}\right) \tag{4.36a}$$

If the nonlinearity in hydraulic conductivity is ignored by setting the series index m to zero, eq. (4.36a) reduces to Terzaghi's original analytical solution [eq. (4.33)]. Convergent solutions are typically obtained by setting m equal to 10. In applying eq. (4.36a), the material parameter β can be determined if either the hydraulic conductivity function or soil-water characteristic curve is measured a priori. Given the former, β can be determined in conjunction with Gardner's (1958) model to find the value giving a best fit to the data. Given the latter, β can be determined by estimating the air-entry head h_a and by recognizing that β may be interpreted as its inverse. The practical range of β for most soil reported in the literature varies from 1.0 cm^{-1} for coarse-grained materials, to 0.001 cm^{-1} or lower for relatively fine-grained materials. The ultimate height of the capillary rise for use in eq. (4.36a) may be approximated using a capillary tube analogy and applying the Young-Laplace equation or by applying the empirical relationships described in the previous section.

Equation (4.36a) may also be written in terms of the dimensionless variables T and Z as

$$T = \sum_{j=0}^{m=\infty}\frac{(\beta h_c)^j}{j!}\left(\ln\frac{1}{1 - Z} - \sum_{s=0}^{j}\frac{Z^{j+1-s}}{j + 1 - s}\right)$$

$$T = \frac{k_s t}{nh_c} \qquad Z = \frac{z}{h_c} \tag{4.36b}$$

By writing the solution in dimensionless space and time, arrival time contours can be predicted if the soil parameters h_c, β, n, and k_s are known. Figure 4.10 shows a series of such contours for T_{50}, T_{60}, T_{70}, T_{80}, and T_{90}. The arrival time T_{50}, for example, is defined as the dimensionless time required to advance the wetting front to the position half of the total height of the maximum capillary rise, that is, $z = 0.5\,h_c$.

4.3.3 Experimental Verification

In 1946, Lane and Washburn reported a systematic experimental study on the height and rate of capillary rise using open-tube column tests. Soils were

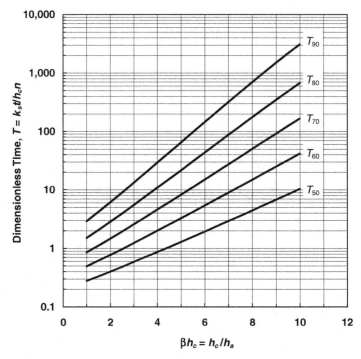

Figure 4.10 Solution for rate of capillary rise in dimensionless time and space.

prepared from natural sandy gravel that was graded and remixed in desired proportions to create eight "classes" of soils representing a wide range in grain size and grain size distribution. Figure 4.11 shows grain size distribution curves for four of these classes. Direct measurements were obtained for saturated hydraulic conductivity k_s, porosity n, soil-water characteristic curves, total height of capillary rise h_c, rate of capillary rise (e.g., the elevation of wetting front as a function of time), and in some cases, the height of the capillary fringe h_a.

Figure 4.12 (page 148) shows height of capillary rise as a function of time from the experimental measurements for class 2 (Fig. 4.12a) and class 4 (Fig. 4.12b) materials, a poorly graded coarse sand and poorly graded fine sand, respectively. Experimental data for class 5 and 6 soils, a well-distributed coarse sand with fines and a sandy silt, respectively, are shown in Figs. 4.13a and 4.13b (page 149). Theoretical solutions for the rate of capillary rise based on the saturated hydraulic conductivity formulation [eq. (4.33)] and the unsaturated hydraulic conductivity formulation [eq. (4.36)] are included for comparison. Note the significant improvement in the prediction when the unsaturated nature of the soil is considered.

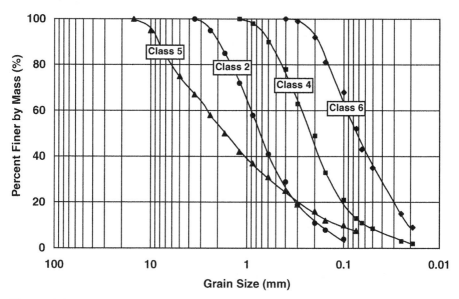

Figure 4.11 Grain size distributions for Lane and Washburn (1946) capillary rise tests.

4.4 CAPILLARY PORE SIZE DISTRIBUTION

4.4.1 Theoretical Basis

The size, shape, and distribution of the pore spaces in soil comprise a critical element of soil fabric and play principal roles in governing the overall engineering behavior of the bulk soil mass. Methodologies to measure or estimate physical properties of the pore space provide significant insight in predicting strength, compressibility, and permeability behavior. This section describes the theoretical basis for evaluating relationships among pore size, pore size distribution, and capillary pressure in unsaturated soil. A step-by-step list of computational procedures and a series of example problems are provided to demonstrate use of the soil-water characteristic curve for estimating pore size distribution.

Kelvin's equation provides the thermodynamic basis to relate relative humidity or matric suction to pore size. As introduced in Section 3.3, capillary radius r can be expressed as a function of surface tension T_s, contact angle α, and relative humidity RH as

$$r = -\frac{2T_s v_w \cos \alpha}{RT \ln(\text{RH})} \qquad (4.37\text{a})$$

or in terms of matric suction $u_a - u_w$ as

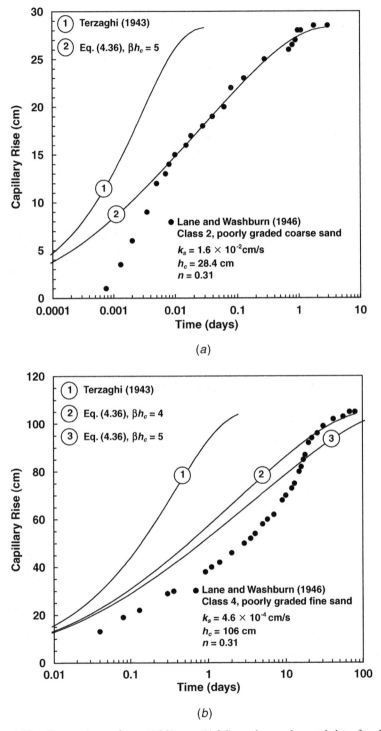

Figure 4.12 Comparison of eq. (4.33), eq. (4.36), and experimental data for the rate of capillary rise in (*a*) coarse sand and (*b*) fine sand.

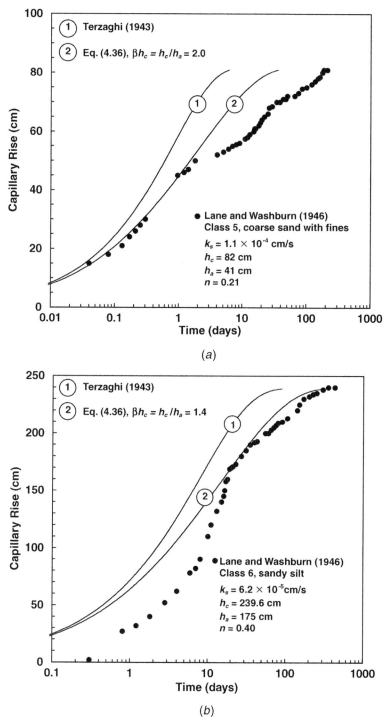

Figure 4.13 Comparison of eq. (4.33), eq. (4.36), and experimental data for the rate of capillary rise in (*a*) coarse sand with fines and (*b*) sandy silt.

$$r = \frac{2T_s \cos \alpha}{u_a - u_w} \qquad (4.37b)$$

Analyses based on eqs. (4.37a) and (4.37b) have been extensively explored to evaluate pore size and pore size distribution in porous media (e.g., Lowell, 1979). Pore fluids commonly used for such analyses most commonly include water, water vapor, nitrogen, and mercury. Water vapor sorption isotherms for use with eq. (4.37a), whereby the relationship between relative humidity and pore size may be established, and soil-water characteristic curves (SWCC) for use with eq. (4.37b), whereby the relationship between matric suction and pore size may be established, are typically considered along drying (desorption or drainage) paths. A zero contact angle is typically assumed. A related type of analysis involves the intrusion of a nonwetting pore fluid (most commonly mercury) into an initially evacuated specimen under externally applied positive pressure. In this case, a more general form of the pore size–capillary pressure relationship can be written in terms of applied intrusion pressure u_p as

$$r = -\frac{2T_s \cos \alpha}{u_p} \qquad (4.37c)$$

where the contact angle α is greater than 90°, or about 130° to 150° for mercury. Diamond (1970) and Sridharan et al. (1971) provide detailed descriptions of mercury intrusion porosimetry (MIP) and its application to the evaluation of pore size and pore size distribution in soil.

By definition, capillary pore size analysis is applicable over the range of pore size for which capillarity remains the dominant pore fluid retention mechanism. As described in Chapter 3, this range is approximately 10^{-9} to 10^{-4} m in terms of pore radius, which corresponds to matric suction ranging from approximately 144,000 to 0 kPa, or relative humidity ranging from approximately 35 to 100%. Below relative humidity of about 35%, pore water adsorption and retention are controlled primarily by surface hydration mechanisms, which cannot be directly described by eq. (4.37). Application of conventional pore size distribution analyses to clayey soil, particularly expansive clay, is also limited because adsorption mechanisms other than capillarity (e.g., hydration and osmotic effects) dominate over an extremely wide and poorly understood range of suction and because the pore fabric is not a constant but rather may radically deform as a function of water content.

4.4.2 Pore Geometry

Several relationships are required to conceptualize the geometry of the soil pores in order for analysis based on capillary pressure measurements to be possible. These include pore volume, average pore radius, the thickness of the adsorbed water film on the soil solids, and the ratio of pore volume to surface area. Because the computational procedures for estimating pore size distribution involve numerical integration, it is convenient to define these quantities in incremental form as functions of relative humidity or matric suction. Each may then be quantified at incremental steps along the sorption isotherm or soil-water characteristic curve under consideration.

The change in the air-filled pore volume or the water-filled pore volume per unit mass of solid, ΔV_p^i (m^3/kg), for the ith increment of relative humidity or suction can be defined as

$$\Delta V_p^i = \frac{\Delta w^i}{\rho_w} \tag{4.38a}$$

or in an integral form

$$V_p^i = \frac{w^i}{\rho_w} \tag{4.38b}$$

The gravimetric water content w^i in the above equations can be directly obtained from the sorption isotherm or soil-water characteristic curve. The density of water, ρ_w, may be considered essentially constant within the capillary adsorption regime, assuming that solid-liquid interaction effects will only cause significant density changes in the thin films located adjacent to the particle surfaces (see Section 2.1.3).

The ratio of pore volume to surface area for a given pore depends on the pore geometry. However, because pore shapes in soil are highly irregular, an exact mathematical expression of the volume-to-area ratio is practically impossible. Alternatively, simple shapes such as cylinders, parallel plates, and spheres may be assumed to provide estimates or bounds on such ratios. The volume-to-area ratio for a cylinder, pair of parallel plates, and sphere are $r/2$, $r/2$, and $r/3$, respectively, where r is the cylinder radius, sphere radius, or the separation distance between parallel plates (Fig. 4.14). The geometry of the air-filled pores under a given relative humidity, matric suction, or degree of saturation can also be conceptualized as the simple shapes depicted in Fig. 4.14 and can be calculated using eq. (4.37).

The Kelvin radius r_k (air-filled pore radius) can be evaluated from eqs. (4.37a) or (4.37b) as

$$r_k^i = \frac{2T_s v_w}{RT \ln(u_v/u_{v0})} = \frac{2T_s}{u_a - u_w} \tag{4.39}$$

The actual pore radius r_p^i is the Kelvin radius plus the thickness of the water film, t^i, adsorbed on the particle surface at the prevailing relative humidity or matric suction, and thus may be written as

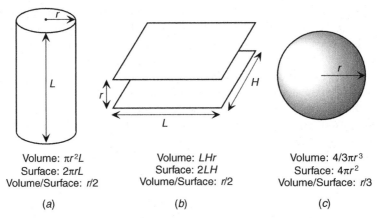

Figure 4.14 Idealized geometries for soil pores: (*a*) cylinder, (*b*) parallel plates, and (*c*) sphere.

$$r_p^i = r_k^i + t^i \qquad (4.40)$$

Several methods have been proposed to estimate adsorbed film thickness t. The Halsey equation (1948) is commonly used for pore size distribution analyses as it has been shown to provide a close fit to experimental data for many porous media and because it is independent of porous media type for relative humidity greater than 30%. The Halsey (1948) equation is written as

$$t^i = \tau \left[-\frac{5}{\ln(\mathrm{RH}^i)} \right]^{1/3} \qquad (4.41)$$

where t^i is the thickness of the water layer on the surface of the soil solid at the ith increment in relative humidity, and τ is the effective diameter of the sorbate molecule. The effective diameter of an adsorbed water molecule may be calculated by considering the area and volume occupied by one mole of water if it were spread over a surface to a depth of one molecular layer. Assuming the occupied cross-sectional area of a liquid water molecule is approximately $A = 10.8$ Å^2 (Livingston, 1949), and given the molar volume of water $v_w = 18 \times 10^{-6}$ m^3/mol, and Avogadro's number $N_A = 6.02 \times 10^{23}$ mol^{-1}, the effective diameter for an adsorbed water molecule may be estimated as

$$\tau = \frac{v_w}{AN_A} = \frac{18 \times 10^{-6}\ \mathrm{m^3/mol}}{(10.8\ \mathrm{Å^2})(6.02 \times 10^{23}\ 1/\mathrm{mol})} = 2.77\ \mathrm{Å} \qquad (4.42)$$

Figure 4.15 shows a plot of adsorbed water film thickness as a function of relative humidity calculated using eqs. (4.41) and (4.42) for an effective diameter of water molecules equal to 2.77 Å.

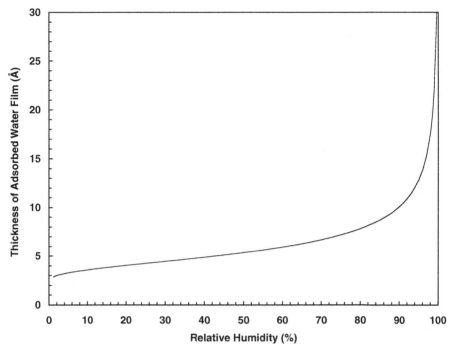

Figure 4.15 Thickness of adsorbed water film as function of relative humidity.

The change in the specific surface area, S, over the ith increment of relative humidity or suction can be determined by the volume-to-area ratio for a given pore geometry. For example, if a cylinder or pair of parallel plates is assumed, the incremental specific surface area is

$$\Delta S^i = \frac{2 \Delta V_p^i}{r_p^i} \qquad (4.43a)$$

If a spherical pore geometry is assumed, then the incremental specific surface area is

$$\Delta S^i = \frac{3 \Delta V_p^i}{r_p^i} \qquad (4.43b)$$

4.4.3 Computational Procedures

Numerical integration procedures for calculating pore size distribution from a sorption isotherm or soil-water characteristic curve are summarized in the following steps.

1. Select data from the sorption isotherm or soil-water characteristic curve for relative humidity greater than about 35% or matric suction less than about 144,000 kPa.
2. Convert volumetric water content to gravimetric water content if the soil-water characteristic curve or vapor sorption isotherm is obtained in terms of volumetric water content.
3. Convert matric suction to relative humidity if the SWCC is obtained in terms of suction.
4. Convert gravimetric water content to the water-filled pore volume per unit mass of solid by dividing the water content by water density, eq. (4.38b).
5. Calculate the Kelvin radius using eq. (4.39).
6. Calculate the thickness of the water film using eq. (4.41).
7. Calculate the pore radius using eq. (4.40).
8. For a given change in relative humidity (i.e., decrement along the desorption curve under consideration), calculate the decrement in the pore volume per unit mass of solid.
9. Calculate the average Kelvin radius during the decrement.
10. Calculate the average pore radius during the decrement.
11. Calculate the incremental surface area for the assumed pore geometry using eq. (4.43).
12. Calculate the cumulative pore volume per unit mass by summing the previous incremental pore volumes.
13. Plot the decrement in pore volume per unit mass versus the average pore radius and plot the cumulative pore volume versus the pore radius.

Example Problem 4.1 Figure 4.16 shows a soil-water characteristic curve in the form of matric suction versus gravimetric water content, $\psi(w)$, for a pulverized specimen of Georgia kaolinite. Given that the surface tension of water, T_s, is 72 mN/m, the gas constant R is 8.314 J/mol · K, and the molar volume of liquid water, v_w, is 0.018 m^3/kmol, develop the pore size and cumulative pore size distribution functions for the clay. Assume the ambient temperature corresponding to the soil-water characteristic curve is 25°C.

Solution The worksheet shown as Table 4.2 was created to follow the general computational procedures described above. Figure 4.17a illustrates the resulting pore size distribution for the kaolinite in terms of pore volume per unit mass versus average pore size. Figure 4.17b illustrates the pore size distribution in terms of cumulative pore volume versus average pore size. The calculated specific surface area is 19.83 m^2/g, which is within the typical range for Georgia kaolinite of about 10 to 20 m^2/g (e.g., Klein and Hurlbut, 1977). The total pore volume calculated for the kaolinite is 0.396 cm^3/g. Note from Fig. 4.17a that pore sizes between about 100 and 10,000 Å dominate

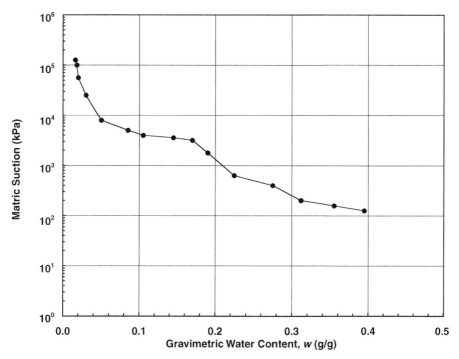

Figure 4.16 Soil-water characteristic curve for Georgia kaolinite.

the total pore volume. Since most of the grain sizes for typical kaolinite are less than 2 μm, it follows that the majority of pores fall within the range of 0.1 μm (1000 Å) and 1 μm (10,000 Å). The valley occurring at about 700 Å reflects the rapid change in matric suction noted in the soil-water characteristic curve at water content between 0.16 and 0.22 g/g.

Example Problem 4.2 Figure 4.18a shows grain size distribution curves for two sandy soil specimens: poorly graded sand with silt (SP-SM) and silty sand (SM). Soil-water characteristic curves for the sands (Fig. 4.18b) were obtained in the laboratory along drying paths using a Tempe pressure cell apparatus (Section 10.3). Develop the pore size and cumulative pore size distribution functions for each material from this data.

Solution Figure 4.19a shows the pore size distribution for each sand in terms of pore volume per unit mass versus average pore size. Figure 4.19b illustrates pore size distributions in terms of cumulative pore volume versus average pore size. Note that the relatively narrow grain size distribution of the SP-SM specimen is reflected in its poorly graded, or "steep," grain size distribution curve, its relatively "flat" soil-water characteristic curve, and by the distinct maximum on the pore size distribution function (Fig. 4.19a) oc-

TABLE 4.2 Computational Worksheet for Determining Pore Size Distribution from Soil-Water Characteristics for Georgia Kaolinite (Fig. 4.16)

$u_a - u_w$ (kPa)	w (g/g)	RH (%)	V_p (cm³/g)	r_k (Å)	t (Å)	r_p (Å)	ΔV_p (cm³/g)	$(r_k)_{avg}$ (mm)	$(r_p)_{avg}$ (mm)	ΔS (m²/g)	$\sum (V_p)$ (cm³/g)
126	0.395	99.91	0.396	11,438.3	48.8	11,487.1					0.040
158	0.355	99.89	0.356	9,085.8	45.2	9,131.0	0.040	10,262.1	10309.1	0.078	0.083
200	0.312	99.86	0.313	7,217.1	41.9	7,259.0	0.043	8,151.4	8195.0	0.105	0.120
398	0.275	99.71	0.276	3,617.1	33.3	3,650.4	0.037	5,417.1	5454.7	0.136	0.170
631	0.225	99.54	0.225	2,282.2	28.5	2,310.8	0.050	2,949.7	2980.6	0.336	0.205
1,778	0.190	98.72	0.190	809.8	20.2	830.0	0.035	1,546.0	1570.4	0.447	0.225
3,162	0.170	97.73	0.170	455.4	16.7	472.0	0.020	632.6	651.0	0.616	0.251
3,548	0.145	97.46	0.145	405.8	16.0	421.9	0.025	430.6	447.0	1.121	0.291
3,981	0.105	97.15	0.105	361.7	15.4	377.1	0.040	383.8	399.5	2.006	0.311
5,012	0.085	96.43	0.085	287.3	14.3	301.6	0.020	324.5	339.4	1.181	0.346
7,943	0.050	94.40	0.050	181.3	12.3	193.5	0.035	234.3	247.6	2.833	0.366
25,119	0.030	83.33	0.030	57.3	8.4	65.7	0.020	119.3	129.6	3.092	0.376
56,234	0.020	66.48	0.020	25.6	6.4	32.0	0.010	41.5	48.8	4.104	0.378
100,000	0.018	48.38	0.018	14.4	5.3	19.7	0.002	20.0	25.8	1.552	0.380
125,893	0.016	40.09	0.016	11.4	4.9	16.3	0.002	12.9	18.0	2.277	0.396
							0.016			$\Sigma = 19.83$	

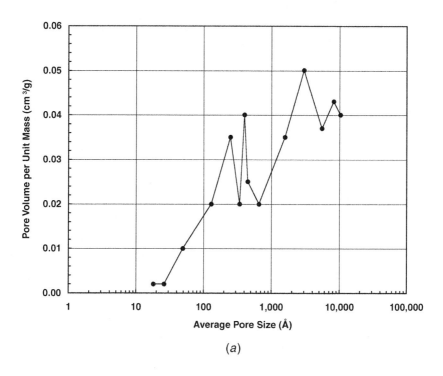

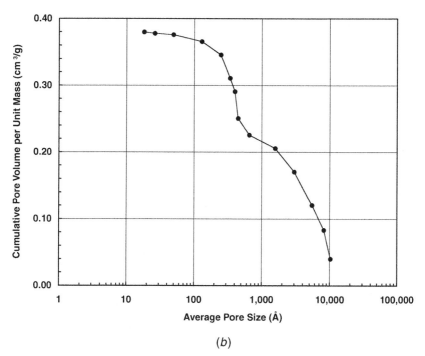

Figure 4.17 Pore size distribution functions for Georgia kaolinite: (*a*) pore volume per unit mass versus average pore size and (*b*) cumulative pore volume per unit mass versus average pore size.

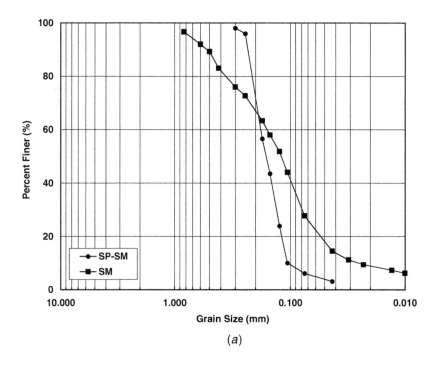

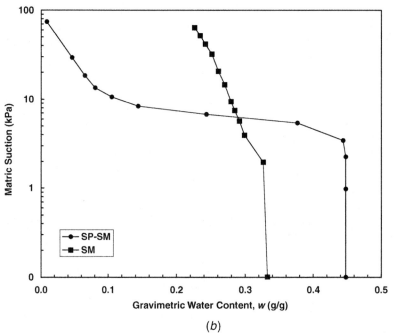

Figure 4.18 (*a*) Particle size distributions (*b*) and soil-water characteristic curves (*c*) for two sandy soil specimens (data from Clayton, 1996).

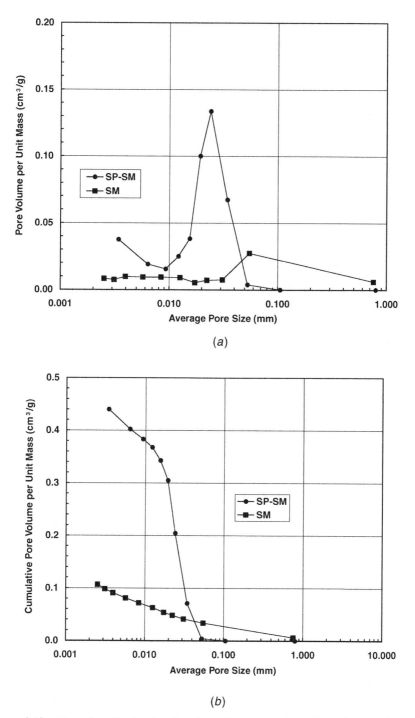

Figure 4.19 Pore size distribution functions for two sandy soil specimens: (*a*) pore volume per unit mass versus average pore size and (*b*) cumulative pore volume per unit mass versus average pore size.

curring at about 0.02 mm. The average predominant pore sizes for both soils are marked by values less than the predominant grain size.

4.5 SUCTION STRESS

4.5.1 Forces between Two Spherical Particles

Suction stress refers to the net interparticle force generated within a matrix of unsaturated granular particles (e.g., silt or sand) due to the combined effects of negative pore water pressure and surface tension. The macroscopic consequence of suction stress is a force that tends to pull the soil grains toward one another, similar in effect and sign convention to an overburden stress or surcharge loading.

One approach to evaluating the magnitude of suction stress is to consider the microscale forces acting between and among idealized assemblies of spherical unsaturated soil particles. Consider, for example, the two-particle system shown on Fig. 4.20. At low degrees of pore water saturation, or the "pendular" regime, interparticle forces arise from the presence of the air-water-solid interface defining the pore water menisci between the particles. The magnitude of the capillary force arising from this so-called liquid bridge between the particles may be analyzed as a function of water content by considering the local geometry of the air-water-solid interface as follows.

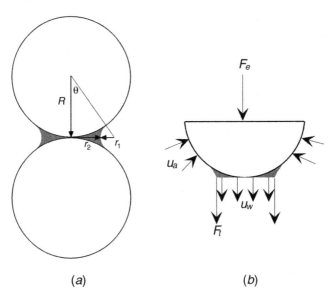

Figure 4.20 Air-water-solid interaction for two spherical particles and water meniscus: (*a*) toroidal geometry of the air-water-solid interface and (*b*) free-body diagram for analysis of interparticle forces.

For monosized particles (Fig. 4.20a), it was established in the previous chapter that the water meniscus formed between them may be described by two radii r_1 and r_2, the particle radius R, and a filling angle θ. A free-body diagram for the relevant system forces, which involves contribution from air pressure u_a, pore water pressure u_w, surface tension T_s, and applied external force or overburden F_e, is shown in Fig. 4.20b.

Positive, isotropic air pressure u_a will exert a compressive force on the soil skeleton. The total force due to air pressure, F_a, is equal to the product of the magnitude of the air pressure and the area of the air-solid interface over which it acts:

$$F_a = u_a(\pi R^2 - \pi r_2^2) \tag{4.44}$$

The total force due to surface tension, F_t, acts along the perimeter of the water meniscus:

$$F_t = -T_s 2\pi r_2 \tag{4.45}$$

The projection of total force due to water pressure acting on the water-solid interface in the vertical direction, F_w, is

$$F_w = u_w \pi r_2^2 \tag{4.46}$$

The resultant capillary force, F_{sum}, is the sum of all three of the above forces:

$$F_{\text{sum}} = u_a \pi R^2 - u_a \pi r_2^2 - T_s 2\pi r_2 + u_w \pi r_2^2 \tag{4.47}$$

Assuming the air pressure is the only contribution to external force leads to the following:

$$F_e = u_a \pi R^2 - (u_a - u_w)\pi r_2^2 - T_s 2\pi r_2 \tag{4.48}$$

which is the net interparticle force due to the interfacial interaction. This force exerts a tensile stress on the soil skeleton as long as the following condition is met:

$$(u_a - u_w)r_2^2 + T_s 2 r_2 > u_a R^2 \tag{4.49}$$

It was demonstrated in Chapter 3 that matric suction $u_a - u_w$ within the water lens formed between two spherical particles may be described independent of contact angle by the spherical radii r_1 and r_2 and surface tension T_s as

$$u_a - u_w = T_s \left(\frac{1}{r_1} - \frac{1}{r_2}\right) \tag{4.50}$$

Substituting the above equation into eq. (4.49) results in

$$\frac{T_s(r_2 - r_1)r_2^2}{r_1 r_2} + T_s 2r_2 > u_a R^2$$

and setting air pressure to a reference value equal to zero leads to

$$T_s r_2 (r_2 + r_1) > 0 \tag{4.51}$$

The above condition will always be satisfied if $r_1 > 0$ because r_2 is always greater than or equal to zero. This implies that suction stress in hydrophilic unsaturated soil is always greater than or equal to zero. Therefore, the force on the soil skeleton will always be tensile, even though r_1 and r_2 have the opposite effect on the sign of the pore water pressure, as shown below.

4.5.2 Pressure in the Water Lens

The water pressure in the lens between two spherical particles can be either positive, zero, or negative. The relationship between the sign of the water pressure and the lens geometry may be illustrated by rearranging eq. (4.50) as

$$u_w = u_a - T_s \left(\frac{1}{r_1} - \frac{1}{r_2}\right) \tag{4.52}$$

Accordingly, the absolute value of pore water pressure depends on both air pressure and the interface geometry. For example, if $r_1 < r_2$, a pore water pressure less than the air pressure will develop within the lens. However, if $r_1 > r_2$, a pore pressure greater than air pressure will develop within the lens. For u_a equal to zero, eq. (4.52) dictates that a decrease in the menisci radius r_1 results in increasingly negative values of pore water pressure, a reflection of radius r_1's relationship to the concave curvature of the water lens. A decrease in r_2, on the other hand, causes the pore water pressure to be less negative, a reflection of its relationship to the convex curvature of the water lens.

Considering the geometry of the contacting spheres and the water lens for zero contact angle, a relationship between R, r_1, and r_2 may be written as follows:

$$(R + r_1)^2 = R^2 + (r_1 + r_2)^2 \tag{4.53}$$

If r_1 is equal to r_2, which must occur at some value of water content, the pressure in the water lens is equal to the air pressure and the matric suction is thus equal to zero. Imposing this condition to eq. (4.53) leads to

$$(R + r_2)^2 = R^2 + (r_2 + r_2)^2 \quad R = \tfrac{3}{2} r_2 \tag{4.54}$$

Considering the geometry shown in Fig. 4.20a, it can be shown that

$$\tan \theta = \frac{r_1 + r_2}{R} = \frac{2r_2}{R} = \frac{2r_2}{(3/r_2 2)} = \frac{4}{3} \tag{4.55a}$$

or

$$\theta = 53.13° \tag{4.55b}$$

Therefore, the water content regime corresponding to a negative pore water pressure corresponds to the range in filling angle described by

$$0 \leq \theta \leq 53.13° \quad r_1 < r_2 \tag{4.56}$$

and the water content regime corresponding to positive pore water pressure is described by

$$53.13° \leq \theta \leq 90° \quad r_1 > r_2 \tag{4.57}$$

For relatively loosely packed particles, such as the simple cubic (SC) order, the filling angle θ may not be greater than 45° because the adjacent water lenses start to overlap each other. The condition described by eq. (4.57) is unlikely to occur in unsaturated soil with zero contact angle, indicating that the pore water pressure in the water lens is likely to be negative. The condition where the contact angle is not zero is often the case in real soil and will be covered in the next chapter.

4.5.3 Effective Stress due to Capillarity

Effective stress owing to the balance of the interfacial forces described above can be evaluated by considering the area over which they act. Figure 4.21 illustrates two such areas for analysis: the area over one spherical soil grain, or πR^2, and a unit area for simple cubic packing order, or $4R^2$. Considering

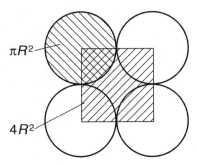

Figure 4.21 Unit areas for analyzing effective stress in simple cubic packing order.

eq. (4.48), the stress contribution due to the capillary interparticle force over the area πR^2 is

$$\sigma_w = u_a - \frac{r_2^2}{R^2}(u_a - u_w) - \frac{2r_2^2 r_1}{R^2(r_2 - r_1)}(u_a - u_w)$$

$$= u_a - \left[\frac{r_2^2}{R^2} + \frac{2r_2^2 r_1}{R^2(r_2 - r_1)}\right](u_a - u_w)$$

$$= u_a - \frac{r_2^2}{R^2} \frac{r_2 + r_1}{r_2 - r_1}(u_a - u_w) \qquad (4.58)$$

and the effective stress under an external total stress σ is

$$\sigma' = \sigma - \sigma_w = \sigma - u_a + \frac{r_2^2}{R^2} \frac{r_2 + r_1}{r_2 - r_1}(u_a - u_w) \qquad (4.59a)$$

which is in the same form as Bishop's (1959) effective stress equation for unsaturated soil, or

$$\sigma' = \sigma - \sigma_w = \sigma - u_a + \frac{r_2^2}{R^2} \frac{r_2 + r_1}{r_2 - r_1}(u_a - u_w) = \sigma - u_a + \chi(u_a - u_w)$$

$$(4.59b)$$

where the *effective stress parameter* χ is in this case equal to

$$\chi = \frac{r_2^2}{R^2} \frac{r_2 + r_1}{r_2 - r_1} \qquad (4.59c)$$

Similarly, for analysis using a cross-sectional area of $4R^2$, the effective stress is

$$\sigma' = \sigma - \sigma_w = \sigma - u_a + \frac{\pi}{4} \frac{r_2^2}{R^2} \frac{r_2 + r_1}{r_2 - r_1} (u_a - u_w) = \sigma - u_a + \chi(u_a - u_w)$$

(4.60a)

where, for this geometry,

$$\chi = \frac{\pi}{4} \frac{r_2^2}{R^2} \frac{r_2 + r_1}{r_2 - r_1} \quad (4.60b)$$

Equations (4.59c) and (4.60b) provide a great deal of insight into the nature of suction stress in unsaturated soil. Physically, the effective stress parameter χ represents the contribution of matric suction to effective stress. The χ parameter clearly depends on water content in these equations via r_1 and r_2. When water content for SC packing order approaches saturation, radius r_2 approaches the particle radius R and radius r_1 approaches zero. Examination of eq. (4.59c) for a unit area of πR^2 demonstrates that χ approaches unity under these conditions, thus reducing eq. (4.59a) to the classical effective stress equation for saturated soil:

$$\sigma' = \sigma - u_w \quad (4.61)$$

On the other hand, if water content approaches zero (i.e., perfectly dry conditions), then r_2 and r_1 both approach zero, thus leading to χ approaching zero and the condition where the effective stress is equal to the total stress minus the air pressure. Matric suction in this case, no matter its value, has no contribution to effective stress. For water content values between the completely dry and completely saturated conditions, the effective stress parameter is dependent on the relationship between r_1 and r_2. In general, and in real soil, the relationship between r_1 and r_2 is complicated and depends on contact angle and the geometric constraints imposed by the soil pores. The analysis below illustrates a special case when the contact angle is zero.

4.5.4 Effective Stress Parameter and Water Content

A specific relationship between effective stress parameter χ and water content can be established by considering the geometry of the water lens. As introduced in Section 3.4, Dallavalle (1943) presented the following approximations relating the parameters r_1, r_2, R, and θ for the case where contact angle is assumed equal to zero:

$$r_1 = R \left(\frac{1}{\cos \theta} - 1 \right) \quad r_2 = R \tan \theta - r_1 \quad 0 \leq \theta \leq 85° \quad (4.62)$$

Substituting the above equation into eq. (4.59c), the effective stress parameter χ may thus be described in terms of filling angle θ for an elementary cross section of πR^2:

$$\chi = \frac{(\sin\theta + \cos\theta - 1)^2}{\cos^2\theta} \frac{\sin\theta}{\sin\theta - 2 + 2\cos\theta} \quad (4.63)$$

or considering eq. (4.60b) for an elementary cross section of $4R^2$:

$$\chi = \frac{\pi}{4}\frac{(\sin\theta + \cos\theta - 1)^2}{\cos^2\theta} \frac{\sin\theta}{\sin\theta - 2 + 2\cos\theta} \quad (4.64)$$

Equation (4.63) or (4.64) can be used to explore a physical interpretation of the effective stress parameter and suction stress, and their dependency on soil water content in terms of filling angle θ. For any filling angle θ, the radii r_1 and r_2 and the effective stress parameter χ can be uniquely defined. The relationship between the effective stress parameter and filling angle is illustrated in Fig. 4.22 for θ less than 45° (corresponding to gravimetric water content, $w < 0.063$). Interestingly, this relationship is independent of the particle size R as inferred from equations (4.63) and (4.64).

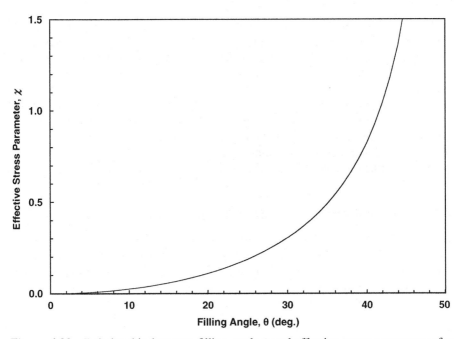

Figure 4.22 Relationship between filling angle θ and effective stress parameter χ for spherical particles in simple cubic packing order with $4R^2$ unit area.

Effective stress due to suction stress can also be studied without introducing the concept of the effective stress parameter χ. Eliminating matric suction in eq. (4.59b) by substituting eq. (4.50) leads to the effective stress due to suction stress, σ_c, for the unit area πR^2:

$$\sigma_c = \frac{r_2^2}{R^2} \frac{r_2 + r_1}{r_1 r_2} T_s \qquad (4.65)$$

and by substituting eq. (4.50) into eq. (4.60a) for a unit area of $4R^2$:

$$\sigma_c = \frac{\pi}{4} \frac{r_2^2}{R^2} \frac{r_2 + r_1}{r_1 r_2} T_s \qquad (4.66)$$

Substituting eq. (4.62) into the above equation to express r_1 and r_2 in terms of θ, suction stress can be expressed in terms of filling angle θ:

$$\sigma_c = \frac{\pi T_s}{4R} \frac{\sin \theta + \cos \theta - 1}{1 - \cos \theta} \tan \theta \qquad (4.67)$$

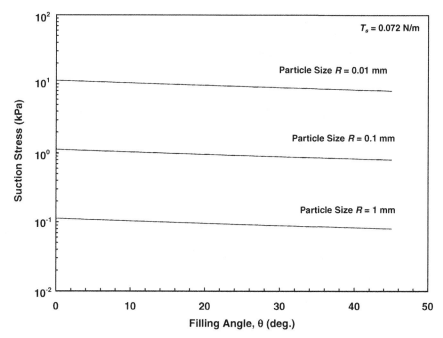

Figure 4.23 Suction stress as function of filling angle for spherical particles in simple cubic packing order.

168 CAPILLARITY

Through eq. (4.67) and as illustrated on Fig. 4.23, one can infer that suction stress is dependent on particle size R and water content but not directly on matric suction. The fundamental question stemming from the above analysis is: Is it necessary to use matric suction to represent effective stress in unsaturated soil? At the present time, this remains an open question.

PROBLEMS

4.1. Compute and compare the equilibrium height of capillary rise in a 5×10^{-5} m diameter capillary tube for free water with surface tension of 0.072 N/m and soapy water with surface tension of 0.010 N/m. Assume zero contact angle and a fluid density equal to 1 g/cm^3 in both cases.

4.2. Water is in a capillary tube at equilibrium. The tube has an inner radius of 2×10^{-5} m, the contact angle is 60°, and the surface tension is 0.072 N/m. What are the pressure in the water and the relative humidity in the tube? If the tube were placed in a spacecraft with zero gravity, water from capillary condensation is likely to spread over the inner wall with a uniform water film thickness. Assume the thickness of the water film at equilibrium is 10^{-5} m. What are the pressure in the water and the relative humidity in the tube?

4.3. Uniform fine sand with particle radius of 0.1 mm is packed in two arrays—simple cubic packing and tetrahedral closest packing—for an open-tube capillary rise test. The contact angle is 50° and surface tension is 0.072 N/m. What is the expected range for height of capillary rise?

4.4. A fine sand specimen was tested for grain size and pore size distribution parameters and the soil-water characteristic curve. Particle size analysis shows $D_{10} = 0.06$ mm. Pore size analysis shows a mean pore radius of 0.05 cm and a void ratio of 0.4. Soil-water characteristic curve testing indicates an air-entry head of 100 cm. Estimate the maximum height of capillary rise for this soil using three different empirical relationships.

4.5. Derive Terzaghi's (1943) solution for the rate of capillary rise [eq. (4.33a)].

4.6. Show that eq. (4.36a) can be reduced to eq. (4.33a) if the summation index m is zero. Reproduce the theoretical curves shown in Fig. 4.12a using the system parameters shown in the figure. Use a summation index $m = 5$.

4.7. Data describing the soil-water characteristic curve for a sand specimen is shown in Table 4.3. If the surface tension is 0.072 N/m, the molar volume of water is 0.018 m^3/kmol, and R is 8.314 J/mol · K, conduct a pore size distribution analysis and provide the following information: specific surface area (m^2/g), total pore volume (cm^3/g), average pore radius vs. pore volume in an *x*-*y* plot, and average pore radius vs. cumulative pore volume in an *x*-*y* plot.

TABLE 4.3 Soil-Water Characteristic Curve Data for Problem 4.7

$u_a - u_w$ (kPa)	RH	w (g/g)
10	0.99993	0.330
16	0.99988	0.310
32	0.99977	0.250
63	0.99954	0.140
158	0.99885	0.070
1259	0.99090	0.040
12589	0.91265	0.035
125893	0.40092	0.034

4.8. Calculate and plot the interparticle force between two spherical particles ($R = 0.1$ mm) as a function of filling angle from $\theta = 0°$ to $\theta = 30°$.

PART II

STRESS PHENOMENA

CHAPTER 5

STATE OF STRESS

5.1 EFFECTIVE STRESS IN UNSATURATED SOIL

5.1.1 Macromechanical Conceptualization

The state of stress in unsaturated soil is fundamentally different from the state of stress in saturated soil. Unlike saturated soils, which are two-phase systems comprised essentially of solids and liquid only (i.e., soil particles and pore water as in a liquid-saturated system) or solids and gas only (i.e., soil particles and pore air as in a gas-saturated, or perfectly dry, system), unsaturated soils are three-phase systems comprised of solids (soil particles), liquid (pore water), and gas (pore air). The relative amounts and corresponding pressures of the pore water and pore air phases in unsaturated soil have a direct impact on the state of stress acting at the particle-particle contacts and, consequently, on the macroscopic physical behavior of the soil mass (e.g., shear strength and volume change). As such, changes in the relative amounts of the pore air and pore water phases, which may occur under natural processes such as precipitation or evaporation, or under anthropogenic processes such as irrigation or imposed changes in the boundary conditions (e.g., water table lowering), have a direct impact on the state of stress and physical behavior of the soil system. Understanding this impact is of critical importance to the design and performance of engineered geotechnical systems comprised of unsaturated soils. An excellent practical example is the common occurrence of precipitation-induced failures in unsaturated earthen slopes.

Early attempts at understanding capillarity and its role in the stress-strain behavior of unsaturated soil recognized that when soil is saturated and the pore water pressure is compressive, the net effect of the water pressure is to reduce the effective stress. At the opposite condition when the soil is relatively dry, it was recognized that the pore water in the voids might sustain very high negative pore pressures, thus creating tensile forces acting to increase the effective stress and pull the soil grains together. The resultant interparticle stress in the range between these extremes was described in a variety of extended forms of Terzaghi's classic effective stress equation modified to account for the negative pore water pressures. Bishop (1959), for example, proposed the following single-valued effective stress equation for unsaturated soil:

$$\sigma' = (\sigma - u_a) + \chi(u_a - u_w) \tag{5.1}$$

where σ' is the effective interparticle stress, σ is total stress, u_a is pore air pressure, u_w is pore water pressure, the quantity $u_a - u_w$ is matric suction, and χ is a material property that depends on the degree of saturation or matric suction. The χ parameter, which was introduced in the previous chapter, is referred to as the *effective stress parameter*.

The first term on the right-hand side of eq. (5.1) $(\sigma - u_a)$ represents the component of net normal stress applicable to bulk soil. The product $\chi(u_a - u_w)$, on the other hand, represents the interparticle stress due to suction, herein referred to as *suction stress*. In the case where capillarity is the sole mechanism contributing to matric suction, suction stress is identical to the microscopically formulated suction stress described in the previous chapter. The effective stress parameter χ is generally believed to vary with degree of saturation, being equal to zero for perfectly dry soil and unity for saturated soil. In either of these extreme cases, eq. (5.1) reduces to the classic effective stress equation.

Understanding suction stress and its dependency on degree of saturation in unsaturated soil has historically been a challenging task from both theoretical and experimental perspectives. Early experimental efforts were primarily concerned with determining χ indirectly as a function of water content or degree of saturation. The majority of the experimental work relied on measurement or independent control of matric suction and total stress in triaxial or direct shear specimens loaded to failure conditions. Figure 1.17 shows a series of relationships between χ and degree of saturation for a wide range of soil types. The figure illustrates the apparent variation in χ between zero and one for perfectly dry and saturated conditions, respectively. Very few studies in the past, either theoretical or experimental, have investigated hysteretic phenomena in the suction stress behavior of unsaturated soil.

5.1.2 Micromechanical Conceptualization

The fundamental physical mechanisms responsible for the retention of pore water by unsaturated soil and the corresponding soil-water characteristic curve include capillary mechanisms, osmotic mechanisms, and short-range solid-liquid interaction, or hydration, mechanisms. It is important that each of these mechanisms be fully considered for analysis of pore water flow phenomena in unsaturated soil. The role of each mechanism in stress and deformation phenomena, however, remains to a great extent uncertain. A comprehensive framework for describing the roles of osmotic and hydration mechanisms on stress and volume change behavior in unsaturated soil has yet to be established. The role of capillarity on the state of stress, on the other hand, is reasonably well understood from a micromechanical standpoint, particularly for relatively coarse-grained materials (e.g., silts and sands) over a finite range of water content.

Bishop's effective stress approach described above is a macroscale interpretation that attempts to describe the microscale contribution of interparticle pore water menisci located between and among soil particles to the net interparticle stress. This suction stress contribution can be more readily understood by examining the forces and fluid pressures that arise in unsaturated soil from a micromechanical particle-scale point of reference for idealized soil particles. Numerous micromechanical studies have focused on the complementary roles of negative pore pressure and surface tension in controlling interparticle forces between and among simple particle systems comprised of spheres, plates, or other idealized geometries. These have included theoretical studies involving consideration of the changing geometry of pore water menisci and the consequent relationships among water content, soil suction, and interparticle forces and stresses (e.g., Fisher, 1926; Dallavalle, 1943; Blight, 1967; Sparks, 1963; Lian et al., 1993; Cho and Santamarina, 2001; Molenkamp and Nazemi, 2003; Likos and Lu, 2004) as well as micromechanical experimental studies involving direct measurement of interparticle forces for two-particle or multiparticle systems (e.g., Mason and Clark, 1965; Dushkin et al., 1996; Rossetti et al., 2003). Together, these types of studies have provided significant insight into the role of capillary forces in governing the basic interaction of unsaturated granular particles. Perhaps more importantly, the studies have provided a rational conceptual link between the microscale physics that govern the state of stress in unsaturated soil and the macroscopic engineering formulations that have been proposed to describe its physical behavior.

5.1.3 Stress between Two Spherical Particles with Nonzero Contact Angle

A micromechanical theoretical development for evaluating interparticle forces and suction stress in contacting, monosized, unsaturated spherical particles with a constant contact angle equal to zero was presented in Chapter 4. This section takes the analysis several steps further by considering contact angle as a nonzero material variable. As before, simple cubic (SC) packing and tetrahedral (TH) packing are considered to represent end members in granular soil fabric. It is presumed that the range in material properties and water retention and suction stress behavior of real soil, particularly coarse-grained material such as silt or sand, falls somewhere between these two idealized scenarios. Figures 5.1a and 5.1b illustrate geometries for uniformly sized spheres coordinated under SC packing and TH packing, respectively. Unit volumes for SC and TH packing have void ratios of 0.91 and 0.34, respectively, corresponding to porosities of 47.6 and 26.0%.

Two quantities are required to analyze suction stress and its dependency on water content in such particle arrangements: the capillary force between the particles and the water content of the particle/pore water system. The capillary force between two contacting spherical particles for a toroidal meniscus geometry and zero contact angle (Fig. 4.20) was derived earlier as

$$F_{\text{sum}} = u_a \pi R^2 - u_a \pi r_2^2 - T_s 2\pi r_2 + u_w \pi r_2^2 \quad (5.2)$$

Figure 5.2 shows a more general system geometry for describing the water content between two particles with a nonzero and variable contact angle α.

Simple cubic: radius R
Coordination number = 6
Layer spacing = $2R$
Unit volume = $8R^3$
Void ratio = 0.91
Porosity = 47.6%

(a)

Tetrahedral: radius R
Coordination number = 12
Layer spacing = $2R(2/3)^{0.5}$
Unit volume = $4(2R^3)^{0.5}$
Void ratio = 0.34
Porosity = 26.0%

(b)

Figure 5.1 Uniform spheres in (a) simple cubic packing order and (b) tetrahedral close packing order.

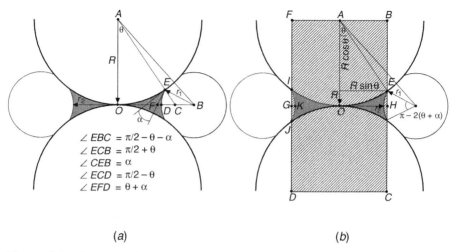

Figure 5.2 Geometrical constraints for defining the water meniscus between contacting spheres with consideration for a variable contact angle: (*a*) system radii and angles and (*b*) two-dimensional surface boundaries of water lens.

Here, the water lens represented by radii r_1 and r_2 can be written in terms of filling angle θ, the common particle radius R, and contact angle α as

$$r_1 = R\frac{1 - \cos\theta}{\cos(\theta + \alpha)} \tag{5.3}$$

$$r_2 = R\tan\theta - r_1\left(1 - \frac{\sin\alpha}{\cos\theta}\right) \tag{5.4}$$

When contact angle is equal to zero, eqs. (5.3) and (5.4) reduce to those proposed by Dallavalle (1943) [eq. (3.38)] as presented in Chapter 3.

The water content of the system may be evaluated by considering the volume of the water lens. In a two-dimensional projection, the water lens is bounded by the three hatched surfaces shown in Fig. 5.2*b*. The rectangle *BFGH*, which is a cylinder in three dimensions having radius $R\sin\theta$ and height R, bounds the bottom half of the symmetrical water lens. The partial circle of radius R defined by *FBEOI* bounds the water lens on the top half. The partial circle defined by *IKJ* bounds the water lens on both sides. Rotated in three dimensions, these surfaces become volumes that can be used to define the total volume of the water lens. The volume of the rotated area *BFGH* for one unit particle is

$$V_c = 2\pi R^3 \sin^2\theta \tag{5.5}$$

The volume of the rotated area *FBEOI* for one unit particle is

$$V_s = 2\pi R^3 \sin^2 \theta \cos \theta + \frac{2\pi}{3} R^3 (1 - \cos \theta)^2 (2 + \cos \theta) \quad (5.6)$$

The volume of the rotated area *IJK* for one unit particle is

$$V_r = 2\pi \left[r_2 + r_1 - \frac{2}{3} \frac{r_1 \cos^3(\theta + \alpha)}{(\pi/2) - (\theta + \alpha) - \sin(\theta + \alpha)\cos(\theta + \alpha)} \right]$$

$$\frac{1}{2} r_1^2 [\pi - 2(\theta + \alpha) - \sin 2(\theta + \alpha)] \quad (5.7)$$

Accordingly, the total volume of the water lens V_l is

$$V_l = V_c - V_s - V_r$$
$$= 2\pi R^3 \sin^2 \theta - 2\pi R^3 \sin^2 \theta \cos \theta$$
$$- \frac{2\pi}{3} R^3 (1 - \cos \theta)^2 (2 + \cos \theta) - V_r \quad (5.8)$$

Determining gravimetric water content for each unit cell of particles in SC packing requires summation of three orthogonal water lens volumes and can be expressed as

$$w_{SC} = \frac{3V_l}{V_{sphere} G_s} \quad (5.9)$$

where V_{sphere} is the volume of one soil particle (i.e., $V_{sphere} = \frac{4}{3}\pi R^3$) and G_s is the specific gravity of the soil solids. It follows that water content can be written in terms of the angles θ and α as

$$w_{SC} = \frac{9}{2G_s} \sin^2 \theta - \frac{9}{2G_s} \sin^2 \theta \cos \theta$$
$$- \frac{3}{2G_s} (1 - \cos \theta)^2 (2 + \cos \theta) - \frac{9V_r}{4G_s \pi R^3} \quad (5.10)$$

In TH packing, gravimetric water content is simply twice that of SC packing for the same filling angle:

$$w_{TH} = 2 w_{SC} \quad (5.11)$$

For zero contact angle, the limits of the pendular water regime in SC and TH packing are 0.063 g/g gravimetric water content and 0.032 g/g, respectively. These values represent water contents where the individual water lenses between neighboring particles begin to touch each other and the meniscus geometry idealized in Fig. 5.2 is no longer valid.

Equations (5.10) and (5.11) are plotted in Fig. 5.3a to show relationships between filling angle θ and gravimetric water content for contact angle equal to zero ($R = 1$ mm, $G_s = 2.65$). Similar analytical solutions for $R = 1$ mm, $G_s = 2.65$, and $\alpha = 0°$ developed previously by Dallavalle (1943) and Cho and Santamarina (2001) are included for comparison. Figure 5.3b shows w_{SC} and w_{TH} as functions of θ for contact angle equal to 0°, 20°, and 40°. It can be seen here that the increase in contact angle has a significant effect on the volume of the pore water lens and the corresponding water content of the two-particle system. Larger contact angles, which may be considered to coincide with a wetting process, result in higher water contents for a given filling angle θ. Zero contact angles, which might correspond to a drying process, result in relatively low water contents. This observation forms the basis for an analysis of *contact angle hysteresis* presented in Section 5.2.

As introduced in Chapter 4, effective stress resulting from suction stress can be evaluated by dividing the interparticle capillary force, that is, eq. (5.2), by the area over which it acts. Taking the cross-sectional area of one particle (πR^2) as an elementary area, and employing eq. (4.50) to describe surface tension T_s in terms of the spherical radii r_1 and r_2, eq. (5.2) can be written in terms of a stress contribution due to capillarity σ_w as

$$\sigma_w = u_a - \frac{r_2^2}{R^2}(u_a - u_w) - \frac{2r_2^2 r_1}{R^2(r_2 - r_1)}(u_a - u_w)$$

$$= u_a - \left[\frac{r_2^2}{R^2} + \frac{2r_2^2 r_1}{R^2(r_2 - r_1)}\right](u_a - u_w)$$

$$= u_a - \frac{r_2^2}{R^2} \frac{r_2 + r_1}{r_2 - r_1}(u_a - u_w) \qquad (5.12)$$

and the effective stress under an external total stress σ is

$$\sigma' = \sigma - \sigma_w = \sigma - u_a + \frac{r_2^2}{R^2} \frac{r_2 + r_1}{r_2 - r_1}(u_a - u_w) \qquad (5.13)$$

which is in the same form as Bishop's (1959) single-valued effective stress equation for unsaturated soil, that is, eq. (5.1). Equating the two leads to

$$\sigma - u_a + \chi(u_a - u_w) = \sigma - u_a + \frac{r_2^2}{R^2} \frac{r_2 + r_1}{r_2 - r_1}(u_a - u_w) \qquad (5.14)$$

where

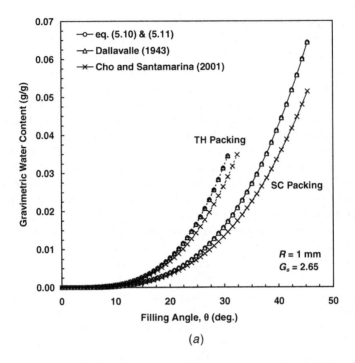

(a)

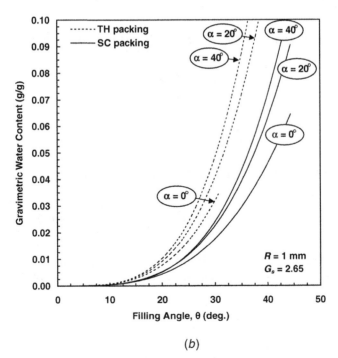

(b)

Figure 5.3 Relationship between filling angle and gravimetric water content for 1-mm spheres in simple cubic (SC) and tetrahedral (TH) close packing: (*a*) for $\alpha = 0°$ and (*b*) for $\alpha = 0°$, $\alpha = 20°$, and $\alpha = 40°$.

$$\chi = \frac{r_2^2}{R^2} \frac{r_2 + r_1}{r_2 - r_1} \tag{5.15}$$

Equation (5.15) can now be used in conjunction with eqs. (5.3) and (5.4) to write the effective stress parameter χ as a function of filling angle θ and contact angle α as

$$\chi = \left[\tan\theta - \frac{1 - \cos\theta}{\cos(\theta + \alpha)} \frac{\cos\theta - \sin\alpha}{\cos\theta} \right]^2$$
$$\frac{\tan\theta + (\sin\alpha/\cos\theta)(1 - \cos\theta)/\cos(\theta + \alpha)}{\tan\theta - (2 - \sin\alpha/\cos\theta)(1 - \cos\theta)/\cos(\theta + \alpha)} \tag{5.16}$$

The above equation can be used to investigate the dependency of χ on water content and contact angle, as shown subsequently.

5.1.4 Pore Pressure Regimes

Contact angle and filling angle both play important roles in the transition between regimes of positive pore pressure and negative pore pressure. These roles can be considered by examining the impact of α and θ on the geometry of the lens between spheres. Matric suction within the water lens for contacting spheres may be described as

$$u_a - u_w = T_s \left(\frac{1}{r_1} - \frac{1}{r_2} \right) \tag{5.17}$$

If $r_1 = r_2$ in the above equation, which must occur at some value of water content, then the matric suction is equal to zero. Considering eqs. (5.3) and (5.4), this occurs when contact angle α and filling angle θ satisfy the condition

$$(1 - \cos\theta)(2\cos\theta - \sin\alpha) = \sin\theta \cos(\theta + \alpha) \tag{5.18}$$

Equation (5.18) is instructive because it identifies the boundary between a negative pore water pressure regime and a positive pore water pressure regime for our idealized two-particle unsaturated soil system. Figure 5.4, for example, is a plot of eq. (5.18) for both angles (θ and α) varying between zero and 60°. If $\alpha = 0°$, which may represent a drying process in soil, the zero matric suction condition occurs at $\theta = 53.13°$. If $\alpha = 60°$, which may represent a wetting process, the zero matric suction condition occurs at $\theta = 20°$. This observation indicates that positive pressures may develop at much lower water contents for soil undergoing wetting. Development of positive pore water pressures may be partially responsible for slaking processes that occur upon the wetting of certain materials, most notably clay shale. Figure 5.4 and eq.

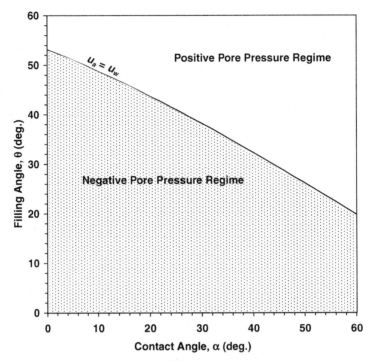

Figure 5.4 Relationship between filling angle and contact angle showing positive and negative pore water pressure regimes.

(5.18) also imply that for soil with relatively dense packing, there is less likelihood to enter the positive pressure regime than for loosely packed soil.

5.2 HYSTERESIS

5.2.1 Hysteresis Mechanisms

Hysteresis is a well-known but poorly understood phenomenon in unsaturated soil behavior. Perhaps the most outstanding example of hysteretic behavior is that between wetting and drying paths of the soil-water characteristic curve. There is no unique equilibrium between moisture content and soil suction. Rather, soil undergoing drying processes such as evaporation or gravity drainage generally tends to retain a greater amount of water than for the same magnitude of suction during wetting processes such as infiltration or capillary rise.

Figure 5.5 shows a conceptualization of hysteresis in the suction-water content relationship for a typical coarse-grained unsaturated porous material. Note that the horizontal dashed line at some suction value ψ_1 intersects the

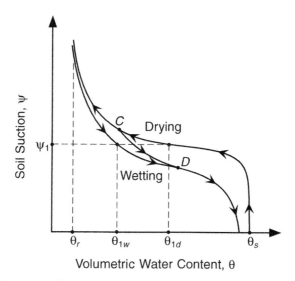

Figure 5.5 Conceptual illustration of hysteresis in soil-water characteristic curve.

curve at different water contents along the wetting loop (θ_{1w}) and the drying loop (θ_{1d}), where $\theta_{1d} > \theta_{1w}$. The breadth of the hysteresis loop across the entire range of water content is most pronounced in the region of relatively rapid pore drainage or adsorption (i.e., the flat portion of the curve) where pore water is retained primarily by capillary mechanisms. In general, hysteresis is less pronounced near the residual water content where pore water retention falls within the pendular regime. The figure also illustrates that full saturation (θ_s) may not be reached during the wetting process due to the entrapment of occluded air bubbles. The portion of the curve from C to D represents a partial rewetting step along a so-called scanning loop, implying that the actual soil-water characteristic curve for soil under fluctuating field conditions will be contained within two boundaries defined by the full wetting and drying loops, but may have a unique form if small wetting and drying cycles occur.

There is strong motivation to understand hysteretic behavior in the soil-water characteristic curve and its consequent impact on the stress, strength, flow, and deformation behavior of unsaturated soil systems. This is particularly true in practical engineering situations where cyclical wetting and drying processes are likely to occur with fluctuations in atmospheric or moisture loading conditions. Some form of rationale is required to predict the expected range of wetting or drying for the system and to then define the boundaries of the soil-water characteristic curve between these two extremes. Because most experimental measurement techniques and models for quantifying the soil-water characteristic curve (Chapters 10 and 12) are path dependent (i.e., specific to either wetting or drying processes), the type of measurement or model should be selected to best match the expected direction of moisture change in the field. By practical constraint, it is common to measure or model the desorption branch of the curve and assume that it represents a true equilibrium relationship.

Although not fully understood, significant insight into soil-water hysteresis has been gained from both experimental and theoretical perspectives (e.g., Haines; 1930; Mualem, 1984; Israelachvili, 1992; Nimmo, 1992; Iwata et al., 1995). Hysteretic behavior has been attributed to several mechanisms that act on both a relatively microscopic (particle) scale and a relatively macroscopic (interparticle) scale. Major theorized mechanisms include: (1) geometrical effects associated with nonhomogenous pore size distribution, often referred to as the "ink-bottle" effect, (2) capillary condensation, which becomes a unique wetting process at relatively low water content (Section 3.4), (3) entrapped air, which refers to the formation of occluded air bubbles in "dead-end" pores during wetting, (4) swelling and shrinkage, which may alter the pore fabric of fine-grained soil differently during wetting and drying processes, and (5) contact angle hysteresis, which is related to the intrinsic difference between drying and wetting contact angles at the soil particle–pore water interface.

The exact roles and relative importance of the various possible hysteresis mechanisms for a wide range of soil types and water content regimes remain unclear. The remainder of this section provides more detailed descriptions of two mechanisms most likely to be important for relatively coarse-grained soil, specifically, ink-bottle hysteresis and contact angle hysteresis. The theoretical development introduced in the previous section for spherical particles and a nonzero contact angle is then applied to illustrate the potential role of contact angle in terms of hysteresis for three aspects of unsaturated soil behavior: (1) the soil-water characteristic curve, (2) the relationship between the effective stress parameter χ and water content, and (3) the relationship between suction stress and water content.

5.2.2 Ink-Bottle Hysteresis

The so-called ink-bottle effect in porous media arises due to nonhomogeneity in pore size and shape distribution. This effect can be better understood through analogy by considering the nonuniform capillary tube system shown as Fig. 5.6. The capillary tube is described by two different radii, R being the larger tube radius and r being the smaller tube radius. During upward capillary flow, which is a wetting process, the maximum height of capillary rise is controlled by the smaller tube radius, ceasing at the point where the larger radius is encountered (Fig. 5.6a). This height is denoted h_w and is a direct function of r. For a zero contact angle, the matric suction at the maximum rise is equal to $2T_s/r$. On the other hand, if the tube is initially filled, then the capillary height h_d during drainage may extend beyond the larger pore radius R (Fig. 5.6b). The matric suction at equilibrium for the drainage condition is also equal to $2T_s/r$; however, much like the soil-water characteristic curve shown in Fig. 5.5, the total water content of the capillary tube during drainage is larger than that during wetting.

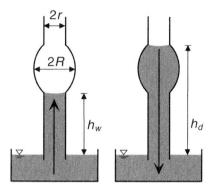

Figure 5.6 Capillary tube model for demonstrating ink-bottle effect.

As illustrated in Fig. 5.7, Childs (1969) presented a hypothetical cross section through a soil specimen that further clarifies ink-bottle hysteresis. The soil system shown in the figure is initially saturated, with the air-water interface standing at some elevation above the soil surface (stage 1). The pore pressure at this stage is positive, having a magnitude equal to the hydrostatic pressure governed by the height of the standing water. The solid lines corresponding to stages 2 through 6 denote progressive positions of the air-water interface as the pore pressure is incrementally decreased into a negative regime, thus causing the pore water to retreat into smaller and smaller pore throats within the specimen. The matric suction at each stage is described by the curvature of the air-water interface, which becomes more severe as the drainage process continues under increasing suction. The dashed lines denoting stages 7 through 9 represent positions of the air-water interface during a

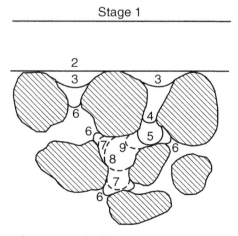

Figure 5.7 Conceptual unsaturated soil system at progressive stages of drainage and rewetting (after Childs, 1969).

186 STATE OF STRESS

subsequent refilling process. In order for the pore structure to refill, however, the air-water interface must proceed through the widest pore throat (within the vicinity of stages 8 and 9). Because the curvature within this pore throat becomes progressively less severe, the suction must be progressively reduced to a point low enough to be in equilibrium with the curvature for the filling process to proceed. The net effect is that the water content of the system during the refilling process is systematically less than that during the drainage process for the same magnitude of suction.

5.2.3 Contact Angle Hysteresis

At many solid-liquid-gas interfaces, the wetting solid-liquid contact angle is substantially larger than the drying contact angle. Figure 5.8 shows a classic conceptual example for a drop of water on an inclined solid surface. As the drop geometry reaches steady-state under the influence of gravity, a wetting front characterized by relatively large contact angle α_w develops at the advancing edge of the drop. A drying front, which is characterized by a much smaller contact angle α_d, develops at the receding edge.

The difference between wetting and drying contact angles in unsaturated soil can be significant. Experimental studies based on capillary rise and horizontal infiltration testing, for example, have shown that wetting contact angles in sands can be as high as 60° to 80° (e.g., Letey et al., 1962; Kumar and Malik, 1990). Drying contact angles, on the other hand, have been estimated to range from 0° to as much as 20° to 30° less than the corresponding wetting angles (e.g., Laroussi and DeBacker, 1979). These differences may have an important impact on the water retention behavior of unsaturated soil and may contribute to hysteresis in the soil-water characteristic curve and suction stress characteristic curve. The micromechanical theoretical development presented in the previous section becomes a useful tool to investigate this notion.

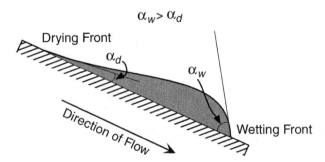

Figure 5.8 Water droplet on inclined surface illustrating difference between wetting and drying contact angles.

5.2.4 Hysteresis in the Soil-Water Characteristic Curve

For filling angle θ ranging from 0° to 45°, eq. (5.10) can be used to calculate gravimetric water content in SC packing. For θ ranging from 0° to 30°, eq. (5.11) can be used to calculate gravimetric water content in TH packing. The upper limits defined by θ = 45° and θ = 30° represent transitions from the pendular regime to the funicular regime for SC and TH packing, respectively. Given some particle diameter R and contact angle α, matric suction corresponding to the calculated water content values can be determined from eqs. (5.3), (5.4), and (5.17).

Figure 5.9 shows theoretical soil-water characteristic curves calculated in this manner for six values of R (0.1 μm to 1.0 mm) and a zero contact angle. It can be observed that the larger the particle size, the less the magnitude of suction for the same value of water content. Figure 5.10a shows characteristic curves for two particle radii (0.1 and 1 mm) in SC packing for α equal to 0°, 20°, and 40°. Figure 5.10b shows characteristic curves for the same radii and contact angles in TH packing order. It is apparent from both figures that the larger contact angles (which may simulate a wetting process) result in less water retained by the soil than at the same value of suction for lower contact angles (simulating a drying process). The hysteresis is similar in behavior to that observed in typical characteristic curves for real soil.

5.2.5 Hysteresis in the Effective Stress Parameter

Equation (5.16) provides insightful information into the constitutive relationships among the effective stress parameter χ, water content, and contact angle. Figure 5.11a and 5.11b show relationships based on this equation for SC and TH packing, respectively. According to eq. (5.16), χ is independent of particle size. For both SC and TH packing, larger contact angles result in larger values of χ for the same water content. Note also that χ in Fig. 5.11a exceeds unity, which is contrary to previous experimental studies (i.e., Fig. 1.17) but similar to earlier theoretical studies (e.g., Sparks, 1963). A recent study by the authors employing a free energy formulation (e.g., Orr et al., 1975) to calculate a more accurate meniscus geometry shows a very similar χ function to the toroidal approximation shown in Fig. 5.11, confirming that χ greater than unity is not due to a manifestation of the toroidal model for the meniscus. Physically, an effective stress parameter greater than unity implies that suction stress $\chi(u_a - u_w)$ can exceed matric suction $u_a - u_w$. Theoretically, increasingly large values of χ can be interpreted as a reflection of the relatively important role of surface tension compared with matric suction on the total capillary interparticle force [i.e., eq. (5.2)]. The relatively large values of χ for large contact angles may reflect the larger resultant of surface tension in the direction of suction stress normal to the particles. At the present time, experimental evidence for χ greater than unity remains unproven.

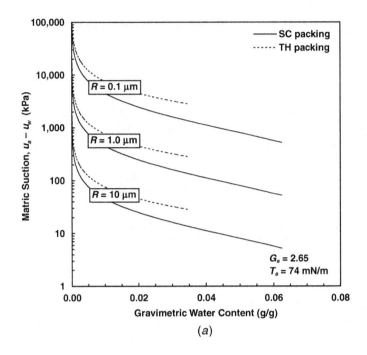

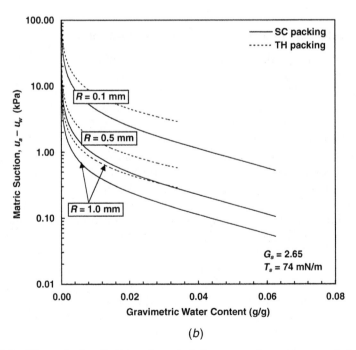

Figure 5.9 Theoretical soil-water characteristic curves for various particle sizes in simple cubic (SC) and tetrahedral (TH) packing order: (*a*) relatively fine-grained materials and (*b*) relatively coarse-grained materials.

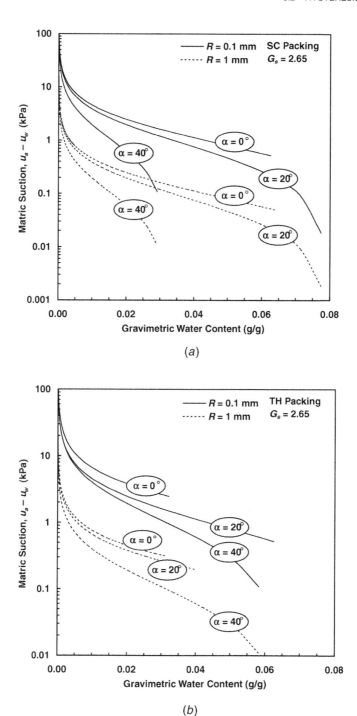

Figure 5.10 Effect of contact angle on hysteresis in soil-water characteristic curve: (*a*) particles in simple cubic (SC) packing and (*b*) particles in tetrahedral (TH) packing.

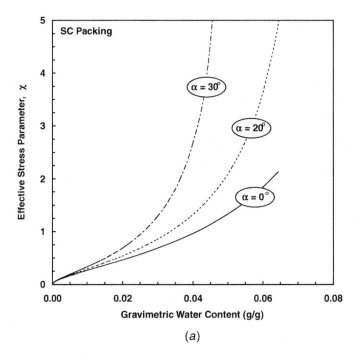

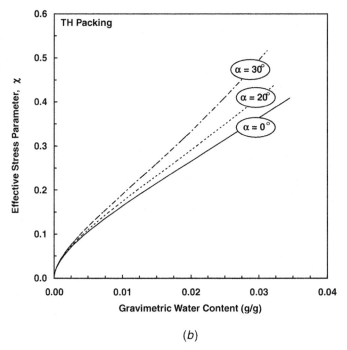

Figure 5.11 Theoretical relationship between water content and effective stress parameter: (*a*) particles in simple cubic (SC) packing and (*b*) particles in tetrahedral (TH) packing.

5.2.6 Hysteresis in the Suction Stress Characteristic Curve

Equation (5.12) describes suction stress due to capillarity. Imposing the Laplace eq. (5.17) and setting air pressure to a reference value of zero leads to the net capillary stress as a function of particle size and water lens radii r_1 and r_2:

$$\sigma_w = -\frac{r_2^2}{R^2}\left(\frac{r_2+r_1}{r_2-r_1}\right)\left(\frac{r_2-r_1}{r_2 r_1}\right)T_s = -\frac{r_2^2}{R^2}\left(\frac{r_2+r_1}{r_1 r_2}\right)T_s \quad (5.19)$$

The total suction force between two particles with a radius R is

$$F_{cap} = \sigma_w \pi R^2 = -\frac{r_2+r_1}{r_1}\pi r_2 T_s \quad (5.20)$$

Figure 5.12 shows corresponding relationships between water content and suction stress for $R = 1$ mm (Fig. 5.12a) and $R = 0.1$ mm (Fig. 5.12b). Note that suction stress increases by an order of magnitude with a decrease in particle size of the same order of magnitude. The greater tendency for large capillary forces to develop between relatively fine-grained soil particles may partially explain the greater tendency of fine-grained soil to shrink during drying. Note also that increasing the contact angle has a significant effect on the magnitude of suction stress. The larger the contact angle, the less the suction stress, a reflection of the decrease in matric suction as water content increases. This observation may have important practical implications. For example, this implies that soil at a certain value of suction undergoing a wetting process (e.g., an unsaturated slope during a precipitation event) may have a lesser contribution to effective stress from capillarity, and thus less shear strength, than at the same value of suction during a drying process.

5.3 STRESS TENSOR REPRESENTATION

5.3.1 Net Normal Stress, Matric Suction, and Suction Stress Tensors

The governing variables for the state of stress in perfectly dry soil are the total principal stress in each coordinate direction and the pore air pressure. The latter is always isotropic. The stress state variable in dry soil is the difference between the total normal stress and the pore air pressure. This difference is designated the *net normal stress*, or $\sigma - u_a$. Figure 5.13a shows the stress state variables for a cubic element of dry soil in three Cartesian coordinate directions. Normal and shear stresses act on every plane in the x, y, and z directions. By convention, positive normal stresses (shown) indicate compression on the cubic element. Negative normal stresses indicate tension.

The governing variables for the state of stress in saturated soil are the total principal stress in each coordinate direction and the pore water pressure.

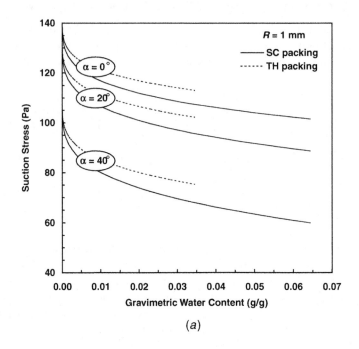

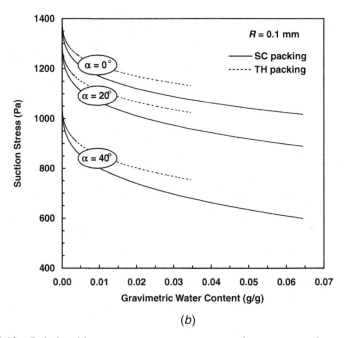

Figure 5.12 Relationships among water content, suction stress, and contact angle: (*a*) 1-mm particles and (*b*) 0.1-mm particles.

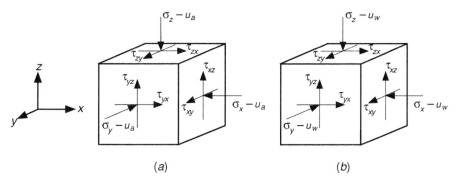

Figure 5.13 Normal and shear stresses on cubical element of soil: (a) perfectly dry soil and (b) saturated soil.

Again, the latter is always isotropic. For incompressible soil particles, the stress state variable is the difference between the total stress and the pore pressure. This is Terzaghi's classic effective stress, typically designated σ' where $\sigma' = \sigma - u_w$. Figure 5.13b shows the stress state variables for a cubic element of saturated soil in three coordinate directions.

Some unsaturated soil mechanics problems may be effectively approached as an extension of saturated soil mechanics if it is assumed that the state of stress can be described by two independent stress state variables. For convenience of analysis and measurement, independent variables are chosen in terms of physically measurable properties. One set of commonly cited independent stress state variables is the net normal stress $\sigma - u_a$ and matric suction $u_a - u_w$. Following the continuum mechanics methodology, each of these independent stress variables in three-dimensional space can be represented by a tensor. The net normal stress tensor in the Cartesian coordinate system is

$$\begin{bmatrix} \sigma_x - u_a & \tau_{yx} & \tau_{zx} \\ \tau_{xy} & \sigma_y - u_a & \tau_{zy} \\ \tau_{xz} & \tau_{yz} & \sigma_z - u_a \end{bmatrix}$$

And the matric suction tensor is

$$\begin{bmatrix} u_a - u_w & 0 & 0 \\ 0 & u_a - u_w & 0 \\ 0 & 0 & u_a - u_w \end{bmatrix}$$

The superimposed net normal stress and matric suction tensors are shown for a cubic element of unsaturated soil in Fig. 5.14a.

Following Bishop's (1959) effective stress formulation, suction stress may be considered an isotropic stress tensor in conjunction with net normal stress to describe the state of stress in unsaturated soil. The suction stress tensor is

194 STATE OF STRESS

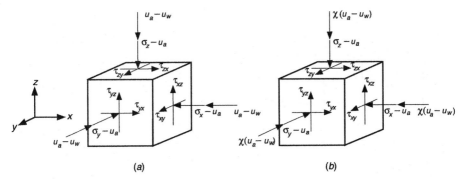

Figure 5.14 Normal and shear stresses on cubical element of unsaturated soil: (*a*) net normal stress and matric suction tensors following independent stress state variable approach and (*b*) net normal stress and suction stress tensors following effective stress approach.

$$\begin{bmatrix} \chi(u_a - u_w) & 0 & 0 \\ 0 & \chi(u_a - u_w) & 0 \\ 0 & 0 & \chi(u_a - u_w) \end{bmatrix}$$

which is illustrated in Fig. 5.14*b* superimposed with the net normal stress tensor for a cubic element of soil. In general, suction stress in anisotropic soil is not the same in all directions and the more general form for suction stress may be considered as an anisotropic stress tensor, but formulation along this line is in its infancy:

$$\begin{bmatrix} \chi_x(u_a - u_w) & 0 & 0 \\ 0 & \chi_y(u_a - u_w) & 0 \\ 0 & 0 & \chi_z(u_a - u_w) \end{bmatrix}$$

The treatment of suction stress as anisotropic completely respects the nature of the characteristic function for suction stress in unsaturated soil.

Under stable equilibrium conditions in unsaturated soil, the total normal stress exceeds the pore air pressure, which in turn exceeds the pore water pressure. By recognizing the relative magnitude of each stress component, the following hierarchy can be established:

$$\sigma > u_a > u_w \tag{5.21}$$

If this hierarchy is indeed followed, then the diagonal components of the net stress tensor $\sigma - u_a$ are positive. This condition is generally true under field conditions, but not always. For example, under unique loading conditions such as blasting during mining operations, the pore air pressure is suddenly increased to a value exceeding the total stress, resulting in a limiting stress state condition and an explosion of the soil skeleton when the tensile strength of the material is reached.

Matric suction ($u_a - u_w$) in unsaturated soil under field conditions generally exhibits positive values, implying that the pore air pressure is greater than the pore water pressure. Typically, the air pressure is set to a reference of zero for convenience, thus implying that the pore water pressure is negative when matric suction is positive. For matric suction to be negative, the water pressure must increase to positive values while maintaining the air pressure as a constant, resulting in a trend toward saturation.

5.3.2 Stress Tensors in Unsaturated Soil: Conceptual Illustration

The state of stress in unsaturated soil may be evaluated with consideration of the subsurface matric suction profile or field. This profile is influenced by several factors, including the type of soil, the thickness of the unsaturated zone, and the fluid fluxes occurring at the subsurface-atmosphere interface. Because atmospheric and moisture loading conditions are inherently subject to natural and anthropogenic fluctuations, the state of stress for unsaturated soil in the field is rarely a constant.

Consider the example shown in Fig. 5.15. For simplicity, the unsaturated layer can be divided into two general zones: an unsteady, or "active," zone

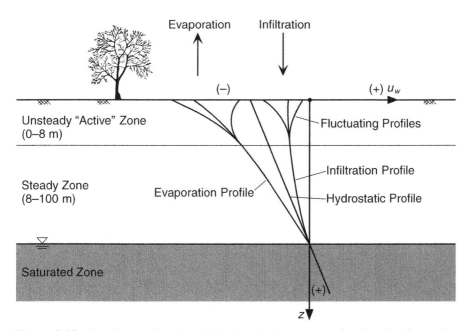

Figure 5.15 Suction profiles in typical deposit of unsaturated soil under fluctuating atmospheric conditions.

and a steady zone. In the steady zone, which comprises the relatively deep portion of the unsaturated zone, the suction profile is independent of time. In the unsteady, or "active," zone, the suction profile is close enough to the ground surface to be influenced by seasonal environmental changes and thus varies with time. In the wintertime, for example, the suction profile may trend toward relatively high values due to the low relative humidity and relative lack of precipitation. In the summertime, the suction profile may trend toward a minimum as a result of the relatively high amount of precipitation. The breadth of the profile at any time during the seasonal cycle can fluctuate widely between these two extremes. The depth of active zone, which defines the depth from the ground surface to the point where the profile is relatively constant, varies considerably from location to location. In the semiarid climate of Colorado, for example, the active zone is generally about 3 to 8 m deep. Under the hydrostatic condition (i.e., no net surface moisture flux), the suction profile is linearly distributed between zero at the water table and some maximum at the ground surface.

Example Problem 5.1: State of Stress in Shallow Unsaturated Soil A 20-m-thick flat unsaturated silty clay layer (Fig. 5.16) has the following properties: total unit weight $\gamma = 18.5$ kN/m³, void ratio $e = 0.40$, Poisson's ratio $\mu = 0.35$, effective stress parameter $\chi = 0.5$, and 10% finer particle size $D_{10} = 1.2$ μm. The pore air pressure is set as a zero reference. Assume that the

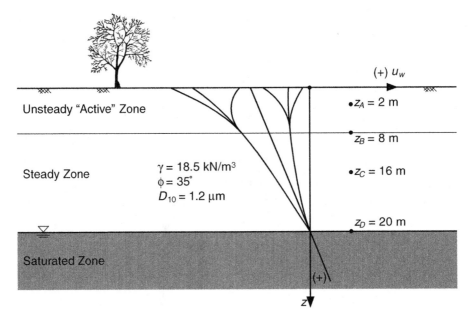

Figure 5.16 Soil profile for Example Problem 5.1.

height of capillary rise may be estimated by considering a minimum pore diameter equal to D_{10}. Calculate the net normal stress and matric suction tensors at points B, C, and D under the hydrostatic condition.

Solution The matric suction profile is governed by the height of capillary rise. Capillary water extends from the water table to some distance h_c above the water table. The height of capillary rise may be estimated from D_{10} using the lower bound of equation [eq. (4.28)] as follows:

$$h_c = \frac{C}{eD_{10}} = \frac{10 \text{ mm}^2}{0.4 \times 0.0012 \text{ mm}} = 20.8 \text{ m}$$

which is slightly greater than the thickness of the soil layer. Accordingly, the capillary suction will cause water to move upward to the ground surface, where the suction is

$$u_a - u_w = \gamma_w H = (9.8 \times \text{kN/m}^3)(20 \text{ m}) = 196 \text{ kPa}$$

and is linearly distributed to a value of zero at the water table. The matric suction at points B, C, and D, therefore, are

$$(u_a - u_w)_B = \gamma_w H_B = (9.8 \times \text{kN/m}^3)(12 \text{ m}) = 117.6 \text{ kPa}$$
$$(u_a - u_w)_C = \gamma_w H_C = -(9.8 \times \text{kN/m}^3)(4 \text{ m}) = 39.2 \text{ kPa}$$
$$(u_a - u_w)_D = \gamma_w H_D = -(9.8 \times \text{kN/m}^3(0 \text{ m}) = 0 \text{ kPa}$$

The total vertical stress is due to the soil's self-weight:

$$\sigma_z = \gamma z \tag{5.22}$$

The horizontal stress components can be approximated under the at-rest, or K_0, condition formulated in Section 7.3:

$$\sigma_x = \sigma_y = \frac{\mu}{1 - \mu} \sigma_z - \frac{1 - 2\mu}{1 - \mu} \chi(u_a - u_w) \tag{5.23}$$

The shear stresses on all three orthogonal planes are zero because the ground surface is flat and loading occurs only in the vertical and horizontal directions. Therefore, at point B, the components of the stress tensors are

$$\sigma_z = \gamma z_B = (18.5 \text{ kN/m}^3)(8 \text{ m}) = 148 \text{ kPa}$$

$$\sigma_x = \frac{0.35}{1 - 0.35}(148 \text{ kPa}) - \frac{1 - (2)(0.35)}{1 - 0.35}(0.5 \times 117.6 \text{ kPa}) = 52.6 \text{ kPa}$$

$$\sigma_y = \sigma_x = 52.6 \text{ kPa}$$

$$u_a - u_w = 0 - (-117.6 \text{ kPa}) = 117.6 \text{ kPa}$$

And the stress state tensors at point B are

$$\begin{bmatrix} \sigma_x - u_a & \tau_{yx} & \tau_{zx} \\ \tau_{xy} & \sigma_y - u_a & \tau_{zy} \\ \tau_{xz} & \tau_{yz} & \sigma_z - u_a \end{bmatrix} = \begin{bmatrix} 52.6 & 0 & 0 \\ 0 & 52.6 & 0 \\ 0 & 0 & 148 \end{bmatrix}$$

$$\begin{bmatrix} u_a - u_w & 0 & 0 \\ 0 & u_a - u_w & 0 \\ 0 & 0 & u_a - u_w \end{bmatrix} = \begin{bmatrix} 117.6 & 0 & 0 \\ 0 & 117.6 & 0 \\ 0 & 0 & 117.6 \end{bmatrix}$$

Similarly, at point C,

$$\begin{bmatrix} \sigma_x - u_a & \tau_{yx} & \tau_{zx} \\ \tau_{xy} & \sigma_y - u_a & \tau_{zy} \\ \tau_{xz} & \tau_{yz} & \sigma_z - u_a \end{bmatrix} = \begin{bmatrix} 150.3 & 0 & 0 \\ 0 & 150.3 & 0 \\ 0 & 0 & 296 \end{bmatrix}$$

$$\begin{bmatrix} u_a - u_w & 0 & 0 \\ 0 & u_a - u_w & 0 \\ 0 & 0 & u_a - u_w \end{bmatrix} = \begin{bmatrix} 39.2 & 0 & 0 \\ 0 & 39.2 & 0 \\ 0 & 0 & 39.2 \end{bmatrix}$$

And at point D,

$$\begin{bmatrix} \sigma_x - u_a & \tau_{yx} & \tau_{zx} \\ \tau_{xy} & \sigma_y - u_a & \tau_{zy} \\ \tau_{xz} & \tau_{yz} & \sigma_z - u_a \end{bmatrix} = \begin{bmatrix} 199.2 & 0 & 0 \\ 0 & 199.2 & 0 \\ 0 & 0 & 370 \end{bmatrix}$$

$$\begin{bmatrix} u_a - u_w & 0 & 0 \\ 0 & u_a - u_w & 0 \\ 0 & 0 & u_a - u_w \end{bmatrix} = \begin{bmatrix} 0 & 0 & 0 \\ 0 & 0 & 0 \\ 0 & 0 & 0 \end{bmatrix}$$

Example Problem 5.2: State of Effective Stress in Shallow Unsaturated Soil For the same problem above, assume the soil-water characteristic curve for the soil follows the Brooks and Corey (1964) model, which, as described in Chapter 12, states the following:

$$\theta = \theta_s \qquad\qquad u_a - u_w < (u_a - u_w)_b \qquad (5.24)$$

$$\theta = \theta_r + (\theta_s - \theta_r)\left(\frac{(u_a - u_w)_b}{u_a - u_w}\right)^\lambda \quad u_a - u_w \geq (u_a - u_w)_b$$

where θ is volumetric water content, θ_s is the saturated water content, θ_r is the residual water content, $(u_a - u_w)_b$ is the air-entry suction, and λ is a pore size distribution index. Assume the soil layer has $\theta_s = 0.5$, $\theta_r = 0.1$, $(u_a - u_w)_b = 5$ kPa, and $\lambda = 0.3$. Also assume that the effective stress parameter χ may be approximated between zero and unity as a function of water content as

$$\chi = \frac{\theta - \theta_r}{\theta_s - \theta_r} \qquad (5.25)$$

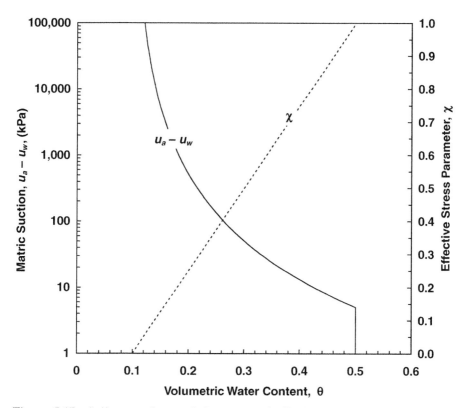

Figure 5.17 Soil-water characteristic curve and effective stress parameter function $\chi(\theta)$ for unsaturated soil layer from Example Problem 5.2.

200 STATE OF STRESS

which implies that when $\theta = \theta_r$, $\chi = 0$ and when $\theta = \theta_s$, $\chi = 1$. Calculate the state of effective stress using the Bishop (1959) formulation at points B, C, and D.

Solution The soil-water characteristic curve and effective stress parameter function $\chi(\theta)$ are plotted in Fig. 5.17. As before, the matric suctions at points B, C, and D are

$$(u_a - u_w)_B = 117.6 \text{ kPa} \qquad (u_a - u_w)_C = 39.2 \text{ kPa} \qquad (u_a - u_w)_D = 0 \text{ kPa}$$

The effective stress parameters at points B, C, and D may be calculated from these values and eqs. (5.24) and (5.25), leading to

$$\chi_B = 0.388 \qquad \chi_C = 0.539 \qquad \chi_D = 1.000$$

The state of effective stress is

$$\sigma'_{ij} = (\sigma_{ij} - u_a) + \chi(u_a - u_w)\delta_{ij} \qquad (5.26)$$

Applying eqs. (5.22), (5.23), and (5.26) leads to effective stresses at B, C, and D,

$$\begin{pmatrix} \sigma'_x & 0 & 0 \\ 0 & \sigma'_y & 0 \\ 0 & 0 & \sigma'_z \end{pmatrix}_B = \begin{pmatrix} 58.6 & 0 & 0 \\ 0 & 58.6 & 0 \\ 0 & 0 & 148 \end{pmatrix}$$

$$+ \begin{pmatrix} (0.388)(117.6) & 0 & 0 \\ 0 & (0.388)(117.6) & 0 \\ 0 & 0 & (0.388)(117.6) \end{pmatrix}$$

$$= \begin{pmatrix} 104.3 & 0 & 0 \\ 0 & 104.3 & 0 \\ 0 & 0 & 193.6 \end{pmatrix} \text{ kPa}$$

$$\begin{pmatrix} \sigma'_x & 0 & 0 \\ 0 & \sigma'_y & 0 \\ 0 & 0 & \sigma'_z \end{pmatrix}_C = \begin{pmatrix} 149.6 & 0 & 0 \\ 0 & 149.6 & 0 \\ 0 & 0 & 296 \end{pmatrix}$$

$$+ \begin{pmatrix} (0.539)(39.2) & 0 & 0 \\ 0 & (0.539)(39.2) & 0 \\ 0 & 0 & (0.539)(39.2) \end{pmatrix}$$

$$= \begin{pmatrix} 170.8 & 0 & 0 \\ 0 & 170.8 & 0 \\ 0 & 0 & 317.1 \end{pmatrix} \text{ kPa}$$

$$\begin{pmatrix} \sigma'_x & 0 & 0 \\ 0 & \sigma'_y & 0 \\ 0 & 0 & \sigma'_z \end{pmatrix}_D = \begin{pmatrix} 199.2 & 0 & 0 \\ 0 & 199.2 & 0 \\ 0 & 0 & 370 \end{pmatrix} + \begin{pmatrix} (1)(0) & 0 & 0 \\ 0 & (1)(0) & 0 \\ 0 & 0 & (1)(0) \end{pmatrix}$$

$$= \begin{pmatrix} 199.2 & 0 & 0 \\ 0 & 199.2 & 0 \\ 0 & 0 & 370 \end{pmatrix} \text{kPa}$$

5.4 STRESS CONTROL BY AXIS TRANSLATION

5.4.1 Rationale for Axis Translation

Matric suction may be considered an important variable in defining the state of stress in unsaturated soil. Control or measurement of matric suction, therefore, becomes necessary in order to evaluate the physical behavior (e.g., fluid flow, strength, and volume change) of unsaturated soil under changing stress conditions. Difficulties associated with the measurement and control of negative pore water pressure, however, present an important practical limitation.

As described in Section 2.5, cavitation in free water under negative pressure occurs as the magnitude of the pressure approaches -1 atm. As cavitation occurs, the water phase in both the soil and measurement system becomes discontinuous, making the measurements unreliable or impossible. Because control of the matric suction variable over a range far greater than 1 atm is required for many soil types and applications, alternatives to measurement or control of negative water pressure are desirable.

The general term *axis translation* refers to the practice of elevating pore air pressure in unsaturated soil while maintaining the pore water pressure at a measurable reference value, typically atmospheric. As such, the matric suction variable $u_a - u_w$ may be controlled over a range far greater than the cavitation limit for water under negative pressure. The origin of reference, or "axis," for the matric suction variable is "translated" from the condition of atmospheric air pressure and negative water pressure to the condition of atmospheric water pressure and positive air pressure. Matric suction may be accurately controlled in this manner because positive air pressure may be easily controlled and measured.

Axis translation is accomplished by separating the air and water phases of the soil through the minute pores of a high-air-entry (HAE) material. When saturated, these materials have the unique capability of restricting the advection of air while allowing free advection of water. If a specimen of soil is placed in good contact with a saturated HAE material, positive air pressure may be applied to the pore air on one side, while allowing the pore water to drain freely through the material under atmospheric pressure maintained on the other side. Separation of the air and water pressure is maintained as long as the applied pressure does not exceed the air-entry pressure of the HAE material, which can be as high as 1500 kPa for sintered ceramics or 10,000 kPa for special cellulose membranes. Chapter 10 contains detailed descriptions of the specific techniques, materials, and limitations of the axis translation concept for investigation of unsaturated soil behavior. The remainder of this section provides insight into the equilibrium condition between pore air and pore water under axis translation through a series of thought experiments. The use of axis translation to investigate net normal stress and matric suction as potential independent stress state variables in unsaturated soil is then illustrated.

5.4.2 Equilibrium for an Air-Water-HAE System

Figure 5.18 shows a closed chamber containing air and water separated by a saturated HAE ceramic disk. Air pressure u_a and water pressure u_w may be independently controlled through ports located on the top and bottom of the chamber, respectively. Assume that there is no communication between the air and water phases along the sides of the chamber, the pore water pressure is less than or equal to atmospheric, and the pore air pressure is greater than or equal to atmospheric. The ceramic disk is described by an air-entry pressure denoted u_{wa}. There are three possible equilibrium positions for the air-water interface depending on the relative magnitudes of u_a and u_{wa}.

1. $u_a > u_{wa}$: at equilibrium $u_a = u_w$ and the position of the air-water interface may be located somewhere in the water compartment (position A),
2. $u_a < u_{wa}$: at equilibrium $u_a = u_w$ and the position of the air-water interface may be located somewhere in the air compartment (position C),

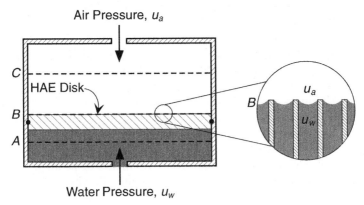

Figure 5.18 Equilibrium positions for air-water interface in air-water-HAE system.

3. $u_a < u_{wa}$: at equilibrium $u_a > u_w$, where the pressure difference is compensated by surface tension at the ceramic-air-water interface (position B).

Axis translation for unsaturated soil testing applications relies on the equilibrium condition described by position B. In effect, the surface tension at the air-water interface in the pores of the ceramic disk acts as a "membrane" to separate the air and water pressure. The maximum sustainable difference between the air pressure and the water pressure is a function of the magnitude of the surface tension and the maximum effective pore size of the HAE material. This maximum pressure, or air-entry pressure, may be captured by the Young-Laplace equation for a capillary tube:

$$u_{wa} = (u_a - u_w)_b = \frac{2T_s}{R_s} \tag{5.27}$$

where $(u_a - u_w)_b$ is the difference between the air and water pressure at air entry, often called the *bubbling* pressure, T_s is the surface tension of the interface, and R_s is the maximum effective radius of the pores of the HAE material. The smaller this radius (i.e., the finer the pores), the larger the air-entry pressure.

5.4.3 Equilibrium for an Air-Water-HAE-Soil System

Figure 5.19 builds upon the previous discussion by considering a specimen of soil placed in good contact on top of the HAE disk. The saturated pores

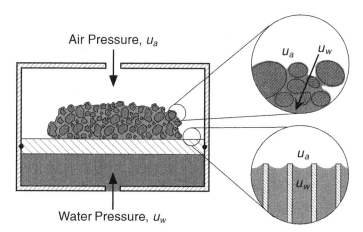

Figure 5.19 Equilibrium positions for air-water interface in air-water-HAE-soil system.

of the HAE disk provide a hydraulic connection between the soil pore water and the water reservoir below the disk. Under any given air and water pressure conditions, the possible equilibrium positions for the air-water interface in the system are similar to those described without soil present (Fig. 5.18). In this case, however, condition B is described by two interfaces: an air-water interface in the pores of the ceramic and an air-water interface in the pores of the soil (see inset figures). At equilibrium, the difference between the air and water pressure acting across both of these interfaces is the same. If drainage from the water reservoir is allowed, the pressure deficit at the air-water-soil interface is accommodated by drainage of pore water from the soil. Drainage continues until the curvature of the interface satisfies equilibrium with the applied suction. The water content of the soil specimen at equilibrium depends on its surface and pore size properties, described by the soil-water characteristic curve.

5.4.4 Characteristic Curve for HAE Material

The water retention model for a system of uniform capillary tubes (Chapter 3) provides insight into the water retention characteristics of HAE materials. Consider, for example, a saturated ceramic disk of sintered kaolin. Ceramic HAE materials generally have a relatively uniform pore size. Accordingly, drainage under increasing suction will occur relatively "rapidly" (i.e., over a narrow range of changing suction) when the suction reaches and exceeds the material's air-entry pressure.

An idealized characteristic curve for a ceramic porous disk with a uniform pore size distribution would follow the simple path shown in Fig. 5.20. The saturated water content and air-entry value (AEV) for the HAE disk, $\theta_{s(HAE)}$ and AEV_{HAE}, respectively, are defined by the two straight-line segments of the characteristic curve. The soil-water characteristic curve for a typical well-graded soil (e.g., sand) is included in the figure for comparison. Note that the soil's characteristic curve is distinguished by sustained drainage over a relatively wide range of suction. The shape of the curve reflects the soil's relatively wide pore size distribution and relatively low air-entry pressure. As illustrated in the figure, matric suction for this soil in contact with the disk may be controlled using axis translation over a significant portion of the soil-water characteristic curve because the air-entry pressure of the HAE material is significantly larger than the air-entry pressure of the soil.

5.4.5 Controlled Stress Variable Testing

The axis translation concept may be used in modified triaxial, direct shear, or oedometer systems for independently controlling the stress state variables $u_a - u_w$ and $\sigma - u_a$ in unsaturated soil. Figure 5.21, for example, illustrates one variation on this testing concept for a cylindrical soil specimen in an isotropic confining cell. The specimen is seated on the pedestal in good con-

5.4 STRESS CONTROL BY AXIS TRANSLATION   205

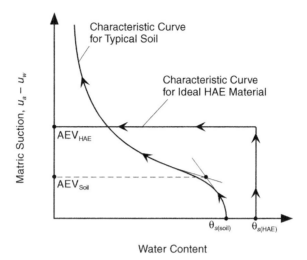

Figure 5.20 Characteristic curves for idealized high-air-entry material and typical coarse-grained soil.

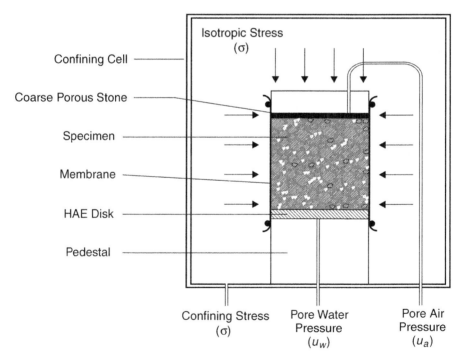

Figure 5.21 Isotropic loading system for controlled stress variable testing. Net normal stress ($\sigma - u_a$) is controlled by independent manipulation of cell pressure σ and air pressure u_a. Matric suction ($u_a - u_w$) is controlled by independent manipulation of pore water pressure u_w and air pressure u_a using axis translation.

tact with a saturated high-air-entry ceramic disk. The ceramic disk maintains communication with the pore water such that the pore pressure may be measured and/or externally controlled through a port located outside the cell. The pore air pressure may be measured or controlled using an external pressure supply in communication with the specimen through a relatively coarse porous stone (i.e., low air entry) located on the specimen top cap. Under equilibrium conditions, the difference between the applied pore air pressure and the pore water pressure is maintained by axis translation. An axial loading ram (not shown) may also be used to apply deviator stress to the specimen for triaxial testing.

Matric suction $u_a - u_w$ and net normal stress $\sigma - u_a$ in such a set up may be independently controlled by manipulating the pore air pressure u_a, confining stress σ, and pore water pressure u_w components. If the magnitudes of these stress components are varied equally and in the same direction (i.e., the changes are either all positive or all negative in sign), then the matric suction and net normal stress variables may be maintained at constant values, thus allowing their roles on the macroscopic behavior of the unsaturated soil system to be effectively isolated, a strategy often referred to as *null testing* (Fredlund, 1973). Experiments can proceed as long as the changes in the component stresses satisfy the general equilibrium hierarchy $\sigma > u_a > u_w$.

Fredlund and Morgenstern (1977) report a series of null tests conducted using modified oedometer and triaxial systems for specimens of compacted kaolinite. Table 5.1 summarizes the initial stress components (σ, u_a, and u_w) and the applied changes in these stresses ($\Delta\sigma$, Δu_a, and Δu_w) for a select series of six tests. Table 5.2 summarizes the test results in terms of the initial and final matric suction and net normal stress values and the corresponding percent change in the overall specimen volume. Each test follows a stress path such that the offsetting changes in the component stresses result in essentially constant matric suction and net normal stress. The very small changes in the overall specimen volume (< 0.4%) under the null test conditions reflect the potential applicability of matric suction and net normal stress as stress

TABLE 5.1 Changes in Component Stress Variables for Null Tests Conducted on Compacted Kaolinite

Test Number	Initial Stresses (kPa)			Changes in Stresses (kPa)		
	σ	u_a	u_w	$\Delta\sigma$	Δu_a	Δu_w
24	359.4	270.9	3.0	+135.9	+135.9	+140.5
30	343.1	270.5	91.2	+68.8	+68.5	+68.8
35	410.9	338.5	208.3	+69.5	+69.3	+69.7
38	615.4	541.2	411.3	−66.0	−64.1	−63.7
39	549.4	477.1	347.6	−70.2	−69.5	−69.8
41	412.6	340.7	211.4	−140.5	−140.3	−139.8

Source: Data from Fredlund (1973).

TABLE 5.2 Initial and Final Net Normal Stress and Matric Suction and Corresponding Specimen Volume Change for Null Tests Conducted on Compacted Kaolinite

Test Number	Initial Stress State Variables (kPa)		Final Stress State Variables (kPa)		Volume Change (%)[a]
	$\sigma - u_a$	$u_a - u_w$	$\sigma - u_a$	$u_a - u_w$	
24	88.5	267.9	88.5	263.3	+0.40
30	72.6	179.3	72.9	179.0	+0.012
35	72.4	130.2	72.6	129.8	+0.033
38	74.2	129.9	72.3	129.5	+0.002
39	72.3	129.5	71.6	129.8	+0.005
41	71.9	129.3	71.7	128.8	+0.007

[a] Overall specimen volume change at elapsed time 1350 min.
Source: Data from Fredlund (1973).

state variables in unsaturated soil. The results also support the applicability of Bishop's (1959) effective stress formulation, whereby the null change in effective stress [$\Delta\sigma' = \Delta(\sigma - u_a) + \chi\Delta(u_a - u_w) = 0 + \chi(0) = 0$] results in little specimen volume change.

5.5 GRAPHICAL REPRESENTATION OF STRESS

5.5.1 Net Normal Stress and Matric Suction Representation

The stress conditions at any point in a mass of soil can be described by the normal and shear stresses acting on a particular plane passing through that point. For an orthogonal element of unsaturated soil subject to both normal and shear stresses and isotropic pore air pressure (Fig. 5.22a), the equilibrium net normal and shear stresses at an angle θ from the vertical direction z are

$$\sigma_\theta - u_a = \tfrac{1}{2}(\sigma_x + \sigma_z) + \tfrac{1}{2}(\sigma_z - \sigma_x)\cos 2\theta + \tau_{xz}\sin 2\theta - u_a \quad (5.28a)$$

$$\tau_\theta = \tfrac{1}{2}(\sigma_z - \sigma_x)\sin 2\theta + \tau_{xz}\cos 2\theta \quad (5.28b)$$

The customary convention in geotechnical engineering is to describe compressive normal stresses as positive. Similarly, shear stresses that tend to cause counterclockwise rotation of the element are defined as positive.

For a reference pore air pressure ($u_a = 0$), eqs. (5.28a) and (5.28b) describe a Mohr circle of stress centered at a point described by

$$\sigma = \tfrac{1}{2}(\sigma_x + \sigma_z) \qquad \tau = 0$$

and at a radius

208 STATE OF STRESS

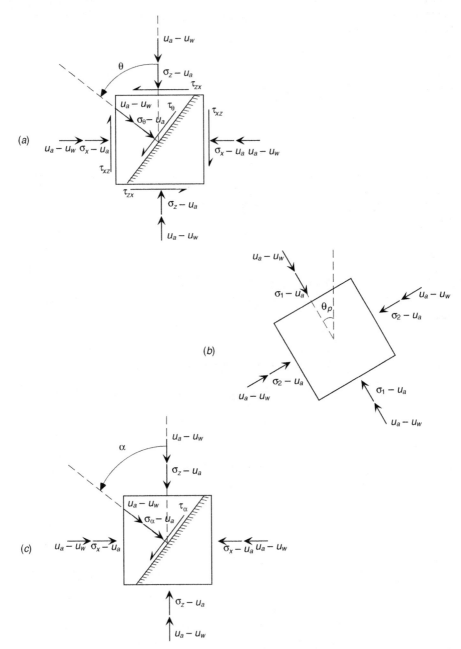

Figure 5.22 State of stress in two-dimensional space: (*a*) orthogonal element with net normal and shear stresses, (*b*) principal stresses and their directions, and (*c*) orthogonal element with principal stresses in horizontal and vertical directions.

$$\sqrt{[\tfrac{1}{2}(\sigma_z - \sigma_x)]^2 + \tau_{xz}^2}$$

Planes on which no shear stresses exist are principal stress planes. These planes are described by points on the Mohr circle where it intersects the normal stress axis. The stresses acting on these planes, the principal stresses (Fig. 5.22b), can be found by the equations

$$\sigma_1 - u_a = \tfrac{1}{2}(\sigma_x + \sigma_z) - u_a + \sqrt{[\tfrac{1}{2}(\sigma_x - \sigma_z)]^2 + \tau_{xz}^2} \quad (5.29)$$
$$\sigma_2 - u_a = \tfrac{1}{2}(\sigma_x + \sigma_z) - u_a - \sqrt{[\tfrac{1}{2}(\sigma_x - \sigma_z)]^2 + \tau_{xz}^2}$$

The principal stress ($\sigma_1 - u_a$) acts on the plane at an angle θ_p from horizontal, where

$$\theta_p = \frac{1}{2} \tan^{-1}\left(\frac{2\tau_{xz}}{\sigma_x - \sigma_z}\right) \quad (5.30)$$

If the vertical and horizontal directions are the principal stress directions (Fig. 5.22c), the normal and shear stresses acting on a plane at angle α to the horizontal direction are

$$\sigma_\alpha - u_a = \tfrac{1}{2}(\sigma_x + \sigma_z) + \tfrac{1}{2}(\sigma_z - \sigma_x)\cos 2\alpha - u_a \quad (5.31)$$
$$\tau_\alpha = \tfrac{1}{2}(\sigma_z - \sigma_x)\sin 2\alpha$$

The Mohr circle diagram for the above state of stress is shown as Fig. 5.23a. For unsaturated soil where matric suction is not zero, matric suction may be included as a description of the state of stress by adding a third

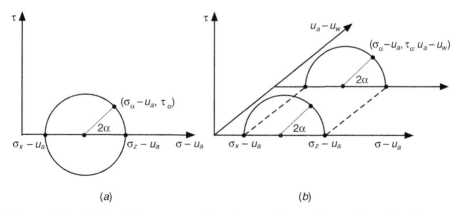

Figure 5.23 Mohr circle representation of state of stress in two-dimensional space: (a) when matric suction is zero and (b) when matric suction is nonzero.

orthogonal axis to the Mohr circle diagram. This extended Mohr circle representation is shown in Fig. 5.23b. As the matric suction approaches zero, the third axis disappears and the Mohr circle diagram projects to the single plane of shear stress and net normal stress. For stress in three-dimensional space, three Mohr circles are generally required to represent the state of stress. Assuming there are only normal stresses on the faces of the orthogonal element (Fig 5.24a), Mohr circles for zero matric suction are shown in Fig. 5.24b and for nonzero matric suction in 5.24c.

Example Problem 5.3 An isotropic loading test using axis translation to control matric suction is performed (e.g., Fig. 5.21). At equilibrium, the total stress σ is 300 kPa, the pore air pressure u_a is 200 kPa, and the pore water pressure u_w is 100 kPa. Draw the Mohr circle for the state of stress and find the maximum shear stress.

Solution The net normal stresses and matric suction are

$$\sigma_1 - u_a = \sigma_2 - u_a = \sigma_3 - u_a = 300 - 200 = 100 \text{ kPa}$$

$$u_a - u_w = 200 - 100 = 100 \text{ kPa}$$

Because all three principal net stresses are equal, the Mohr circle for the state of stress is the single point shown as point A on Fig. 5.25. Accordingly, the maximum shear stress is zero.

Example Problem 5.4 A triaxial loading test using axis translation to control matric suction is performed. The vertical total stress is $\sigma_1 = 300$ kPa, the horizontal stresses are $\sigma_2 = \sigma_3 = 200$ kPa, the applied pore air pressure u_a is 100 kPa, and the pore water pressure u_w is zero. Draw the Mohr circles for the state of stress. Find the maximum shear stress and the plane where it acts.

Solution The net normal stresses and matric suction are

$$\sigma_1 - u_a = 300 - 100 = 200 \text{ kPa}$$

$$\sigma_2 - u_a = \sigma_3 - u_a = 200 - 100 = 100 \text{ kPa}$$

$$u_a - u_w = 100 - 0 = 100 \text{ kPa}$$

The Mohr circle for the principal stresses ($\sigma_1 - u_a$ and $\sigma_2 - u_a$ or $\sigma_3 - u_a$) is the circle shown as B on Fig. 5.25. The Mohr circle for the principal stresses ($\sigma_2 - u_a$ and $\sigma_3 - u_a$) is the point denoted A. The maximum shear stress is equal to the radius of circle B, or $\frac{1}{2}(200 - 100) = 50$ kPa, and acts on a plane $2\alpha = 90°$ or $\alpha = 45°$ from the major principal plane.

5.5 GRAPHICAL REPRESENTATION OF STRESS

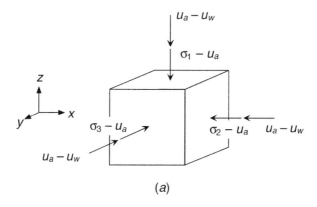

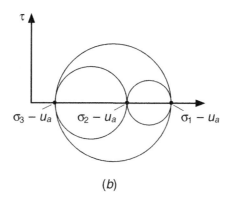

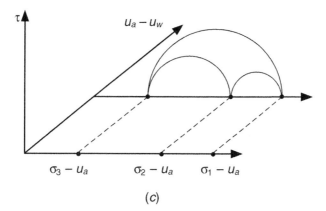

Figure 5.24 Independent stress state variable representation of state of stress in three-dimensional space: (*a*) elementary stress diagram, (*b*) Mohr circle representation at matric suction equal to zero, and (*c*) Mohr circle in shear stress, net normal stress, and matric suction space.

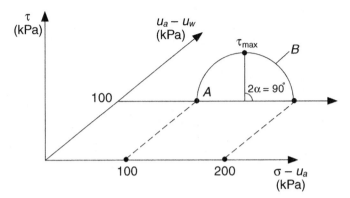

Figure 5.25 Mohr circles for Example Problems 5.3 and 5.4.

Example Problem 5.5 Analyze the state of stress at a point A beneath the slope shown in Fig. 5.26a. The total stress $\sigma_z = \sigma_1$ due to the self-weight of soil is about 300 kPa, the horizontal stresses are $\sigma_y = \sigma_2 = 150$ kPa and $\sigma_x = \sigma_3 = 10$ kPa. During the dry season, the pore water pressure at point A is measured to be -200 kPa. During the wet season, the pore pressure is 0 kPa. Draw the Mohr circles for the state of stress at point A during both seasons. Set the air pressure to a zero reference value.

Solution The net normal stresses and matric suction at point A during the dry season are

$$\sigma_1 - u_a = 300 - 0 = 300 \text{ kPa}$$

$$\sigma_2 - u_a = 150 - 0 = 150 \text{ kPa}$$

$$\sigma_3 - u_a = 10 - 0 = 10 \text{ kPa}$$

$$u_a - u_w = 0 - (-200) = 200 \text{ kPa}$$

The Mohr circle representations for the state of stress during the dry season are shown as the solid circles shown in Fig. 5.26b. During the wet season, all the net normal stress components remain the same, but the matric suction becomes

$$u_a - u_w = 0 - 0 = 0 \text{ kPa}$$

The corresponding Mohr circles for the wet season are the dashed circles shown in Fig. 5.26b.

5.5 GRAPHICAL REPRESENTATION OF STRESS

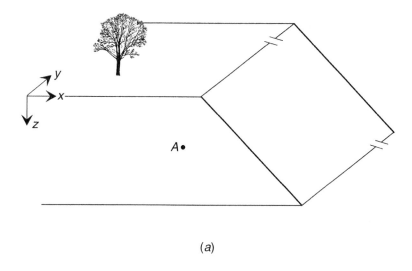

(a)

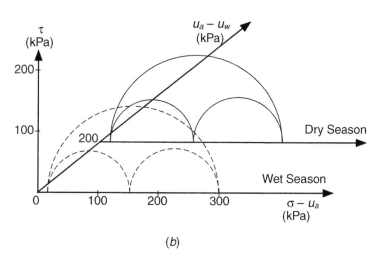

(b)

Figure 5.26 State of stress beneath slope for Example Problem 5.5: (a) field condition and (b) Mohr circle representation.

5.5.2 Effective Stress Representation

In three dimensions, Bishop's effective stress [eq. (5.1)] can be expressed in tensor notation as

$$\sigma'_{ij} = (\sigma_{ij} - u_a) + \chi(u_a - u_w)\delta_{ij} \quad (5.32)$$

214 STATE OF STRESS

where δ_{ij} is the Kronecker delta. Mohr circle representation of effective stress can be accomplished by assigning the suction stress defined by the second term on the right side of the above equation as the third axis shown in Fig. 5.27.

Alternatively, the state of stress can be represented in stress invariant space. One commonly used plot is the $p' - q'$ plane, where p' is the mean effective stress and q' is the deviatoric stress defined as

$$p' = \tfrac{1}{2}(\sigma_1' + \sigma_3') \tag{5.33a}$$

$$q' = \sigma_1' - \sigma_3' \tag{5.33b}$$

The state of stress in the space of mean effective stress p' and half-deviatoric stress $\tfrac{1}{2}q'$ is illustrated in Fig. 5.28.

Example Problem 5.6 Two triaxial tests were conducted for a silty soil specimen under a constant matric suction of 100 kPa. The tests were conducted using axis translation with an externally applied air pressure of 100 kPa. Pore water pressure was maintained at an atmospheric reference pressure (0 kPa). Test 1 showed that failure occurred when the applied stresses were $\sigma_1 = 1200$ kPa and $\sigma_3 = 300$ kPa. Test 2 showed that failure occurred when the applied stresses were $\sigma_1 = 1850$ kPa and $\sigma_3 = 450$ kPa. The effective stress parameter χ is assumed to be 0.5. Represent the state of stress at failure for the tests in $(\sigma - u_a) - \tau - \chi(u_a - u_w)$ space and $p' - \tfrac{1}{2}q'$ space.

Solution The net normal stress and suction stress tensors for test 1 at failure are

$$\sigma_{ij} - u_a = \begin{bmatrix} \sigma_1 - u_a & 0 & 0 \\ 0 & \sigma_2 - u_a & 0 \\ 0 & 0 & \sigma_3 - u_a \end{bmatrix}$$

$$= \begin{bmatrix} 1200 - 100 & 0 & 0 \\ 0 & 300 - 100 & 0 \\ 0 & 0 & 300 - 100 \end{bmatrix}$$

$$= \begin{bmatrix} 1100 & 0 & 0 \\ 0 & 200 & 0 \\ 0 & 0 & 200 \end{bmatrix}$$

$$\chi(u_a - u_w)\delta_{ij} = \begin{bmatrix} \chi(u_a - u_w) & 0 & 0 \\ 0 & \chi(u_a - u_w) & 0 \\ 0 & 0 & \chi(u_a - u_w) \end{bmatrix}$$

$$= \begin{bmatrix} 0.5(100) & 0 & 0 \\ 0 & 0.5(100) & 0 \\ 0 & 0 & 0.5(100) \end{bmatrix} = \begin{bmatrix} 50 & 0 & 0 \\ 0 & 50 & 0 \\ 0 & 0 & 50 \end{bmatrix}$$

5.5 GRAPHICAL REPRESENTATION OF STRESS

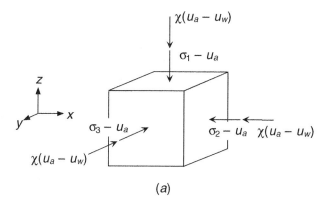

(a)

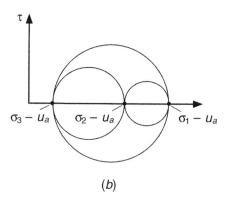

(b)

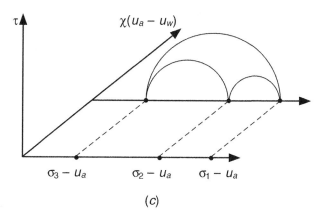

(c)

Figure 5.27 Effective stress representation of state of stress in three-dimensional space: (a) elementary stress diagram, (b) Mohr circle representation at matric suction equal to zero, and (c) Mohr circle in shear stress, net normal stress, suction stress space.

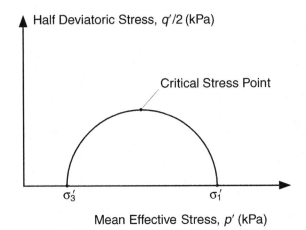

Figure 5.28 State of stress in mean effective stress and half deviatoric stress ($p' - q'/2$) space.

The net normal stress and suction stress tensors for test 2 at failure are

$$\sigma_{ij} - u_a = \begin{bmatrix} \sigma_1 - u_a & 0 & 0 \\ 0 & \sigma_2 - u_a & 0 \\ 0 & 0 & \sigma_3 - u_a \end{bmatrix}$$

$$= \begin{bmatrix} 1850 - 100 & 0 & 0 \\ 0 & 450 - 100 & 0 \\ 0 & 0 & 450 - 100 \end{bmatrix} = \begin{bmatrix} 1750 & 0 & 0 \\ 0 & 350 & 0 \\ 0 & 0 & 350 \end{bmatrix}$$

$$\chi(u_a - u_w)\delta_{ij} = \begin{bmatrix} \chi(u_a - u_w) & 0 & 0 \\ 0 & \chi(u_a - u_w) & 0 \\ 0 & 0 & \chi(u_a - u_w) \end{bmatrix}$$

$$= \begin{bmatrix} 0.5(100) & 0 & 0 \\ 0 & 0.5(100) & 0 \\ 0 & 0 & 0.5(100) \end{bmatrix} = \begin{bmatrix} 50 & 0 & 0 \\ 0 & 50 & 0 \\ 0 & 0 & 50 \end{bmatrix}$$

The above stress tensors are plotted in Fig. 5.29. The state of effective stress at failure for both tests can be calculated from eq. (5.32). For test 1, the effective stresses are

$$\sigma'_1 = 1200 - 100 + (0.5)(100) = 1150 \text{ kPa}$$
$$\sigma'_3 = 300 - 100 + (0.5)(100) = 250 \text{ kPa}$$

and for test 2,

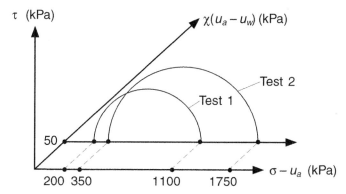

Figure 5.29 Mohr circle representation of states of stress at failure for triaxial tests from Example Problem 5.6.

$$\sigma'_1 = 1850 - 100 + (0.5)(100) = 1800 \text{ kPa}$$
$$\sigma'_3 = 450 - 100 + (0.5)(100) = 400 \text{ kPa}$$

The mean effective stress and deviatoric stress from eqs. (5.33a) and (5.33b) are as follows. For test 1,

$$p' = \tfrac{1}{2}(1150 + 250) = 700 \text{ kPa}$$
$$q' = 1150 - 250 = 900 \text{ kPa}$$

and for test 2,

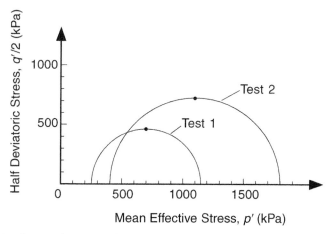

Figure 5.30 States of stress at failure for triaxial tests from Example Problem 5.6 in $p' - q/2$ space.

$$p' = \tfrac{1}{2}(1800 + 400) = 1100 \text{ kPa}$$
$$q' = 1800 - 400 = 1400 \text{ kPa}$$

The state of stress in the space of mean effective stress p' and half-deviatoric stress $\tfrac{1}{2}q'$ is illustrated in Fig. 5.30.

PROBLEMS

5.1. What is the soil-water characteristic curve? Draw, semiquantitatively, the characteristic curve for sand, silt, and clay. For the same water content, which of these three soils has the highest value of matric suction? For the same value of matric suction, which soil has the highest water content?

5.2. Assume that a soil's specific gravity G_s is 2.65, the void ratio is 0.91, and the effective stress parameter varies linearly with saturation from $\chi = 0$ at $S = 0$ to $\chi = 1.0$ at $S = 100\%$. What is the relationship between χ and gravimetric water content w? What is the relationship between χ and volumetric water content θ_w? Hint: $Se = G_s w$, and $\theta_w = Se/(1 + e)$.

5.3. a. Calculate a theoretical soil-water characteristic curve for spheres in simple cubic packing order with a uniform particle radius $R = 1$ mm using the following parameters: surface tension $T_s = 74$ mN/m, contact angle $\alpha = 0°$, and $G_s = 2.8$. Filling angle θ ranges from $0°$ to $45°$.
b. Repeat the above procedures for particles with a radius of 0.1 mm.
c. Plot your results in a figure showing matric suction vs. gravimetric water content.

5.4. a. Calculate the theoretical relationship between the effective stress parameter χ and water content for spheres in simple cubic packing order using the following parameters: surface tension $T_s = 74$ mN/m, contact angle $\alpha = 0°$, and $G_s = 2.8$. Filling angle θ ranges from $0°$ to $45°$.
b. Repeat the above procedures for contact angle $\alpha = 20°$.
c. Repeat the above procedures for contact angle $\alpha = 60°$.
d. Does the material variable χ depend on particle size? Is the material variable χ sensitive to contact angle α?

5.5. What are the major mechanisms to explain soil-water hysteresis? Given these mechanisms, explain why the same soil has higher soil suction when drying than when wetting? Will suction stress be higher during a drying process or a wetting process?

5.6. Explain why matric suction cannot be controlled using axis translation for values higher than the air-entry pressure of the HAE material.

5.7. The following stress components are measured at a point in an unsaturated soil:

$$\sigma_x = 200 \text{ kPa} \quad \sigma_y = 250 \text{ kPa} \quad \sigma_z = 400 \text{ kPa}$$

$$u_a = 100 \text{ kPa} \quad u_w = -100 \text{ kPa}$$

where x and y are the horizontal directions, z is the vertical direction, positive stress or pressure is compressive, and negative stress or pressure is tensile.

a. What is the matric suction?
b. What are the stress tensors?
c. Draw Mohr circles in shear stress, matric suction, and net normal stress space.
d. On which plane (i.e., x-y, y-z, or z-x) does the shear stress reach its maximum?
e. If the effective stress parameter $\chi = 0.5$, what are the effective stress components in the x, y, and z directions, respectively?

5.8. If the stress components at the same point as Problem 5.7 change to the following values:

$$\sigma_x = 400 \text{ kPa} \quad \sigma_y = 450 \text{ kPa} \quad \sigma_z = 600 \text{ kPa}$$

$$u_a = 300 \text{ kPa} \quad u_w = 100 \text{ kPa}$$

a. What are the new stress tensors?
b. Does the state of stress change?
c. Would you expect the volume to change?

CHAPTER 6

SHEAR STRENGTH

6.1 EXTENDED MOHR-COULOMB (M-C) CRITERION

6.1.1 M-C for Saturated Soil

Measuring, modeling, and predicting the shear strength of soil are hallmarks of soil mechanics and geotechnical engineering. A solid understanding of shear strength behavior is required for addressing numerous engineering problems where stability of a given soil mass under load is of concern. Examples of these types of problems include bearing capacity, slope stability, lateral earth pressure, pavement design, and foundation design, among many others.

The shear strength of soil, whether saturated or unsaturated, may be defined as the maximum internal resistance per unit area the soil is capable of sustaining along the failure plane under external or internal stress loading. For saturated soil, shear strength is commonly described by the M-C failure criterion, which defines shear strength in terms of the material variables ϕ' and c' and the stress state variable effective stress as

$$\tau_f = c' + (\sigma - u_w)_f \tan \phi' \tag{6.1}$$

where τ_f is the shear stress on the failure plane at failure, c' is the effective cohesion, $(\sigma - u_w)_f$ is the effective normal stress on the failure plane at failure, and ϕ' is the effective angle of internal friction.

As shown in Fig. 6.1, the M-C failure criterion defines a straight line with a slope equal to $\tan \phi'$ and an intercept equal to c' in the space of effective normal stress and shear stress. For cohesionless soil (i.e., $c' = 0$), eq. (6.1) reduces to a line passing through the origin. The M-C criterion is commonly

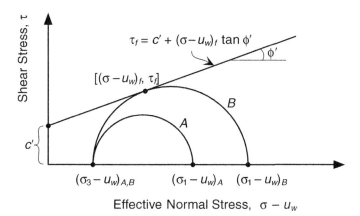

Figure 6.1 Mohr-Coulomb failure envelope for saturated soil. State of stress described by Mohr's circle A is stable. State of stress described by Mohr's circle B represents a failure condition.

referred to as a *failure envelope* because any combination of effective normal stress and shear stress defined by the points along the line corresponds to a failure condition. Accordingly, the shear stress along the failure envelope describes the shear strength of the soil under the corresponding effective normal stress.

Mohr's circles can be drawn to represent the state of normal and shear stress acting on any plane in a soil element. If the Mohr circle for some state of stress falls entirely below the M-C failure envelope, the shear strength has not been exceeded and the soil mass remains stable. Consider, for example, the states of stress defined by circles A and B in Fig. 6.1. Under condition A, the combination of minor and major effective principal stresses $\sigma_3 - u_w$ and $\sigma_1 - u_w$ is such that the soil element remains stable. However, if the major principal stress is increased to the condition described by Mohr's circle B, then failure occurs under the normal and shear stress conditions $(\sigma - u_w)_f$ and τ_f. The orientation of the failure plane may be evaluated by considering the geometry of the Mohr circle.

6.1.2 Experimental Observations of Unsaturated Shear Strength

Modern experimental studies regarding the shear strength of unsaturated soil date back to the 1950s and 1960s. Laboratory tests have most commonly been conducted using triaxial or direct shear testing equipment modified to incorporate pore air pressure control and a high-air-entry (HAE) ceramic disk for control of matric suction by axis translation (Section 5.4). By directly controlling or measuring total normal stress σ, pore air pressure u_a, and pore water pressure u_w under various stress paths and drainage conditions, the dependency of shear strength and volume change behavior on the stress state variables net normal stress $\sigma - u_a$ and matric suction $u_a - u_w$ may be evaluated.

Figure 6.2 illustrates one variation of the basic experimental setup for triaxial testing of unsaturated soil. Similar to conventional triaxial testing, a cylindrical soil specimen is placed on a pedestal in a fluid-filled confining cell, separated from the confining fluid by a flexible membrane. A saturated HAE ceramic disk is placed in good contact with the bottom of the specimen to establish an external hydraulic connection with the pore water. A low-air-entry (coarse) porous disk is placed between the specimen and the specimen top cap to establish a similar connection for external control of the pore air pressure. Filter papers, fibers, or other low-air-entry materials may also be placed along the sides of the specimen to create additional contact area for pore air pressure control. Isotropic stress may be applied by pressurizing the confining fluid. An axial loading ram allows application of deviator stress for shear loading.

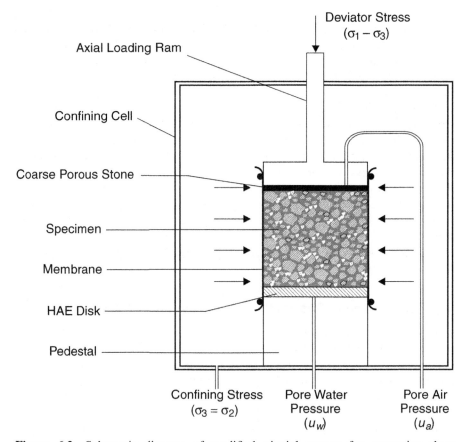

Figure 6.2 Schematic diagram of modified triaxial system for measuring shear strength of unsaturated soil.

6.1 EXTENDED MOHR-COULOMB (M-C) CRITERION

Specimens are typically initially saturated by applying pore water backpressure increments at constant effective stress. The specimen may then be consolidated under isotropic effective confining pressure if desired. Some desired level of matric suction is imposed prior to the shearing phase by elevating the pore air pressure while allowing pore water to drain through the ceramic disk. The corresponding matric suction is measured at equilibrium as the difference between the applied air pressure and the pore water pressure $(u_a - u_w)$, where the latter is directly controlled or measured at some external location. Deviator stress $\sigma_1 - \sigma_3$ and axial or volumetric strain are measured as the specimen is loaded in compression to failure under drained or undrained conditions. Numerous specimens are prepared and tested at various levels of matric suction and net normal stress in order to develop a reliable failure envelope. Alternatively, multistage tests may be conducted for a single specimen whereby the applied stresses are maintained constant or released just prior to failure. Subsequent test stages are conducted by reloading the specimen under different magnitudes of net normal stress and matric suction, thereby maximizing the amount of information that may be gained for one specimen and eliminating the effect of specimen variability on the results.

Figure 6.3 shows an example of data obtained from a series of six consolidated-drained (CD) triaxial tests for specimens of unsaturated silt (Blight, 1967). All specimens were initially compacted by the standard American Association for State Highway and Transportation Organization (AASHTO) method at a molding gravimetric water content of 16.5%. Figure 6.3a shows results in terms of deviator stress $\sigma_1 - \sigma_3$ versus axial strain for three specimens at net confining stress $\sigma_3 - u_a$ of 13.8 kPa and three levels of matric suction $u_a - u_w$: 6.9, 68.9, and 137.9 kPa. Figure 6.3b shows results for three additional specimens at a slightly higher net confining stress of 27.6 kPa and the same three levels of matric suction.

The state of stress at failure for each test can be analyzed following an independent stress tensor approach (Section 5.3). For example, consider the results from the test at net confining stress of 13.8 kPa and matric suction of 6.9 kPa (labeled A on Fig. 6.3a). If the deviator stress at failure is interpreted to be 30 kPa, then the corresponding net normal stress tensor at failure for test A is as follows:

$$\begin{bmatrix} \sigma_1 - u_a & 0 & 0 \\ 0 & \sigma_2 - u_a & 0 \\ 0 & 0 & \sigma_3 - u_a \end{bmatrix}_A = \begin{bmatrix} 43.8 & 0 & 0 \\ 0 & 13.8 & 0 \\ 0 & 0 & 13.8 \end{bmatrix} \text{ kPa}$$

The matric suction tensor at failure for test A is

$$\begin{bmatrix} u_a - u_w & 0 & 0 \\ 0 & u_a - u_w & 0 \\ 0 & 0 & u_a - u_w \end{bmatrix}_A = \begin{bmatrix} 6.9 & 0 & 0 \\ 0 & 6.9 & 0 \\ 0 & 0 & 6.9 \end{bmatrix} \text{ kPa}$$

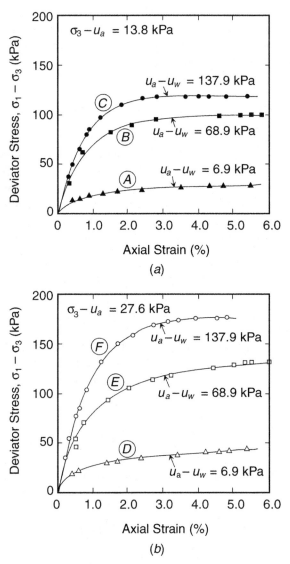

Figure 6.3 Results of consolidated-drained triaxial tests for unsaturated silt (data from Blight, 1967). Series of six tests were conducted: (*a*) tests labeled *A*, *B*, and *C* were conducted at net confining stress ($\sigma_3 - u_a$) of 13.8 kPa and three levels of matric suction ($u_a - u_w$), and (*b*) tests labeled *D*, *E*, and *F* were conducted at net confining stress of 27.6 kPa and the same levels of matric suction as tests *A*, *B*, and *C*.

Similarly, the net normal stress tensor at failure for test D, which was conducted at the same matric suction but increased net confining stress (27.6 kPa), may be interpreted as

$$\begin{bmatrix} \sigma_1 - u_a & 0 & 0 \\ 0 & \sigma_2 - u_a & 0 \\ 0 & 0 & \sigma_3 - u_a \end{bmatrix}_D = \begin{bmatrix} 75 & 0 & 0 \\ 0 & 27.6 & 0 \\ 0 & 0 & 27.6 \end{bmatrix} \text{kPa}$$

and the matric suction tensor for test D is

$$\begin{bmatrix} u_a - u_w & 0 & 0 \\ 0 & u_a - u_w & 0 \\ 0 & 0 & u_a - u_w \end{bmatrix}_D = \begin{bmatrix} 6.9 & 0 & 0 \\ 0 & 6.9 & 0 \\ 0 & 0 & 6.9 \end{bmatrix} \text{kPa}$$

The net normal stress tensors at failure for tests B and E at matric suction equal to 68.9 kPa are

$$\begin{bmatrix} \sigma_1 - u_a & 0 & 0 \\ 0 & \sigma_2 - u_a & 0 \\ 0 & 0 & \sigma_3 - u_a \end{bmatrix}_B = \begin{bmatrix} 113.8 & 0 & 0 \\ 0 & 13.8 & 0 \\ 0 & 0 & 13.8 \end{bmatrix} \text{kPa}$$

$$\cdots \begin{Bmatrix} u_a - u_w \\ u_a - u_w \\ u_a - u_w \end{Bmatrix} = \begin{Bmatrix} 68.9 \\ 68.9 \\ 68.9 \end{Bmatrix} \text{kPa}$$

and

$$\begin{bmatrix} \sigma_1 - u_a & 0 & 0 \\ 0 & \sigma_2 - u_a & 0 \\ 0 & 0 & \sigma_3 - u_a \end{bmatrix}_E = \begin{bmatrix} 155.6 & 0 & 0 \\ 0 & 27.6 & 0 \\ 0 & 0 & 27.6 \end{bmatrix} \text{kPa}$$

$$\cdots \begin{Bmatrix} u_a - u_w \\ u_a - u_w \\ u_a - u_w \end{Bmatrix} = \begin{Bmatrix} 68.9 \\ 68.9 \\ 68.9 \end{Bmatrix} \text{kPa}$$

Finally, the net normal stress tensors at failure for tests C and F at matric suction equal to 137.9 kPa are

$$\begin{bmatrix} \sigma_1 - u_a & 0 & 0 \\ 0 & \sigma_2 - u_a & 0 \\ 0 & 0 & \sigma_3 - u_a \end{bmatrix}_C = \begin{bmatrix} 130.8 & 0 & 0 \\ 0 & 13.8 & 0 \\ 0 & 0 & 13.8 \end{bmatrix} \text{kPa}$$

$$\cdots \begin{Bmatrix} u_a - u_w \\ u_a - u_w \\ u_a - u_w \end{Bmatrix} = \begin{Bmatrix} 137.9 \\ 137.9 \\ 137.9 \end{Bmatrix} \text{kPa}$$

and

$$\begin{bmatrix} \sigma_1 - u_a & 0 & 0 \\ 0 & \sigma_2 - u_a & 0 \\ 0 & 0 & \sigma_3 - u_a \end{bmatrix}_F = \begin{bmatrix} 202.6 & 0 & 0 \\ 0 & 27.6 & 0 \\ 0 & 0 & 27.6 \end{bmatrix} \text{kPa}$$

$$\cdots \begin{Bmatrix} u_a - u_w \\ u_a - u_w \\ u_a - u_w \end{Bmatrix} = \begin{Bmatrix} 137.9 \\ 137.9 \\ 137.9 \end{Bmatrix} \text{kPa}$$

The state of stress at failure for each test may be plotted by considering an extended M-C diagram as shown in Fig. 6.4. The matric suction axis extends orthogonally from the conventional plane of shear stress versus net normal stress. Tests A, B, and C are described by a constant net confining pressure of 13.8 kPa. Tests D, E, and F are described by a constant net confining pressure of 27.6 kPa. The Mohr circles for tests A and D are seated on the matric suction axis at 6.9 kPa. Tests B and E are seated on the matric suction axis at 68.9 kPa. Tests C and F are seated on the matric suction axis at 137.9 kPa.

Figure 6.5 illustrates a variation of the basic experimental setup for shear strength testing using direct shear testing equipment modified for control of matric suction. Here, the specimen is confined by a split box that allows the top half of the specimen to be displaced relative to the bottom half along a prescribed horizontal failure plane. A saturated HAE ceramic disk is installed in the base of the shear box and the entire box is enclosed in an air-tight

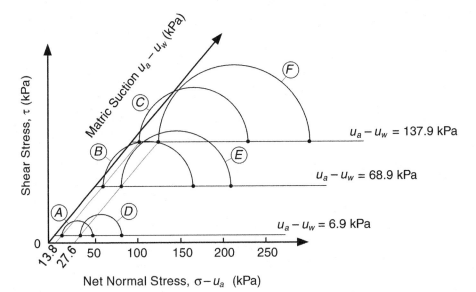

Figure 6.4 Extended Mohr-Coulomb diagram showing state of stress at failure interpreted from Blight's (1967) results for triaxial tests on unsaturated silt.

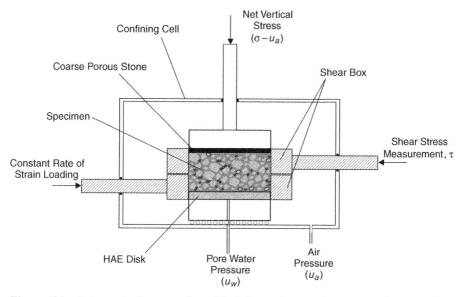

Figure 6.5 Schematic diagram of modified direct shear testing system for measuring shear strength of unsaturated soil.

chamber such that elevated air pressure may be applied. A coarse porous stone in contact with the top of the specimen allows communication between the specimen and the chamber pressure. Pore water pressure is maintained at a lower pressure than the air pressure by axis translation through the HAE ceramic disk, thus allowing matric suction to be controlled and maintained.

For testing, the specimen is usually initially saturated and then consolidated under a vertical normal stress supplied by the axial loading ram. Prior to the shearing phase, matric suction is increased to a desired value by elevating the pore air pressure and measuring/controlling the pore water pressure. Net normal stress and matric suction are measured at equilibrium. Shear stress is imparted by applying horizontal load to the lower half of the shear box at a constant rate of strain. The buildup of shear stress and the shear stress at failure are recorded by monitoring the force mobilized to the top half of the shear box as a function of horizontal strain. As in triaxial testing, numerous specimens may be tested under different confining or matric suction conditions or multistage tests may be conducted whereby shearing is ceased just prior to failure and subsequently loaded under different stress state conditions.

Figure 6.6 shows experimental data reported by Escario (1980) for a series of consolidated-drained direct shear tests on unsaturated Madrid gray clay. Figure 6.6a shows peak shear stress as a function of applied net normal stress for four levels of normal stress at four levels of applied matric suction. Figure 6.6b shows the same data in terms of peak shear stress as a function of matric suction.

228 SHEAR STRENGTH

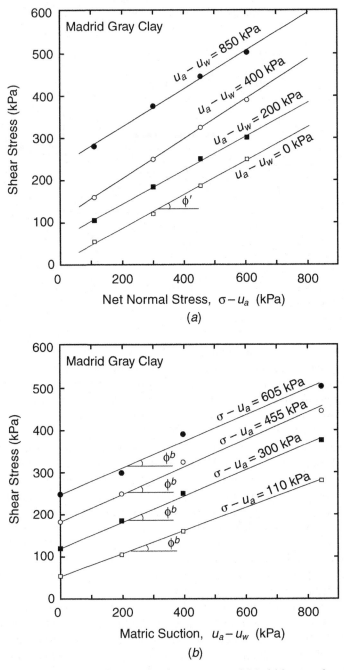

Figure 6.6 Direct shear testing results for unsaturated Madrid gray clay (data from Escario, 1980): (a) peak shear stress as function of net normal stress and (b) peak shear stress as function of matric suction.

6.1.3 Extended M-C Criterion

Inspection of Blight's triaxial testing results (Fig. 6.4) and Escario's direct shear test results (Fig. 6.6) demonstrates two general trends in the shear strength behavior of unsaturated soil. First, as in saturated soil, the shear strength of unsaturated soil generally increases as net normal stress increases. This trend is readily seen by comparing the Mohr circles at failure for triaxial tests A and D, B and E, or C and F (Fig. 6.4) or by direct examination of the direct shear results in Fig. 6.6a. For analysis, this trend is captured in the classical M-C failure criterion by introducing the shear strength parameters cohesion c' and internal friction angle ϕ'. The angle of internal friction may be evaluated from the slope of the failure envelope at zero matric suction, as shown on Fig. 6.6a. Note the failure envelopes at the various levels of applied matric suction for this data are essentially parallel, suggesting that ϕ' is effectively independent of matric suction. The second trend emerging from the triaxial and direct shear testing results is that shear strength increases as applied matric suction increases. This trend is readily seen by comparing the Mohr circles at failure for triaxial tests A, B, and C or for tests D, E, and F (Fig. 6.4) or by direct examination of the direct shear results on Fig. 6.6b. For the range of suction measured in the direct shear testing series, the increase in strength with increasing suction appears to be linear.

As shown in Fig. 6.6b, Fredlund et al. (1978) introduced an additional variable ϕ^b to capture the increase in shear strength with increasing matric suction. Comparison between Figs. 6.6a and 6.6b would indicate that the slopes describing the shear strength versus net normal stress envelopes are larger than those describing shear strength versus matric suction, that is, $\phi' > \phi^b$. Fredlund and Rahardjo (1993) summarized the results of ϕ^b measurements for a variety of soils and showed that ϕ^b indeed appears to be generally smaller than or equal to the internal friction angle ϕ' (Table 6.1).

Fredlund et al. (1978) formulated an extended M-C criterion to describe the shear strength behavior of unsaturated soil. The failure envelope is a planar surface in the space of the stress state variables $\sigma - u_a$ and $u_a - u_w$ and shear stress τ and may be written as

$$\tau_f = c' + (\sigma - u_a)_f \tan \phi' + (u_a - u_w)_f \tan \phi^b \qquad (6.2)$$

where c' is the cohesion at zero matric suction and zero net normal stress, $(\sigma - u_a)_f$ is the net normal stress on the failure plane at failure, ϕ' is the angle of internal friction associated with the net normal stress variable, $(u_a - u_w)_f$ is the matric suction at failure, and ϕ^b is an internal friction angle associated with matric suction that discribes the rate of increase in shear strength relative to matric suction. Table 6.1 summarizes the results of experimental tests for a variety of soils in the literature, indicating the approximate range and variability of these shear strength parameters.

The first two terms on the right-hand side of eq. (6.2) describe the conventional M-C criterion for the strength of saturated soil. The third term

TABLE 6.1 Shear Strength Parameters Measured for Wide Variety of Soil Types

Soil Type	c' (kPa)	ϕ' (deg)	ϕ^b (deg)	References
Compacted shale; $w = 18.6\%$	15.8	24.8	18.1	Bishop et al. (1960)
Boulder clay; $w = 11.6\%$	9.6	27.3	21.7	Bishop et al. (1960)
Dhanauri clay; $w = 22.2\%$, $\rho_d = 1580$ kg/m³	37.3	28.5	16.2	Satija (1978)
Dhanauri clay; $w = 22.2\%$, $\rho_d = 1478$ kg/m³	20.3	29.0	12.6	Satija (1978)
Madrid gray clay; $w = 29\%$	23.7	22.5	16.1	Escario (1980)
Undisturbed decomposed granite	28.9	33.4	15.3	Ho and Fredlund (1982)
Tappen-Notch Hill silt; $w = 21.5\%$, $\rho_d = 1590$ kg/m³	0.0	35.0	16.0	Krahn et al. (1989)
Compacted glacial till; $w = 12.2\%$, $\rho_d = 1810$ kg/m³	10.0	25.3	7–25.5	Gan et al. (1988)

Source: Modified from Fredlund and Rahardjo (1993).

captures the increase in shear strength with increasing matric suction in unsaturated soil. The corresponding failure surface for the extended M-C criterion is illustrated in three-dimensional stress space in Fig. 6.7. The projection of the failure surface for constant matric suctions leads to a series of straight lines in the space of net normal stress and shear stress, illustrated by the direct shear data of Fig. 6.6a. The projection of the failure surface for constant net normal stresses leads to a series of straight lines in the space of matric suction and shear stress, illustrated by the direct shear data of Fig. 6.6b.

For a projection of the failure surface to the shear stress versus net normal stress plane, the extended M-C criterion may be written as

$$\tau_f = c_1' + (\sigma - u_a)_f \tan \phi' \qquad (6.3a)$$

where

$$c_1' = \tau_f|_{(\sigma - u_a) = 0} = c' + (u_a - u_w)_f \tan \phi^b \qquad (6.3b)$$

Similarly, for a projection of the failure surface to the shear stress versus matric suction plane, the extended M-C criterion may be written as

$$\tau_f = c_2' + (u_a - u_w)_f \tan \phi^b \qquad (6.4a)$$

where

$$c_2' = \tau_f|_{(u_a - u_w) = 0} = c' + (\sigma - u_a)_f \tan \phi' \qquad (6.4b)$$

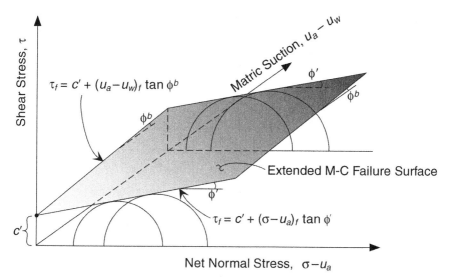

Figure 6.7 Extended Mohr-Coulomb failure surface for unsaturated soil.

6.1.4 Extended M-C Criterion in Terms of Principal Stresses

It is desirable to express the extended M-C criterion in terms of the principal net normal stresses when triaxial tests are used to characterize shear strength behavior. The state of stress at failure for a triaxial test specimen is illustrated in Fig. 6.8. The failure plane is defined at an angle θ from the principal net normal stress in the vertical direction. The net normal stress and shear stress components acting on the failure plane are

$$\sigma_\theta - u_a = \tfrac{1}{2}(\sigma_{1f} + \sigma_{3f}) - u_a + \tfrac{1}{2}(\sigma_{1f} - \sigma_{3f}) \cos 2\theta \qquad (6.5)$$

$$\tau_\theta = \tfrac{1}{2}(\sigma_{1f} - \sigma_{3f}) \sin 2\theta$$

where σ_{1f} and σ_{3f} are the principal total stresses at failure.

From the geometric considerations shown in Fig. 6.9, the angle θ can be related to the angle of internal friction ϕ' as

$$\theta = \tfrac{1}{4}\pi + \tfrac{1}{2}\phi' \qquad (6.6)$$

The extended M-C criterion can also be rewritten in terms of the principal net normal stresses. From Fig. 6.9, it can be shown that

$$\sin \phi' = \frac{DC}{AC} = \frac{(\sigma_1 - \sigma_3)/2}{c_1' \cot \phi' + (\sigma_1 - u_a + \sigma_3 - u_a)/2} \qquad (6.7)$$

Rearranging the above equation leads to

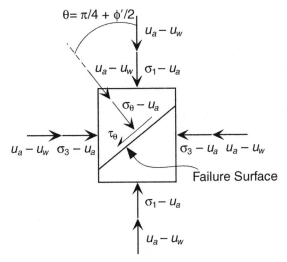

Figure 6.8 State of stress at failure for triaxial test on unsaturated soil.

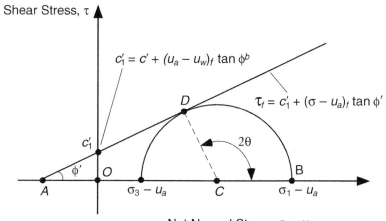

Figure 6.9 Mohr circle representation of failure envelope in space of net normal stress and shear stress.

$$\sigma_1 - u_a = \sigma_3 - u_a + 2c'_1 \sin \phi' \cot \phi' + (\sigma_1 - u_a + \sigma_3 - u_a) \sin \phi'$$

$$\sigma_1 - u_a = \frac{1 + \sin \phi'}{1 - \sin \phi'} (\sigma_3 - u_a) + \frac{\cos \phi'}{1 - \sin \phi'} 2c'_1 \quad (6.8)$$

and since

$$\frac{1 + \sin \phi'}{1 - \sin \phi'} = \tan^2\left(\frac{\pi}{4} + \frac{\phi'}{2}\right) \qquad \frac{\cos \phi'}{1 - \sin \phi'} = \tan\left(\frac{\pi}{4} + \frac{\phi'}{2}\right) \quad (6.9)$$

The extended M-C criterion becomes

$$(\sigma_1 - u_a) = (\sigma_3 - u_a) \tan^2(\tfrac{1}{4}\pi + \tfrac{1}{2}\phi') + 2c'_1 \tan(\tfrac{1}{4}\pi + \tfrac{1}{2}\phi') \quad (6.10a)$$

where

$$c'_1 = c' + (u_a - u_w) \tan \phi^b \quad (6.10b)$$

thus allowing the shear strength parameters ϕ', ϕ^b, and c' to be solved analytically from the results of laboratory tests.

6.2 SHEAR STRENGTH PARAMETERS FOR THE EXTENDED M-C CRITERION

6.2.1 Interpretation of Triaxial Testing Results

The shear strength parameters c', ϕ', and ϕ^b can be determined by conducting laboratory direct shear or triaxial tests. This section contains two example

234 SHEAR STRENGTH

problems to illustrate the interpretation of triaxial and direct shear results and to demonstrate the subsequent construction of the extended M-C failure surface.

Example Problem 6.1 A series of triaxial tests was conducted for four identically prepared specimens of unsaturated silty soil. The following matrices describe the state of stress at failure for each test.

Test 1:

$$\begin{Bmatrix} \sigma_1 - u_a \\ \sigma_2 - u_a \\ \sigma_3 - u_a \end{Bmatrix} = \begin{Bmatrix} 60 \\ 14 \\ 14 \end{Bmatrix} \text{kPa} \cdots \begin{Bmatrix} u_a - u_w \\ u_a - u_w \\ u_a - u_w \end{Bmatrix} = \begin{Bmatrix} 10 \\ 10 \\ 10 \end{Bmatrix} \text{kPa}$$

Test 2:

$$\begin{Bmatrix} \sigma_1 - u_a \\ \sigma_2 - u_a \\ \sigma_3 - u_a \end{Bmatrix} = \begin{Bmatrix} 108 \\ 28 \\ 28 \end{Bmatrix} \text{kPa} \cdots \begin{Bmatrix} u_a - u_w \\ u_a - u_w \\ u_a - u_w \end{Bmatrix} = \begin{Bmatrix} 10 \\ 10 \\ 10 \end{Bmatrix} \text{kPa}$$

Test 3:

$$\begin{Bmatrix} \sigma_1 - u_a \\ \sigma_2 - u_a \\ \sigma_3 - u_a \end{Bmatrix} = \begin{Bmatrix} 115 \\ 14 \\ 14 \end{Bmatrix} \text{kPa} \cdots \begin{Bmatrix} u_a - u_w \\ u_a - u_w \\ u_a - u_w \end{Bmatrix} = \begin{Bmatrix} 70 \\ 70 \\ 70 \end{Bmatrix} \text{kPa}$$

Test 4:

$$\begin{Bmatrix} \sigma_1 - u_a \\ \sigma_2 - u_a \\ \sigma_3 - u_a \end{Bmatrix} = \begin{Bmatrix} 160 \\ 28 \\ 28 \end{Bmatrix} \text{kPa} \cdots \begin{Bmatrix} u_a - u_w \\ u_a - u_w \\ u_a - u_w \end{Bmatrix} = \begin{Bmatrix} 70 \\ 70 \\ 70 \end{Bmatrix} \text{kPa}$$

Determine the shear strength parameters c', ϕ', and ϕ^b from the results of the testing series and construct the extended M-C failure surface.

Solution From eq. (6.10a) and the results of test 1, it can be shown that

$$60 = 14 \tan^2(\tfrac{1}{4}\pi + \tfrac{1}{2}\phi') + 2c_1' \tan(\tfrac{1}{4}\pi + \tfrac{1}{2}\phi') \quad (6.11a)$$

and from test 2 it can be shown that

$$108 = 28 \tan^2(\tfrac{1}{4}\pi + \tfrac{1}{2}\phi') + 2c_1' \tan(\tfrac{1}{4}\pi + \tfrac{1}{2}\phi') \quad (6.11b)$$

Because c_1' depends on matric suction but is independent of the net normal stress, tests 1 and 2 conducted under the same matric suction of 10 kPa have the same c_1'. Therefore, eliminating c_1' from the above equations leads to

6.2 SHEAR STRENGTH PARAMETERS FOR THE EXTENDED M-C CRITERION

$$\frac{48}{14} = \tan^2(\tfrac{1}{4}\pi + \tfrac{1}{2}\phi') \qquad \phi' = 33.25°$$

Substituting this value of ϕ' into eq. (6.11a) leads to

$$c_1' = \frac{60 - 14 \tan^2(61.6°)}{2 \tan(61.6°)} = 3.2 \text{ kPa} \qquad (6.12)$$

Similarly, tests 3 and 4 at matric suction equal to 70 kPa lead to

$$115 = 14 \tan^2(\tfrac{1}{4}\pi + \tfrac{1}{2}\phi') + 2c_1' \tan(\tfrac{1}{4}\pi + \tfrac{1}{2}\phi') \qquad (6.13a)$$

and

$$160 = 28 \tan^2(\tfrac{1}{4}\pi + \tfrac{1}{2}\phi') + 2c_1' \tan(\tfrac{1}{4}\pi + \tfrac{1}{2}\phi') \qquad (6.13b)$$

where eliminating the common parameter c_1' leads to

$$\frac{45}{14} = \tan^2(\tfrac{1}{4}\pi + \tfrac{1}{2}\phi') \qquad \phi' = 31.70°$$

The fact that similar values of ϕ' are computed for both levels of matric suction supports the previous observation from Escario's direct shear data (Fig. 6.6a) that the angle of internal friction is essentially independent of matric suction. The average friction angle is $(33.25° + 31.70°) = 32.47°$. Substituting this value into eq. (6.13a) leads to

$$c_1' = \frac{115 - 14 \tan^2(61.2°)}{2 \tan(61.2°)} = 18.8 \text{ kPa} \qquad (6.14)$$

Substituting eqs. (6.12) and (6.14) and the corresponding matric suction values of 10 and 70 kPa back to eq. (6.10b) results in

$$3.2 = c' + 10 \tan \phi^b \qquad 18.8 = c' + 70 \tan \phi^b$$

Solving the above equations leads to

$$\phi^b = 14.6° \qquad c' = 0.6 \text{ kPa}$$

The states of stress at failure may be plotted as Mohr's circles in the three-dimensional space of shear stress, matric suction, and net normal stress. These circles, as well as the corresponding planar failure surface described by the material variables ϕ', ϕ^b, and c' are shown as an extended M-C diagram in Fig. 6.10.

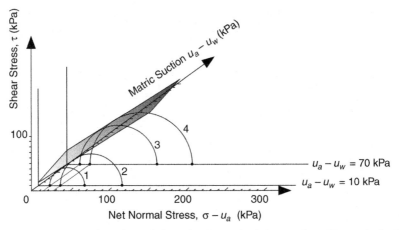

Figure 6.10 States of stress at failure for four triaxial tests from Example Problem 6.1 and corresponding extended Mohr-Coulomb failure surface.

6.2.2 Interpretation of Direct Shear Testing Results

Example Problem 6.2 A series of direct shear tests modified for matric suction control was conducted for four identically prepared specimens of unsaturated silty soil. The states of stress at failure for each test are as follows:

Test 1:

$$u_a - u_w = 0 \text{ kPa} \qquad \tau_f = 65 \text{ kPa} \qquad (\sigma - u_a)_f = 110 \text{ kPa}$$

Test 2:

$$u_a - u_w = 0 \text{ kPa} \qquad \tau_f = 160 \text{ kPa} \qquad (\sigma - u_a)_f = 300 \text{ kPa}$$

Test 3:

$$u_a - u_w = 400 \text{ kPa} \qquad \tau_f = 185 \text{ kPa} \qquad (\sigma - u_a)_f = 110 \text{ kPa}$$

Test 4:

$$u_a - u_w = 400 \text{ kPa} \qquad \tau_f = 285 \text{ kPa} \qquad (\sigma - u_a)_f = 300 \text{ kPa}$$

Determine the shear strength parameters c', ϕ', and ϕ^b from the results of the testing series and construct the extended M-C failure surface.

Solution Substituting the failure states of stress for tests 1 and 2 into eq. (6.3a) results in

$$65 = c_1' + 110 \tan \phi' \qquad 160 = c_1' + 300 \tan \phi'$$

which may be solved to give values of ϕ' and c_1' at $u_a - u_w = 0$ as follows:

$$\phi' = 26.6° \qquad c_1' = 10.0 \text{ kPa}$$

Similarly, substituting the failure stress states for tests 3 and 4 into eq. (6.3a) results in

$$185 = c_1' + 110 \tan \phi' \qquad 285 = c_1' + 300 \tan \phi'$$

which may be solved to give values of ϕ' and c_1' at $u_a - u_w = 400$ kPa as

$$\phi' = 27.8° \qquad c_1' = 127.1 \text{ kPa}$$

Therefore, the average friction angle is as follows: $\phi' = \frac{1}{2}(26.6° + 27.8°) = 27.2°$. Substituting $\phi' = 27.2°$, $c_1' = 10.0$ kPa for $u_a - u_w = 0$, and $c_1' = 127.1$ kPa for $u_a - u_w = 400$ kPa back into eq. (6.3b) leads to

$$10.0 = c' + 0 \tan \phi^b \qquad 127.1 = c' + 400 \tan \phi^b$$

which leads to

$$c' = 10.0 \text{ kPa} \qquad \phi^b = 16.3°$$

The Mohr's circle representation of the state of stress for each test at failure and the corresponding extended M-C failure surface are shown in Fig. 6.11.

Alternatively, the angle ϕ^b may be obtained by considering the state of stress at failure for a constant net normal stress by considering eqs. (6.4a) and (6.4b).

Substituting the stresses at failure for tests 1 and 3 into these equations leads to

$$65 = c_2' + 0 \qquad 185 = c_2' + 400 \tan \phi^b$$

or

$$c_2' = 65 \text{ kPa} \qquad \phi^b = 16.7°$$

Substituting the stresses at failure for tests 2 and 4 leads to

$$160 = c_2' \qquad 285 = c_2' + 400 \tan \phi^b$$

or

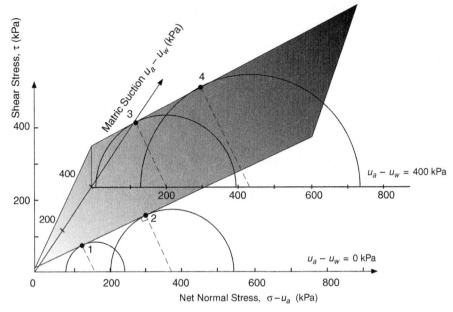

Figure 6.11 States of stress at failure for four direct shear tests from Example Problem 6.2 and corresponding extended Mohr-Coulomb failure surface.

$$c'_2 = 160 \text{ kPa} \qquad \phi^b = 17.4°$$

Equation (6.4b) and the results of tests 1 and 2 lead to

$$65 = c' + 110 \tan \phi' \qquad 160 = c' + 300 \tan \phi'$$

or

$$c' = 9.7 \text{ kPa} \qquad \phi' = 26.6°$$

6.3 EFFECTIVE STRESS AND THE M-C CRITERION

6.3.1 Nonlinearity in the Extended M-C Envelope

Despite the simplicity of the extended M-C criterion for describing the strength of unsaturated soil, several important factors may limit its general validity over a wide range of matric suction. Most notably, there is significant experimental evidence showing that the angle describing the increase in shear strength with respect to matric suction ϕ^b is a highly nonlinear function of matric suction (e.g., Gan et al., 1988; Escario et al., 1989; Vanapalli et al., 1996). The value of ϕ^b for a given soil can vary from a value equal or close to the internal friction angle ϕ' for suctions near zero (i.e., near the saturated condition) to as low as 0° or even negative values for suctions approaching the residual saturation state.

Figures 6.12a and 6.12b show examples of experimental results demonstrating nonlinear behavior in ϕ^b. Both sets of data are in the form of shear strength (peak shear stress) as a function of matric suction, each determined using direct shear systems modified for control of matric suction by axis translation. The parameter ϕ^b describes the slope of the failure envelopes. Note that a regime of apparent softening behavior is observed for both soils where the value of ϕ^b either decreases (Fig. 6.12a) or becomes negative (Fig. 6.12b).

There is a direct correspondence between the nonlinear nature of the shear strength envelope with respect to increasing matric suction and the behavior of the soil-water characteristic curve. Figure 6.13, for example, shows a con-

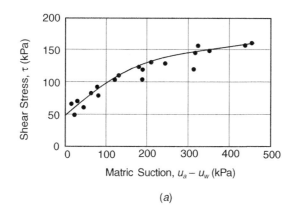

(a)

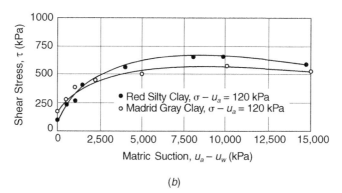

(b)

Figure 6.12 Examples of nonlinear behavior in relationship between shear strength and matric suction: (a) data from modified direct shear tests for unsaturated glacial till (Gan et al., 1988) and (b) data from modified direct shear tests for two clayey materials (Escario et al., 1989).

ceptualized soil-water characteristic curve along a drainage path (Fig. 6.13a) and the corresponding shear strength envelope with respect to increasing matric suction (Fig. 6.13b) for a typical soil. Within the regime of relatively low matric suction and prior to the air-entry pressure, the soil pores remain essentially saturated, the shear strength envelope is approximately linear, and ϕ^b is effectively equal to the angle of internal friction ϕ'. The contribution of matric suction to shear strength within this range may be treated within the conventional M-C framework for saturated soil where the pore pressure is in this case a negative value. Beyond the air-entry pressure, however, a regime of nonlinear behavior commences that corresponds to drainage of the soil pores. As drainage continues through this zone of desaturation, the geometries of the interparticle pore water menisci dramatically change, thus affecting the resultant interparticle forces that contribute to stress on the soil

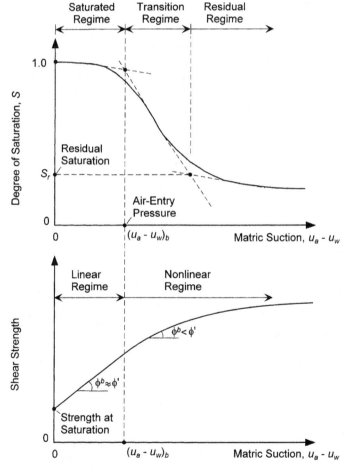

Figure 6.13 Conceptual relationship between soil-water characteristic curve and unsaturated shear strength envelope (modified from Vanapalli et al., 1996).

skeleton and ultimately contribute to shear strength. The reduction in the volume of pore water within this regime effectively reduces the contribution that matric suction has toward increasing shear strength. The micromechanical analyses presented in Chapters 4 and 5 for idealized two-particle systems provide basic insight into the evolution of interparticle forces for a limited portion of the desaturation regime. At the present time, however, the interparticle forces contributing to shear strength over a wide range of the desaturation regime for real soil systems remain poorly understood.

In engineering scenarios where anticipated suction values are expected to extend beyond the regime where ϕ^b may be considered independent of suction, the general validity and applicability of the extended M-C approach begins to come into question. For analysis purposes, Fredlund et al. (1987) suggest that the nonlinearity in the relationship between shear strength and matric suction may be handled in one of several ways: (1) by dividing the failure envelope into two linear portions, the first extending from the point of saturation (zero suction) to the air-entry pressure with a slope equal to ϕ', and the second extending beyond the air-entry pressure with a slope equal to ϕ^b, (2) by neglecting the nonlinearity and adopting a conservative envelope over the entire suction range with a slope equal to ϕ^b, where $\phi^b < \phi'$, or (3) by discretizing the envelope into several linear segments with varying ϕ^b angles.

6.3.2 Effective Stress Approach

One logical way to describe the dependency of shear strength on matric suction is to follow the classical soil mechanics formalism using effective stress and the conventional M-C failure criterion. A practical advantage of the effective stress approach is that it remains firmly within the context of classical soil mechanics, thus requiring minimum modification to the existing elastoplastic theories of stress-strain or constitutive laws that have been implemented in many numerical codes.

As initially proposed by Bishop (1959), effective stress in unsaturated soil can be defined by jointly using the independent state variables: net normal stress $\sigma - u_a$ and matric suction $u_a - u_w$, and one material variable: the effective stress parameter χ. As introduced in previous chapters, Bishop's effective stress is

$$\sigma' = (\sigma - u_a) + \chi(u_a - u_w) \tag{6.15}$$

The effective stress parameter χ is a function of the degree of saturation of the soil mass and reflects the contribution of matric suction to effective stress. For saturated soil, the air pressure is equal to zero, the water pressure is compressive or positive, χ is equal to one, and eq. (6.15) reduces to Terzaghi's classical effective stress equation: $\sigma' = \sigma - u_w$. For completely dry soil, χ is equal to zero and the effective stress is the difference between total stress and air pressure: $\sigma' = \sigma - u_a$. For partially saturated soil, χ is some function of the degree of saturation or matric suction.

Research over the past 30 years has demonstrated that capturing the dependency of the effective stress parameter χ on the degree of saturation or suction is an extremely challenging task. Theoretical endeavors, for example, have focused primarily on considerations of meniscus geometry for simple capillary models employing the Young-Laplace equation to connect an idealized meniscus geometry and volume (or degree of saturation) to capillary stress and matric suction. One such model was developed in Sections 4.5, 5.1, and 5.2 for spherical coarse-grained particles in simple cubic and tetrahedral packing order. An important limitation of these types of theoretical analyses, however, is that they are only valid for the regime of saturation where the meniscus geometry is well defined. For the development in Chapters 4 and 5, this regime corresponds to the pendular regime of discontinuous water menisci and a degree of saturation less than about 25%. For higher degrees of saturation, some empirical relationships have been suggested, but no analytical relation between the χ parameter and the degree of saturation has been reported.

The effective stress parameter χ may not be directly measured or controlled through experiments. However, Bishop (1954) proposed an indirect way to obtain χ from the stresses measured in soil specimens at failure. The traditional M-C criterion was used to represent the failure conditions:

$$\tau_f = c' + \sigma' \tan \phi' \tag{6.16a}$$

which, after substitution of the effective stress expression (6.15), leads to

$$\tau_f = c' + [(\sigma - u_a)_f + \chi_f (u_a - u_w)_f] \tan \phi' \tag{6.16b}$$

where τ_f is shear strength and c' and ϕ' are the effective cohesion and friction angle, respectively.

6.3.3 Measurements of χ at Failure

In a typical direct shear test the net total stress $(\sigma - u_a)$ is known and the net effective stress can be deduced from the shear stress at failure. Hence, by measuring or controlling the matric suction variable $u_a - u_w$, the effective stress parameter χ can be evaluated by rearranging eq. (6.16b)

$$\chi_f = \frac{\tau_f - c' - (\sigma - u_a)_f \tan \phi'}{(u_a - u_w)_f \tan \phi'} \tag{6.17}$$

In a typical triaxial test, the principal net normal stresses $\sigma_1 - u_a$ and $\sigma_3 - u_a$ and matric suction $u_a - u_w$ are known and the M-C criterion can be written as

$$(\sigma_1 - u_a)_f = (\sigma_3 - u_a)_f \tan^2(\tfrac{1}{4}\pi + \tfrac{1}{2}\phi') + 2c_1' \tan(\tfrac{1}{4}\pi + \tfrac{1}{2}\phi') \quad (6.18a)$$

where

$$c_1' = c' + \chi_f(u_a - u_w) \tan \phi' \quad (6.18b)$$

Rearranging eq. (6.18a) leads to

$$\chi_f = \frac{(\sigma_1 - u_a)_f - (\sigma_3 - u_a)_f \tan^2(\pi/4 + \phi'/2) - 2c' \tan(\pi/4 + \phi'/2)}{2(u_a - u_w) \tan(\pi/4 + \phi'/2) \tan \phi'} \quad (6.19)$$

Since the matric suction at failure may be used to indirectly define the degree of saturation by way of the soil-water characteristic curve, a one-to-one relationship between χ and degree of saturation can be established. Following this general strategy, Bishop (1959) proposed a nonlinear form of χ based on direct shear tests taken to failure, shown as a function of degree of saturation in Fig. 6.14.

Other measurements and mathematical representations of χ have been reported as a function of the degree of saturation or as a function of matric suction. Based on a best fit to the experimental data presented in Fig. 6.15a, for example, Khalili and Khabbaz (1998) proposed a form of χ as a function of *suction ratio* $(u_a - u_w)/u_e$ as follows:

$$\chi = \begin{cases} \left(\dfrac{u_a - u_w}{u_e}\right)^{-0.55} & \text{for } u_a - u_w > u_e \\ 1 & \text{for } u_a - u_w \leq u_e \end{cases} \quad (6.20)$$

where u_e is a suction value marking the transition between saturated and unsaturated states, being the air-expulsion pressure for a wetting process and the air-entry pressure for a drying process. A fit of eq. (6.20) to the experimental data shown in Fig. 6.15a is illustrated in Fig. 6.15b.

The validity of several forms of χ as a function of the degree of saturation was also examined by Vanapalli and Fredlund (2000) using a series of shear strength test results for statically compacted mixtures of clay, silt, and sand from Escario and Juca (1989). For matric suction ranging between 0 and 1500 kPa, the following two forms showed good fit to the experimental results:

$$\chi = S^\kappa = \left(\frac{\theta}{\theta_s}\right)^\kappa \quad (6.21)$$

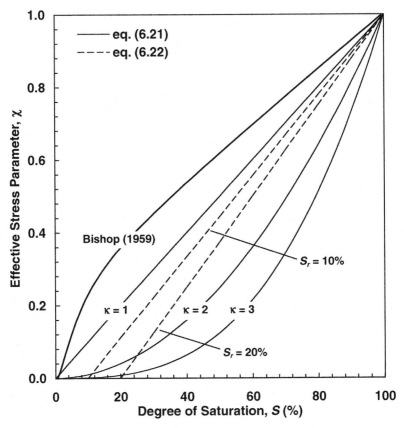

Figure 6.14 Various forms for effective stress parameter χ as function of degree of saturation.

where S is the degree of saturation, θ is volumetric water content, θ_s is the saturated volumetric water content, and κ is a fitting parameter optimized to obtain a best fit between measured and predicted values, and

$$\chi = \frac{S - S_r}{1 - S_r} = \frac{\theta - \theta_r}{\theta_s - \theta_r} \tag{6.22}$$

where θ_r is residual volumetric water content and S_r is residual degree of saturation. The nature of eqs. (6.21) and (6.22) is illustrated in Fig. 6.14 for several values of κ and S_r.

6.3.4 Reconciliation between ϕ^b and χ_f

From eq. (6.15), the effective stress at failure for unsaturated soil can be expressed as

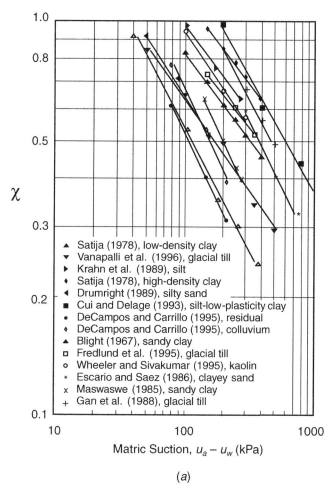

Figure 6.15 Effective stress parameter χ as function of (*a*) matric suction and (*b*) matric suction ratio (after Khalili and Khabbaz, 1998).

$$\sigma'_f = (\sigma_n - u_a)_f + \chi_f(u_a - u_w)_f \tag{6.23}$$

If effective stress is considered the state variable for shear strength, the M-C criterion can be rewritten as

$$\begin{aligned}\tau_f &= c' + \sigma'_n \tan \phi' \\ &= c' + [(\sigma_n - u_a)_f + \chi_f(u_a - u_w)] \tan \phi' \\ &= c' + (\sigma_n - u_a)_f \tan \phi' + \chi_f(u_a - u_w) \tan \phi'\end{aligned} \tag{6.24}$$

Comparing the above equation with the extended M-C failure criterion [eq. (6.21)] in terms of the angle ϕ^b leads to

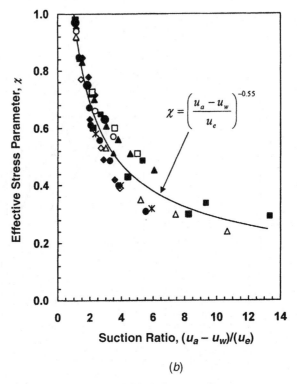

(b)

Figure 6.15 (*Continued*).

$$\tan \phi^b = \chi_f \tan \phi' = f_1(u_a - u_w) = f_2(S) \quad (6.25)$$

where f_1 and f_2 represent functional relationships between ϕ^b and $u_a - u_w$ and ϕ^b and S, respectively. Taking eqs. (6.21) and (6.22), for example, it can be shown that

$$\tau_f = c' + (\sigma_n - u_a)_f \tan \phi' + (u_a - u_w)_f \left(\frac{\theta}{\theta_s}\right)^\kappa \tan \phi' \quad (6.26)$$

or

$$\tau_f = c' + (\sigma_n - u_a)_f \tan \phi' + (u_a - u_w)_f \frac{\theta - \theta_r}{\theta_s - \theta_r} \tan \phi' \quad (6.27)$$

Comparing the above two equations with the extended M-C criterion [eq. (6.2)]

$$\tau_f = c' + (\sigma - u_a)_f \tan \phi' + (u_a - u_w)_f \tan \phi^b \qquad (6.2)$$

it becomes clear that

$$\tan \phi^b = \left(\frac{\theta}{\theta_s}\right)^\kappa \tan \phi' \qquad (6.28)$$

or

$$\tan \phi^b = \frac{\theta - \theta_r}{\theta_s - \theta_r} \tan \phi' \qquad (6.29)$$

Substituting eq. (6.25) into the above equations leads to two specific functional forms for the effective stress parameter:

$$\chi_f = \left(\frac{\theta}{\theta_s}\right)^\kappa \qquad (6.30)$$

$$\chi_f = \frac{\theta - \theta_r}{\theta_s - \theta_r} \qquad (6.31)$$

which are eqs. (6.21) and (6.22).

Note that for χ equal to 1.0, eq. (6.25) dictates that the angle ϕ^b be equal to the angle ϕ', which corresponds to conditions near saturation and the linear portion of the shear strength envelope with respect to matric suction (Fig. 6.13b). For χ less than 1.0, eq. (6.25) dictates that the angle ϕ^b be less than ϕ', which corresponds to unsaturated conditions and the nonlinear portion of the shear strength envelope with respect to matric suction.

6.3.5 Validity of Effective Stress as a State Variable for Strength

Much experimental evidence such as the null tests described in Section 5.4.5 indicates that the behavior of unsaturated soil can be effectively described by the effective stress principle. Application of the effective stress principle to unsaturated soil was recently examined by Khalili et al. (2004), where several sets of experimental data were interpreted to assess the validity of the effective stress equation [eq. (6.15)] proposed by Bishop (1959) and the effective stress parameter function [eq. (6.20)] proposed by Khalili and Khabbaz (1998). It was shown that the critical state line (CSL) can be used to uniquely describe the state of stress at failure in the mean effective stress–deviatoric stress ($p' - q'$) plane for both saturated and unsaturated states. Mean effective stresses at failure were calculated from suction-controlled triaxial shear strength testing data using eqs. (6.15) and (6.20) and the known matric suction values. Deviatoric stresses were calculated using eq. (6.15) and the expression

$$q' = a + [(p - u_a) + \chi(u_a - u_w)]M \qquad (6.32)$$

where

$$a = \frac{6c' \cos \phi'}{3 - \sin \phi'} \qquad (6.33a)$$

$$p = \tfrac{1}{2}(\sigma_1 + \sigma_3) \qquad (6.33b)$$

$$M = \frac{6 \sin \phi'}{3 - \sin \phi'} \qquad (6.33c)$$

Figure 6.16 illustrates the state of stress at failure in $p' - q'$ space for four different soils: kaolin (Wheeler and Sivakumar, 1995), Trois-Rivieres silt (Maâtouk et al., 1995), Sion silt (Geiser, 1999), and Jossigny silt (Cui and Delage, 1996). The results show maximum deviatoric stress (peak shear strength) versus the mean effective stress at several different values of matric suction. As shown, most data points plot close to the critical state line, defined using saturated test data, indicating that a unique failure envelope exists for each of these soils if the effective stress approach is employed.

6.4 SHEAR STRENGTH PARAMETERS FOR THE M-C CRITERION

6.4.1 Interpretation of Direct Shear Testing Results

This section describes two example problems to illustrate procedures for determining shear strength parameters from unsaturated shear strength experiments and the conventional (nonextended) M-C failure criterion.

Example Problem 6.3 A series of direct shear and soil-water characteristic curve tests were conducted for an unsaturated clay-sand mixture. Results are tabulated in Table 6.2. Determine the shear strength parameters c' and ϕ'. Determine the functional relationship between the effective stress parameter χ and the degree of saturation S. Plot matric suction and the effective stress parameter as functions of saturation.

Solution From tests 1 and 2, the cohesion intercept and friction angle under saturated conditions may be calculated as $c' = 40$ kPa and $\phi' = 40°$, respectively. Equation (6.17) can be used to calculate χ for each test, as tabulated in Table 6.3. It is assumed that c' and ϕ' remain constant for all degrees of saturation. The soil-water characteristic curve is plotted in Fig. 6.17a. The effective stress parameter χ is shown as a function of saturation in Fig. 6.17b. As shown, the effective stress parameter function $\chi(S)$ follows the concave

6.4 SHEAR STRENGTH PARAMETERS FOR THE M-C CRITERION   249

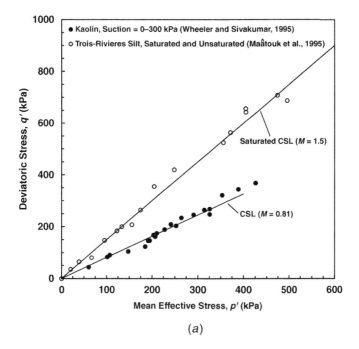

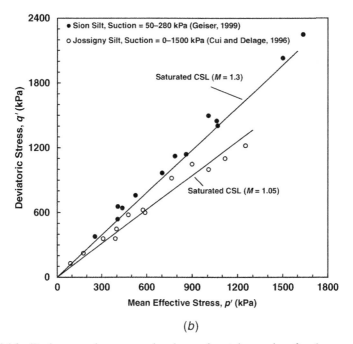

Figure 6.16 Peak strength at several values of matric suction for four unsaturated soils in $p' - q'$ space: (*a*) kaolin (Wheeler and Sivakumar, 1995) and Trois-Rivieres silt (Maatouk et al., 1995) and (*b*) Sion silt (Geiser, 1999) and Jossigny silt (Cui and Delage, 1996). (Data from Khalili et al., 2004.)

TABLE 6.2 Results of Shear Strength and Soil-Water Characteristic Curve Tests for Example Problem 6.3

Test	$u_a - u_w$ (kPa)	S	$\sigma - u_a$ (kPa)	τ (kPa)
1	0	1.0	300	292
2	0	1.0	120	141
3	25	0.98	120	160
4	500	0.78	120	300
5	1,000	0.71	120	400
6	2,000	0.64	120	460
7	5,000	0.45	120	500
8	11,000	0.38	120	590
9	15,000	0.35	120	530
10	50,000	0.2	120	525

upward pattern noted previously in Fig. 6.14 and is closely matched by eq. (6.21) if $\kappa = 3.5$.

6.4.2 Interpretation of Triaxial Testing Results

Example Problem 6.4 A series of triaxial tests was conducted on specimens of unsaturated sandy soil. Results for tests under different levels of matric suctions are tabulated in Table 6.4. Determine c' and ϕ' and plot the functional relationship between the effective stress parameter and matric suction.

Solution The cohesion and friction angle under saturated conditions may be calculated from test 1 as $c' = 0$ kPa and $\phi' = 37°$. Again, it is assumed that c' and ϕ' remain constant for all degrees of saturation. Equation (6.19) can

TABLE 6.3 Effective Stress Parameters Calculated for Example Problem 6.3

Test	$u_a - u_w$ (kPa)	S	$\sigma - u_a$ (kPa)	τ (kPa)	χ
1	0	1.0	300	292	1.000
2	0	1.0	120	141	1.000
3	25	0.98	120	160	0.920
4	500	0.78	120	300	0.380
5	1,000	0.71	120	400	0.309
6	2,000	0.64	120	460	0.190
7	5,000	0.45	120	500	0.086
8	11,000	0.38	120	590	0.049
9	15,000	0.35	120	530	0.031
10	50,000	0.2	120	525	0.009

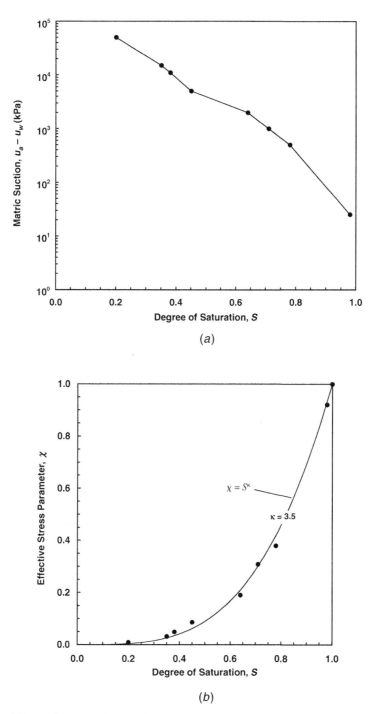

Figure 6.17 Soil-water characteristic curve and effective stress parameter function for Example Problem 6.3: (a) $\psi(S)$ and (b) $\chi(S)$.

TABLE 6.4 Results of Shear Strength Tests for Example Problem 6.4

Test	$u_a - u_w$ (kPa)	$\sigma_1 - u_a$ (kPa)	$\sigma_3 - u_a$ (kPa)
1	0	200	50
2	20	250	50
3	50	280	50
4	200	290	50
5	300	300	50
6	400	310	50
7	500	320	50
8	800	340	50

be used to calculate χ for each test, as plotted in Fig. 6.18. As shown, the relationship is generally linear in log-log scale, as also noted by the data in Fig. 6.15a.

6.5 UNIFIED REPRESENTATION OF FAILURE ENVELOPE

6.5.1 Capillary Cohesion as a Characteristic Function for Unsaturated Soil

In the extended M-C criterion presented in Section 6.1, matric suction is used as an independent stress state variable along with net normal stress for describing unsaturated shear strength. On one hand, introducing the constant internal friction angle with respect to matric suction ϕ^b provides a relatively simple mathematical and graphical representation of the shear strength envelope. On the other hand, the experimental evidence described in Section 6.3.1 necessitates consideration of the strong functional dependency between ϕ^b and matric suction for suctions greater than the air-entry pressure. In this light, the practical advantages of the extended M-C criterion may no longer be sufficient to warrant its applicability for the expected range of matric suction in practical problems. To fully describe the shear strength behavior of unsaturated soil over a realistically wide range, one must appreciate the nonlinear nature of the relationship between strength and suction and construct a nonplanar failure surface in three-dimensional space of shear strength, matric suction, and net normal stress. It has become increasingly clear that material variables are required to do so.

Reconciliation between Bishop's effective stress concept and the extended M-C criterion can be achieved by an alternative approach for describing the state of stress and strength in unsaturated soil. The M-C criterion incorporating both Bishop's effective stress and the suction stress concept expressed by eq. (6.16b) can be rearranged as

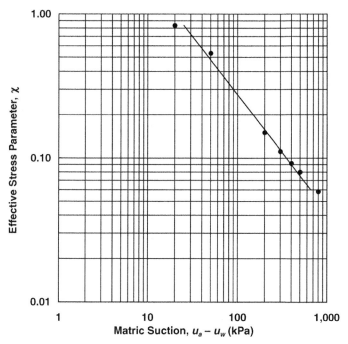

Figure 6.18 Effective stress parameter as a function of matric suction from Example Problem 6.4.

$$\tau_f = c' + \chi_f(u_a - u_w)_f \tan \phi' + (\sigma - u_a)_f \tan \phi'$$
$$= c' + c'' + (\sigma - u_a)_f \tan \phi' \quad (6.34a)$$

where

$$c'' = \chi_f(u_a - u_w)_f \tan \phi' \quad (6.34b)$$

The first two terms in eq. (6.34a), c' and c'', represent shear strength due to so-called *apparent cohesion* in unsaturated soil. As in saturated soil, the third term represents frictional shearing resistance provided by the effective normal force at the grain contacts. The apparent cohesion captured by the first two terms includes the classical cohesion c' representing shearing resistance arising from interparticle physicochemical forces such as van der Waals attraction, and a second term c'' describing shearing resistance arising from capillarity effects. The term c'' is defined as *capillary cohesion* hereafter. Physically, capillary cohesion describes the mobilization of suction stress $[\chi(u_a - u_w)]$ in terms of shearing resistance. The relationship between capillary cohesion and the maximum suction stress at failure, $\chi_f(u_a - u_w)_f$, is defined by eq. (6.34b).

The concepts of suction stress and capillary cohesion may be better illustrated by plotting eq. (6.34) in the three-dimensional space of shear stress, net normal stress, and suction stress, as shown in Fig. 6.19. Net normal stress in this regard is an independent stress state variable and suction stress is a material variable. One unique feature of graphical representation in this space is that the failure surface remains planar no matter whether the soil is unsaturated or saturated. This feature makes it possible to represent the entire failure surface in the net normal stress and shear stress plane by plotting constant suction stress lines, leading to a series of parallel lines with different values of suction stress. For example, Fig. 6.20 shows this projection for two values of suction stress: one for zero suction stress and one for an arbitrary nonzero value. Capillary cohesion associated with the nonzero suction stress (the upper envelope) is apparent from the intersection of the envelope with the shear stress axis. The total intercept value is equal to $c' + c''$, where c' is defined by the intercept of the failure envelope at zero suction stress, and c'' is the additional resistance evident in the nonzero suction stress envelope. The intersections with the net normal stress axis in either case define the tensile strength of the soil. The parallel nature of these two failure lines and the simple relationship between the maximum suction stress and capillary cohesion c'' make the graphical representation instructive for shear strength interpretation.

The definitions of suction stress and capillary cohesion are logical extensions of the classical M-C criterion and Terzaghi's effective stress principle. Physically, suction stress is an internal stress that results specifically from the partial saturation of the soil. Suction stress is independent of external loading

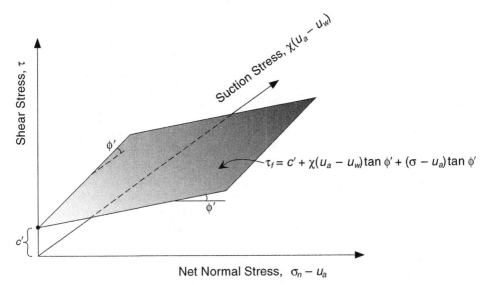

Figure 6.19 Shear strength surface in space of net normal stress, suction stress, and shear stress.

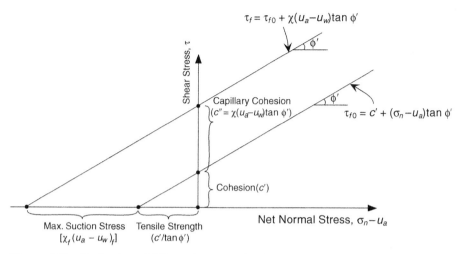

Figure 6.20 Projection of failure shown in Fig. 6.19 on shear stress–net normal stress plane.

or overburden pressure. Rather, suction stress originates from the combined effects of negative pore water pressure and surface tension, as was formally introduced and derived from a micromechanical framework in Sections 4.5 and 5.1. Capillary cohesion is the contribution of the maximum suction stress to the apparent cohesion. As an analog to the soil-water characteristic curve, capillary cohesion may be considered as a strength characteristic curve for unsaturated soil.

There are several advantages to introducing the concept of suction stress over independently considering the variables that define it (i.e., the effective stress parameter and matric suction). First, it is clear that not all matric suction contributes directly to stress acting on the soil skeleton. For instance, 400 kPa of suction may contribute a negligible amount of effective stress to relatively coarse sands, which are likely to be nearly dry at this level of suction, but may be contributed in its entirety to effective stress in clay, which is likely near saturation at this point. Unless the soil remains a two-phase saturated system (i.e., remains at some value of suction less than the air-entry pressure), matric suction cannot be strictly considered a material variable independent stress state variable. Second, in the framework established thus far for unsaturated fluid flow, matric suction is used as a governing state variable and is dependent on water content or the degree of saturation as cast in the soil-water characteristic curve. From a thermodynamic viewpoint, matric suction represents the energy level or potential of the soil pore water. Matric suction is by nature a variable controlled by state variables such as temperature and water content. In this regard, it is inconsistent to define matric suction as an independent state variable. Third, the nonlinear nature of the friction angle with respect to matric suction ϕ^b makes it practically difficult to construct a comprehensive failure surface either experimentally, graphically, or mathematically. Even though it may be accomplished, such surfaces are very difficult to incorporate into analyses as has been readily done for the classical linear M-C failure criterion. Fourth, suction stress is fundamentally a stress that physically exists between soil particles. Much like a spring in tension, suction stress is a force that pulls particles together. The magnitude of suction stress contributes directly to effective stress with no reservation, as demonstrated quantitatively in Sections 4.5 and 5.1. Consideration of suction stress or capillary cohesion as a function of water content and defining it as a stress or strength characteristic curve echoes the conceptualization of the soil-water characteristic curve.

Three additional points speak to the practical advantages of adopting suction stress or capillary cohesion as material variables. First, the suction stress representation requires neither matric suction, nor χ, nor ϕ^b for describing the state of stress or shear strength. Uncertainties and ambiguities in the theoretical formulation and experimental determination of χ and ϕ^b are avoided. Second, the suction stress concept preserves the simplicity and linearity of the classical M-C criterion. The shear strength of soil under unsaturated soil conditions may, as illustrated in the following, be analyzed entirely within the classical framework of saturated soil mechanics. And finally, by remaining within the classical M-C framework, modifications to the existing limit analyses that form the basis for most geotechnical design and analysis are minimized. This notion will be demonstrated throughout Chapter 7 in the context of active and passive earth pressure theory.

6.5.2 Determining the Magnitude of Capillary Cohesion

A key question following the unified representation of the M-C criterion for unsaturated soil becomes how to determine the magnitude of capillary cohesion from the results of a given shear strength test. The following demonstrates these procedures for direct shear and triaxial testing techniques.

Capillary cohesion can be expressed in a form suitable for the direct shear test from the M-C criterion in terms of Bishop's effective stress [eq. (6.17)]:

$$c''(S) = c''(\theta) = \chi_f (u_a - u_w)_f \tan \phi' = \tau_f - c' - (\sigma - u_a)_f \tan \phi' \quad (6.35)$$

Note that c'' is defined directly in terms of the shear stress measured at failure, τ_f; cohesion, c'; net normal stress, $(\sigma - u_a)_f$; and friction angle, ϕ', thus indicating that one can circumvent the necessity to define matric suction or χ.

Similarly, the capillary cohesion function under principal stresses at failure in a triaxial test can be developed from eq. (6.19) as

$$c''(S) = c''(\theta) = \chi_f (u_a - u_w)_f \tan \phi'$$
$$= \frac{(\sigma_1 - u_a)_f - (\sigma_3 - u_a)_f \tan^2(\pi/4 + \phi'/2) - 2c' \tan(\pi/4 + \phi'/2)}{2 \tan(\pi/4 + \phi'/2)}$$

$$(6.36)$$

6.5 UNIFIED REPRESENTATION OF FAILURE ENVELOPE 257

Equations (6.35) and (6.36) can both be used as the theoretical basis for designing and interpreting unsaturated shear strength tests and quantifying the associated shear strength parameters χ_f, c', c'', and ϕ'. If a series of direct shear tests is conducted for specimens prepared at different water contents, the dependency of the capillary cohesion function on water content can be determined. Consider Fig. 6.21. The intercepts c_1'', c_2'', and c_3'' are the capillary cohesions due to the corresponding suction stresses at different water contents. The shear strength under the saturated condition is τ_{f0}. If triaxial tests are used to determine shear strength parameters, the failure states of stress and M-C failure envelopes for tests at various water contents are also a series of parallel lines, as illustrated in Fig. 6.22.

Example Problem 6.5 Results from a series of direct shear tests for a clayey soil are shown in Table 6.5. Plot the M-C failure envelopes for each test. Calculate and plot capillary cohesion c'' as a function of the degree of saturation.

Solution Tests 1 and 2, which were conducted at different values of net normal stress and under saturated conditions, may be interpreted to determine friction angle and cohesion. They are $\phi' = 40°$ and $c' = 40$ kPa. Figure 6.23a illustrates a series of parallel failure envelopes corresponding to the subsequent tests conducted at different degrees of saturation, constructed by drawing lines parallel to the saturated failure envelope and through the states of stress at failure for each test. The failure envelope under saturated conditions (tests 1 and 2) is used as a benchmark for calculating capillary cohesion at the other degrees of saturation. Using the shear stress at failure from test 2

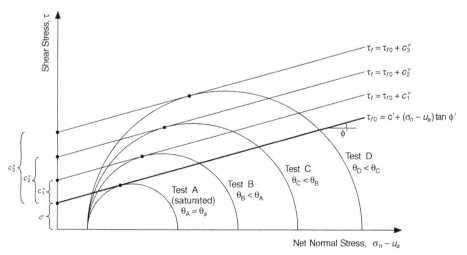

Figure 6.21 Unified representation of Mohr-Coulomb failure criterion for unsaturated soil specimens under direct shear tests.

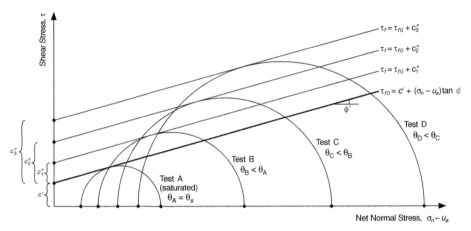

Figure 6.22 Unified representation of Mohr-Coulomb failure criterion for unsaturated soil specimens under triaxial tests.

as the benchmark, the capillary cohesion for each condition can be calculated by eq. (6.35), or can be obtained graphically from the interception of the failure envelopes with the shear stress axis and the known value of c'. The capillary cohesion function $c''(S)$ obtained in this manner is plotted in Fig. 6.23b and tabulated in the last column of Table 6.5. It can be seen that the capillary cohesion of this soil appears to reach a peak value of 449 kPa at a degree of saturation of 0.38. The curve shown in Fig. 6.23b is a characteristic curve for the soil. Much like the soil-water characteristic curve and hydraulic conductivity function may be applied to predict flow phenomena in unsaturated soil, the capillary cohesion characteristic curve may be applied to predict shear strength phenomena.

TABLE 6.5 Direct Shear Testing Results for Example Problem 6.5

Test	$u_a - u_w$ (kPa)	S	$\sigma - u_a$ (kPa)	τ (kPa)	χ	c'' (kPa)
1	0	1.0	120	141	1.000	0
2	0	1.0	300	292	1.000	0
3	25	0.98	120	160	0.920	19
4	500	0.78	120	300	0.380	159
5	1,000	0.71	120	400	0.309	259
6	2,000	0.64	120	460	0.190	319
7	5,000	0.45	120	500	0.086	359
8	11,000	0.38	120	590	0.049	449
9	15,000	0.35	120	530	0.031	389
10	50,000	0.2	120	525	0.009	384

6.5 UNIFIED REPRESENTATION OF FAILURE ENVELOPE

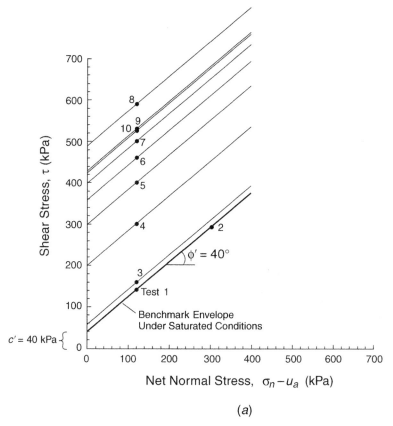

(a)

Figure 6.23 Analysis of direct shear testing data for Example Problem 6.5: (a) unified Mohr-Coulomb failure envelopes for each test and (b) capillary cohesion as function of degree of saturation.

Example Problem 6.6 The results from a series of triaxial tests conducted on a sand are reported in the first four columns of Table 6.6. Calculate and plot capillary cohesion as a function of matric suction.

Solution Because the specimen is sand, the cohesion intercept c' may be assumed to be equal to zero. Accordingly, the results of test 1, which was conducted under saturated conditions, may be interpreted to show that the internal friction angle ϕ' is equal to 37°. The effective stress parameter χ and capillary cohesion c'' may be calculated using eq. (6.36), as shown in the last two columns of Table 6.6, respectively. Mohr circles and the M-C failure envelopes for tests 1, 4, and 8 are plotted in Fig. 6.24a. Capillary cohesion for each test may be evaluated from the intercept of the failure envelopes with the shear stress axis. For example, the capillary cohesion for test 8 conducted at matric suction equal to 800 kPa is 35 kPa. Capillary cohesion

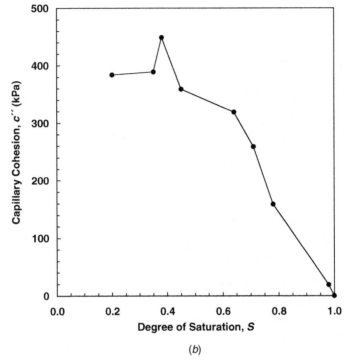

(b)

Figure 6.23 (Continued).

TABLE 6.6 Triaxial Testing Results for Example Problem 6.6

Test	$u_a - u_w$ (kPa)	$\sigma_1 - u_a$ (kPa)	$\sigma_3 - u_a$ (kPa)	χ	c'' (kPa)
1	0	200	50	1.000	0
2	20	250	50	0.833	13
3	50	280	50	0.533	20
4	200	290	50	0.150	23
5	300	300	50	0.111	25
6	400	310	50	0.092	28
7	500	320	50	0.080	30
8	800	340	50	0.058	35

for the entire testing series is depicted as a function of matric suction in Fig. 6.24b.

Example Problem 6.7 Triaxial test results for a silty soil are shown in the first four columns of Table 6.7. Calculate the friction angle and cohesion intercept from tests 1 and 2 conducted under saturated conditions. Plot Mohr circles for tests 1 through 6 and construct the failure envelopes for the test pairs conducted at the same suction. Graphically determine the capillary cohesion for each suction level. Calculate the effective stress parameter, capillary cohesion, and apparent cohesion as functions of matric suction.

Solution The friction angle and cohesion intercept may be determined from the results of tests 1 and 2 conducted under saturated conditions using eq. (6.10), where $u_a - u_w = 0$ and $c_1' = c'$. They are $\phi' = 33°$ and $c' = 20.8$ kPa. Mohr circles for each test and failure envelopes for the tests conducted at the same suction values are shown in Fig. 6.25a. The capillary cohesion c'' function and apparent cohesion $c'' + c'$ function are shown in Fig. 6.25b.

6.5.3 Concluding Remarks

The physical meaning of Bishop's effective stress for application to unsaturated soil appears to be consistent with Terzaghi's original definition of effective stress in saturated soil, which states the following: [effective stress] "represents that part of the total stress which produces measurable effects such as compaction or an increase of the shearing resistance" (Terzaghi, 1943, p. 12). Effective stress for saturated soil does not need to involve material variables because the *neutral stress* that Terzaghi suggests must be subtracted from the total stress to define effective stress is an isotropic stress equal to the pore water pressure. This is no longer the case in unsaturated soil. Rather, the so-called neutral stress is no longer "neutral," but indeed has a profound effect on macroscopic physical behavior such as volume change or shearing resistance. The characteristic stress identified here as suction stress $\chi(u_a - u_w)$ directly represents the stress between unsaturated soil particles. Suction stress depends on the degree of saturation, water content, matric suction, or whatever other quantity is identified as most suitable for defining the unsaturated nature of soil. Thus, suction stress inevitably becomes a material variable. Defining material variables as part of stress state variables is by no means a violation of classical continuum mechanics. According to Fung (1965), and as summarized in Chapter 1, state variables are those that are required to completely describe a system for the phenomenon at hand. In a multiphase system such as unsaturated soil, material variables that describe the relative amounts of each phase are necessary.

Nevertheless, one approach proposed by Fredlund et al. (1978) is to advocate net normal stress and matric suction as independent and nonmaterial

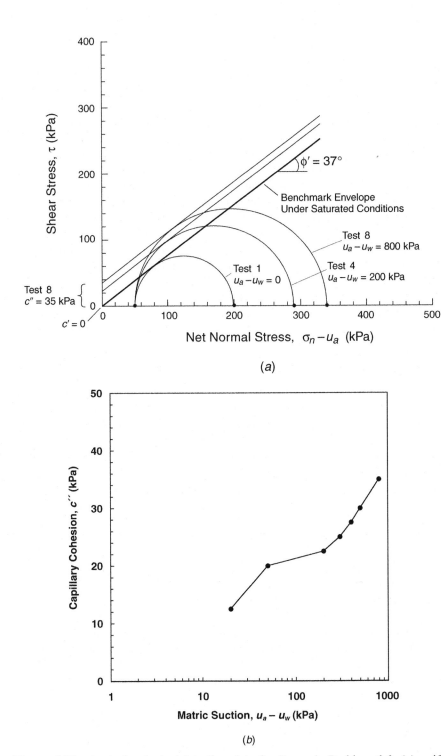

Figure 6.24 Analysis of triaxial testing data for Example Problem 6.6: (*a*) unified Mohr-Coulomb failure envelopes for each test and (*b*) capillary cohesion as function of matric suction.

TABLE 6.7 Triaxial Testing Results and Calculations for Example Problem 6.7

Test	$(u_a - u_w)_f$ (kPa)	$(\sigma_3 - u_a)_f$ (kPa)	$(\sigma_1 - u_a)_f$ (kPa)	χ	c'' (kPa)	$c' + c''$ (kPa)
1	0	23	153	1.000	0.00	20.8
2	0	64	289	1.000	0.00	20.8
3	200	50	410	0.342	44.46	65.26
4	200	100	580	0.343	44.57	65.37
5	400	50	570	0.338	87.90	108.70
6	400	100	740	0.339	88.01	108.81
7	800	50	611	0.191	99.03	119.83
8	800	100	780	0.190	98.86	119.66
9	1500	50	780	0.149	144.91	165.71
10	1500	100	950	0.149	145.02	165.82

dependent stress state variables. Consequently, shear strength is described using a modified M-C criterion. To capture shear strength behavior over a realistically wide range, this requires the introduction of new material variables (e.g., ϕ^b, κ) as functions of matric suction, water content, or degree of saturation. The material variable ϕ^b was considered earlier as a constant but later expanded to reflect its nonlinear nature with respect to matric suction. The material variable κ is empirical and does not possess direct physical meaning. As illustrated in Section 6.3.4, both the ϕ^b and κ approaches can be reconciled with Bishop's effective stress approach. However, approaches employing ϕ^b and χ encounter uncertainties and difficulties because the material parameters required for their formulation (i.e., ϕ^b or χ) are highly dependent on matric suction or water content. This uncertainty is particularly pronounced when matric suction is either very high (practically important in expansive soil) or very low (practically important in sandy soil). Equations (6.17) and (6.19) illustrate theoretically that the uncertainty in determining χ from laboratory experiments on specimens taken to failure is due to the fact that matric suction appears in the denominator of the equations from which it is calculated.

The parameter of most relevance to physical behavior and of most practical significance is neither matric suction nor the effective stress parameter, but rather, the product of the two, the quantity identified and defined as suction stress. This quantity, which was established within a micromechanical framework by considering equilibrium between idealized particles in Chapters 4 and 5, directly contributes to shearing resistance and can be directly quantified from laboratory shear strength tests. Capillary cohesion, which represents the mobilization of suction stress in terms of shearing resistance, can be used as a material variable to reconcile the two competing theories for the strength of unsaturated soil. Thus, the necessity to calculate χ, ϕ^b, or κ can be avoided.

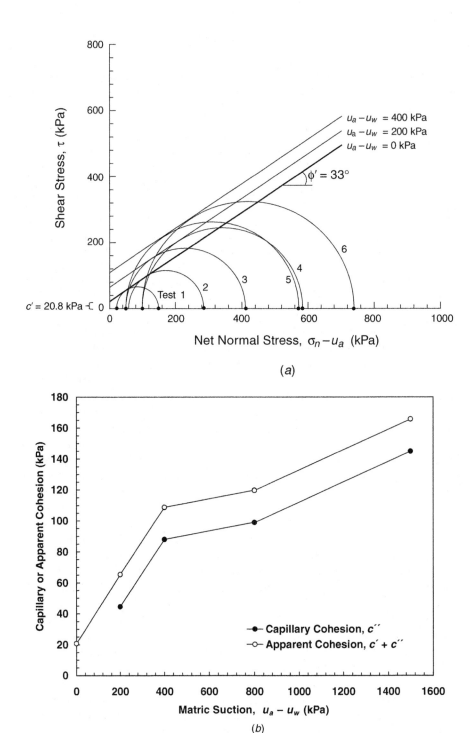

Figure 6.25 Analysis of triaxial testing data for Example Problem 6.7: (*a*) unified Mohr-Coulomb failure envelopes for tests 1 through 6 and (*b*) capillary cohesion and apparent cohesion as functions of matric suction.

PROBLEMS

6.1. Under the same external loading condition, which state, saturated or unsaturated, has a higher strength? What are the possible state variables that control changes in soil strength? Among the controlling variables matric suction, the effective stress parameter, and suction stress, which one do you think will be more representative to describe the strength of unsaturated soil?

6.2. Show that the material variables ϕ^b, κ, and χ can be related mathematically under the framework of Terzaghi's effective stress and the M-C failure criterion.

6.3. A series of direct shear tests was conducted to determine the unsaturated shear strength properties of a glacial till. Results are shown in Table 6.8. Determine the internal friction angle ϕ' and cohesion c' at saturation, the effective stress parameter as a function of saturation and matric suction, and the suction stress characteristic curve, $\chi(u_a - u_w)$ versus S. Plot matric suction, the effective stress parameter, and suction stress as functions of the degree of saturation.

TABLE 6.8 Direct Shear Testing Results for Problem 6.3

Test	$u_a - u_w$ (kPa)	S	$\sigma - u_a$ (kPa)	τ (kPa)
1	0	1.0	50	33.31
2	0	1.0	25	21.66
3	50	0.9	25	40.5
4	100	0.6	25	50.33
5	150	0.4	25	59.72
6	250	0.2	25	74.36
7	350	0.15	25	85.11
8	500	0.12	25	100.91
9	1000	0.1	25	110.24

6.4. A series of saturated and unsaturated triaxial tests were conducted on a sand-clay mixture. Results are shown in Table 6.9. Assume that the cohesion c' is zero and determine the internal friction angle ϕ', effective stress parameter function, χ versus $u_a - u_w$, and capillary cohesion function, c'' versus $u_a - u_w$. Plot the effective stress parameter and capillary cohesion functions.

TABLE 6.9 Triaxial Testing Results for Problem 6.4

Test	$(u_a - u_w)_f$ (kPa)	$(\sigma_1 - u_a)_f$ (kPa)	$(\sigma_3 - u_a)_f$ (kPa)
1	0	486	150
2	200	600	120
3	400	950	200
4	850	1300	260

6.5. Using Fig. 6.20, show mathematically that the relationship between suction stress and capillary cohesion can be established.

6.6. Triaxial test results for a silty soil are shown Table 6.10. The friction angle and cohesion intercept were determined by tests conducted under saturated conditions as $\phi' = 25°$ and $c' = 20.8$ kPa. Plot Mohr circles for tests 1 through 6 and construct failure envelopes for the test pairs conducted at the same suction. Graphically determine capillary cohesion for each suction level. Calculate the effective stress parameter, capillary cohesion, and apparent cohesion as functions of matric suction.

TABLE 6.10 Triaxial Testing Results for Problem 6.6

Test	$(u_a - u_w)_f$ (kPa)	$(\sigma_3 - u_a)_f$ (kPa)	$(\sigma_1 - u_a)_f$ (kPa)
1	200	200	850
2	200	400	1350
3	400	200	1050
4	400	400	1560
5	1500	200	1380
6	1500	400	1900

CHAPTER 7

SUCTION AND EARTH PRESSURE PROFILES

7.1 STEADY SUCTION AND WATER CONTENT PROFILES

7.1.1 Suction Regimes in Unsaturated Soil

Profiles of stress in soil are often used as the theoretical basis for foundation design and analysis. The vertical distribution of matric suction in a natural deposit of unsaturated soil generally depends on several factors: in particular, the soil's hydrologic properties as given by the soil-water characteristic curve (SWCC) and hydraulic conductivity function, environmental factors which control infiltration and evaporative fluxes at the surface, and geometrical boundary or drainage conditions such as the depth of the water table. The combination of these material properties, environmental influences, and geometrical factors results in different matric suction profiles with depth, illustrated for a homogeneous unsaturated soil deposit in Fig. 7.1.

The unsaturated zone can be conceptually divided into two subzones: a seasonally unsteady-state zone and a steady-state zone. Time-dependent environmental factors including precipitation, evaporation, relative humidity, temperature, and airflow conditions cause the soil suction near the ground surface to fluctuate. The depth of this "active" zone varies significantly from place to place and from time to time and is highly dependent on the local geological and environmental conditions. Below the active zone, soil suction is relatively independent of time. The suction profile in this steady zone is controlled by factors including the type of soil, the steady recharge rate (net surface influx), the surface topography, and the location of the water table, as described in detail in Chapter 9.

The flux of water within the unsaturated zone is a complex function of the soil properties and transient infiltration, evaporation, and storage processes.

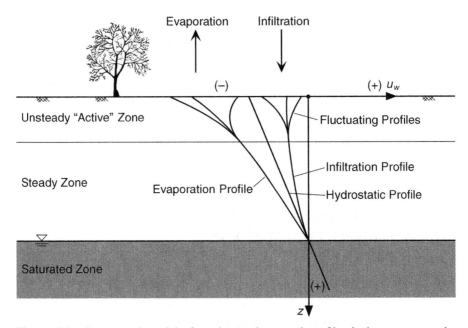

Figure 7.1 Conceptual model of suction regimes and profiles in homogeneous deposit of unsaturated soil under various surface flux boundary conditions.

Insight regarding the impact of fluid flow on the suction profile may be gleaned by considering two very simple cases of steady downward flow (e.g., infiltration) and steady upward flow (e.g., evaporation) processes. Conceptual illustrations of pressure head (suction head) and water content profiles under steady downward (q_{-z}) and steady upward (q_z) unsaturated fluid flow are shown in Figs. 7.2a and 7.2b. At the no-flow, or hydrostatic, condition, suction head is distributed linearly because total head is a constant everywhere. The corresponding water content distribution is the soil-water characteristic curve $\theta(h)$. A minimum water content value occurs at the ground surface, and the 100% saturation condition occurs at the water table. The air-entry head is the elevation above the water table at which desaturation commences (i.e., the height of the capillary fringe). As shown by the profiles for downward flow, an increase in the rate of infiltration, perhaps through a precipitation event that reaches steady state, leads to a decrease in suction head along the profile and a corresponding increase in water content. Conversely, an increase in the upward flow rate, such as during evaporation, leads to an increase in suction and a corresponding decrease in water content. The rate of the infiltration or evaporation process, q, controls the extent to which the suction profile is shifted from the hydrostatic condition.

Vertical profiles of matric suction and how they are influenced by infiltration or evaporation have been important areas of study for many years. The

7.1 STEADY SUCTION AND WATER CONTENT PROFILES   269

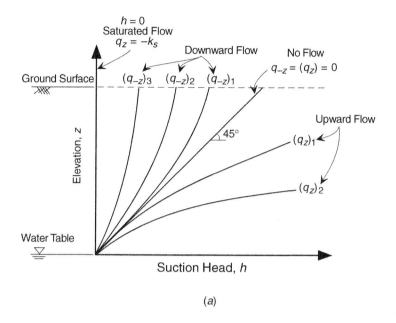

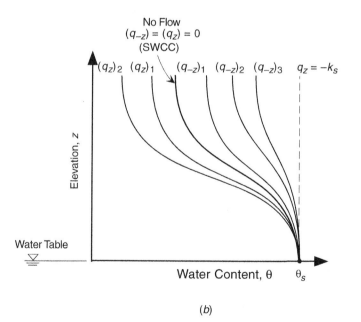

Figure 7.2 Conceptual pressure-head (*a*) and water-content (*b*) distributions in homogeneous layer of unsaturated soil under steady vertically downward and steady vertically upward flow processes: $(q)_3 > (q)_2 > (q)_1$ (after Bear, 1972).

following sections present analytical solutions for evaluating matric suction profiles within the steady zone. Although the matric suction profile within the unsteady zone is indeed important to the stability of many shallow geotechnical structures and critically important for understanding the behavior of unsaturated expansive soil, quantitative description of transient suction fluctuation requires explicit and extensive solutions of the governing equation for transient unsaturated flow. This development follows in Chapter 9.

7.1.2 Analytical Solutions for Profiles of Matric Suction

Mathematical prediction of matric suction profiles can be established by solving the governing flow equation with appropriate initial and boundary conditions. For steady-state profiles, Darcy's law may be generally applied to describe vertical unsaturated flow. Following an upward-positive sign convention, vertical specific discharge may be written as

$$q = -k \left[\frac{d(u_w - u_a)}{\gamma_w \, dz} + 1 \right] \tag{7.1a}$$

where k is the unsaturated hydraulic conductivity dependent on matric suction (m/s) and γ_w is the unit weight of water. The specific discharge q (m/s) can also be written in terms of matric suction head h_m as

$$q = -k \left(\frac{dh_m}{dz} + 1 \right) \tag{7.1b}$$

where h_m is equal to $(u_w - u_a)/\gamma_w$ in units of length (m).

A number of models may be adopted to capture the characteristic dependency of hydraulic conductivity on matric suction. A number of these models are described in Chapter 12. For the current development, consider Gardner's (1958) one-parameter, exponential model. Gardner's model has been widely used to obtain many analytical solutions of unsaturated flow problems (e.g., Philip, 1987) and is written in terms of matric suction head as

$$k = k_s e^{(\beta h_m)} \tag{7.2a}$$

where k_s is the saturated hydraulic conductivity and β (m^{-1}, cm^{-1}) is a parameter capturing the rate of decrease in hydraulic conductivity with increasing suction. Equation (7.2a) may also be written in terms of suction pressure and a parameter α (kPa^{-1}) as

$$k = k_s e^{[-\alpha(u_a - u_w)]} \tag{7.2b}$$

7.1 STEADY SUCTION AND WATER CONTENT PROFILES

Given eqs. (7.1) and (7.2a), and imposing the boundary condition of zero suction head at the water table ($z = 0$), an analytical solution for the one-dimensional suction profile can be derived. Substituting eq. (7.2a) into eq. (7.1) leads to the following:

$$q = -k_s e^{\beta h_m}\left(\frac{dh_m}{dz} + 1\right) \quad (7.3)$$

Integrating the above equation and imposing the suction condition $h_m = h_0$ at the lower boundary $z = 0$ leads to (Yeh, 1989)

$$h_m = \frac{1}{\beta}\ln\left[\left(1 + \frac{q}{k_s}\right)e^{-\beta(z-h_0)} - \frac{q}{k_s}\right] \quad (7.4)$$

or in terms of matric suction $u_a - u_w$ and the parameter α

$$u_a - u_w = \frac{-1}{\alpha}\ln\left[\left(1 + \frac{q}{k_s}\right)e^{-\alpha\gamma_w(z-h_0)} - \frac{q}{k_s}\right] \quad (7.5a)$$

If the lower boundary is set at the water table where the suction is zero, the above equation becomes

$$u_a - u_w = \frac{-1}{\alpha}\ln\left[\left(1 + \frac{q}{k_s}\right)e^{-\alpha\gamma_w z} - \frac{q}{k_s}\right] \quad (7.5b)$$

By mathematical definition, the bracketed quantity on the right-hand side of eq. (7.5b) should be greater than zero. By physical constraint, the quantity should be less or equal to unity to ensure that matric suction is positive or zero, that is,

$$0 < \left(1 + \frac{q}{k_s}\right)e^{-\gamma_w\alpha z} - \frac{q}{k_s} \leq 1.0 \quad (7.6a)$$

The upper bound leads to the constraint that the flux q must be less than or equal to the saturated hydraulic conductivity, reasoned as follows ($q \leq k_s$):

$$\left(1 + \frac{q}{k_s}\right)e^{-\gamma_w\alpha z} - \frac{q}{k_s} \leq 1.0$$

$$\frac{q}{k_s} \geq \frac{1 - e^{-\gamma_w\alpha z}}{e^{-\gamma_w\alpha z} - 1} = -1 \quad (7.6b)$$

If the lower bound is considered,

$$0 < \left(1 + \frac{q}{k_s}\right)e^{-\gamma_w \alpha z} - \frac{q}{k_s} \qquad (7.7a)$$

when $1.0 \geq q/k_s > 0$, the above condition leads to

$$q > \frac{-k_s e^{-\gamma_w \alpha z}}{e^{-\gamma_w \alpha z} - 1} \qquad (7.7b)$$

For the analytical solution to eq. (7.5) to be valid, the above inequality must be satisfied. When this condition is not satisfied, the permissible solution becomes trivial, that is,

$$u_a - u_w = 0 \qquad (7.8)$$

For the hydrostatic condition of $q = 0$, the solution of eq. (7.5) describes a linear suction distribution:

$$u_a - u_w = z\gamma_w \qquad (7.9)$$

Equation (7.5b) can also be rewritten in terms of dimensionless matric suction $\alpha(u_a - u_w)$, depth $\gamma_w \alpha z$, and flow ratio q/k_s as

$$\alpha(u_a - u_w) = -\ln\left[\left(1 + \frac{q}{k_s}\right)e^{-\alpha \gamma_w z} - \frac{q}{k_s}\right] \qquad (7.10)$$

7.1.3 Hydrologic Parameters for Representative Soil Types

The remainder of this chapter includes several analyses that demonstrate the impact of steady-state pore water flow on suction profiles and corresponding states of stress in idealized homogeneous deposits of unsaturated soil. Identification of representative soil properties and flow conditions for different soil types (e.g., sand, silt, and clay) allows the impacts of suction stress and fluid flow on the general behavior of the various soil types to be demonstrated.

A range of representative soil parameters for homogenous deposits of sand, silt, and clay has been identified and is listed in Table 7.1. Specifically, these parameters are the soil-water characteristic curve and hydraulic conductivity

TABLE 7.1 Representative Hydrologic Parameters for Sand, Silt, and Clay

Soil Type	n (dimensionless)	α (kPa^{-1})	$S_r(\%)$	k_s (m/s)
Sand	4–8.5	0.1–0.5	5–10	10^{-2}–10^{-5}
Silt	2–4	0.01–0.1	8–15	10^{-6}–10^{-9}
Clay	1.1–2.5	0.001–0.01	10–20	10^{-8}–10^{-13}

function modeling constants n and α, residual degree of saturation S_r, and saturated hydraulic conductivity k_s. As described in Chapter 12, the n parameter is required in many SWCC and hydraulic conductivity function models to capture the pore size distribution of the soil.

Ranges of steady-state infiltration and evaporation rates commonly encountered in the field under natural environmental conditions are listed in Table 7.2. Flow rates greater than zero correspond to an upward flow process (e.g., evaporation). Flow rates less than zero correspond to a downward flow process (e.g., infiltration).

7.1.4 Profiles of Matric Suction for Representative Soil Types

Analytical profiles of matric suction with depth for three representative soils are illustrated in Fig. 7.3. These profiles were calculated for the various upward and downward flow rates bounded in Table 7.2 using the solutions derived in the previous section and the hydrologic parameters described in Table 7.1. In all cases, a 10-m-thick, homogeneous layer has been considered. The water table is located at $z = 0$, or 10 m below the ground surface.

Investigation of Fig. 7.3 provides insight into the role of soil type on matric suction profiles under steady flow conditions. It is shown in Fig. 7.3a, for example, that in a sandy soil layer, the various flow rates only start to have an influence on matric suction at elevations relatively far from the water table. For the sandy soil parameters modeled here, this height is about 8 m or more. Matric suction at the ground surface varies by about 20 kPa, ranging from a maximum value corresponding to an evaporation condition ($q = 1.15 \times 10^{-8}$ m/s) to a minimum value corresponding to an infiltration condition ($q = -3.14 \times 10^{-8}$ m/s). The suction distribution corresponding to the hydrostatic condition ($q = 0$) is a linear extrapolation of hydrostatic pore pressure from zero at the water table into the negative pressure regime.

For the silty soil (Fig. 7.3b), the flow conditions have a greater impact on the matric suction profile. In this case, the zone of suction variation extends all the way to the water table with the maximum influence corresponding to the highest infiltration rate ($q = -3.14 \times 10^{-8}$ m/s). This general trend continues for the clay, as shown in Fig. 7.3c. Here, the maximum matric suction variation reaches about 60 kPa under the highest infiltration rate. The relatively large impact of the flow condition on the suction profile in increas-

TABLE 7.2 Representative Infiltration (Negative Flux) and Evaporation (Positive Flux) Rates Used for Modeling Steady-State Flow Conditions

Direction of Flow	q (m/s)	q (mm/day)	q (m/yr)
Infiltration	-3.14×10^{-8}	-2.73	-1.00
Hydrostatic (no flow)	0	0	0
Evaporation	1.15×10^{-8}	1.00	0.365

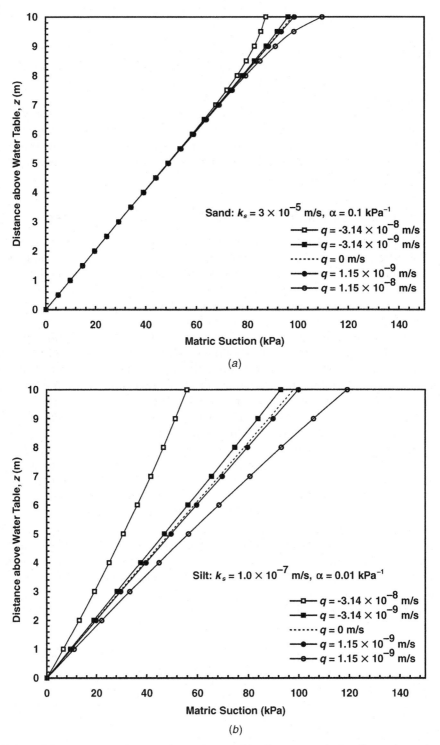

Figure 7.3 Representative matric suction profiles for (*a*) sand, (*b*) silt, and (*c*) clay under hydrostatic and steady-state vertical flow conditions.

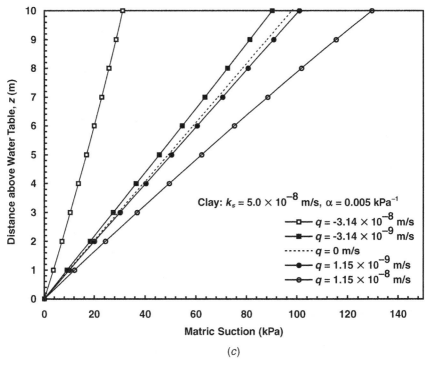

(c)

Figure 7.3 (*Continued*).

ingly fine-grained soil seen here is consistent with physical reasoning and previous field and laboratory observations.

7.1.5 Profiles of Water Content for Representative Soil Types

Evaluating the profile of water content in an unsaturated soil layer requires a constitutive link with the suction profile using the SWCC. Air-entry pressure and residual degree of saturation are often used as benchmark points in mathematical models developed to describe the SWCC (Chapter 12). For the current development, consider a form of the van Genuchten (1980) model that may be written in terms of a relationship between effective saturation S_e and matric suction as

$$S_e = \frac{S - S_r}{1 - S_r} = \left\{\frac{1}{1 + [\alpha(u_a - u_w)]^n}\right\}^{1 - 1/n} \qquad (7.11)$$

where α and n are fitting parameters. The α parameter approximates the inverse of the air-entry pressure and typically falls within the range $0 < \alpha < 0.5$ kPa^{-1}. The n parameter is related to the breadth of the soil's pore size distribution. Relatively large values of n reflect a relatively narrow pore size distribution where the majority of the pore water drains over a relatively narrow range of suction. In general, n has been shown to fall within the range $1.1 < n < 8.5$ for most natural soil types (e.g., van Genuchten, 1980). Assumed values of these parameters for representative sand, silt, and clay are included in Table 7.1.

An expression for the vertical profile of effective saturation as a function of dimensionless depth $\gamma_w \alpha z$, and dimensionless flow ratio q/k_s can be derived by substituting eq. (7.5b) into eq. (7.11) as

$$\frac{S - S_r}{1 - S_r} = \left(\frac{1}{1 + \{-\ln[(1 + q/k_s)e^{-\gamma_w \alpha z} - q/k_s]\}^n} \right)^{1 - 1/n} \quad (7.12)$$

Corresponding profiles for sand, silt, and clay are illustrated in Fig. 7.4. Note that the effective degree of saturation for the sandy soil (Fig. 7.4a) is relatively insensitive to flow rate. Here, saturation falls toward zero quite rapidly at a relatively small elevation above the water table, reaching zero at an elevation of about 4 m. Following capillary theory, the relatively short capillary rise for the hydrostatic condition ($q = 0$) reflects the relatively large pores of the sand. As shown in Figs. 7.4b and 7.4c, the reduction in saturation with increasing elevation from the water table in the silt and clay is much less pronounced. Clay has the smallest variation in saturation, with less than a 20% reduction occurring within the 10-m layer. For all three soils, the reduction in saturation is greatest with evaporation and least with infiltration. A small change in the degree of saturation in clay may imply a considerable change in gravimetric water content.

As discussed in Chapter 6, Vanapalli and Fredlund (2000) showed that the following two forms of χ may fit reasonably well to experimental results from unsaturated shear strength tests:

$$\chi = \left(\frac{\theta}{\theta_s} \right)^\kappa \quad (7.13)$$

and

$$\chi = \frac{\theta - \theta_r}{\theta_s - \theta_r} = \frac{S - S_r}{1 - S_r} \quad (7.14)$$

Applying eq. (7.11) to model the SWCC and substituting it into eq. (7.14) leads to a direct relationship between the effective stress parameter χ and matric suction:

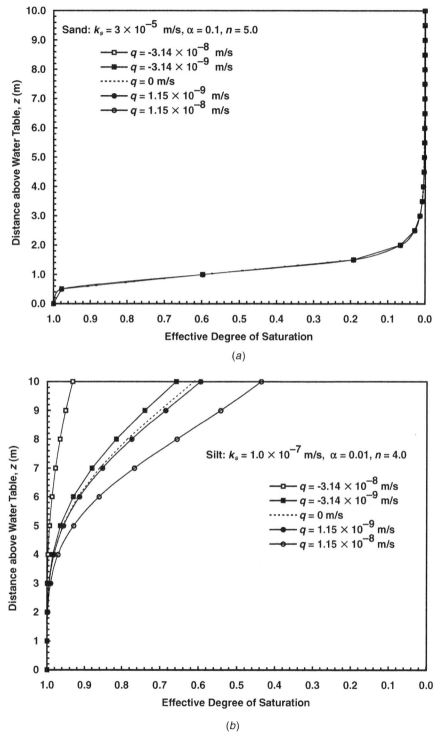

Figure 7.4 Effective degree of saturation profiles under various vertical unsaturated flow rates for representative deposits of (*a*) sand, (*b*) silt, and (*c*) clay.

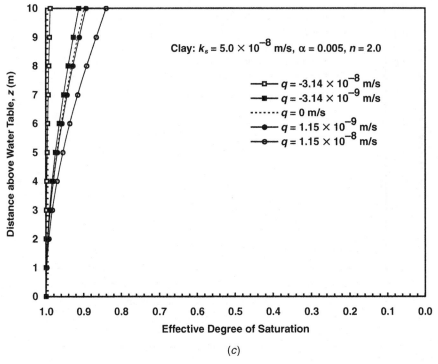

(c)

Figure 7.4 (*Continued*).

$$\chi = S_e = \left\{ \frac{1}{1 + [\alpha(u_a - u_w)]^n} \right\}^{1-1/n} \tag{7.15}$$

Figure 7.5 shows soil-water characteristic curves and corresponding effective stress parameter functions $\chi[(u_a - u_w)]$ for sand, silt, and clay modeled using eq. (7.15). Note that the functions vary widely for the different soil types. Physically, the magnitude of χ reflects the percentage of matric suction at a given degree of saturation that contributes to suction stress. For example, at a suction of 20 kPa, Fig. 7.5a indicates that sand can convert only a few percent of matric suction into suction stress. Silt-sized particles, on the other hand, can convert about 40 to 90% percent of matric suction into suction stress (Fig. 7.5b). Finally, clay can convert essentially 100% of matric suction into suction stress (Fig. 7.5c). The disparities in these values reflect the differences in the geometry of the air-water-solid interface for the increasingly small particle sizes. In all cases, the value of χ decreases with increasing suction.

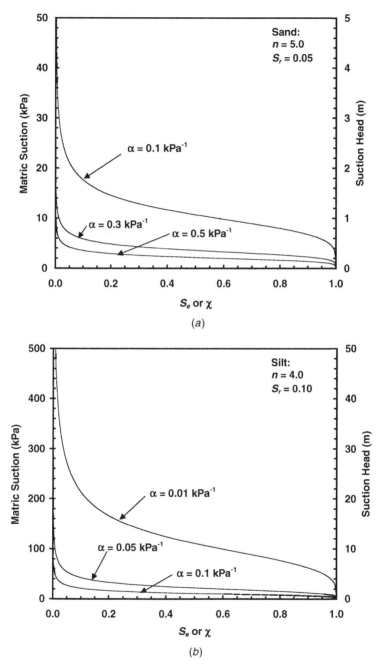

Figure 7.5 Soil-water characteristic curves and effective stress parameter functions for (*a*) sand, (*b*) silt, and (*c*) clay.

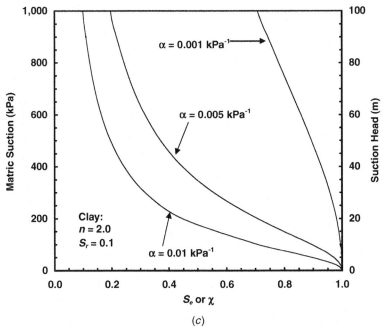

Figure 7.5 (*Continued*).

7.2 STEADY EFFECTIVE STRESS PARAMETER AND STRESS PROFILES

7.2.1 Profiles of the Effective Stress Parameter χ

The matric suction profiles and effective stress parameter functions developed in the preceding section provide an analytical basis to study stress development and shear strength behavior in simulated deposits of unsaturated soil. This section and the following section build upon the previous developments by providing a series of analytical expressions that may be used to evaluate vertical profiles of the effective stress parameter χ and the corresponding suction stress $\chi(u_a - u_w)$ under steady-state flow conditions.

Profiles of the effective stress parameter as a function of dimensionless depth $\gamma_w \alpha z$ and dimensionless flow ratio q/k_s can be derived by substituting eq. (7.5) into eq. (7.15):

$$\chi = \left(\frac{1}{1 + \{-\ln[(1 + q/k_s)e^{-\gamma_w \alpha z} - q/k_s]\}^n} \right)^{1-1/n} \quad (7.16)$$

Effective stress parameter profiles for sand, silt, and clay are illustrated in Fig. 7.6. It is clear from Fig. 7.6a that the effective stress parameter for sandy

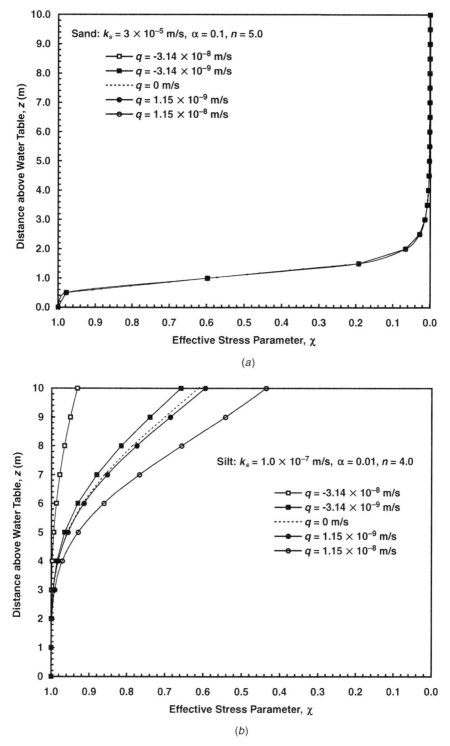

Figure 7.6 Effective stress parameter profiles under various surface flux boundary conditions in representative soils: (*a*) sand, (*b*) silt, and (*c*) clay.

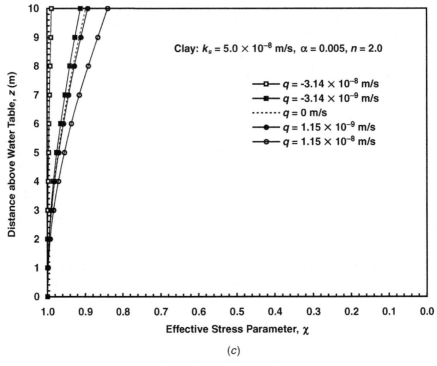

Figure 7.6 (*Continued*).

soil is relatively insensitive to flow rate, decreasing quite rapidly above the water table and reaching zero at an elevation of about 4 m. The reduction in the effective stress parameter for silt and clay is much less pronounced, as shown in Figs. 7.6b and 7.6c, respectively. The clay shows the smallest variation, with less than 20% reduction occurring within the 10-m-thick layer. In all cases, the effective stress parameter reduces with height above the water table. The reduction is greatest for evaporation processes and least for infiltration processes.

7.2.2 Profiles of Suction Stress and Their Solution Regimes

By definition, the absolute magnitude of suction stress $\chi(u_a - u_w)$ depends on both the magnitude of the effective stress parameter and matric suction itself. Because $u_a - u_w$ generally increases with increasing height above the water table but χ generally decreases (Fig. 7.6), the product of the two reaches some maximum between the water table and the ground surface. For steady-state flow conditions, the position and magnitude of this maximum is a function of the soil's hydrologic properties and the flow direction and rate.

Suction stress profiles may be obtained by combining eqs. (7.16) and (7.10) as follows:

$$\chi(u_a - u_w) = \frac{-1}{\alpha} \frac{\ln[(1 + q/k_s)e^{-\gamma_w \alpha z} - q/k_s]}{(1 + \{-\ln[(1 + q/k_s)e^{-\gamma_w \alpha z} - q/k_s]\}^n)^{(n-1)/n}} \quad (7.17)$$

or in a dimensionless form

$$\chi(u_a - u_w)\alpha = \frac{-\ln[(1 + q/k_s)e^{-\gamma_w \alpha z} - q/k_s]}{(1 + \{-\ln[(1 + q/k_s)e^{-\gamma_w \alpha z} - q/k_s]\}^n)^{(n-1)/n}} \quad (7.18)$$

There are only three soil parameters involved in the above equations; α, n, and k_s. Together, these three parameters define the SWCC and the unsaturated hydraulic conductivity function. The dimensionless flow ratio varies in the range of $-1 < q/k_s < 1$.

Although eq. (7.18) is a smooth and continuous function, it displays distinct characteristics in terms of its shape, maxima and minima, and asymptotes. Its solution can be conveniently subdivided into four regimes as a function of the dimensionless flow ratio and the n parameter, as shown in Fig. 7.7. A brief description of the properties and physical implications of each

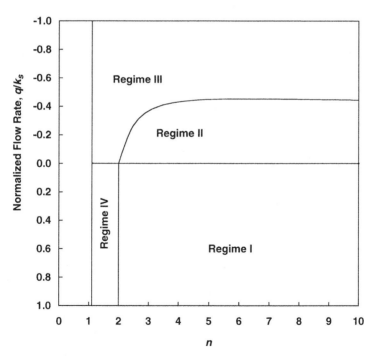

Figure 7.7 Regimes of suction stress profiles: Regime I: maximum suction stress with zero asymptotic postmaximum suction stress; Regime II: maximum suction stress with finite asymptotic postmaximum suction stress; Regime III: monotonically increasing suction stress; and Regime IV: maximum suction stress with rapidly decreasing asymptotic postmaximum suction stress of zero.

regime are summarized in the following sections. For further detailed analysis, refer to Lu and Griffiths (2004).

Regime I: $q/k_s \geq 0$ and $n > 2.0$ This is a steady-state evaporation case. Profiles of suction stress exhibit a constant maximum value for all normalized flow rates q/k_s. The maximum suction stress depends only on the parameter n as follows:

$$[\alpha\chi(u_a - u_w)]_{max} = \frac{(n-2)^{(n-2)/n}}{(n-1)^{(n-1)/n}} \qquad (7.19)$$

and occurs at the following elevation from the water table:

$$(\alpha\gamma_w z)_{max} = \ln\left\{\frac{(1 + q/k_s)e^{(n-2)^{-1/n}}}{(1 + q/k_s)e^{(n-2)^{-1/n}}}\right\} \qquad (7.20)$$

After passing the maximum value, suction stress decreases and tends to zero as the normalized depth $\alpha\gamma_w z$ approaches $\ln(1 + k_s/q)$. For $\alpha\gamma_w z < \ln(1 + k_s/q)$, a solution to eq. (7.5) for matric suction does not exist. Normalized suction stress profiles for various normalized flow ratios (evaporation rates) and $n = 5$ are provided in Fig. 7.8a for illustration.

For the hydrostatic case ($q/k_s = 0$), the suction stress profile has a maximum value expressed by eq. (7.19). The location of the maximum suction stress can be obtained by imposing a zero value of flow ratio on eq. (7.20); hence

$$(\alpha\gamma_w z)_{max} = (n-2)^{-1/n} \qquad (7.21)$$

After passing this maximum, suction stress decreases and tends to zero as normalized depth $\alpha\gamma_w z$ tends to infinity. Normalized suction profiles for the hydrostatic case and for various values of the soil parameter n are illustrated in Fig. 7.8b. Note that soils with relatively large n values (i.e., poorly graded soils) display a sharp maximum at points relatively close to the water table.

Regime II: $-e^{(n-2)^{-1/n}} < q/k_s < 0$ This is the "small" steady-state infiltration case. Suction stress reaches the same maximum at the same elevation as in regime I:

$$[\alpha\chi(u_a - u_w)]_{max} = \frac{(n-2)^{(n-2)/n}}{(n-1)^{(n-1)/n}} \qquad (7.22)$$

and the maximum suction stress occurs at the following elevation above the water table:

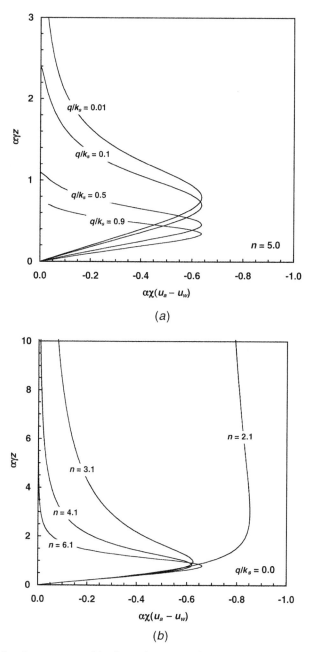

Figure 7.8 Suction stress profiles in regime I: (*a*) for various normalized evaporation rates and $n = 5$ and (*b*) for the hydrostatic condition and various values of n.

$$(\alpha\gamma_w z)_{max} = \ln\left\{\frac{(1 + q/k_s)e^{(n-2)^{-1/n}}}{1 + (q/k_s)e^{(n-2)^{-1/n}}}\right\} \tag{7.23}$$

Normalized suction profiles for various infiltration ratios q/k_s and $n = 4$ within this regime are illustrated in Fig. 7.9a. Normalized suction profiles for $q/k_s = -0.1$ and various n values are shown in Fig. 7.9b.

An interesting characteristic of the behavior in regime II is that after passing the maximum value, suction stress decreases and asymptotically approaches a value that is dependent on both soil parameter n and flow ratio q/k_s:

$$[\alpha\chi(u_a - u_w)]_{\alpha\gamma_w z \to \infty} = \frac{-\ln(-q/k_s)}{\{1 + [-\ln(-q/k_s)]^n\}^{(n-1)/n}} \tag{7.24}$$

Regime III: $-1 < q/k_s \leq -e^{(n-2)^{-1/n}}$, and $n > 1.1$ This is the "large" steady-state infiltration case. Suction stress increases as the distance above the water table increases and asymptotically approaches the following value:

$$[\alpha\chi(u_a - u_w)]_{\alpha\gamma_w z \to \infty} = \frac{-\ln(-q/k_s)}{\{1 + [-\ln(-q/k_s)]^n\}^{(n-1)/n}} \tag{7.25}$$

which is equal to the asymptotic value of postmaximum suction stress in regime II [eq. (7.24)]. Normalized suction stress profiles for various values of infiltration ratio q/k_s are illustrated in Fig. 7.10a.

A limiting case within this regime is $q/k_s = -e^{(n-2)^{-1/n}}$ where suction stress profiles begin to vary monotonically with distance from the water table. Here, suction stress increases as the distance from the water table increases and asymptotically approaches the following value:

$$[\alpha\chi(u_a - u_w)]_{\alpha\gamma_w z \to \infty} = \frac{(n - 2)^{(n-2)/n}}{(n - 1)^{(n-1)/n}} \tag{7.26}$$

which is equal to the maximum suction stress value in regime II [eq. (7.22)]. Under this limiting case, the maximum suction stress is no longer a function of the normalized flow ratio but is solely a function of the soil parameter n. Normalized suction profiles for various n values are illustrated in Fig. 7.10b. A smooth transition between regimes II and III is also illustrated in Fig. 7.10b for flow ratio of -0.4, where $n = 4.1$ is in regime II with a maximum suction stress $\alpha\chi(u_a - u_w) = 0.62$ at $\alpha\gamma_w z = 2.2$.

Regime IV: $0 \leq q/k_s$ and $1.1 < n \leq 2.0$ This corresponds to a "dry clayey soil" evaporation case. The maximum suction stress always occurs at $\alpha\gamma_w z =$

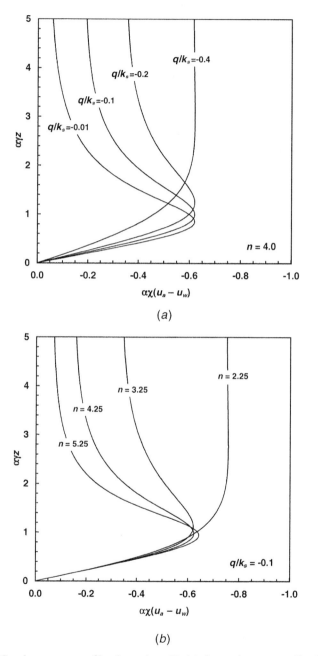

Figure 7.9 Suction stress profiles in regime II: (*a*) for various normalized infiltration rates and $n = 4$ and (*b*) for $q/k_s = -0.1$ and various values of n.

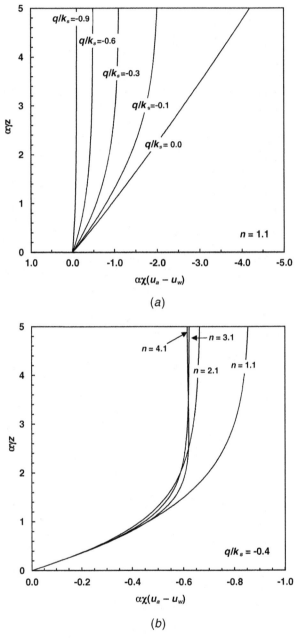

Figure 7.10 Suction stress profiles in regime III: (*a*) for various normalized infiltration rates and $n = 1.1$ and (*b*) for $q/k_s = -0.4$ and various values of n.

$\ln(1 + k_s/q)$ (Fig. 7.11a). After passing this maximum, suction stress quickly approaches infinity as $\alpha\gamma_w z \to \ln(1 + k_s/q)$. For $\alpha\gamma_w z \geq \ln(1 + k_s/q)$, the solution for suction stress does not exist. Another feature of the suction stress profile in this regime is that it is very sensitive to evaporation rate as shown in Fig. 7.11b. This feature is due to the strong dependency of the suction profile on evaporation rate [eq. (7.10)], illustrated previously in Fig. 7.2 and noted in previous work (e.g., Bear, 1972; Marshall and Holmes, 1988; Stephens, 1996).

7.2.3 Profiles of Suction Stress for Representative Soil Types

Application of the analytical solution given by eq. (7.18) to representative sand, silt, and clay provides insight into the magnitude and general patterns of the suction stress profiles in unsaturated soil.

First consider saturation and suction stress profiles under the hydrostatic condition ($q = 0$). As described previously, hydrostatic suction stress profiles follow the characteristics of regimes I and II for sand and silt but follow the characteristics of regime III for clay. Despite soil type, the suction profile under hydrostatic conditions follows the linear distribution predicted by eq. (7.10). Corresponding saturation profiles for the various representative soil parameters can be calculated using eq. (7.15), shown in Fig. 7.5, as well as the suction stress profiles by employing eq. (7.17), shown in Fig. 7.12.

The soil parameter α strongly controls the shape of the soil saturation profiles and the magnitude and shape of the hydrostatic suction stress profiles. Large α values represent relatively large pore sizes, resulting in a relatively small zone of water retention above the water table and hence relatively small suction stresses. It was observed previously in Fig 7.5a that the retention of water by sand becomes relatively insignificant at a distance of about 2 – 4 m above the water table. Figure 7.12a illustrates the consequent effect on the maximum suction stress, which reaches a value of only 6.4 kPa at 1.2 m above the water table. In the unsaturated silt layer (Fig. 7.5b), on the other hand, the zone of significant water retention extends to about 20 m. The corresponding maximum suction stress reaches about 64 kPa at 9 m above the water table (Fig 7.12b). In the unsaturated clay layer, the zone of significant water retention could be greater than 100 m (Fig. 7.5c). Here, the maximum suction stress could reach 700 kPa (Fig 7.12c). Unlike the cases for sand and silt, there is no distinct maximum for the profile in the clay.

Suction stress profiles modeled for sand, silt, and clay under various steady-state infiltration and evaporation conditions are shown in Fig. 7.13. Under steady infiltration conditions, the behavior of most clayey soil follows the characteristics of regime III, and suction stress increases nearly linearly as elevation from the water table increases (Fig. 7.13c). Under the steady evaporation condition, most clayey soil follows the characteristics of regime IV. Since clayey soil has very small values of α, typically between 0.001 and 0.01 kPa^{-1}, the value of $\alpha\gamma_w z$ is very small for a soil layer less than 10 m

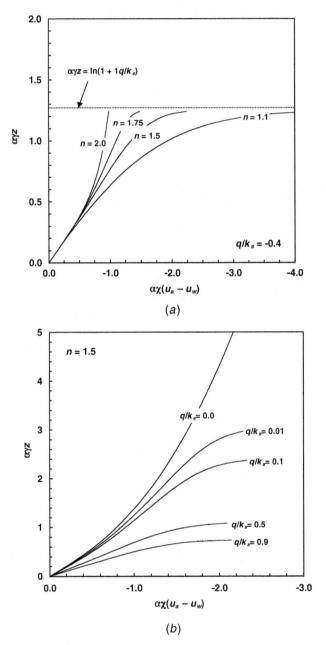

Figure 7.11 Suction stress profiles in regime IV: (a) for $q/k_s = -0.4$ and various values of n and (b) for various normalized evaporation rates and $n = 1.5$.

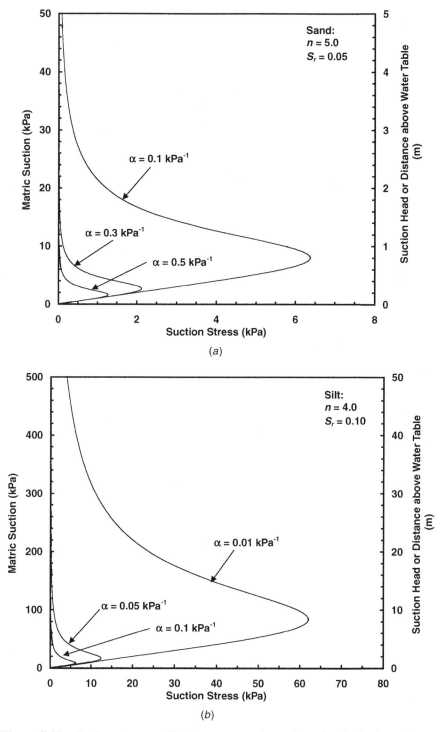

Figure 7.12 Suction stress profiles in representative soils under the hydrostatic condition: (*a*) sand, (*b*) silt, and (*c*) clay.

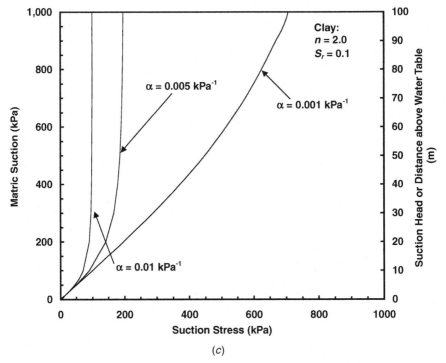

Figure 7.12 (*Continued*).

thick. The small value of $\alpha\gamma_w z$ leads to suction stress profile varying nearly linearly. The maximum suction stress could reach 110 kPa in clay, 60 kPa in silt, and 6 kPa in sand. This range of values may be an important consideration in many shallow foundation analyses or retaining wall design and slope stability.

7.2.4 Concluding Remarks

The models described in this section provide a theoretical framework for predicting vertical profiles of suction stress in near-surface unsaturated soil deposits under hydrostatic or steady-state flow conditions. Analysis for a 10-m-thick layer of representative sand, silt, and clay shows that in relatively coarse-grained materials, suction stresses will modify the effective stress profile mostly near the water table. In finer grained soil such as silt or clay, suction stress can have a significant influence on effective stresses at large distances from the water table. Under steady-state infiltration or evaporation conditions, four characteristic regimes of behavior can occur.

The theory retains the simplicity and generality of the principle of effective stress. Accordingly, it may be used in conjunction with existing analysis tools

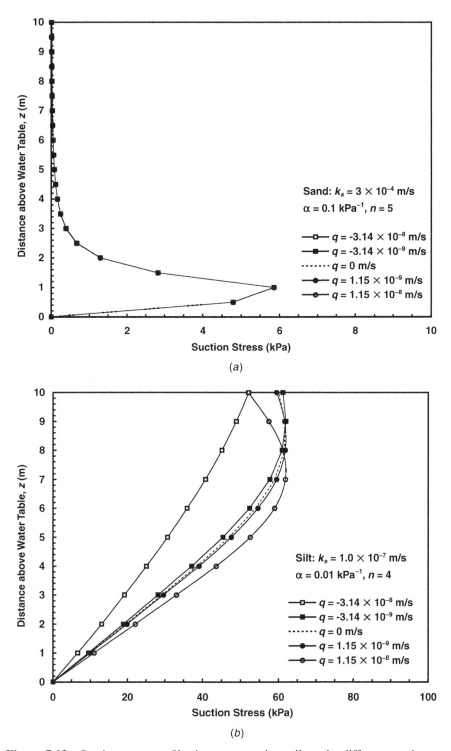

Figure 7.13 Suction stress profiles in representative soils under different steady-state flow conditions.

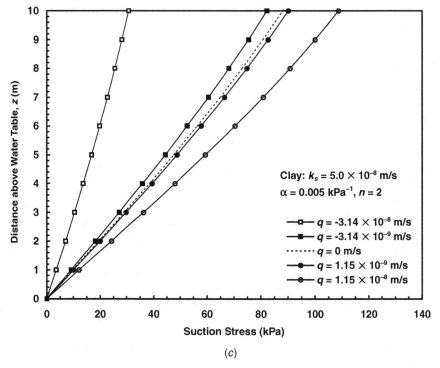

(c)

Figure 7.13 (*Continued*).

and readily applied to the study of classical geotechnical problems such as earth pressure, slope stability, and bearing capacity. The remaining sections of this chapter demonstrate the applicability of the suction stress profile model to the quantitative evaluation of lateral earth pressure, specifically earth pressure at rest (Section 7.3), active earth pressure (Section 7.4), and passive earth pressure (Section 7.5). Profiles of each are modeled for representative deposits of sand, silt, and clay under hydrostatic and various steady-state unsaturated flow conditions.

7.3 EARTH PRESSURE AT REST

7.3.1 Extended Hooke's Law

Establishing the relationships among different stress components such as horizontal and vertical earth pressures requires stress-strain constitutive laws. The most commonly used linear stress-strain equation in elasticity is Hooke's law, that is,

7.3 EARTH PRESSURE AT REST

$$\varepsilon_x = \frac{\sigma'_x}{E} - \frac{\mu}{E}(\sigma'_y + \sigma'_z) \qquad (7.27a)$$

$$\varepsilon_y = \frac{\sigma'_y}{E} - \frac{\mu}{E}(\sigma'_x + \sigma'_z) \qquad (7.27b)$$

$$\varepsilon_z = \frac{\sigma'_z}{E} - \frac{\mu}{E}(\sigma'_y + \sigma'_x) \qquad (7.27c)$$

where ε_x, ε_y, and ε_z are the principal strain components in the horizontal and vertical directions (Fig. 7.14), σ'_x, σ'_y, and σ'_z are the principal effective stress components in the horizontal and vertical directions, E is Young's modulus, and μ is Poisson's ratio.

For unsaturated soil, an extended Hooke's law in light of the suction stress concept can be derived by substituting the effective stress components in the above equations with eq. (5.1):

$$\varepsilon_x = \frac{\sigma_x - u_a}{E} - \frac{\mu}{E}(\sigma_y + \sigma_z - 2u_a) + \frac{(1-2\mu)\chi(u_a - u_w)}{E} \qquad (7.28a)$$

$$\varepsilon_y = \frac{\sigma_y - u_a}{E} - \frac{\mu}{E}(\sigma_x + \sigma_z - 2u_a) + \frac{(1-2\mu)\chi(u_a - u_w)}{E} \qquad (7.28b)$$

$$\varepsilon_z = \frac{\sigma_z - u_a}{E} - \frac{\mu}{E}(\sigma_x + \sigma_y - 2u_a) + \frac{(1-2\mu)\chi(u_a - u_w)}{E} \qquad (7.28c)$$

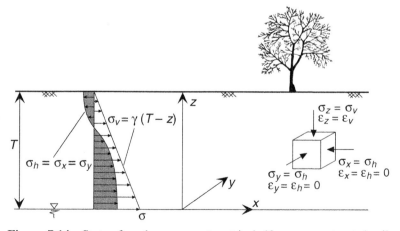

Figure 7.14 State of earth pressure at rest in half-space unsaturated soil.

For the half-space problem of a homogeneous unsaturated soil layer, two general conditions can be imposed: (1) the horizontal stresses $\sigma_x = \sigma_y = \sigma_h$, and (2) the horizontal strains $\varepsilon_x = \varepsilon_y = \varepsilon_h = 0$. Imposing the first condition leads to

$$\varepsilon_v = \frac{\sigma_v - u_a}{E} - \frac{2\mu}{E}(\sigma_h - u_a) + \frac{(1 - 2\mu)\chi(u_a - u_w)}{E} \quad (7.29a)$$

$$\varepsilon_h = \frac{\sigma_h - u_a}{E} - \frac{\mu}{E}(\sigma_v + \sigma_h - 2u_a) + \frac{(1 - 2\mu)\chi(u_a - u_w)}{E} \quad (7.29b)$$

Imposing the second condition leads to

$$\sigma_h - u_a = \frac{\mu}{1 - \mu}(\sigma_v - u_a) - \frac{1 - 2\mu}{1 - \mu}\chi(u_a - u_w) \quad (7.30)$$

Rearranging the above equation results in

$$\frac{\sigma_h - u_a}{\sigma_v - u_a} = K_0 = \frac{\mu}{1 - \mu} - \frac{1 - 2\mu}{1 - \mu}\frac{\chi(u_a - u_w)}{(\sigma_v - u_a)} \quad (7.31)$$

The above equation provides a theoretical framework to assess the dependency of the coefficient of earth pressure at rest, or K_0, on matric suction and overburden stress.

7.3.2 Profiles of Coefficient of Earth Pressure at Rest

An expression to evaluate profiles of suction stress under various unsaturated flow conditions and unsaturated soil material parameters was established in the previous section as

$$\chi(u_a - u_w) = \frac{-1}{\alpha}\frac{\ln[(1 + q/k_s)e^{-\gamma_w \alpha z} - q/k_s]}{(1 + \{-\ln[(1 + q/k_s)e^{-\gamma_w \alpha z} - q/k_s]\}^n)^{(n-1)/n}} \quad (7.17)$$

The above equation, together with eq. (7.31), provides a general quantitative description for profiles of the coefficient of earth pressure at rest. Substituting the above equation into eq. (7.31), one can arrive at a general K_0 profile expression:

$$\frac{\sigma_h - u_a}{\sigma_v - u_a} = K_0 = \frac{\mu}{1 - \mu}$$
$$+ \frac{1 - 2\mu}{1 - \mu}\frac{\ln[(1 + q/k_s)e^{-\gamma_w \alpha z} - q/k_s]}{\alpha(\sigma_v - u_a)(1 + \{-\ln[(1 + q/k_s)e^{-\gamma_w \alpha z} - q/k_s]\}^n)^{(n-1)/n}} \quad (7.32)$$

When soil is saturated, suction stress is zero, the second term drops from eq. (7.32), and K_0 is a constant equal to $\mu/(1 - \mu)$. When soil is unsaturated, suction stress may have an important impact on earth pressure. It is instructive to first consider the suction stress profiles for different soils as plotted in Fig. 7.13. The profiles of matric suction shown in Fig. 7.3 will be used for the purpose of illustration in the following to obtain the corresponding profiles of K_0 using eq. (7.32) and assuming μ is 0.35 and the unit weight of the unsaturated soil is 18 kN/m^3.

Figure 7.15 shows K_0 profiles for representative 10-m deposits of sand, silt, and clay. Several observations can be made. At the water table, the coefficient of earth pressure is a constant equal to 0.538. Physically, this means that if soil is fully saturated, the horizontal earth pressure is 53.8% of the vertical earth pressure or overburden stress and is an invariant in space. If soil is partially saturated (i.e., as the distance from the water table increases), the coefficient of earth pressure decreases as matric suction increases. For the sandy soil (Fig. 7.15a), a minimum value of K_0 occurs at about 1 m above the water table and is insensitive to the magnitude of steady-state flow rate. The K_0 values remain positive and fairly constant over the entire unsaturated zone.

For finer grained soils (Figs. 7.15b and 7.15c), the coefficient of earth pressure reaches zero at some distance above the water table. For the hydrostatic condition, this occurs in both the silt and clay at a similar location of about 7.5 m above the water table or 2.5 m below the ground surface. The K_0 profiles in the silt and clay follow similar decreasing patterns as the elevation from the water table increases. Under nonzero flow rates, the location of the zero K_0 condition appears to be sensitive to flow rate, being closest to the ground surface for clayey soil when infiltration occurs.

For any given elevation, infiltration causes K_0 to increase throughout the soil layer. Conversely, evaporation causes K_0 to decrease. For example, at 3 m below the ground surface in the clay ($z = 7$ m), an evaporation rate of 1.15×10^{-8} m/s reduces the value of K_0 to -0.18 from nearly zero at the hydrostatic condition, whereas an infiltration rate of -3.14×10^{-8} m/s increases K_0 to about 0.4. Negative values of K_0 indicate the existence of tensile stress in soil. Physically, the fact that an increasing evaporation rate increases the depth of the zero K_0 condition results in deeper tension cracks in unsaturated soil under evaporative conditions.

7.3.3 Depth of Cracking

Because soil has relatively low strength in tension, cracking may develop when the value of the coefficient of earth pressure K_0 approaches zero and the tansile strength is reached. Figure 7.16 shows a conceptual model for tension cracking in the unsaturated zone, occurring over a depth where K_0 is zero or negative.

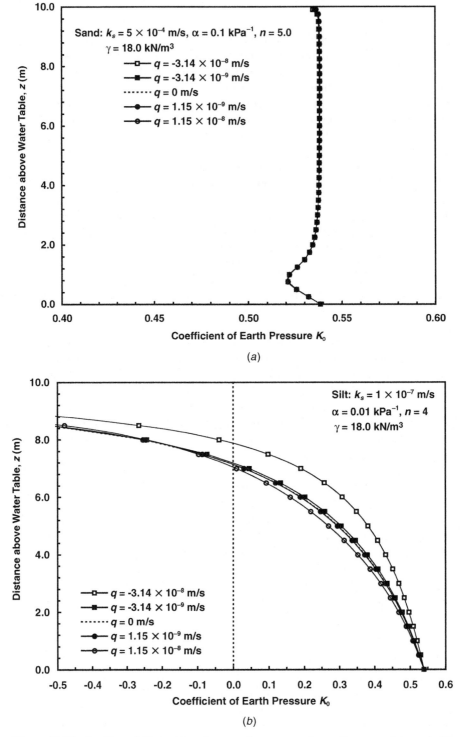

Figure 7.15 Profiles of K_0 in 10-m layers of representative soils types: (a) sand, (b) silt, and (c) clay.

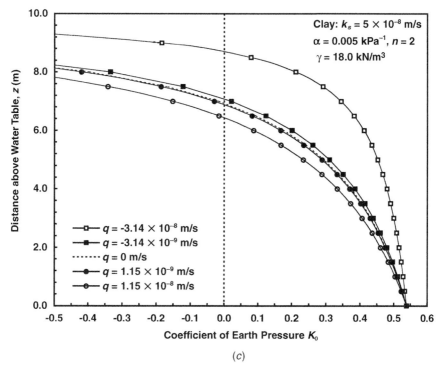

(c)

Figure 7.15 (*Continued*).

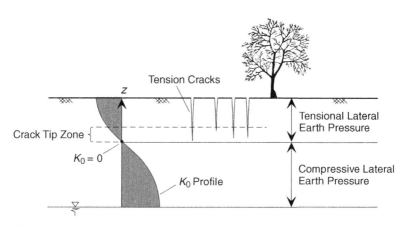

Figure 7.16 Conceptual illustration of tension crack development in unsaturated soil.

If it is assumed that cracking indeed occurs at $K_0 = 0$, then eq. (7.32) becomes

$$\sigma_v - u_a = \frac{1 - 2\mu}{\mu\alpha} \frac{-\ln[(1 + q/k_s)e^{-\gamma_w\alpha z} - q/k_s]}{(1 + \{-\ln[(1 + q/k_s)e^{-\gamma_w\alpha z} - q/k_s]\}^n)^{(n-1)/n}} \quad (7.33a)$$

Assuming that the vertical normal stress $\sigma_v - u_a$ is due to the soil's self-weight and is equal to $\gamma(z_0 - z)$, where z_0 is the depth of the water table from the ground surface and γ is the unit weight, the depth of a tension crack from the ground surface for a given flow condition becomes

$$z_0 - z = \frac{1 - 2\mu}{\gamma\mu\alpha} \frac{-\ln[(1 + q/k_s)e^{-\gamma_w\alpha z} - q/k_s]}{(1 + \{-\ln[(1 + q/k_s)e^{-\gamma_w\alpha z} - q/k_s]\}^n)^{(n-1)/n}} \quad (7.33b)$$

If one defines the dimensionless variables

$$Q = \frac{q}{k_s} \qquad Z = \alpha\gamma_w z \qquad Z_0 = \alpha\gamma_w z_0 \qquad G = \frac{1 - 2\mu}{\mu} \frac{\gamma_w}{\gamma} \quad (7.34)$$

then eq. (7.33b) can be written as

$$Z_0 - Z = G \frac{-\ln[(1 + Q)e^{-Z} - Q]}{(1 + \{-\ln[(1 + Q)e^{-Z} - Q]\}^n)^{(n-1)/n}} \quad (7.35)$$

where the dimensionless quantity G represents the deformability of the soil. For most soil the range of G falls between 0.4 and 1.5. Large G values indicate relatively deformable materials.

Under the hydrostatic condition ($Q = q/k_s = 0$), eq. (7.35) leads to

$$Z_0 - Z = G \frac{Z}{(1 + Z^n)^{(n-1)/n}} \quad (7.36)$$

Equation (7.36) is plotted in Fig. 7.17 for a fixed G value of 0.47. It can be seen that the depth of cracking under the hydrostatic condition is a function of the depth of the unsaturated layer. Based on this analysis, the possible depth of cracking for a wide variety of soil types (i.e., α and n values) lies somewhere between 0.0 and 1.6 m. The depth of cracking remains fairly constant when α is less than 0.01 kPa^{-1} and is sensitive to the pore size distribution parameter n when α is greater than 0.01 kPa^{-1} (i.e., for relatively coarse-grained soil). Within this sensitive regime, the cracking depth is greater for smaller values of n.

Equation (7.36) is plotted in Fig. 7.18 as a function of α and G for a fixed n value equal to 4. Here, higher values of G result in relatively large depths for crack development. In other words, the more deformable the soil, the greater the tendency to develop cracking.

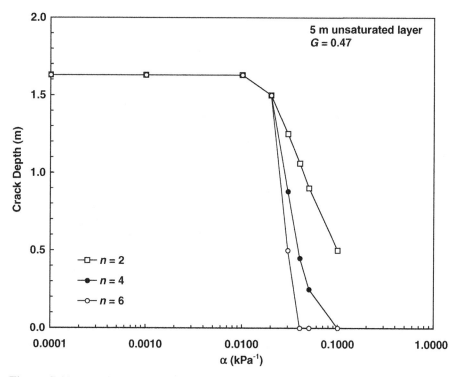

Figure 7.17 Predicted depth of tension cracking from ground surface under hydrostatic conditions in 5-m unsaturated soil layer as function of material parameters α and n. Relatively large values of α describe relatively coarse-grained materials. Relatively large values of n describe materials with relatively narrow pore size distribution.

Figure 7.19 illustrates the influence of the air-entry pressure ($\approx 1/\alpha$) and steady-state infiltration or evaporation rate on the depth of tension cracking. For soil with the same deformability G and pore size distribution parameter n, the cracking depth for relatively fine soil (small α) tends to be greater. An increase in evaporation rate (higher values of Q) causes deeper crack development.

7.4 ACTIVE EARTH PRESSURE

7.4.1 Mohr-Coulomb Failure Criteria for Unsaturated Soil

As described in Chapter 6, the shear strength of unsaturated soil can be described by the Mohr-Coulomb criterion and Bishop's (1959) effective stress as

$$\tau_f = c' + [(\sigma - u_a)_f + \chi_f(u_a - u_w)_f] \tan \phi' \qquad (7.37)$$

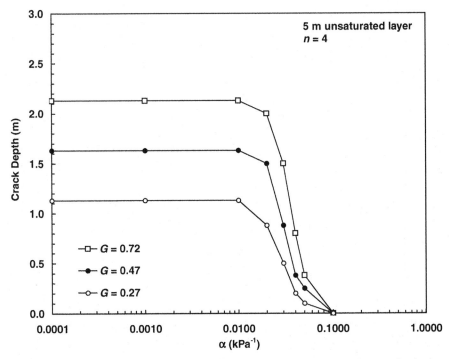

Figure 7.18 Predicted depth of tension cracking from ground surface under hydrostatic conditions in 5-m unsaturated soil layer as function of material parameters α and G. Relatively large values of α describe relatively coarse-grained materials. Relatively large values of G describe relatively deformable materials.

A graphical representation of the failure criterion defined by eq. (7.37) is shown in Fig. 7.20. The point of failure on the Mohr circle (Fig. 7.20a) represents the plane in the soil element (Fig. 7.20b) such that the angle formed by the failure point, the major principal stress, and the point at the center of the circle represents twice the angle of the actual failure plane. Referring to Fig. 7.20b, the angle 2θ between σ'_{1f} and the point defined by the interception of the Mohr-Coulomb failure envelope is twice the angle between σ'_{1f} and the normal direction of the failure plane.

7.4.2 Rankine's Active State of Failure

Rankine's active state of failure is illustrated in Fig. 7.21. The term *active* falls from the fact that if soil fails as shown in Fig. 7.21a, the cause for failure is due to the stress generated from soil self-weight rather than an external load.

In a setting such as Fig. 7.21a, several unique features led Rankine to simplify the failure or limit state analysis. The free-stress boundary induces

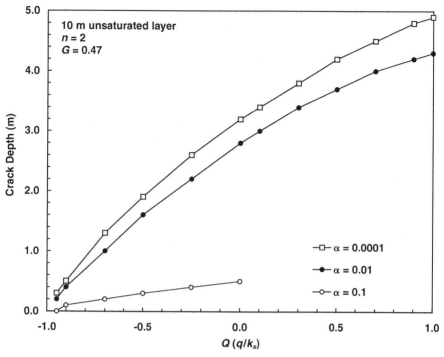

Figure 7.19 Predicted depth of tension cracking from ground surface in 10-m unsaturated soil layer as function of normalized flow rate and material parameter α.

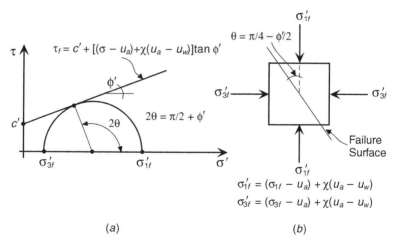

Figure 7.20 State of stress at failure in unsaturated soil following the effective stress concept: (*a*) Mohr's circle and Mohr-Coulomb failure envelope and (*b*) states of stress on failure surface.

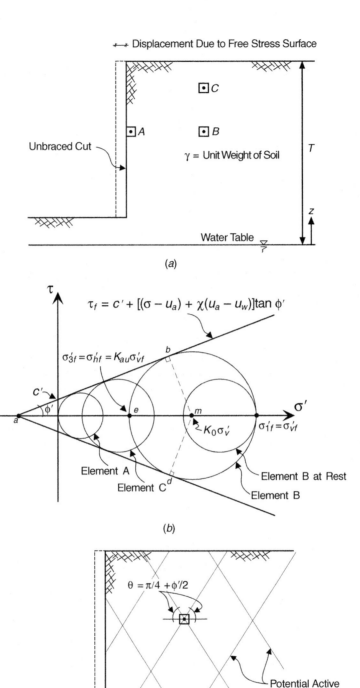

Figure 7.21 Rankine's active earth pressure: (*a*) system geometry, (*b*) Mohr circles and failure envelopes, and (*c*) active failure planes.

lateral movement of the soil mass and a reduction in horizontal stress. It is also a reasonably good assumption to consider the principal stresses along the vertical and horizontal directions, with a maximum occurring vertically and a minimum occurring horizontally. Under these simplifications, the limit state at failure can be described by the group of Mohr circles shown in Fig. 7.21b.

If, for example, failure occurs at point A near the free-stress surface in Fig. 7.21a, the horizontal stress is zero and the vertical stress is equal to the overburden stress $\gamma(T - z_A)$. At points B and C, failure occurs with different magnitudes of horizontal and vertical stresses. However, the failure planes are parallel to each other since the limit state stress points occur at the same angle with respect to the maximum principal plane (Figs. 7.21b and 7.21c).

Since the maximum principal stress is due to the soil self-weight and can be estimated as $\gamma(T - z)$, the minimum principal stress at failure in the horizontal direction can be inferred from the limit state shown in Fig. 7.21b or eq. (7.37). Knowing the principal stresses in both directions, it is logical to rewrite eq. (7.37) in terms of the principal stresses rather than Coulombian stresses. The triangle abm for the geometry shown in Fig. 7.21b leads to

$$\sin \phi' = \frac{bc}{ac} = \frac{(\sigma'_1 - \sigma'_3)/2}{(\sigma'_1 + \sigma'_3)/2 + c' \cot \phi'}$$

$$= \frac{\sigma'_1 - \sigma'_3}{\sigma'_1 + \sigma'_3 + 2c' (\cos \phi'/\sin \phi')} \qquad (7.38)$$

$$= \sin \phi' \frac{\sigma'_1 - \sigma'_3}{\sigma'_1 \sin \phi' + \sigma'_3 \sin \phi' + 2c' \cos \phi}$$

and rearranging leads to

$$\sigma'_1 \sin \phi' + \sigma'_3 \sin \phi' + 2c' \cos \phi' = \sigma'_1 - \sigma'_3 \qquad (7.39a)$$

$$\sigma'_3(\sin \phi' + 1) = \sigma'_1(1 - \sin \phi') - 2c' \cos \phi' \qquad (7.39b)$$

$$\sigma'_3 = \frac{1 - \sin \phi'}{1 + \sin \phi'} \sigma'_1 - 2c' \frac{\cos \phi'}{1 + \sin \phi'} = \sigma'_1 \tan^2\left(\frac{\pi}{4} - \frac{\phi'}{2}\right) - 2c' \tan\left(\frac{\pi}{4} - \frac{\phi'}{2}\right) \qquad (7.39c)$$

When cohesion c' is zero, the above equation becomes

$$\frac{\sigma'_3}{\sigma'_1} = \tan^2\left(\frac{\pi}{4} - \frac{\phi'}{2}\right) = K_a \qquad (7.40)$$

where K_a is the coefficient of Rankine's active earth pressure.

For soil at a state of active failure, the vertical effective stress σ'_v is the maximum principal stress σ'_1 and the horizontal effective stress σ'_h is the minimum principal stress σ'_3. Following the suction stress concept, each may be expressed for unsaturated soil as

$$\sigma'_1 = \sigma'_v = (\sigma_v - u_a) + \chi(u_a - u_w) \quad (7.41\text{a})$$

$$\sigma'_3 = \sigma'_h = (\sigma_h - u_a) + \chi(u_a - u_w) \quad (7.41\text{b})$$

Substituting eqs. (7.40) and (7.41) into eq. (7.39) leads to

$$(\sigma_h - u_a) + \chi(u_a - u_w) = [(\sigma_v - u_a) + \chi(u_a - u_w)]K_a - 2c'\sqrt{K_a} \quad (7.42\text{a})$$

$$\sigma_h - u_a = (\sigma_v - u_a)K_a + \chi(u_a - u_w)(K_a - 1) - 2c'\sqrt{K_a} \quad (7.42\text{b})$$

$$\sigma_h - u_a = (\sigma_v - u_a)K_a - 2c'\sqrt{K_a} - \chi(u_a - u_w)(1 - K_a) \quad (7.42\text{c})$$

It is evident that the first two terms on the right-hand side of eq. (7.42c) are included in Rankine's original theory. The third term represents the contribution of suction stress that arises in unsaturated soil. The first term induces a compressive earth pressure in the soil mass or on adjacent retaining structures. The second and third terms are tensional stresses. If, for example, a retaining wall was installed next to the unsaturated soil mass shown in Fig. 7.21c, the total lateral earth pressure can be estimated from eq. (7.42c). Given eq. (7.42), the coefficient of active earth pressure for unsaturated soil K_{au} can be defined as

$$K_{au} = \frac{\sigma_h - u_a}{\sigma_v - u_a} = K_a - \frac{2c'\sqrt{K_a}}{\sigma_v - u_a} - \chi \frac{u_a - u_w}{\sigma_v - u_a}(1 - K_a) \quad (7.43)$$

7.4.3 Active Earth Pressure Profiles for Constant Suction Stress

If suction stress is assumed constant (i.e., the same at all depths from the ground surface) and the soil is homogeneous, the relative contribution of each term in eq. (7.42c) can be conceptualized in Fig. 7.22. The first term (shown as a stress increasing linearly with depth) is the classical Rankine's linear earth pressure due to the soil's self-weight. The second term reflects the mobilized cohesion at the failure state, a constant value with depth. The third term is the constant suction stress due to the existence of suction stress in unsaturated soil.

The combined effect of these three components of stress results in a linear lateral earth pressure profile that divides the unsaturated soil zone into zones of resultant tensional stress and compressional stress. Tensional stress will act to cause soil to crack, thus nullifying the lateral earth stress within the zone near the surface.

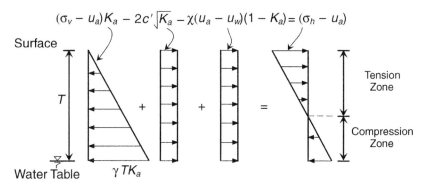

Figure 7.22 Decomposition of total active earth pressure components for condition of constant suction stress with depth.

The resultant earth pressure profile moves from the tension zone to the compression zone at a coefficient of active earth pressure K_{au} equal to zero. This condition allows eq. (7.43) to be rewritten as

$$K_{au} = \frac{2c'\sqrt{K_a}}{\sigma_v - u_a} + \chi \frac{u_a - u_w}{\sigma_v - u_a}(1 - K_a) \qquad (7.44a)$$

Assuming the vertical stress $\sigma_v - u_a$ is equal to $\gamma(T - z)$, eq. (7.43) becomes

$$K_a \gamma(T - z) = 2c'\sqrt{K_a} + \chi(u_a - u_w)(1 - K_a) \qquad (7.44b)$$

The depth of the zero tension stress condition, Z, is then

$$Z = T - z = \frac{2c'}{\gamma\sqrt{K_a}} + \frac{\chi(u_a - u_w)}{\gamma}\left(\frac{1}{K_a} - 1\right) \qquad (7.44c)$$

The above equation can be used to estimate the depth of tensional zone under various suction stress or matric suction conditions. The special case for a constant suction stress is demonstrated in the following example.

Example Problem 7.1 A silty unsaturated soil deposit with a 10-m vertical cut is at the active limit state. The soil has the following properties: internal friction angle $\phi' = 30°$, cohesion $c' = 10$ kPa, unit weight $\gamma = 18$ kN/m³, and suction stress $\chi(u_a - u_w) = 20$ kPa. These values are constant throughout the entire depth of the unsaturated zone. Calculate and plot the lateral active earth pressure profile. Determine the depth of the tension zone.

Solution From eq. (7.40), the coefficient of Rankine's active earth pressure is

308 SUCTION AND EARTH PRESSURE PROFILES

$$K_a = \tan^2\left(45° - \frac{30°}{2}\right) = 0.33 \qquad \sqrt{K_a} = 0.58$$

The lateral earth pressure due to soil's self-weight is zero at the ground surface and the following at 10 m:

$$\sigma = \gamma(T - z)K_a = (18 \text{ kN/m}^3)(10 \text{ m} - 0)(0.33) = 59.9 \text{ kPa}$$

The cohesion-induced tension at both the surface and at 10 m depth is

$$\sigma = -2c'\sqrt{K_a} = -(2)(10 \text{ kPa})(0.58) = -11.6 \text{ kPa}$$

The tension induced by suction stress at the surface and at 10 m depth is

$$\sigma = -\chi(u_a - u_w)(1 - K_a) = -(20 \text{ kPa})(1 - 0.33) = -13.4 \text{ kPa}$$

Therefore, from eq. (7.42c), the total lateral active earth pressure at the ground surface is

$$0 - 11.6 - 13.4 = -25.0 \text{ kPa}$$

and at 10-m depth

$$59.9 - 11.6 - 13.4 = 34.9 \text{ kPa}$$

The depth of tension zone may be estimated using eq. (7.44c):

$$Z = T - z = \frac{2c'}{\gamma\sqrt{K_a}} + \frac{\chi(u_a - u_w)}{\gamma}\left(\frac{1}{K_a} - 1\right) = \frac{(2)(10 \text{ kPa})}{(18 \text{ kN/m}^3)(0.58)}$$

$$+ \frac{20 \text{ kPa}}{18 \text{ kN/m}^3}\left(\frac{1}{0.333} - 1\right) = 4.2 \text{ m}$$

Lateral earth pressure profiles resulting from each component well as the total earth pressure profile are shown in Fig. 7.23. The profile based on classical Rankine theory is included for comparison.

7.4.4 Active Earth Pressure Profiles for Variable Suction Stress

While the assumption of constant suction stress with depth provides a simple approximation of active earth pressure profiles in unsaturated soil, a more general and accurate solution can be evaluated by describing a suction stress profile that incorporates environmental factors (fluid flow) and material factors (soil-water characteristic curve, hydraulic conductivity function). It was

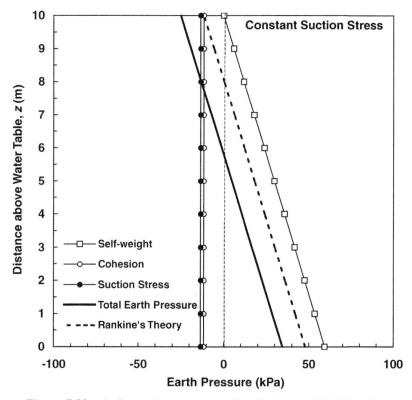

Figure 7.23 Active earth pressure profiles for Example Problem 7.1.

shown in Section 7.2 that a suction stress profile accounting for these factors can be described analytically as follows:

$$\chi(u_a - u_w)\alpha = \frac{-\ln[(1 + q/k_s)e^{-\gamma_w \alpha z} - q/k_s]}{(1 + \{-\ln[(1 + q/k_s)e^{-\gamma_w \alpha z} - q/k_s]\}^n)^{(n-1)/n}} \quad (7.45)$$

Suction stress profiles for representative sand, silt, and clay under various flow rates predicted by eq. (7.45) were shown previously in Fig. 7.13. Note that the suction stress profiles are far from constant or linear, and the magnitude of suction stress can vary from several kilopascals in sand to over 110 kPa in clay.

Fig. 7.24 conceptualizes the corresponding behavior of active earth pressure profiles in unsaturated silty or clayey soil. The nonlinearity in the suction stress profile causes not only a relatively complex lateral active earth pressure distribution but also increases the depth of the tensional stress zone. It is entirely possible that no compressive earth pressure acts along the entire depth of a retaining wall adjacent to unsaturated soil.

310 SUCTION AND EARTH PRESSURE PROFILES

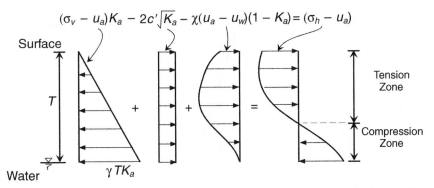

Figure 7.24 Decomposition of total active earth pressure components for condition of suction stress varying with depth.

Example Problem 7.2 Consider a silty soil with the following material properties: $k_s = 1 \times 10^{-7}$ m/s, $\alpha = 0.01$ kPa^{-1}, $\gamma = 18$ kN/m³, $\phi' = 30°$, and $c' = 15$ kPa. Calculate the active earth pressure profiles for a 10-m-thick unsaturated deposit of this soil ($T = 10$ m) under the following conditions: (a) steady infiltration rate of 1 m/yr ($q = -3.14 \times 10^{-8}$ m/s), (b) the hydrostatic condition, and (c) steady evaporation rate of 0.36 m/yr ($q = 1.15 \times 10^{-8}$ m/s).

Solution Suction stress profiles can be calculated using eq. (7.45). These results are shown in Fig. 7.13b. The suction stress profiles may then be used to calculate earth pressure profiles using eq. (7.42c). Earth pressure profiles for the three different flow conditions are shown in Fig. 7.25. As shown, evaporation (Fig. 7.25c) results in a deeper zone of tension throughout the soil mass. Because the suction stress profiles are nonlinear, the corresponding profiles of total earth pressure are also nonlinear.

7.4.5 Active Earth Pressure Profiles with Tension Cracks

When the depth of the tension-cracking zone is known behind a retaining wall, total earth pressure profiles must be reassessed to account for the fact that no earth pressure, either in tension or compression, exists. This involves two adjustments to the earth pressure calculation: (1) all the stresses must initiate from the depth of deepest crack, and (2) the weight of overburden soil is accounted for as a surcharge load.

The magnitude of the surcharge load q_s can be calculated as

$$q = \gamma Z_c \qquad (7.46)$$

where Z_c is the depth of the crack zone. The corresponding active earth pressure according to Rankine's active earth pressure concept is modified as

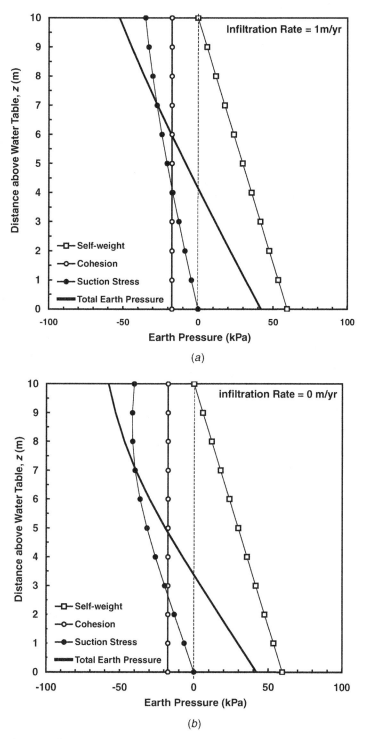

Figure 7.25 Earth pressure profiles for Example Problem 7.2: (*a*) under steady infiltration rate of 1 m/yr, (*b*) under hydrostatic condition, and (*c*) under steady evaporation rate of 0.36 m/yr.

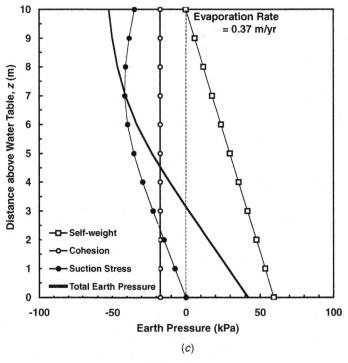

Figure 7.25 (*Continued*).

$$\sigma_h - u_a = (\sigma_v - u_a)K_a + q_s K_a - 2c'\sqrt{K_a} - \chi(u_a - u_w)(1 - K_a) \quad (7.47)$$

and the coefficient of active earth pressure becomes

$$K_{au} = \frac{\sigma_h - u_a}{\sigma_v - u_a} = K_a + \frac{q_s K_a}{\sigma_v - u_a} - \frac{2c'\sqrt{K_a}}{\sigma_v - u_a} - \frac{\chi(u_a - u_w)}{\sigma_v - u_a}(1 - K_a) \quad (7.48)$$

The surcharge term causes earth pressure in compression. For the case of constant suction stress, the component and total active earth pressure profiles are illustrated conceptually in Fig. 7.26.

7.5 PASSIVE EARTH PRESSURE

7.5.1 Rankine's Passive State of Failure

In situations such as a soil mass located in front of a failing retaining wall or an expansive soil mass located behind a retaining wall, the horizontal earth

7.5 PASSIVE EARTH PRESSURE

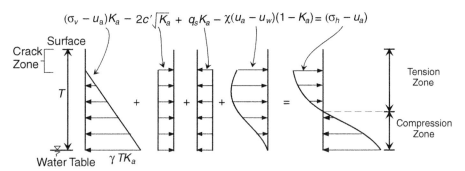

Figure 7.26 Conceptual profiles of active earth pressure when tension cracks are considered.

pressure could be greater than the vertical stress induced by the overburden. Failure occurs when the horizontal stress develops to a magnitude such that the state of stress reaches the Coulombian failure stress. This general condition is referred to as the passive limit state.

Points A, B, and C shown in Fig. 7.27a, for example, could be subjected to a passive limit state. Point D is located at the same distance from the ground surface as points A and B, and thus has the same vertical stress. Displacement of the retaining wall may cause the state of stress at any or all of these points to change from the at-rest condition to their respective limit state.

States of stress for points A, B, C, and D are illustrated in terms of Mohr's circles in Fig. 7.27b. At rest, the Mohr circles for points A, B, and D are identical. The fundamental difference between the active failure state, such as that occurring at point D, and the passive failure state, such as that occurring at points A, B, and C, is the direction of the principal stresses. In the passive state, the minimum principal stress is the overburden stress acting in the vertical direction and the maximum principal stress acts in the horizontal direction.

The relationship between the minimum principal stress and maximum principal stress at failure can be derived by considering the geometry of triangle abm in Fig. 7.27b, which leads to

$$\sin \phi' = \frac{bm}{am} = \frac{(\sigma'_1 - \sigma'_3)/2}{(\sigma'_1 + \sigma'_3)/2 + c' \cot \phi'} = \frac{\sigma'_1 - \sigma'_3}{\sigma'_1 + \sigma'_3 + 2c'(\cos \phi'/\sin \phi')}$$

$$= \sin \phi' \frac{\sigma'_1 - \sigma'_3}{\sigma'_1 \sin \phi' + \sigma'_3 \sin \phi' + 2c' \cos \phi'} \quad (7.49a)$$

where rearrangement leads to

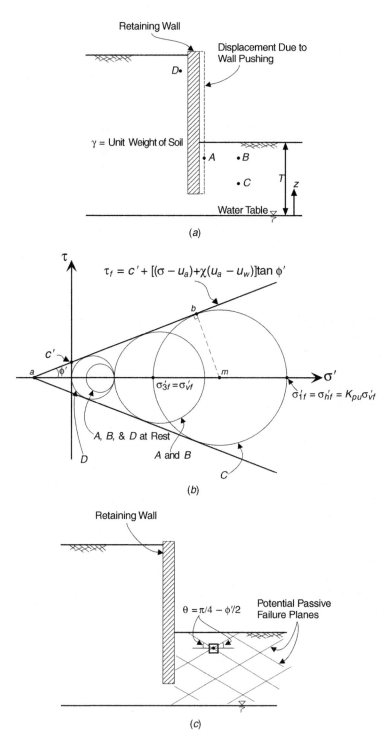

Figure 7.27 Rankine's passive earth pressure: (*a*) system geometry, (*b*) Mohr circles and failure envelopes, and (*c*) passive failure planes.

$$\sigma'_1 = \frac{1 + \sin \phi'}{1 - \sin \phi'} \sigma'_1 + 2c' \frac{\cos \phi'}{1 - \sin \phi'} = \sigma'_3 \tan^2\left(\frac{\pi}{4} + \frac{\phi'}{2}\right)$$
$$+ 2c' \tan\left(\frac{\pi}{4} + \frac{\phi'}{2}\right) \tag{7.49b}$$

When cohesion c' is zero, the above equation becomes

$$\frac{\sigma'_1}{\sigma'_3} = \tan^2\left(\frac{\pi}{4} + \frac{\phi'}{2}\right) = K_p \tag{7.50}$$

where K_p is the coefficient of Rankine's passive earth pressure.

For soil at a state of passive failure, the vertical effective stress σ'_v is the minimum principal stress σ'_3 and the horizontal effective stress σ'_h is the maximum principal stress σ'_1. According to the concept of suction stress, each can be expressed for unsaturated soil as

$$\sigma'_1 = (\sigma_v - u_a) + \chi(u_a - u_w) \tag{7.51a}$$

$$\sigma'_3 = (\sigma_h - u_a) + \chi(u_a - u_w) \tag{7.51b}$$

Substituting eqs. (7.50) and (7.51) into eq. (7.49) leads to

$$(\sigma_h - u_a) + \chi(u_a - u_w) = [(\sigma_v - u_a) + \chi(u_a - u_w)]K_p + 2c'\sqrt{K_p} \tag{7.52a}$$

$$\sigma_h - u_a = (\sigma_v - u_a)K_p + 2c'\sqrt{K_p} + \chi(u_a - u_w)(K_p - 1) \tag{7.52b}$$

The first two terms on the right-hand side of eq. (7.52b) fall from classical Rankine theory. The third term represents the suction stress component arising in unsaturated soil. All three terms induce compressive earth pressure. The coefficient of passive earth pressure for unsaturated soil K_{pu} can also be defined from eq. (7.52b) as

$$K_{pu} = \frac{\sigma_h - u_a}{\sigma_v - u_a} = K_p + \frac{2c'\sqrt{K_p}}{\sigma_v - u_a} + \frac{\chi(u_a - u_w)}{\sigma_v - u_a}(K_p - 1) \tag{7.53}$$

For soil under the passive limit state, the failure planes form a group of parrallel lines with respect to the horizontal plane at $\pi/4 - \phi'/2$, as illustrated in Fig. 7.27c.

7.5.2 Passive Earth Pressure Profiles for Constant Suction Stress

If suction stress is assumed constant with depth and the soil is homogeneous, the relative contribution of each term in eq. (7.52b) can be conceptualized in Fig. 7.28. The first term, shown as a linearly increasing stress with depth, is

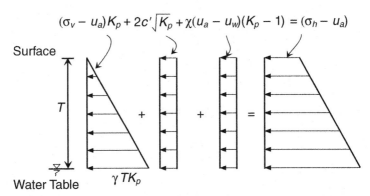

Figure 7.28 Decomposition of passive earth pressure for condition of constant suction stress with depth.

the limit state horizontal stress resulting from overburden. This contribution is usually 3 to 6 times the overburden stress. The second term reflects internal resistance arising from cohesion. The third term is the contribution from suction stress, equal to about 2 to 5 times the magnitude of the suction stress. In contrast with the active limit state described in section 7.4, both the cohesion and the suction stress contribute positively to the total lateral earth pressure. The combined effect of all three components leads to a linearly distributed earth pressure profile. Accordingly, there is no zone of tension.

Example Problem 7.3 Consider a 10-m-thick layer of silty unsaturated soil having a 5-m vertical cut in front of a retaining wall at the passive limit state. The distance from the ground surface to the water table in front of the retaining wall is 5 m. The soil has the following properties: internal friction angle $\phi' = 30°$, cohesion $c' = 10$ kPa, unit weight $\gamma = 18$ kN/m³, and a constant suction stress $\chi(u_a - u_w) = 20$ kPa throughout the entire unsaturated zone. Calculate and plot the passive earth pressure profile.

Solution From eq. (7.50), the coefficient of Rankine's passive earth pressure is

$$K_p = \tan^2\left(45° + \frac{30°}{2}\right) = 3.00 \quad \sqrt{K_p} = 1.73$$

Lateral earth pressure due to the soil's self-weight is zero at the ground surface and the following at 5 m depth:

$$\sigma = \gamma(T - z)K_p = (18 \text{ kN/m}^3)(5 \text{ m})(3.00) = 270 \text{ kPa}$$

The cohesion-induced earth pressure is

$$\sigma = 2c'\sqrt{K_p} = (2)(10 \text{ kN/m}^3)(1.73) = 34.6 \text{ kPa}$$

The suction stress-induced earth pressure is

$$\sigma = \chi(u_a - u)(K_p - 1) = (20 \text{ kN/m}^2)(3 - 1) = 40.0 \text{ kPa}$$

Therefore, from eq. (7.52b), the total passive earth pressure at the ground surface is

$$0 + 34.6 + 40.0 = 74.6 \text{ kPa}$$

and at 5 m depth it is

$$270 + 34.6 + 40.0 = 344.6 \text{ kPa}$$

The component and total lateral earth pressure profiles are plotted in Fig. 7.29. The profile based on classical Rankine theory is included for comparison. Clearly, suction stress acts to increase the earth pressure by acting as an additional cohesion term.

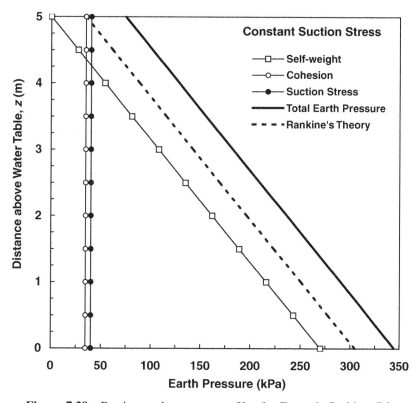

Figure 7.29 Passive earth pressure profiles for Example Problem 7.3.

7.5.3 Passive Earth Pressure Profiles for Variable Suction Stress

The impact of a variable suction stress profile on passive earth pressure is conceptually illustrated in Fig. 7.30. The major impacts of the nonlinear suction stress are a nonlinear distribution of passive earth pressure and an overall increase in the earth pressure magnitude. Each of these effects can be better appreciated by considering the following two examples.

Example Problem 7.4 Consider a silty unsaturated soil layer having the following properties: $k_s = 1 \times 10^{-7}$ m/s, $n = 4.0$, $\alpha = 0.01$ kPa^{-1}, $\gamma = 18$ kN/m^3, $T = 5$ m, $\phi' = 30°$, and $c' = 15$ kPa. Calculate the passive earth pressure profiles for this soil situated in front of the retaining wall shown in Fig. 7.27a under: (a) a steady infiltration rate of 1 m/yr (-3.14×10^{-8} m/s), (b) the hydrostatic condition, and (c) a steady evaporation rate of 0.36 m/yr (1.15×10^{-8} m/s).

Solution Suction stress profiles can be calculated using eq. (7.45) and are shown in Fig. 7.31a. Note that suction stress for the hydrostatic condition distributes slightly nonlinearly. Evaporation causes suction stress to increase, whereas infiltration leads to a decrease in suction stress. These suction stress profiles may then be used to calculate earth pressure profiles using eq. (7.52b), as shown in Figs. 7.31b through 7.31d. It can be seen that evaporation (Fig. 7.31d) results in a higher lateral earth pressure and that the total earth pressure profiles are slightly nonlinear. The difference between Rankine's classical theory and the current one becomes more and more pronounced as the distance from the water table increases.

Example Problem 7.5 Consider a clayey soil deposit having the following properties: $k_s = 5 \times 10^{-8}$ m/s, $n = 2.0$, $\alpha = 0.001$ kPa^{-1}, $\gamma = 18$ kN/m^3, $T = 5$ m, $\phi' = 30°$, and $c' = 15$ kPa. Calculate the passive earth pressure

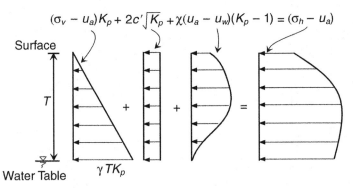

Figure 7.30 Decomposition of total passive earth pressure components when variable suction stress profile is considered.

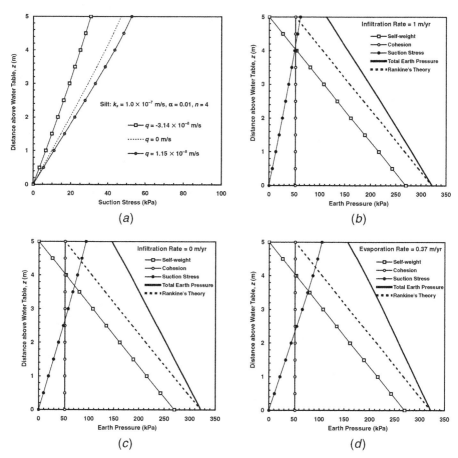

Figure 7.31 Passive earth pressure profiles for Example Problem 7.4: (a) suction stress profiles, (b) earth pressure profiles for infiltration, (c) hydrostatic condition, and (d) evaporation.

profiles for this soil situated in front of the retaining wall shown in Fig. 7.27a under: (a) a steady infiltration rate of 1 m/yr (-3.14×10^{-8} m/s), (b) the hydrostatic condition, and (c) a steady evaporation rate of 0.36 m/yr (1.15×10^{-8} m/s).

Solution Suction stress profiles can be calculated using eq. (7.45) and are shown in Fig. 7.32a. The suction stress profile for the hydrostatic condition is essentially linearly distributed. The difference between the infiltration and evaporation conditions becomes greater in comparison with the previous example for silt. The suction stress profiles may then be used to calculate earth pressure profiles using eq. (7.52b) as illustrated in Figs. 7.32b through 7.32d. Compared to the previous case for silty soil, the suction stress profile under

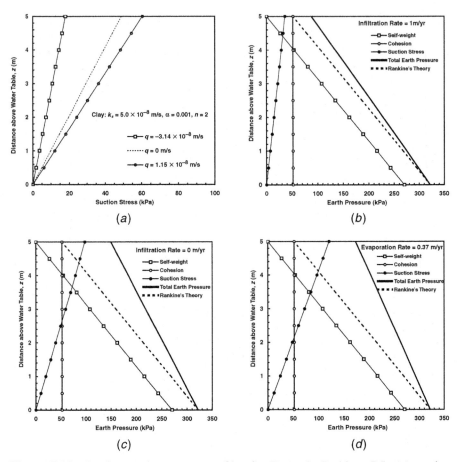

Figure 7.32 Passive earth pressure profiles for Example Problem 7.5: (a) suction stress profiles, (b) earth pressure profiles for infiltration, (c) hydrostatic condition, and (d) evaporation.

infiltration is generally less. The consequent total earth pressure at the ground surface is smaller (95 kPa in clay vs. 120 kPa in silt). At the hydrostatic condition, the earth pressure profiles in the clay and silt layers are nearly identical. Under evaporation conditions, the suction stress is higher in clay than in silt, leading to a higher total lateral earth pressure.

7.5.4 Concluding Remarks

The fundamental difference in the limit state between classical Rankine theory and the theory presented here for unsaturated soil is the contribution of pore pressure. Since pore pressure or suction stress is considered an isotropic stress tensor, the size (diameter) of Mohr's circle for a given point will remain the same, no matter the value and sign of the pore pressure or suction stress. The center of the Mohr circle representing the first invariant of the effective stress tensor, however, depends on the magnitude and sign of the pore pressure or suction stress. According to continuum mechanics and plasticity theory, the isotropic and invariant nature of pore pressure or suction stress leads to the same failure planes for both saturated and unsaturated conditions. When the state of stress reaches its limit, the failure plane will occur at $\pi/4 + \phi'/2$ with respect to the horizontal plane for the active limit state and $\pi/4 - \phi'/2$ for the passive limit state. However, the stresses on these failure planes for saturated and unsaturated soil conditions will be quite different.

The semiquantitative analysis procedures presented in this chapter clearly demonstrate the significant impact of suction stress on the distribution of lateral earth pressure in unsaturated soil. The suction stress profile has been shown to be nonlinear with depth and is highly dependent on soil type and steady-state fluid flow conditions.

The existence of suction stress under active state conditions causes a deeper zone of tension and an overall reduction in lateral earth pressure. This deeper tension zone may lead to the development of cracks at significant depth from the ground surface. Infiltration causes the soil profile to become relatively moist in comparison with the hydrostatic condition and less apt to develop tension cracks. Evaporation causes the soil to be relatively dry and more apt to develop tension cracks. The reduction of lateral earth pressure arising from the suction stress component may increase the factor of safety for retaining wall applications. However, the tendency to form deeper tension cracks may promote more drastic pore pressure changes behind retaining walls during and after precipitation events. Under passive conditions, suction stress may significantly contribute to the overall cohesion of the soil mass, an effect that may also lead to higher factors of safety compared with classical design based on Rankine theory. However, the higher passive earth pressures resulting from suction stress could have a concurrent impact on the force and moment balance of the retaining wall system, possibly resulting in different optimum design configurations. Quantitative assessment of the impact of suction stress on the design of retaining structures has yet to be established.

Earth pressure distributions calculated using Rankine theory and the current theory show almost no difference for depths near the water table. The difference becomes more and more pronounced, however, as the distance from the water table increases. For representative sand, silt, and clay soil types examined here, this difference was as much as 100 kPa. It appears that the suction stress developed in sandy soil has a very limited impact on the total lateral earth pressure distribution. The impact is significantly greater for silty and clayey soil. Although the current design procedures in engineering practice typically follow an assumption of zero pore pressure in the unsaturated zone, the observations presented in this chapter should be borne in mind as refinements that incorporate unsaturated soil mechanics principles continue to emerge.

PROBLEMS

7.1. Consider a 10-m-thick unsaturated silty soil layer with the following properties:

$$k_s = 10^{-7} \text{ m/s} \qquad \alpha = 0.01 \text{ kPa}^{-1} \qquad n = 4.0$$

Use the analytical procedures developed in this chapter to estimate and plot matric suction and effective degree of saturation profiles with depth under the following steady conditions: (a) hydrostatic, (b) infiltration rate of 1 m/yr, and (c) evaporation rate of 0.365 m/yr.

7.2. Plot the suction stress profiles for the unsaturated soil layer from Problem 7.1.

7.3. Assume that the unsaturated soil layer from Problem 7.1 is bounded by the retaining wall shown in Fig. 7.27a. Given the following properties: $\gamma = 18 \text{ kN/m}^3$, $T = 5$ m, $\phi' = 30°$, and $c' = 15$ kPa, calculate the passive earth pressure profiles in front of the retaining wall under: (a) steady infiltration rate of 1 m/yr (-3.14×10^{-8} m/s), (b) hydrostatic condition, and (c) steady evaporation rate of 0.36 m/yr (1.15×10^{-8} m/s).

7.4. When suction stress is present under the unsaturated condition, will the lateral earth pressure at rest be less than or greater than that under a saturated condition? What are the possible practical implications for unsaturated soil under nonconstant value of K_0? Which climatic condition, evaporation or infiltration, will cause lateral earth pressure at rest to be smaller?

7.5. A silty unsaturated soil deposit with an 8-m vertical cut is at the active limit state. The soil has the following properties: internal friction angle $\phi' = 33°$, cohesion $c' = 15$ kPa, unit weight $\gamma = 18 \text{ kN/m}^3$, and suction stress $\chi(u_a - u_w) = 30$ kPa throughout the entire depth of unsaturated zone. Calculate and plot the lateral active earth pressure profile.

7.6. Consider the soil from Problem 7.5 with the following additional material properties: $k_s = 2 \times 10^{-7}$ m/s, $\alpha = 0.01 \text{ kPa}^{-1}$. Calculate active earth pressure profiles under the following conditions: (a) steady infiltration rate of 1 m/yr ($q = -3.14 \times 10^{-8}$ m/s), (b) hydrostatic condition, and (c) steady evaporation rate of 0.36 m/yr ($q = 1.15 \times 10^{-8}$ m/s).

7.7. Consider the soil from Problem 7.6. Calculate passive earth pressure profiles for this soil situated in front of the retaining wall with a depth of 3 m shown in Fig. 7.27a under: (a) a steady infiltration rate of 1 m/yr (-3.14×10^{-8} m/s), (b) hydrostatic condition, and (c) a steady evaporation rate of 0.36 m/yr (1.15×10^{-8} m/s).

PART III

FLOW PHENOMENA

PART III

FLOW PHENOMENA

CHAPTER 8

STEADY FLOWS

8.1 DRIVING MECHANISMS FOR WATER AND AIRFLOW

8.1.1 Potential for Water Flow

The fundamental thermodynamic quantity governing the flow of liquid water in unsaturated soil is the total potential of the pore water, most commonly described in terms of total suction or total head. As described in Section 1.6, total potential (in J/kg) may be expressed in terms of total suction (kPa) ψ_t as follows:

$$\mu_t = \frac{\psi_t}{\rho_w} \tag{8.1}$$

or in terms of total head h_t (m):

$$\mu_t = gh_t \tag{8.2}$$

For example, soil described by a total suction of 200 kPa has a pore water potential of 200 J/kg or a driving head for fluid flow of about 20.4 m.

Total suction is sufficiently defined for most practical geotechnical engineering seepage problems by considering the suction due to gravity, ψ_g, together with matric suction ψ_m and osmotic suction ψ_o as follows:

$$\psi_t = \psi_g + \psi_m + \psi_o \tag{8.3a}$$

where the gravitational component represents the change in elevation, z, from one point under consideration to another ($\psi_g = \rho_w g z$). Similarly, the total driving head is

$$h_t = h_g + h_m + h_o = z + h_m + h_o \tag{8.3b}$$

It should be noted that total potential as defined in the above two equations lumps osmotic potential, which refers only to the free water component of the soil-water solution, with gravitational and pressure potential, which refer to the entire soil-water solution. Corey and Klute (1985) argue that, strictly speaking, total potential defined in this manner cannot be a valid potential and that distinction should be made between elements of the soil solution and the soil solution as a whole. For most practical seepage problems occurring on a relatively macroscopic scale, however, defining total potential as the algebraic sum of the pressure, gravitational, and osmotic components appears to be sufficient in governing pore water equilibrium and transport.

8.1.2 Mechanisms for Airflow

The flow of pore air in unsaturated soil is governed by the total potential (absolute pressure) of the air phase. If the pore air is assumed to follow ideal gas behavior, changes in total air potential can be captured through the ideal gas law (Chapter 2). The largest changes in air potential are the result of changes in pressure and temperature due to variation of the prevailing atmospheric conditions. As illustrated schematically in Fig. 8.1, the major mechanisms responsible for airflow in unsaturated soil include the following:

1. Daily, weekly, and seasonal barometric pressure fluctuation (e.g., Stallman, 1967; Stallman and Weeks, 1969)
2. Daily, weekly, and seasonal atmospheric temperature fluctuation (e.g., Weeks, 1978, 1979)
3. Fluctuations in wind conditions (e.g., Weeks, 1991)
4. Temperature gradients due to topographic relief (e.g., Ross et al., 1992)
5. Heat sources such as pyrite oxidation and radioactive materials (e.g., Lu and Zhang, 1997)

8.1.3 Regimes for Pore Water Flow and Pore Airflow

Neglecting vapor phase transport, the flow of pore water in unsaturated soil may occur only through the pore space occupied by a continuous liquid phase. Neglecting diffusive transport in liquid, the flow of pore air may occur only through the pore space occupied by a continuous gas phase. Water flow and airflow, therefore, occur under specific regimes, depending primarily on the type of soil and the water content or degree of saturation.

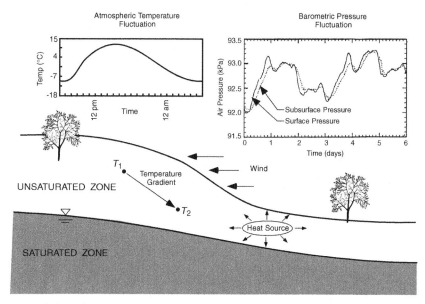

Figure 8.1 Primary driving mechanisms for flow of pore air in unsaturated soil.

Fig. 8.2 is a conceptual model illustrating three distinct regimes in unsaturated fluid flow corresponding to pore airflow only, concurrent pore air and pore water flow, and pore water flow only. Each regime is delineated as a function of water content and particle (pore) size by boundaries that establish a residual water content regime, an occluded-air-bubble water content regime, and a saturated water content regime. The magnitude of water content separating each regime decreases with increasing pore size.

At water content less than or equal to the residual condition (hatched area in Fig. 8.2), the pore water exists in a pendular state of isolated pockets, thin films, or disconnected menisci among the soil grains. Here, there is essentially no continuous liquid phase and pore water transport occurs primarily by vapor transport mechanisms (Section 8.6). The flow of pore air, on the other hand, readily occurs and is driven by gradients in total air potential.

When water content is some value greater than the residual water content but less than the occluded-air-bubble water content (shaded region in Fig. 8.2), simultaneous pore air and pore water flow is permissible. The driving potential for airflow is the gradient in total air potential, and the driving potential for liquid flow is the gradient in total pore water potential. Hydraulic conductivity is some value less than its value at saturation and is a function of the degree of saturation of the soil matrix. The hydraulic conductivity function (Section 8.3) describes the characteristic relationship between hydraulic conductivity and saturation, water content, or matric suction.

The occluded-air-bubble water content regime describes a condition where the water content of the system is such that the remaining pore air exists

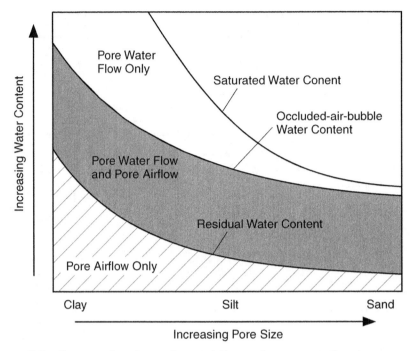

Figure 8.2 Conceptual regimes of pore airflow and pore water flow in unsaturated soil.

primarily as isolated bubbles among the soil grains. Here, continuous paths for the flow of air are essentially cut off and the primary mechanism for pore air transport becomes gaseous diffusion through the water phase (Section 8.7). Hydraulic conductivity in this regime is relatively high, approaching its maximum value at full saturation.

8.1.4 Steady-State Flow Law for Water

Darcy's law states that the discharge velocity of fluid from a porous medium, v, is linearly proportional to the gradient in the relevant driving head, ∇h, written as

$$v = -k\, \nabla h \tag{8.4}$$

where k is a proportionality term describing the conductivity of the porous medium (m/s). The negative sign preceding the right-hand side of eq. (8.4) indicates that fluid flow occurs from a locale of relatively high head to a locale of relatively low head. Seepage velocity v_s, which describes the average actual flow velocity through the pores of the medium, is the discharge velocity divided by the medium porosity, that is, $v_s = v/n$.

The proportionality term k in eq. (8.4) describes the ability for a specific porous medium under specific conditions to transmit a specific fluid. For the flow of pore water in soil, the driving gradient is the water potential or hydraulic head, and the constant of proportionality is the hydraulic conductivity (k or k_w). For the flow of pore air in soil, the driving gradient is the total head with respect to the air phase (absolute air pressure) and the constant of proportionality is the air conductivity (k_a).

In principle, the total water potential [eq. (8.3)] can be employed in Darcy's law to describe steady liquid flow in unsaturated soil. Darcy's law in three-dimensional space for unsaturated soil is written in terms of total head as

$$\mathbf{q} = -k_x(h_m)\frac{\partial h_t}{\partial x}\mathbf{i} - k_y(h_m)\frac{\partial h_t}{\partial y}\mathbf{j} - k_z(h_m)\frac{\partial h_t}{\partial z}\mathbf{l} \qquad (8.5)$$

where $h_t = \psi_t/\rho_w g$, $\mathbf{i}$, $\mathbf{j}$, and $\mathbf{l}$ are unit vectors in the x, y, and z directions, respectively, and $k_x(h_m)$, $k_y(h_m)$, and $k_z(h_m)$ are the hydraulic conductivity functions in each coordinate direction.

Solving the above equation under appropriate boundary conditions provides a quantitative description of the total head field and forms the classical approach to analyzing a rich body of seepage problems. Some of these classical approaches will be illustrated in Section 8.5. Modern approaches often involve imposing the principle of mass conservation to Darcy's law. For steady flow, the principle of mass conservation states that the net flow through any element at a fixed point in space is zero and independent of time:

$$\nabla \cdot (k \, \nabla h_t) = 0 \qquad (8.6a)$$

In most solutions of the total head field described by eq. (8.6a), matric suction and gravitational heads have been considered, whereas osmotic head has been ignored. By imposing this omission ($h_t = h_m + z$), eq. (8.6a) in two-dimensional (horizontal and vertical) scalar form becomes

$$\frac{\partial k_x}{\partial x}\frac{\partial h_m}{\partial x} + \frac{\partial k_z}{\partial z}\left(\frac{\partial h_m}{\partial z} + 1\right) + k_x \frac{\partial^2 h_m}{\partial x^2} + k_z \frac{\partial^2 h_m}{\partial z^2} = 0 \qquad (8.6b)$$

8.2 PERMEABILITY AND HYDRAULIC CONDUCTIVITY

8.2.1 Permeability versus Conductivity

Discharge velocity is proportional to the viscosity and density of the permeant fluid, being higher for relatively high density or low viscosity fluids. These proportionalities may be captured mathematically as

$$v \propto \frac{\rho g}{\mu}$$

where ρ is the fluid density (kg/m³), g is gravitational acceleration (m/s²), and μ is the dynamic (absolute) fluid viscosity (N · s/m²) (Section 2.1).

Experimental results and theoretical considerations also reveal that discharge velocity is highly dependent on pore size and pore size distribution. Following Poiseuille's law, the discharge velocity is proportional to the square of the pore diameter d, or

$$v \propto d^2$$

Combining the above two proportionalities with Darcy's original observation that discharge velocity is linearly proportional to the total head gradient leads to

$$v = -C \frac{d^2 \rho g}{\mu} \nabla h_t \tag{8.7}$$

where C is a dimensionless constant related to the geometry of the soil pores. Comparing eq. (8.7) with eq. (8.4) leads to

$$k = (Cd^2)\left(\frac{\rho g}{\mu}\right) \tag{8.8}$$

If intrinsic permeability K is defined as

$$K = Cd^2 \tag{8.9}$$

then, together with eq. (8.8), the relationship between intrinsic permeability and hydraulic conductivity becomes

$$k = \frac{\rho g}{\mu} K \tag{8.10}$$

Intrinsic permeability, often simply referred to as permeability, has units of length squared (m²) and is dependent only on the pore size, pore geometry, and pore size distribution. Permeability may also be expressed in units of darcies where one darcy is approximately equal to 10^{-12} m². Permeability is the same for any given soil regardless of the properties of the fluid being transmitted as long as the pore structure remains unaltered.

Consider two identical columns of soil having intrinsic permeability $K = 10^{-12}$ m², a possible value for silty material. One column is completely saturated with water and the other is completely dry (saturated with air). An intriguing question arises as to which pore fluid, air or water, will have the higher discharge velocity for the same applied gradient with respect to that fluid?

Assuming that the density of water is about 1000 kg/m³ and the dynamic viscosity is 1.0×10^{-3} N · s/m² at 20°C, the hydraulic conductivity for the soil column may be calculated from eq. (8.10) as

$$k_w = \frac{\rho_w g}{\mu_w} K = \frac{(1000 \text{ kg/m}^3)(9.81 \text{ m/s}^2)}{1 \times 10^{-3} \text{ N} \cdot \text{s/m}^2} (10^{-12} \text{ m}^2) = 9.6 \times 10^{-6} \text{ m/s}$$

For a density of air equal to 1 kg/m³ and dynamic viscosity equal to 1.785×10^{-5} N · s/m², the air conductivity is

$$k_a = \frac{\rho_a g}{\mu_a} K = \frac{(1 \text{ kg/m}^3)(9.81 \text{ m/s}^2)}{1.785 \times 10^{-5} \text{ N} \cdot \text{s/m}^2} (10^{-12} \text{ m}^2) = 0.55 \times 10^{-6} \text{ m/s}$$

The above calculation indicates that, contrary to some intuitive guesses, the conductivity of water is about 17.5 times higher than the conductivity of air. Consequently, the discharge velocity is 17.5 times greater.

The determining factors in the above analysis are clearly the material properties of the fluid, described through the quantity $\rho g/\mu$. As introduced in Section 2.1, the viscosity and density of both air and water are functions of temperature, thus illustrating the potentially important impact that variations in state variables may have on the material variables governing flow phenomena in soil. In fine-grained soil, fluid properties other than viscosity and density may also significantly influence hydraulic conductivity. These include chemical and electrical fluid characteristics that act to either alter the pore fabric (e.g., swelling, shrinkage, dispersion) or to induce fluid flow under induced or applied gradients in electric or chemical potential. Changes in the concentration and type of chemical species in the pore fluid can significantly affect the soil structure, particularly for materials comprised of or containing expansive clay. Significant deviations from Darcy's law have been observed for flow through fine-grained materials as a result of these and other effects.

8.2.2 Magnitude, Variability, and Scaling Effects

The magnitude of permeability varies extraordinarily from one type of soil to another, ranging from as high as perhaps 10^{-7} m² for gravel to as low as 10^{-20} m² for heavily overconsolidated clay. Figure 8.3 illustrates the wide range of permeability values for a variety of soil types and the relationships between permeability, hydraulic conductivity, and air conductivity. Note that permeability can vary several orders of magnitude for the same nominal type of soil. Permeability is also sensitive to changes in pore structure, may display anisotropic behavior, and is scale dependent, generally increasing as the rep-

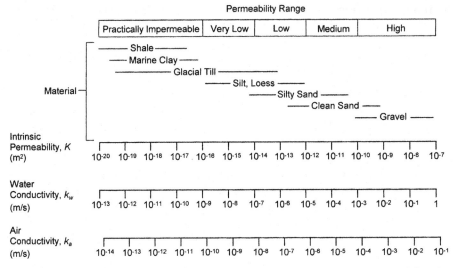

Figure 8.3 Intrinsic permeability, hydraulic conductivity, and air conductivity for variety of soil types.

resentative volume of soil under consideration increases. Fabric and scaling sensitivity is particularly significant in fine-grained clayey soil.

The fabric of soil is often described in terms of three levels of increasing scale by the microfabric, the minifabric, and the macrofabric. Mitchell (1993) describes the *microfabric* as consisting of aggregations, or "clusters," of fine-grained particles, including the micron- or submicron-scale pores formed between neighboring particles comprising these aggregations. Fluid flow through these very small pores is relatively limited and is markedly influenced by the physical and physicochemical interactions occurring at the soil solid–pore water interface. The *minifabric* includes particle clusters defining the microfabric as well as the assemblage of larger pores formed between and among these clusters. Minifabric pores may be up to several tens of micrometers in size. Consequently, fluid flow in this regime (i.e., around the clusters) is much more significant than fluid flow in the microfabric regime (i.e., through the clusters). On a yet larger scale, the *macrofabric* contains cracks, fissures, anisotropic formations and bedding, or other relatively large-scale features. The dissimilar and directionally dependent permeability of these large-scale features may completely obscure the much smaller microfabric and minifabric flow and dominate the overall permeability of the soil mass. As a result, it is critical to ensure representative testing elements (specimens) when permeability or hydraulic conductivity is measured in either the laboratory or field. In attempting to capture permeability in terms of a single number obtained from laboratory tests, field tests, or theoretical or empirical considerations, it is important to recognize the limitations of the measurement and to appreciate the consequent impact on the engineering problem at hand.

8.3 HYDRAULIC CONDUCTIVITY FUNCTION

8.3.1 Conceptual Model for the Hydraulic Conductivity Function

The hydraulic conductivity of unsaturated soil is a function of material variables describing the pore structure (e.g., void ratio and porosity), the pore fluid properties (e.g., density and viscosity), and the relative amount of pore fluid in the system (e.g., water content and degree of saturation). The *unsaturated hydraulic conductivity function* describes the characteristic dependence on the relative amount of pore fluid in the system. The hydraulic conductivity function is typically described in terms of matric suction head $k(h_m)$, matric suction $k(\psi)$, degree of saturation $k(S)$, or volumetric water content $k(\theta)$.

To better understand the hydraulic conductivity function, consider the conceptual model illustrated as Fig. 8.4, which shows a series of cross-sectional areas for a rigid mass of relatively coarse-grained soil (e.g., sand). The soil is initially saturated at condition 8.4(*a*) and allowed to drain under increasing suction through conditions 8.4(*b*) and 8.4(*c*) to a residual condition at point 8.4(*d*). The soil-water characteristic curve $\theta(\psi)$ and hydraulic conductivity function $k(\psi)$ corresponding to these four saturation conditions are conceptualized as Figs. 8.5*a* and 8.5*b*, respectively.

At condition (*a*) in Fig. 8.5, the soil matrix is completely saturated and the matric suction is zero. The saturated volumetric water content θ_s is equal to about 0.34 (Fig. 8.5*a*) and the saturated hydraulic conductivity k_s is equal to about 2×10^{-3} cm/s (Fig 8.5*b*), both reasonable values for sand. The saturated hydraulic conductivity is a maximum for the system because the cross-sectional area of pore space available for the conduction of water is at its maximum. Conversely, the air conductivity at condition (*a*) is effectively zero. Between points (*a*) and (*b*), the soil matrix sustains a finite amount of suction prior to desaturation, which commences at the air-entry pressure. The soil remains saturated within this regime and the hydraulic conductivity may decrease slightly as the air-entry pressure is approached. Condition (*b*) represents the air-entry pressure, corresponding to the point where air begins to enter the largest pores. A further increase in suction from this point results in continued drainage of the system. At point (*c*), drainage under increasing suction has resulted in a significant decrease in both the water content and hydraulic conductivity. The reduction in conductivity continues with increasing suction as the paths available for water flow continue to become smaller and more tortuous. The reduction is initially relatively steep because the first pores to empty are the largest and most interconnected and, consequently, the most conductive to water. At point (*d*), which occurs near the residual water

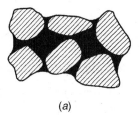

(a)

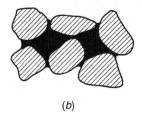

(b)

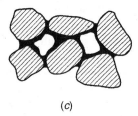

(c)

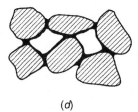

(d)

Figure 8.4 Conceptual distributions of pore water and pore air in a cross-sectional area of rigid soil matrix during incremental drainage process.

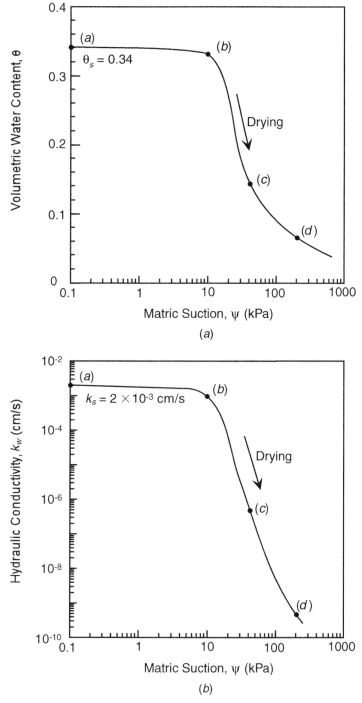

Figure 8.5 (*a*) Conceptual soil-water characteristic curve and (*b*) hydraulic conductivity function corresponding to saturation conditions for rigid soil matrix shown in Fig. 8.4.

content, the pore water exists primarily in the form of disconnected menisci among the soil grains. Here, the hydraulic conductivity reduces effectively to zero and pore water is transported primarily through the vapor phase. Typical of many soils, the total change in the magnitude of hydraulic conductivity from point (*a*) to point (*d*) is over six orders of magnitude.

8.3.2 Hysteresis in the Hydraulic Conductivity Function

If the previous thought experiment were to continue along a rewetting path starting from point (*d*), hysteresis would be observed in both the soil-water characteristic curve and the hydraulic conductivity function. Because the soil-water characteristic curve exhibits hysteresis (Fig. 8.6*a*), and because hydraulic conductivity is directly related to the soil-water content, hysteresis becomes evident when hydraulic conductivity is plotted as a function of suction or suction head (Fig. 8.6*b*). Hydraulic conductivity is generally greater along a drying path (where the volume fraction of liquid-filled pores is greater) than for the same magnitude of suction along a wetting path.

On the other hand, only minor hysteresis is noted in the relationship between hydraulic conductivity and volumetric water content $k_w(\theta)$ or degree of saturation $k_w(S)$. This observation is commonly attributed to the fact that hydraulic conductivity is directly related to the volume fraction of the pore space available for liquid flow, which is directly described by either θ or S. Childs (1969), however, cautions that although volumetric water content and degree of saturation are indeed direct descriptions of the fraction of liquid-filled pores, neither can specifically identify the characteristics of those pores that are in fact filled. Pores that are filled during drying may certainly be different in size and shape than those that are filled during wetting, having a consequent effect on the hydraulic conductivity. In the majority of cases, these possible hysteretic effects are neglected in light of the advantages afforded by expressing k as a unique function of either θ or S in simplifying the prediction and modeling of unsaturated fluid flow phenomena.

8.3.3 Relative Conductivity

It is common to normalize the air and water conductivity of unsaturated soil with respect to their maximum values at complete air saturation and water saturation, respectively. These normalized values, referred to as *relative conductivity*, may be written as

$$k_{rw} = \frac{k_w}{k_{sw}} \tag{8.11a}$$

$$k_{ra} = \frac{k_a}{k_{sa}} \tag{8.11b}$$

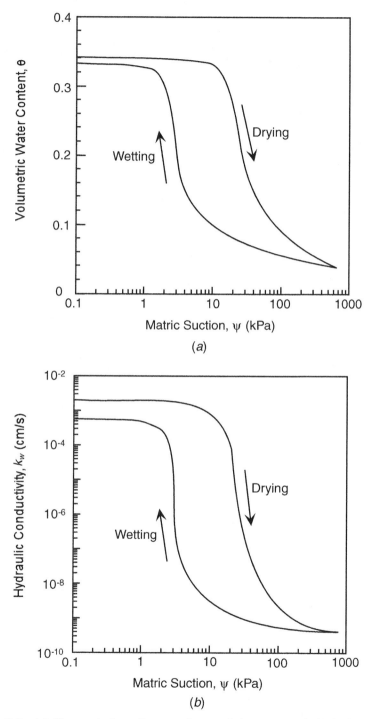

Figure 8.6 (a) Hysteresis in soil-water characteristic curve and (b) hydraulic conductivity function $k_w(\psi)$.

where k_{rw} and k_{ra} are the relative conductivity of water and air, respectively, and k_{sw} and k_{sa} are conductivities at 100% water saturation and 100% air saturation, respectively. Relative conductivity is a dimensionless scalar ranging from 0 to 1. Figure 8.7 illustrates the offsetting nature of relative air conductivity and relative water conductivity as functions of degree of water saturation for a hypothetical soil modeled using the Brooks and Corey (1964) model described Chapter 12.

8.3.4 Effects of Soil Type

Figure 8.8 compares hydraulic conductivity functions for a relatively coarse-grained material (Superstition sand) and a relatively fine-grained material (Yolo light clay) in the form $k_w(\psi)$ (Fig. 8.8a) and $k_{rw}(\psi)$ (Fig. 8.8b). The saturated hydraulic conductivity for the sand and clay are 1.83×10^{-3} cm/s and 1.23×10^{-5} cm/s, respectively. The dramatic reduction in the sand's hydraulic conductivity at approximately 2 to 3 kPa suction corresponds to drainage of the pores at the air-entry pressure. Following capillary theory, the relatively low air-entry value reflects the relatively large pores of the sand. The relatively sharp transition into the regime of decreasing hydraulic con-

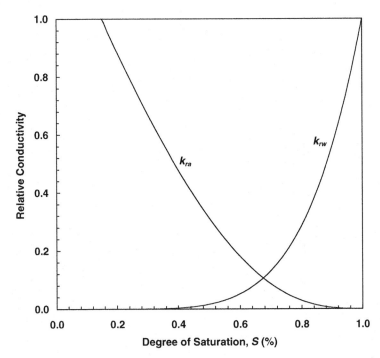

Figure 8.7 Relationships among relative air conductivity, relative water conductivity, and degree of saturation.

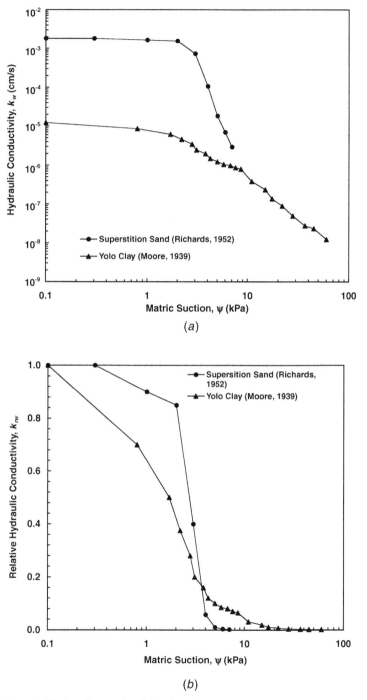

Figure 8.8 (*a*) Hydraulic conductivity function and (*b*) relative hydraulic conductivity function for Superstition sand (Richards, 1952) and Yolo light clay (Moore, 1939).

ductivity reflects the sand's relatively well-defined air-entry pressure. The steep decline in conductivity beyond the air-entry value reflects the sand's relatively uniformly distributed pore size, resulting in relatively "rapid" drainage over a narrow range of suction (i.e., the majority of the pores are drained over a narrow range of suction). By comparison, the smoother behavior of the conductivity function for the clay reflects its relatively poorly defined air-entry pressure, small pores, and well-distributed array of pore sizes.

Figure 8.9 shows hydraulic conductivity functions in the form $k_w(h)$ for a silty material and sandy material together on the same plot. The hydraulic conductivity of the sand at saturation ($h = 0$) is more than one order of magnitude greater than that of the silt, a direct reflection of the relatively large pores formed among the larger sand grains. As both soils desaturate under increasing suction, however, a crossover point is reached where the silt becomes significantly more conductive to water than the sand. The crossover point occurs because the sand's relatively large pores desaturate quite completely and uniformly with only a small increase in suction, whereas a larger fraction of the silt's smaller pores remain available to conduct water at increasingly large values of suction. This observation is a direct reflection of

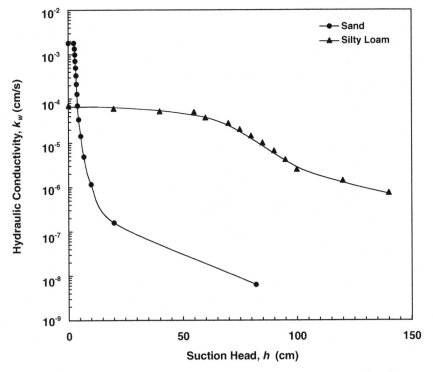

Figure 8.9 Hydraulic conductivity functions for silty loam and sand showing crossover point (data from Hillel, 1982).

capillary theory and the Young-Laplace equation described in Chapters 3 and 4. The capillary barrier systems described in the following section rely on this interesting and fundamental disparity in the hydraulic conductivity and water retention characteristics of relatively coarse-grained and fine-grained unsaturated soils.

8.4 CAPILLARY BARRIERS

8.4.1 Natural and Engineered Capillary Barriers

Capillary barriers are formed at the interface of hydrologically dissimilar unsaturated soil strata where a relatively fine soil layer overlies a relatively coarse soil layer. Figure 8.10 shows a conceptual diagram for a two-layer capillary barrier system located near the ground surface. Under unsaturated conditions, the capillary tension at the interface between the soil layers prohibits the movement of water from the fine layer into the coarse layer. Percolating groundwater can be hydrostatically suspended, stored, or rerouted within or above the fine layer. If the rate of subsequent evaporation, lateral drainage, or vegetative uptake out of the fine layer exceeds the influx, then leaching into the underlying coarse layer can be prevented.

Significant interest has been shown in recent years regarding the use of engineered or natural capillary barriers for isolating buried waste from water

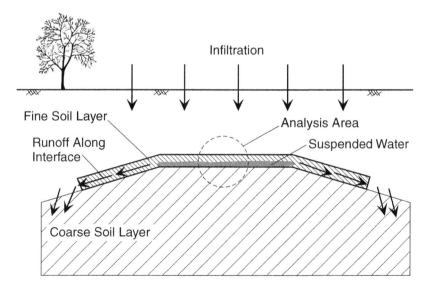

Figure 8.10 Suspension and diversion of infiltrating water by two-layer capillary barrier system. Suspended water located above flat interface is at hydrostatic equilibrium. Runoff diverted along inclined interface is under steady-state flow conditions.

342 STEADY FLOWS

percolating through the near-surface unsaturated soil zone. In semiarid or arid regions, these types of barriers are often an effective alternative as a part of the final cover system for municipal solid waste landfills. Understanding the working principles of capillary barriers provides the necessary guidelines for engineering design of variously configured barrier systems under different soil and climatic conditions.

8.4.2 Flat Capillary Barriers

The working principles of capillary barriers can be illustrated through a relatively simple equilibrium analysis of head within a two-layer system under hydrostatic conditions. Consider the progressive accumulation of water in an area within the fine layer of Fig. 8.10 located above the coarse layer, shown within the dashed circle. Figure 8.11 conceptualizes the interface at this location as a thin transitional zone where equilibrium considerations may be applied. The transition between the fine soil layer and the coarse soil layer is idealized as a cone-shaped pore with radii on either side corresponding to the average pore sizes of the fine and coarse layers. In each stage of water accumulation, the hydrostatic, or no-flow, equilibrium condition is assumed. This leads to a constant head in the vertical direction where the water phase is continuous. If it is assumed that the solid-liquid contact angle in both soil layers is zero ($\alpha = 0$) and the air pressure is a zero reference value ($u_a = 0$), then mechanical equilibrium at the air-water-solid interface in the fine soil requires the following to be true when the suspended water lens is infinitesimally thin (Fig. 8.11a):

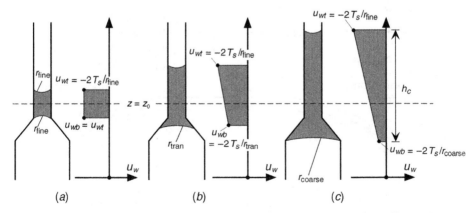

Figure 8.11 Hydrostatic equilibrium of capillary water near interface of fine soil–coarse soil capillary barrier system: (a) thin suspended water layer, (b) intermediate suspended water layer, and (c) water layer at threshold of breakthrough.

$$u_{wt} = -\frac{2T_s}{r_{\text{fine}}} \qquad (8.12)$$

where u_{wt} is the pore water pressure at a point near the air-water interface in the fine soil, and r_{fine} is a representative pore radius for the fine soil. Similarly, mechanical equilibrium at the air-water-solid interface near the bottom of the water lens leads to

$$u_{wb} = -\frac{2T_s}{r_{\text{fine}}} \qquad (8.13)$$

which is equal to the pore pressure near the top since the water lens is infinitesimally thin (i.e., $u_{wt} = u_{wb}$).

As the overlying water lens becomes thicker as shown in Fig. 8.11b, the total head buildup due to gravity requires the pore water to move slightly into the transitional zone between the fine and coarse layers. The pore pressure near the bottom of the lens is greater than that near the top by an amount proportional to the thickness of the water lens and $\rho_w g$. At mechanical equilibrium, the pore pressure near the bottom of the water lens becomes

$$u_{wb} = -\frac{2T_s}{r_{\text{tran}}} \qquad (8.14)$$

where r_{tran} is the equilibrium radius in the transition zone. Since r_{tran} is generally smaller than the representative radius of the coarse soil r_{coarse}, but larger than the representative radius of the fine soil r_{fine}, the pore pressure described by eq. (8.14) is less than the water-entry pressure of the coarse soil. The water-entry pressure of the coarse soil, defined as the pressure at which water begins to enter the coarse soil layer, can be expressed as

$$u_w = -\frac{2T_s}{r_{\text{coarse}}} \qquad (8.15)$$

As the thickness of the water lens progressively increases under infiltration from the ground surface, the pore pressure near the bottom of the water lens increases and the wetting front progressively advances to a new equilibrium position. When this pressure is equal to the water-entry pressure of the coarse layer, the wetting front advances to the position at the end of the transition zone (Fig. 8.11c). Here, the mechanical equilibrium is at a breakthrough threshold that leads to

$$h_c \rho_w g = u_{wb} - u_{wt} = \frac{2T_s}{r_{fine}} - \frac{2T_s}{r_{coarse}} \tag{8.16}$$

where h_c is the "breakthrough," or critical, head. At this point, the capillary barrier fails and water flows into the coarse soil if any additional water is added into the system, persisting for as long as the following condition is satisfied:

$$h_c \geq \frac{2T_s}{\rho_w g r_{fine}} - \frac{2T_s}{\rho_w g r_{coarse}} \tag{8.17}$$

It is instructive to recognize that in the proceeding stages of Figs. 8.11a, 8.11b, and 8.11c, the total head (or pore pressure at $z = z_0$) increases as the thickness of water lens increases. The first term on the right-hand side of the above condition controls the magnitude of the minimum pore water pressure. The second term controls the maximum pore water pressure (see pressure profile in Fig. 8.11c). Accordingly, the larger the difference between the two terms, the more effective the capillary barrier. It follows that the ideal capillary barrier consists of two soils with a sharp disparity in pore size or particle size. Each should also have a relatively uniform particle size distribution to minimize the possibility of overlapping pore sizes.

The soil-water characteristic curve becomes a useful constitutive parameter for the design of capillary barrier systems. For a perfectly wetting material, the representative pore radius r_{fine} may be related to the air-entry pressure value u_b or the parameter α used to model the soil-water characteristic curve in many mathematical formulations (Chapter 12):

$$r_{fine} = \frac{2T_s}{u_b} = 2T_s \alpha_{fine} \tag{8.18}$$

Similarly, the representative pore radius for the coarse soil r_{coarse} may be related to the water-entry pressure u_w, which has been suggested to be half of the air-entry pressure (Bouwer, 1966):

$$r_{coarse} = \frac{2T_s}{u_w} = T_s \alpha_{coarse} \tag{8.19}$$

Given eqs. (8.18) and (8.19), eq. (8.17) can be written as

$$h_c \geq \frac{1}{\rho_w g \alpha_{fine}} - \frac{2}{\rho_w g \alpha_{coarse}} \tag{8.20}$$

Example Problem 8.1 If the overlying fine soil layer in a flat two-layer capillary barrier system is a silty sand with $\alpha_{fine} = 0.29 \text{ kPa}^{-1}$, the coarse soil layer is a sandy gravel with $\alpha_{coarse} = 29$ kPa^{-1}, and both soils are considered perfectly wetting materials, estimate the maximum thickness of the water lens prior to breakthrough failure.

Solution From eq. (8.20), the maximum thickness of the water lens is

$$h_c = \frac{1}{\rho_w g \alpha_{fine}} - \frac{2}{\rho_w g \alpha_{coarse}} = \frac{1}{(1000 \text{ kg/m}^3)(9.8 \text{ m/s}^2)(0.00029 \text{ m}^2/\text{N})}$$

$$- \frac{2}{(1000 \text{ kg/m}^3)(9.8 \text{ m/s}^2)(0.029 \text{ m}^2/\text{N})}$$

$$= 0.352 \text{ m} - 0.007 \text{ m} = 0.345 \text{ m}$$

The above calculation shows that the pressure head in the coarse soil is negligible in comparison with the pressure head in the fine soil. Thus, the maximum thickness of the suspended water in the fine soil can be simply estimated as

$$h_c = \frac{1}{\rho g \alpha_{fine}} \qquad (8.21)$$

If a nonzero contact angle α is considered, the maximum height of the suspended water will be decreased by a factor of $\cos \alpha$

$$h_c = \frac{\cos \alpha}{\rho g \alpha_{fine}} \qquad (8.22)$$

For example, if the soil considered in Example Problem 8.1 has a contact angle of 60°, a possible value for initially dry silty soil under wetting conditions, the maximum thickness of the suspended water would be decreased from 0.345 to 0.173 m.

8.4.3 Dipping Capillary Barriers

Flat capillary barriers may not be effective if the coarse soil is prewetted, if there is a continuous downward infiltration, or if fingering flow exists due to heterogeneity in the overlying fine soil. One effective way to significantly improve the performance of capillary barriers is to dip the interface between the two layers.

In the regions of Fig. 8.10 where the interface between the two soils is inclined, the flow of water may be confined within the fine layer and subsequently forced to run along the dipping portions of the interface under the combined forces of capillarity and gravity. Figure 8.12 illustrates the flow field in a natural dipping capillary barrier system under steady-state flow

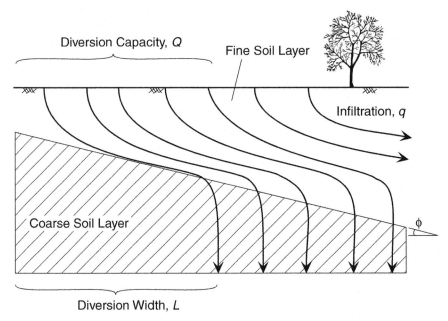

Figure 8.12 Steady-state flow field in dipping two-layer capillary barrier system.

conditions. To preserve the notion used in the literature, the angle of the dip is denoted ϕ.

Consider the pore pressure variation along the dipping interface within the fine soil. Near the top of the dipping interface, the suction in the fine soil is higher (more negative) than that in the underlying coarse soil. Accordingly, water flows along the interface. The magnitude of lateral flow increases along the down dip direction because the volume of infiltration increases, leading to a decrease in suction in the fine soil. At a sufficiently far location along the interface, the suction is equal to the water-entry pressure of the coarse soil and the water begins to break through the capillary barrier and flow downward into the coarse soil. The lateral distance from the highest portion of the dipping barrier (left side of Fig. 8.12) to the point of breakthrough is referred to as the *diversion width* (L). The total flux of water that is diverted is called the *diversion capacity* Q. Diversion capacity Q is equal to qL, where q is the steady infiltration flux.

Both diversion width and diversion capacity depend on the unsaturated hydrologic properties of the fine and coarse soil, the system geometry including the distance to the ground surface and the water table, the angle of dip ϕ, and the uniform steady flux at a position far from the interface. Quantitative relationships among these quantities can be established by solving the steady flow equation introduced in the previous section. Ross (1990), for example, assumed that the unsaturated hydraulic conductivity functions for both the coarse and fine soil could be described as exponential functions in the form

$$k_{\text{fine}} = k_{s,\text{fine}} \exp(\gamma_w \alpha_{\text{fine}} h_m) \qquad k_{\text{coarse}} = k_{s,\text{coarse}} \exp(\gamma_w \alpha_{\text{coarse}} h_m)$$

Ross (1990) then analytically solved eq. (8.6) to arrive at a upper bound for the diversion capacity and diversion width of a dipping capillary barrier system as follows:

$$Q_{\max} < \frac{k_{s,\text{fine}} \tan \phi}{\gamma_w \alpha_{\text{fine}}} \tag{8.23}$$

$$L < \frac{k_{s,\text{fine}} \tan \phi}{\gamma_w \alpha_{\text{fine}} q} \tag{8.24}$$

This approach was broadened by Steenhuis et al. (1991) to include the air-entry head of the fine soil layer $h_{a,\text{fine}}$ and the water-entry head $h_{w,\text{coarse}}$ of the coarse layer using the following model for the hydraulic conductivity function:

$$k = \begin{cases} k_s & |h_m| < h_{a,\text{fine}} \\ k_s \exp[\gamma_w \alpha_{\text{fine}}(h_m + h_{a,\text{fine}})] & |h_m| \geq h_{a,\text{fine}} \end{cases} \tag{8.25}$$

which allowed the following solution for diversion width:

$$L \leq \frac{k_s \tan \phi}{q} \left[\frac{1}{\gamma_w \alpha_{\text{fine}}} + (h_{a,\text{fine}} - h_{w,\text{coarse}}) \right] \tag{8.26}$$

where the diversion capacity from eq. (8.26) is

$$Q_{\max} \leq k_s \tan \phi \left[\frac{1}{\gamma_w \alpha_{\text{fine}}} + (h_{a,\text{fine}} - h_{w,\text{coarse}}) \right] \tag{8.27}$$

The above expressions are more appropriate when the air-entry head is nonzero. If the exchangeability between the term $1/\gamma_w \alpha$ and the parameter h_a is recognized, the above expressions become

$$L \leq \frac{k_s \tan \phi}{q} (2h_{a,\text{fine}} - h_{w,\text{coarse}}) \tag{8.28}$$

$$Q_{\max} \leq k_s \tan \phi (2h_{a,\text{fine}} - h_{w,\text{coarse}}) \tag{8.29}$$

The water-entry head h_w has also been related to air-entry head h_a, but general consensus on this relationship has not been reached. Bouwer (1966), for example, suggested that the air-entry head is twice the water-entry head, while Walter et al. (2000) suggest that the air-entry head is equal to the water-entry head. For $h_w = h_a/2$, expressions (8.28) and (8.29) become

$$L \leq \frac{k_s \tan \phi}{2q} (4h_{a,\text{fine}} - h_{a,\text{coarse}}) \tag{8.30}$$

$$Q_{\max} \leq \frac{k_s \tan \phi}{2} (4h_{a,\text{fine}} - h_{a,\text{coarse}}) \tag{8.31}$$

and for $h_w = h_a$, expressions (8.28) and (8.29) become

$$L \leq \frac{k_s \tan \phi}{q} (2h_{a,\text{fine}} - h_{a,\text{coarse}}) \tag{8.32}$$

$$Q_{\max} \leq k_s \tan \phi (2h_{a,\text{fine}} - h_{a,\text{coarse}}) \tag{8.33}$$

For an ideal capillary barrier, the air-entry head of the coarse soil is much smaller than that of the fine soil and the maximum diversion width becomes

$$L_{\max} \leq \frac{2k_s h_{a,\text{fine}} \tan \phi}{q} \tag{8.34}$$

For design purposes, the efficiency of a capillary barrier may be defined as $\omega = L/L_{\max}$. Combining eq. (8.32) and eq. (8.34), therefore, leads to

$$\frac{h_{a,\text{fine}}}{h_{a,\text{coarse}}} = \frac{1}{2 - 2\omega} \tag{8.35}$$

A less conservative expression for the ratio of $h_{a,\text{fine}}/h_{a,\text{coarse}}$ can be derived by combining eq. (8.30) and eq. (8.34):

$$\frac{h_{a,\text{fine}}}{h_{a,\text{coarse}}} = \frac{1}{4 - 4\omega} \tag{8.36}$$

Both eq. (8.35) and eq. (8.36) provide practical design guidelines for soil involved in capillary barrier systems. For example, if 90% barrier efficiency were required, the ratio of the air-entry head between the fine and coarse soil according to eq. (8.35) would be at least 5.0. The ratio of the air-entry head between the fine and coarse soil according to eq. (8.36) would be at least 2.5.

Example Problem 8.2 A waste package is to be disposed and isolated from infiltration using a dipping two-layer capillary barrier system. If the steady infiltration rate at the site is 0.2 m/yr, the saturated hydraulic conductivity of the overlying fine soil layer is 10^{-7} m/s, the air-entry head of the fine soil is 0.5 m, and the width of the waste package underlying the fine soil layer is 10 m, what is the required dip angle for the barrier?

Solution If a symmetrical capillary barrier is chosen, the diversion width L may be specified as half the width of the waste package, that is, 5.0 m in the horizontal direction. From eq. (8.34), the required dip angle is

$$\phi \geq \tan^{-1}\left(\frac{qL}{2k_s h_{a,\text{fine}}}\right) = \tan^{-1}\frac{(0.2 \times 3.171 \times 10^{-8} \text{ m/s})(5.0 \text{ m})}{(2 \times 10^{-7} \text{ m/s})(0.5 \text{ m})} = 17.6°$$

8.5 STEADY INFILTRATION AND EVAPORATION

8.5.1 Horizontal Infiltration

The fundamental difference between saturated and unsaturated steady-state flow is that the head distribution is linearly distributed in the former and nonlinearly distributed in the latter. For saturated flow in the one-dimensional, homogenous soil column shown in Fig. 8.13, for example, the hydraulic gradient anywhere in the flow domain is a constant. Accordingly, the head loss over a unit distance is equal to the flow rate divided by the hydraulic conductivity. This simple scaling rule is no longer valid for unsaturated fluid flow because the hydraulic conductivity depends on the absolute value of the driving head.

In this section, the nonlinear nature of head distribution and hydraulic conductivity in unsaturated soil will be illustrated through a one-dimensional horizontal flow problem. Horizontal flow is selected to illustrate these basic principles because it may be considered free from the influence of gravity. As such, the elevation head can be ignored and flow is driven solely by the gradient in suction head.

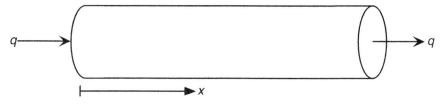

Figure 8.13 One-dimensional, steady-state horizontal flow system for comparing head distribution between saturated and unsaturated conditions.

Head distribution under steady saturated conditions is first developed for comparison. For the one-dimensional, steady saturated flow system illustrated in Fig. 8.13, the governing flow equation is Darcy's law:

$$q = -k_s \frac{dh}{dx} \qquad (8.37)$$

Integrating the above equation and imposing the boundary condition of $h = 0$ at $x = 0$ leads to the following analytical solution for head as a function of position x along the column:

$$h = -\frac{q}{k_s} x \qquad (8.38)$$

The thick line in Fig. 8.14 is a plot of this linear head distribution for $q/k_s = 0.4$. The slope of the distribution is a constant equal to $-q/k_s = 0.4$, which is the hydraulic gradient.

If the unsaturated hydraulic conductivity function is modeled using Gardner's (1958) one-parameter model (Chapter 12):

$$k = k_s \exp(\beta h_m)$$

then Darcy's law may be written in the form

$$q = -k_s \exp(\beta h_m) \frac{dh}{dx} \qquad (8.39)$$

Recognizing $h_m = h$ for zero gravity and integrating the above equation and imposing the same boundary conditions as those for the saturated flow condition results in the following head distribution along the length of the column:

$$h = \frac{1}{\beta} \ln\left(1 - \frac{q\beta x}{k_s}\right) \qquad (8.40)$$

If the solution of eq. (8.40) is confined by the condition of $-1.0 < q\beta x/k_s \leq 1.0$, which is the case for most practical seepage problems, it may be rewritten using a Taylor series expansion as

$$h = -\frac{1}{\beta}\left[\frac{q\beta x}{k_s} - \frac{1}{2}\left(\frac{q\beta x}{k_s}\right)^2 + \frac{1}{3}\left(\frac{q\beta x}{k_s}\right)^3 - \cdots\right] = \sum_{n=1}^{m=\infty} (-1)^n \frac{\beta^{n-1}}{n}\left(\frac{qx}{k_s}\right)^n \qquad (8.41)$$

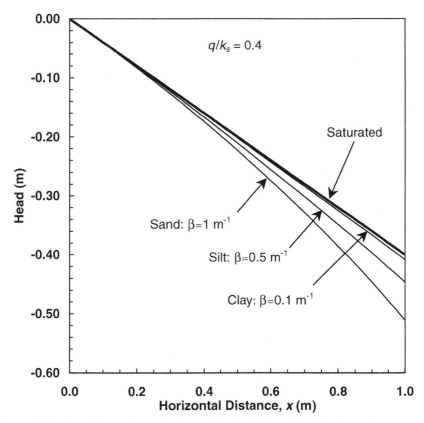

Figure 8.14 One-dimensional, horizontal hydraulic head distribution under saturated and unsaturated conditions.

where the first term in the expansion is identical to the saturated solution [eq. (8.38)] and the remaining terms capture the nonlinear nature of the head distribution. The physical consequence of these additional terms is that the unsaturated hydraulic conductivity becomes smaller and smaller as the suction head increases. Accordingly, the head loss under a constant flow rate becomes larger.

Equation (8.40) is plotted along with the saturated solution in Fig. 8.14 for three representative types of soil and $q/k_s = 0.4$. It can be observed that the nonlinear behavior of head under unsaturated steady flow is most pronounced for sand (modeled using $\beta = 1$ m^{-1}). This observation is readily explained by the sand's characteristically rapid decay in hydraulic conductivity with increasing matric suction head as water drains from the soil. The slopes of the lines shown in Fig. 8.14 also illustrate that the absolute value of the hydraulic gradient for unsaturated flow increases as the distance from the influent boundary increases. By contrast, the hydraulic gradient is a constant for saturated flow.

8.5.2 Vertical Infiltration and Evaporation

Gravity provides an additional driving force for vertical fluid flow and has a consequent influence on the spatial distribution of total head. Here, the governing flow equation for the one-dimensional, vertical case is

$$q = -k \frac{dh_t}{dz} \qquad (8.42)$$

where h_t is the total head, thus representing the sum of the matric suction head and the elevation head: $h_t = h_m + z$.

It is important to recognize that it is the gradient of the total head, not the matric head, that drives fluid flow in unsaturated soil. For example, if two points in the subsurface have the same matric suction value, one cannot draw a definite conclusion regarding the existence or direction of fluid flow between the two points without knowledge of the elevation for each. Using the total head concept, eq. (8.42) can be rewritten in terms of matric suction head h_m and the gravity gradient ($dz/dz = 1$) as

$$q = -k \left(\frac{dh_m}{dz} + 1 \right) \qquad (8.43)$$

Equation (8.43) provides a general basis to quantitatively assess the vertical distribution of total head under the steady flow condition. Two characteristic functions of the soil are required: the soil-water characteristic curve and the hydraulic conductivity function. For vertical, steady, unsaturated flow in the flat soil layer shown in Fig. 8.15, q is a constant throughout the entire vertical space. Thus, eq. (8.43) may be written as

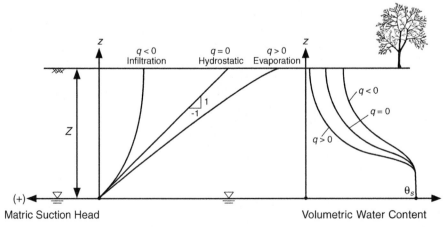

Figure 8.15 Steady vertical flow in unsaturated soil layer under force of gravity.

$$dz = -\frac{dh_m}{1 + q/k} \tag{8.44}$$

Integrating the above equation and imposing the boundary conditions of $h_m = 0$ at $z = 0$ and $h_m = h$ at $z = Z$ yields

$$\int_0^Z dz = Z = -\int_0^h \frac{dh_m(\theta)}{1 + q/k(\theta)} \tag{8.45}$$

which can be solved to yield profiles of the matric or total head profiles if the steady flux q, the soil-water characteristic curve, and the hydraulic conductivity function are known. Because the latter two functions are often measured or modeled as a series of discrete points, it is useful to write eq. (8.45) as a form amenable to numerical integration as

$$z = -\sum_{i=1}^{j} \frac{\Delta h_m(\theta_i)}{1 + q/k(\theta_i)} \tag{8.46}$$

where j is the number of discrete data points selected from the soil-water characteristic curve and hydraulic conductivity function. According to eq. (8.45), the selection of the data points should start from zero suction at the water table. The number of these points can be constrained by several factors, including the total number of points in the data set, the magnitude of the steady flux, and the range of hydraulic conductivity. Thse constraints are illustrated by examining eq. (8.43) in the following form:

$$k = -\frac{q}{dh_m/dz + 1} \tag{8.47}$$

For an infiltration problem, q is negative, and the hydraulic gradient under gravity varies within the following range:

$$-1 \leq \frac{dh_m}{dz} \leq 0 \tag{8.48}$$

The above range can be inferred from the fact that when the infiltration is zero or under the hydrostatic condition, eq. (8.44) leads to the lower bound of -1. When the soil is nearly saturated by infiltration, the matric suction and the gradient in matric suction both approach zero. Combining eq. (8.47) and eq. (8.48) leads to the following condition for downward steady infiltration:

$$k_s \geq k \geq -q \tag{8.49}$$

Equation (8.49) implies that one can discard the experimental data for unsaturated hydraulic conductivity less than the constant flux q. In other words, numerical integration of eq. (8.46) need only be conducted for water contents where the corresponding unsaturated hydraulic conductivity is greater than the steady flux q under consideration. Mathematically, as k approaches $-q$, the denominator in eq. (8.46) approaches infinity and eq. (8.47) becomes nonintegratable.

Example Problem 8.3 A silty unsaturated soil layer is under steady vertical infiltration of 10^{-8} m/s (315 mm/yr), an annual recharge likely to occur in relatively humid regions. The soil-water characteristic curve and hydraulic conductivity function for the soil under both wetting and drying processes are tabulated in Table 8.1 and plotted in Fig. 8.16. Calculate and plot the steady-state matric suction head and total head profiles.

Solution Since the steady flux rate q is 10^{-8} m/s, only data for the first seven water contents ($\theta > 0.2$) is useful for the numerical integration of eq. (8.46). At $k = -q = 1.0 \times 10^{-9}$ m/s, the volumetric water content is about 20% for the wetting state, and the corresponding matric suction head is about -9 m. Steady-state matric suction head and total head profiles under wetting and drying states are plotted in Figs. 8.17a and 8.17b, respectively.

The matric suction head at points near the water table in the preceding example follows the hydrostatic profile (Fig. 8.17a) but starts to deviate from the hydrostatic profile as the distance from the water table increases. For a fixed elevation, the matric suction head under the wetting state is smaller in

TABLE 8.1 Soil-Water Characteristic Curve and Hydraulic Conductivity Function for Silty Soil from Example Problem 8.3

θ	h_m (m) (wetting)	h_m (m) (drying)	k (m/s) (wetting)	k (m/s) (drying)
0.41	−0.001	−0.001	5.00×10^{-7}	5.00×10^{-7}
0.38	−0.5	−0.5	4.00×10^{-7}	5.00×10^{-7}
0.35	−1	−1.5	2.00×10^{-7}	3.00×10^{-7}
0.32	−1.5	−2.5	1.25×10^{-7}	2.00×10^{-7}
0.29	−2	−4	6.00×10^{-8}	9.00×10^{-8}
0.26	−3	−6	3.00×10^{-8}	5.00×10^{-8}
0.23	−5	−10	1.10×10^{-8}	2.00×10^{-8}
0.20	−8	−20	1.10×10^{-9}	3.00×10^{-9}
0.17	−20	−60	8.00×10^{-11}	3.00×10^{-10}
0.15	−60	−120	1.00×10^{-12}	1.00×10^{-11}
0.13	−180	−250	1.00×10^{-13}	5.00×10^{-13}

8.5 STEADY INFILTRATION AND EVAPORATION 355

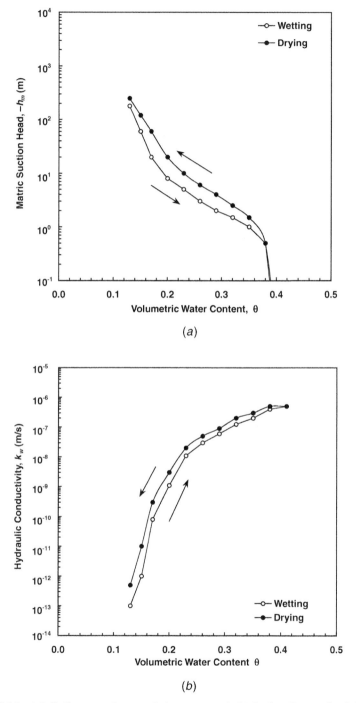

Figure 8.16 (*a*) Soil-water characteristic curve and (*b*) hydraulic conductivity function for silty soil from Example Problem 8.3.

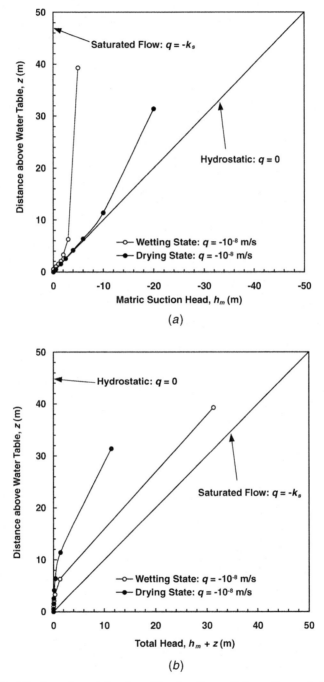

Figure 8.17 Matric and total head profiles in silty soil from Example Problem 8.3 under steady infiltration rate of 10^{-8} m/s: (*a*) matric suction head and (*b*) total head.

absolute value than that under the drying state. This is because the soil undergoing wetting has less matric suction and lower hydraulic conductivity than under drying for the same water content. The matric suction head for the hydrostatic condition ($q = 0$; $h_m = -z$) is shown as the diagonal line for comparison. The saturated flow condition ($q = k_s$) is shown as the vertical axis. Here, the suction head is zero throughout the entire soil profile.

In the soil near the water table, the hydraulic conductivity approaches its maximum value at full saturation. As shown in Fig. 8.17b, the corresponding gradient in total head is at a minimum near the water table because the product of the hydraulic gradient and the hydraulic conductivity must be equal to the constant infiltration rate (evidenced by the nearly vertical slopes in the total head profiles). As the distance from the water table increases, the slopes decrease and approach an asymptotic value of 1.0 when the soil is sufficiently far away from the water table.

The slope of the total suction profile is the hydraulic gradient. The distance over which the gradient varies from zero (i.e., a vertical line) to 1.0 (i.e., a 45° line) depends on the infiltration rate, the soil-water characteristic curve, and the hydraulic conductivity function. The tendency for the total head profile to approach the unit hydraulic gradient at elevations far from the water table is more evident for this soil in the wetting state. In the area where the hydraulic gradient is indeed equal to one, the matric suction head is a constant because $dh_t/dz = dh_m/dz + 1 = 1$, leading to $dh_m/dz = 0$, or h_m = constant. The constant value of h_m at a distance away from the water table is evident for the wetting state profile shown in Fig. 8.17a. Within this regime, the unsaturated hydraulic conductivity is equal to the absolute value of the steady flux q, the water content is a constant, and fluid flow is driven solely by gravity.

For an evaporation process, the steady upward flux is positive ($q > 0$), so the hydraulic conductivity in the integration of eq. (8.46) has no lower bound. Matric suction head and total head profiles for the soil from Example Problem 8.3 and an evaporation rate of 10^{-8} m/s (~1 mm/day) are computed numerically and plotted in Figs. 8.18a and 8.18b, respectively. Because the soil is dryer under the evaporation process than for the hydrostatic condition, the absolute value of the matric suction head (Fig. 8.18a) is higher. Deviation from the hydrostatic profile increases exponentially for distances far from the water table. For the wetting state, the matric suction head approaches infinity as the soil approaches an elevation of about 4 m above the water table, indicating that an evaporation rate greater than 10^{-8} m/s is not physically possible for this soil to maintain a 4-m unsaturated zone. For the drying state, the head approaches infinity at an elevation of about 10 m above the water table.

If the suction head at the ground surface is a finite value (perhaps several hundreds of kilopascals), the possible regime for steady evaporation is confined within several meters, whereas steady infiltration can occur in several tens of meters. Marshall and Holmes (1988) showed that eq. (8.46) could be

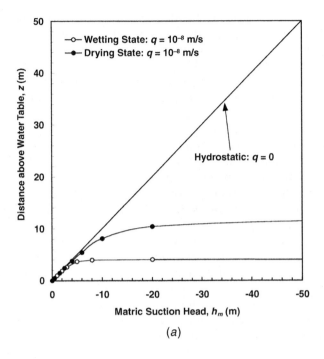
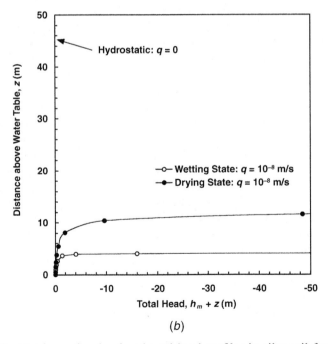

Figure 8.18 Matric suction head and total head profiles in silty soil from Example Problem 8.3 under steady evaporation rate of 10^{-8} m/s: (*a*) matric suction head and (*b*) total head.

used to infer the steady thickness of the unsaturated soil zone if the steady evaporation rate, the soil-water characteristic curve, and the hydraulic conductivity function at a site could be well quantified.

8.6 STEADY VAPOR FLOW

8.6.1 Fick's Law for Vapor Flow

The fundamental driving mechanism for vapor transport in unsaturated soil is the chemical potential of water vapor, typically expressed in terms of vapor concentration or density. Quantitative description of steady vapor flux q_v (in kg/m² · s) is captured by Fick's first law (e.g., de Vries, 1958; Cass et al., 1984) as follows:

$$q_v = -D_v \nabla \rho_v \quad (8.50)$$

where D_v (m²/s) is the diffusion coefficient for water vapor transport in unsaturated soil, and ρ_v (kg/m³) is the vapor density or absolute relative humidity of the pore water vapor (Chapter 2).

The vapor diffusion coefficient for transport in the pores of unsaturated soil is generally smaller than that for transport in free air. Macroscopic mechanisms responsible for this reduction include the limited pore space and tortuous flow path available for vapor movement. The vapor diffusion coefficient is typically approximated in terms of the free air diffusion coefficient D_0 and the air-filled porosity n_a of the soil as follows (Penman, 1940):

$$D_v = \tau \eta n_a D_0 \quad (8.51)$$

where the free air diffusion coefficient D_0 depends on temperature and pressure, ranging from approximately 10^{-9} to 10^{-6} m²/s, τ is a dimensionless tortuosity factor and may be considered to be 0.66 (Penman, 1940), and η is an enhancement factor ranging from 3 at low water content to 16 at saturation (Philip and de Vries, 1957; Cass et al., 1984).

8.6.2 Temperature and Vapor Pressure Variation

Gradients in temperature and vapor pressure (relative humidity) are the two major driving mechanisms for vapor transport in unsaturated soil. Variations in temperature and vapor pressure create gradients in vapor density that, following eq. (8.50), drive vapor flow. Vapor flux may occur across the atmosphere-soil interface in either direction or from one point to another within the subsurface pore space. Vapor transport often becomes the primary mechanism for pore water transport in relatively dry unsaturated soil.

Temperature and vapor pressure can vary significantly in time or space within the atmosphere and to significant depth in the subsurface. Figure 8.19,

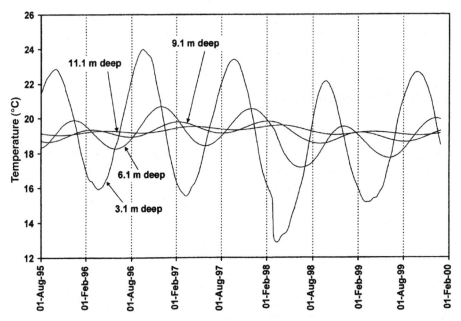

Figure 8.19 Seasonal temperature variation in alluvium soil deposit at different depths—Yucca Mountain, Nevada (Lu and LeCain, 2003).

for example, shows the seasonal temperature variation at various depths within an alluvial soil at Yucca Mountain, Nevada, over a 5-year monitoring period (Lu and LeCain, 2003). Daily and weekly temperature cycles were also detected at the site but were typically fully attenuated within 1 m from the ground surface. Natural variation in vapor pressure occurs on similar daily cycle and seasonal cycle.

As introduced in Chapter 2, the impact of such changes in temperature and vapor pressure on the corresponding vapor density can be quantitatively assessed through the ideal gas law:

$$\rho_v = \frac{\omega_w}{RT} u_v \qquad (8.52)$$

The dependency of vapor pressure on temperature can be expressed by Tetens' (1930) empirical equation (Section 2.2), which relates vapor pressure to temperature (K), saturated vapor pressure $u_{v,\text{sat}}$, and relative humidity RH as follows:

$$u_v = u_{v,\text{sat}} \, \text{RH} = 0.611 \exp\left(17.27 \frac{T - 273.2}{T - 36}\right) \text{RH} \qquad (8.53)$$

The above equation, which is plotted in Fig. 8.20, is accurate to within 1 Pa of exact values from −5 to 45°C. Examination of Fig 8.20 demonstrates the relatively high sensitivity of vapor pressure to temperature. For example, at 20°C the saturated vapor pressure (RH = 100%) is about 2.3 kPa, increasing to about 7.4 kPa at 40°C. At 90°C, the saturated vapor pressure can reach 75 kPa.

8.6.3 Vapor Density Gradient

The vapor density gradient resulting from temperature and vapor pressure gradients can be obtained using the chain rule of differentiation:

$$\nabla \rho_v = \frac{\omega_w u_{v,\text{sat}}}{RT} \nabla \text{RH} + \frac{\omega_w}{R} \frac{\text{RH}}{} \nabla \frac{u_{v,\text{sat}}}{T}$$

$$= \frac{\omega_w u_{v,\text{sat}}}{RT} \nabla \text{RH} - \frac{\omega_w}{R} \frac{\text{RH}}{} \left(u_{v,\text{sat}} \frac{\nabla T}{T^2} - \frac{\nabla u_{v,\text{sat}}}{T} \right) \quad (8.54)$$

Monteith and Unsworth (1990) showed that the gradient of the saturated vapor pressure in eq. (8.54) can be expressed as

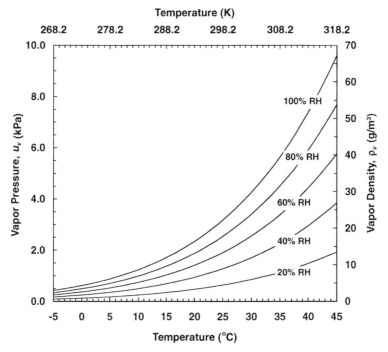

Figure 8.20 Vapor pressure and vapor density as functions of temperature and relative humidity.

$$\nabla u_{v,\text{sat}} = \frac{\lambda \omega_w u_{v,\text{sat}}}{RT^2} \nabla T \qquad (8.55)$$

where λ is the latent heat of vaporization for water, equal to about 2.48 kJ/g at 10°C.

Accordingly, the diffusive vapor flux can be derived using eqs. (8.54), (8.55), and (8.50) as

$$q_v = -D_v \rho_{v,\text{sat}} \left(\frac{\nabla \text{RH}}{\text{RH}} - \frac{\nabla T}{T} + \frac{\lambda \omega_w \nabla T}{RT^2} \right) \qquad (8.56)$$

The negative signs in front of the first and third terms on the right-hand side of eq. (8.56) imply that vapor flows from locales of high relative humidity to locales of low relative humidity and from locales of high temperature to locales of low temperature. The positive sign for the second term implies a counteracting flux from low temperature to high temperature. This offsetting term arises from the fact that low temperature causes air to contract, resulting in a higher vapor density, whereas high temperature causes air to expand, resulting in a lower vapor density.

Example Problem 8.4 The relative humidity is 40% in the atmosphere and 95% at a subsurface depth of 0.1 m, the average temperature is 25°C, the diffusion coefficient of vapor D_v is a constant equal to 10^{-7} m²/s, and the vapor density is 20 g/m³ (from Fig. 8.20). Calculate the vapor flux due to the difference in humidity between the atmosphere and subsurface.

Solution The vapor flux is the first term on the right-hand side of eq. (8.56):

$$q_v = -D_v \rho_{v,\text{sat}} \frac{\nabla \text{RH}}{\text{RH}} = -(10^{-7} \text{ m}^2/\text{s})(20 \text{ g/m}^2) \frac{-0.55}{(0.1 \text{ m})(0.68)}$$

$$= 162 \times 10^{-7} \text{ g/m}^2 \cdot \text{s}$$

The positive flux implies that the vapor movement is from high vapor pressure to low vapor pressure (i.e., from the soil to the atmosphere). The amount of flux (162×10^{-7} g/m² · s) is equivalent to a water flux of 510 g/yr · m², or 0.51 mm/yr of upward vapor flow.

Example Problem 8.5 For the system from the previous example, the relative humidity in the atmosphere and soil are the same but the atmospheric temperature is 5°C and 25°C at a depth of 0.1 m. Calculate the consequent vapor flux.

Solution The temperature difference is -20 K, occurring over a path length of 0.1 m. The corresponding vapor flux is captured by the second and third terms of eq. (8.56):

$$q_v = D_v \rho_{v,\text{sat}} \frac{\nabla T}{T} - D_v \rho_{v,\text{sat}} \frac{\lambda \omega_w}{RT^2} \nabla T$$

$$= -(10^{-7} \text{ m}^2/\text{s}) \frac{20 \text{ g/m}^3}{293 \text{ K}} \frac{20 \text{ K}}{0.1 \text{ m}}$$

$$+ (10^{-7} \text{ m}^2/\text{s}) \frac{20 \text{ g/m}^3 (2.48 \times 10^3 \text{ J/g})(18.016 \text{ g/mol})}{(8.13 \text{ J/mol} \cdot \text{K})(293 \text{ K})^2} \frac{20 \text{ K}}{0.1 \text{ m}}$$

$$= -13.7 \times 10^{-7} \text{ g/m}^2 \cdot \text{s} + 256 \times 10^{-7} \text{ g/m}^2 \cdot \text{s}$$

$$= 242.3 \times 10^{-7} \text{ g/m}^2 \cdot \text{s}$$

The negative $13.7 \times 10^7 \text{g/m}^2 \cdot \text{s}$ flux implies that the vapor movement is from low temperature to high temperature (i.e., from the atmosphere to the soil). The positive $256 \times 10^7 \text{ g/m}^2 \cdot \text{s}$ flux represents vapor movement from the warm soil to the cool atmosphere. The net $242.3 \times 10^{-7} \text{ g/m}^2 \cdot \text{s}$ vapor flux is equivalent to a water flux of 763 mm/yr $\cdot$ m^2, or 0.763 m/yr of upward vapor movement.

It can be seen from the above two examples that vapor fluxes due to temperature and vapor pressure changes in typical field scenarios are on the same order of magnitude. In addition, the second term on the right-hand side of eq. (8.56) is practically negligible compared to the first and third terms.

Vapor density changes due to large variations in relative humidity are more likely to occur on a seasonal time scale. Equally significant temperature changes, on the other hand, can occur on either daily or seasonal cycles. Vapor flux due to relative humidity variation moves primarily from the soil to the atmosphere because the relative humidity in soil is very close to 100% the majority of the time. Vapor flux due to temperature variation, on the other hand, can move in either direction because the temperature in the atmosphere can be higher or lower than the soil temperature for sustained periods of time. Unusually large temperature gradients responsible for vapor flow in unsaturated soil can occur under special circumstances such as underground radioactive waste repositories, where the subsurface temperature can reach several hundred degrees Celsius, or in mine waste rocks, where pyrite oxidation processes may elevate temperature as much as 50°C above ambient (e.g., Lu and Zhang, 1997).

8.7 STEADY AIR DIFFUSION IN WATER

8.7.1 Theoretical Basis

The transport of dissolved air in quiescent water can be described by a diffusion process, which is governed by Fick's law as follows:

$$q_{air} = \frac{\partial M_{air}}{A \, \partial t} = -D \, \nabla C \tag{8.57}$$

where q_{air} is the air flux (mol/s · m^2), M_{air} is the mass of air (mol), A is the cross-sectional area for the flow process (m^2), C is the air concentration (mol/m^3), and D is the diffusion coefficient (m^2/s). The diffusion coefficient is an intrinsic material property of a specific gas in a specific liquid under specific thermodynamic conditions. For example, the diffusion coefficient of oxygen (O$_2$) in water is about 2.6×10^{-9} m^2/s at 25°C and 1 bar, increasing to about 3.24×10^{-9} m^2/s at 40°C, and nearly doubling at 60°C ($D = 4.82 \times 10^{-9}$ m^2/s) (Wilke and Chang, 1955).

The time required for a diffusive process to occur is directly proportional to the square of the length of the diffusion path and inversely proportional to the diffusion coefficient, or $t \sim L^2/D$, where L is the length of the diffusion path (m).

Example Problem 8.6 If the diffusion coefficient for air in water is 2.0×10^{-9} m^2/s, estimate the time required for air to diffuse through water films 0.01 and 0.02 m in thickness.

Solution The amount of time required is proportional to L^2 and D. For a water film of 0.01 m, the amount of time may be estimated as

$$t \sim \frac{L^2}{D} = \frac{(0.01 \text{ m})^2}{2.0 \times 10^{-9} \text{ m}^2/\text{s}} = 0.5 \times 10^5 \text{ s} = 13.88 \text{ h}$$

If the thickness of the water layer is doubled to 0.02 m, the diffusion time increases fourfold:

$$t \sim \frac{L^2}{D} = \frac{(0.02 \text{ m})^2}{2.0 \times 10^{-9} \text{ m}^2/\text{s}} = 2 \times 10^5 \text{ s} = 55.56 \text{ h}$$

To assess diffusive air flux in a body of water, one would need to know the gradient of the air concentration, an extremely challenging measurement task. However, an alternative way to assess the air concentration gradient is to estimate the air pressure gradient. Considering a one-dimensional air diffusion problem, for example, eq. (8.57) can be rewritten as

$$q_{air} = -D \frac{\partial C}{\partial z} = -D \frac{\partial C}{\partial u_a} \frac{\partial u_a}{\partial z} \tag{8.58}$$

where u_a is the partial pressure of dissolved air in water. The change in air concentration with respect to the change in the partial air pressure can be obtained by assuming the dissolved air follows ideal gas behavior:

8.7 STEADY AIR DIFFUSION IN WATER

$$\frac{M_{air}}{V_{air}} = \frac{\omega_{air}}{RT} u_a \qquad (8.59)$$

where M_{air} is the mass of the dissolved air, V_{air} is the volume of the dissolved air, and ω_{air} is the molecular mass of the dissolved air. The change in air concentration with respect to the change in the partial air pressure, therefore, is

$$\frac{\partial C}{\partial u_a} = \frac{\partial \left(\frac{M_{air}}{V_l} \right)}{\partial u_a} = \frac{\partial \left(\frac{V_{air}}{V_l} \frac{\omega_{air}}{RT} u_a \right)}{\partial u_a} \qquad (8.60)$$

where V_l is the volume of the liquid (i.e., water) within which the air is dissolved. According to Henry's law (Section 3.1):

$$\frac{V_{air}}{V_l} = h_{air} \qquad (8.61)$$

where h_{air} is the volumetric coefficient of solubility. Substituting eq. (8.61) into eq. (8.60) yields

$$\frac{\partial C}{\partial u_a} = \frac{\partial [h_{air} (\omega_{air}/RT) u_a]}{\partial u_a} = \frac{h_{air} \omega_{air}}{RT} \qquad (8.62)$$

Inserting eq. (8.62) and eq. (8.59) into eq. (8.58) yields

$$q_{air} = \frac{\partial M_{air}}{A \, \partial t} = \frac{u_a \omega_{air}}{RT} \frac{\partial V_{air}}{A \, \partial t} = -D \frac{h_{air} \omega_{air}}{RT} \frac{\partial u_a}{\partial z} \qquad (8.63)$$

or

$$q_{air} = \frac{u_a \omega_{air}}{RT} \frac{\partial V_{air}}{A \, \partial t} = \frac{u_a \omega_{air}}{RT} v_{air}$$

$$v_{air} = -\frac{Dh_{air}}{u_a} \frac{\partial u_a}{\partial z} = -\frac{Dh_{air} \rho_{air} g}{u_a} \frac{\partial h_a}{\partial z} \qquad (8.64)$$

where v_{air} is the discharge velocity of the dissolved air (m/s), and h_a is the head of the dissolved air (m).

Comparing eq. (8.64) with Darcy's law for water flow:

$$v_{air} = -k_{air} \frac{\partial h_a}{\partial z} = -\frac{Dh_{air}\rho_{air}g}{u_a}\frac{\partial h_a}{\partial z}$$

$$k_{air} = \frac{Dh_{air}\rho_{air}g}{u_a}$$

(8.65)

where k_{air} is the conductivity of dissolved air in water (m/s).

Example Problem 8.7 If the air diffusivity $D = 5 \times 10^{-9}$ m²/s, the volumetric coefficient of solubility $h_{air} = 0.02$, the density of air $\rho_{air} = 1$ kg/m³, and the air pressure $u_a = 100$ kPa, what is the air conductivity k_{air}?

Solution From eq. (8.65):

$$k_{air} = \frac{Dh_{air}\rho_{air}g}{u_a} = \frac{(5 \times 10^{-9} \text{ m}^2/\text{s})(0.02)(1 \text{ kg/m}^3)(9.8 \text{ m/s}^2)}{100 \times 10^3 \text{ N/m}^2} \approx 10^{-14} \text{ m/s}$$

8.7.2 Air Diffusion in an Axis Translation System

Equation (8.64) or (8.65) can be used to calculate the flux of dissolved air in unsaturated soil testing systems. For example, the pressure of dissolved air on either side of the high-air-entry ceramic porous stone in the axis translation system shown in Fig. 8.21 will be different due to the difference in the total pressure between the air chamber and water chamber. If the applied air pressure in the air chamber is 1000 kPa and the total water pressure in the water chamber is 100 kPa, then the dissolved air pressure at equilibrium will be 1000 kPa on top of the ceramic porous stone and 100 kPa at the bottom of the ceramic porous stone. Assuming the thickness of the ceramic disk is 5 mm, the diffusive air discharge velocity into the water chamber can be estimated by eq. (8.63) as

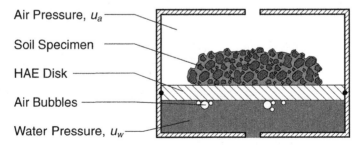

Figure 8.21 Air diffusion and bubble formation through high-air-entry (HAE) ceramic disk in axis translation setup for laboratory soil-water characteristic curve testing.

$$v_{air} = k_{air} \frac{\partial h_{air}}{\partial z} = -\frac{k_{air}}{\rho_{air} g} \frac{\partial u_{air}}{\partial z}$$

$$= \frac{10^{-14} \text{ m/s}}{(1 \text{ kg/m}^3)(9.8 \text{ m/s}^2)} \frac{900 \times 10^3 \text{ N/m}^2}{5 \times 10^{-3} \text{ m}}$$

$$= 1.84 \times 10^{-7} \text{ m/s} = 1.59 \text{ cm/day}$$

If the diameter of the ceramic disk is 5 cm, the area is then about 19.63 cm². The total volume of diffused air per day for a disk with porosity $n = 0.35$ is then

$$Q = v_{air} A n t = (1.59 \text{ cm/day})(19.63 \text{ cm}^2)(0.35)(1 \text{ day}) = 10.92 \text{ cm}^3$$

One should realize that if there is no residual water on top of the ceramic porous stone and the system is at equilibrium, there would be no pressure gradient in the dissolved air, and thus no diffusive air through the saturated porous stone. If the system is not at equilibrium (such as during an incremental increase in applied pressure and for some time after that), there will be additional air dissolved into the pore water. Some of the pore water will be moving out of the system in order for the applied suction and water content of the soil to reach equilibrium. The expelled water will release air bubbles in the water chamber since the pressure there is typically smaller than in the inflow water. The air diffusion mechanism described here is the most likely process for bubble formation in the water chamber.

PROBLEMS

8.1. Does intrinsic permeability depend on fluid properties such as fluid density and viscosity? What are the approximate ranges of intrinsic permeability, water conductivity, and air conductivity for sandy silt? For the same soil, which one, water conductivity or air conductivity, is higher?

8.2. Explain how the unsaturated hydraulic conductivity of a silt could be greater than the hydraulic conductivity of a sand.

8.3. Why can a given unsaturated soil have different values of hydraulic conductivity for the same suction?

8.4. Consider a horizontal capillary barrier consisting of a silt layer overlying a coarse sand layer. If the silt has a pore size distribution index $\alpha = 1$ kPa^{-1} and the sand has $\alpha = 10$ kPa^{-1}, what is the maximum height of water that can be suspended in the overlying silt?

8.5. What is the efficiency of the capillary barrier described in Problem 8.4?

8.6. Assume that the diffusivity of water vapor D_v in a soil layer is 10^{-8} m²/s, the relative humidity of the atmospheric air is 40%, the relative humidity of the soil pore water at the water table is 100%, the thickness of the unsaturated layer is 1 m, and the average temperature is 25°C. Calculate the vapor flux (in mm/day) between the soil and air. What is the mass of vapor moving across 1.0 m² of the ground surface every year?

8.7. The steady net infiltration rate at a two-soil capillary barrier overlying a waste package is 0.1 m/yr, the saturated hydraulic conductivity of the overlying fine soil is 10^{-7} m/s, the air-entry head of the fine soil is 0.1 m, and the width of the waste package is 6 m. What is the required dip of the capillary barrier if a symmetric capillary cap is used in the design?

8.8. An unsaturated soil layer (2 m from the surface to the water table) is described by the soil-water characteristic curve and hydraulic conductivity function tabulated in Table 8.2. Estimate and plot the soil moisture and matric suction profiles for a steady downward infiltration rate of 1×10^{-8} m/s. What are the matric suction and water content at 0.5 m below the ground surface? For soil less than 0.5 m from the surface, what is the dominating driving force for the downward flow? For soil greater than 0.5 m depth, what is the dominating driving force?

TABLE 8.2 Soil-Water Characteristic Curve and Hydraulic Conductivity Function for Silty Soil from Problem 8.8

θ	h_m (m)	k (m/s)
0.43	0.00	3.10×10^{-7}
0.42	−0.25	2.80×10^{-7}
0.37	−0.50	4.20×10^{-8}
0.34	−0.75	1.50×10^{-8}
0.33	−1.00	1.20×10^{-8}
0.325	−1.10	6.00×10^{-9}
0.29	−3.00	7.00×10^{-10}
0.25	−8.00	5.00×10^{-11}
0.21	−21.00	5.00×10^{-12}
0.17	−70.00	2.00×10^{-13}
0.12	−1000.00	5.00×10^{-15}

CHAPTER 9

TRANSIENT FLOWS

9.1 PRINCIPLES FOR PORE LIQUID FLOW

9.1.1 Principle of Mass Conservation

Fluid flow and moisture content in unsaturated soil may vary both spatially and temporally as the result of two basic mechanisms: (1) time-dependent changes in environmental conditions and (2) the storage capacity of soil. For flow prediction purposes, environmental changes are often cast into prescribed boundary conditions for the soil domain under consideration. The storage capacity effect on moisture redistribution is captured in the governing flow equations or laws.

The governing equation for transient water flow in soil under isothermal conditions can be derived by applying the principle of mass conservation. The principle of mass conservation states that for a given elemental volume of soil, the rate of water loss or gain is conservative and is equal to the net flux of inflow and outflow. The mass conservation principle is also called the continuity principle. For the elemental volume of soil shown in Fig. 9.1, with porosity n and volumetric water content θ, the total inflow of water along the positive coordinate direction is

$$q_{in} = \rho(q_x \, \Delta y \, \Delta z + q_y \, \Delta x \, \Delta z + qz \, \Delta x \, \Delta y) \quad (9.1)$$

and the total flux out of the volume is

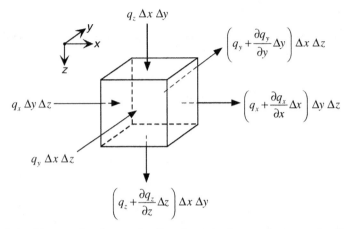

Figure 9.1 Elemental volume of soil and continuity requirements for fluid flow.

$$q_{out} = \rho\left[\left(q_x + \frac{\partial q_x}{\partial x}\Delta x\right)\Delta y\,\Delta z + \left(q_y + \frac{\partial q_y}{\partial y}\Delta y\right)\Delta x\,\Delta z \right.$$
$$\left. + \left(q_z + \frac{\partial q_z}{\partial z}\Delta z\right)\Delta x\,\Delta y\right] \tag{9.2}$$

where ρ is the density of water (kg/m³) and q_x, q_y, and q_z are fluxes in the x, y, and z directions, respectively (m/s).

The rate at which water mass is lost or gained by the element during a transient process is as follows:

$$\frac{\partial(\rho\theta)}{\partial t}\Delta x\,\Delta y\,\Delta z \tag{9.3}$$

For mass conservation, the storage term captured by eq. (9.3) must be equal to the net flux, leading to

$$-\rho\left(\frac{\partial q_x}{\partial x}\Delta x\,\Delta y\,\Delta z + \frac{\partial q_y}{\partial y}\Delta y\,\Delta x\,\Delta z + \frac{\partial q_z}{\partial z}\Delta z\,\Delta x\,\Delta y\right) = \frac{\partial(\rho\theta)}{\partial t}\Delta x\,\Delta y\,\Delta z \tag{9.4}$$

or

$$-\rho\left(\frac{\partial q_x}{\partial x} + \frac{\partial q_y}{\partial y} + \frac{\partial q_z}{\partial z}\right) = \frac{\partial(\rho\theta)}{\partial t} \tag{9.5}$$

Equation (9.5) is the governing equation for unsteady or transient fluid flow in soil and is generally applicable to both saturated and unsaturated conditions.

9.1.2 Transient Saturated Flow

For saturated conditions, the volumetric water content θ is equal to the porosity n, and Eq. (9.5) becomes

$$-\rho\left(\frac{\partial q_x}{\partial x} + \frac{\partial q_y}{\partial y} + \frac{\partial q_z}{\partial z}\right) = \frac{\partial(\rho n)}{\partial t} \tag{9.6}$$

The fluid stored or released by an element of soil due to the fluid volume change ρn can be related to the total head h_t (h for simplicity in this chapter). For saturated soil, it can be shown that the right-hand side of eq. (9.6) can be defined as

$$\frac{\partial(\rho n)}{\partial t} = \rho S_s \frac{\partial h}{\partial t} \tag{9.7a}$$

where S_s is the specific storage, defined as follows (e.g., Freeze and Cherry, 1979):

$$S_s = \rho g(\alpha_s + n\beta_w) \tag{9.7b}$$

where α_s is the bulk compressibility of the soil (m²/N or Pa⁻¹) and β_w is the compressibility of the pore water (m²/N). The bulk soil compressibility depends to some extent on soil type, ranging from approximately 10^{-6} to 10^{-8} m²/N for clay, from 10^{-7} to 10^{-9} m²/N for sand, and from 10^{-8} to 10^{-10} m²/N for gravel (Freeze and Cherry, 1979). The compressibility of water β_w is fairly constant at about 4.4×10^{-10} m²/N.

Darcy's law allows one to write fluid flow in terms of hydraulic conductivity and hydraulic gradient in each coordinate direction as follows:

$$q_x = -k_x \frac{\partial h}{\partial x} \qquad q_y = -k_y \frac{\partial h}{\partial y} \qquad q_z = -k_z \frac{\partial h}{\partial z} \tag{9.8}$$

Thus, for transient saturated flow in isotropic and homogeneous soil (i.e., $k = k_x = k_y = k_z$), eq. (9.6) becomes

$$\frac{\partial^2 h}{\partial x^2} + \frac{\partial^2 h}{\partial y^2} + \frac{\partial^2 h}{\partial z^2} = \frac{S_s}{k}\frac{\partial h}{\partial t} \tag{9.9}$$

which may also be written in the form of the standard diffusion equation

$$D\left(\frac{\partial^2 h}{\partial x^2} + \frac{\partial^2 h}{\partial y^2} + \frac{\partial^2 h}{\partial z^2}\right) = \frac{\partial h}{\partial t} \qquad (9.10)$$

where *hydraulic diffusivity* D has units of length squared over time (m²/s) and is equal to the ratio of hydraulic conductivity to specific storage, that is, $D = k/S_s$.

9.1.3 Transient Unsaturated Flow

For practical applications, Darcy's law may be generalized to unsaturated fluid flow problems by considering hydraulic conductivity as a function of soil suction or suction head as follows (e.g., Buckingham, 1907; Richards, 1931):

$$q_x = -k_x(h_m)\frac{\partial h}{\partial x} \qquad q_y = -k_y(h_m)\frac{\partial h}{\partial y} \qquad q_z = -k_z(h_m)\frac{\partial h}{\partial z} \qquad (9.11)$$

where h_m is matric suction head and $k(h_m)$ is the unsaturated hydraulic conductivity function. In the absence of an osmotic pressure head, the total head in unsaturated soil is the sum of the matric suction head and the elevation head ($h = h_m + z$). Thus, substituting eq. (9.11) into eq. (9.5) and assuming a constant water density leads to

$$\frac{\partial}{\partial x}\left[k_x(h_m)\frac{\partial h_m}{\partial x}\right] + \frac{\partial}{\partial y}\left[k_y(h_m)\frac{\partial h_m}{\partial y}\right] + \frac{\partial}{\partial z}\left[k_z(h_m)\left(\frac{\partial h_m}{\partial z}+1\right)\right] = \frac{\partial \theta}{\partial t} \qquad (9.12)$$

where the additional term in the z coordinate direction arises from the presence of the elevation head.

The right-hand side of eq. (9.12) can be rewritten in terms of the matric suction head by applying the chain rule:

$$\frac{\partial \theta}{\partial t} = \frac{\partial \theta}{\partial h_m}\frac{\partial h_m}{\partial t} \qquad (9.13)$$

where the quantity $\partial\theta/\partial h_m$ is the slope of the relationship between volumetric water content and suction head, which can be obtained directly from the soil-water characteristic curve. This slope is referred to as the *specific moisture capacity*, typically designated C. Because the soil-water characteristic curve is nonlinear, it is necessary to describe specific moisture capacity as a function of suction or suction head, the latter expressed as follows:

$$C(h_m) = \frac{\partial \theta}{\partial h_m} \tag{9.14}$$

Figure 9.2 shows an example of the specific moisture capacity function for a typical unsaturated silty material. The soil-water characteristic curve $\theta(h_m)$ is shown as Fig. 9.2a. The corresponding specific moisture capacity function $C(\theta)$ is shown as Fig. 9.2b. Note that the magnitude of specific moisture capacity reaches a maximum of 0.0042 cm^{-1} at a water content of 0.318. This reflects the fact that for each unit change in head, the change in volumetric water content is 0.0042. Figure 9.3 demonstrates differences in specific moisture capacity functions for sand and clayey materials having relatively narrow and relatively wide pore size distributions, respectively. The relatively sharp maximum for the sand reflects its narrow pore size distribution where the majority of the pores are drained over a narrow range of suction.

Substituting eqs. (9.13) and (9.14) into eq. (9.12), a governing equation for transient unsaturated fluid flow may be written as

$$\frac{\partial}{\partial x}\left[k_x(h_m)\frac{\partial h_m}{\partial x}\right] + \frac{\partial}{\partial y}\left[k_y(h_m)\frac{\partial h_m}{\partial y}\right] + \frac{\partial}{\partial z}\left[k_z(h_m)\left(\frac{\partial h_m}{\partial z}+1\right)\right] = C(h_m)\frac{\partial h_m}{\partial t}$$

(9.15a)

The above equation is the Richards' equation. Solution of the Richards' equation with appropriate boundary and initial conditions gives the suction field in space and time. As implied in eq. (9.15), three characteristic functions are required for its solution: the hydraulic conductivity function, the soil-water characteristic curve, and the specific moisture capacity function.

The Richards' equation may also be written in terms of volumetric water content, as this is often done in soil physics. Following the chain rule, Darcy's law can be expressed in the horizontal direction

$$q_x = -k_x(\theta)\frac{\partial h_m}{\partial x} = -k_x(\theta)\frac{\partial h_m}{\partial \theta}\frac{\partial \theta}{\partial x} = -D_x\frac{\partial \theta}{\partial x} \tag{9.15b}$$

Similarly, the fluxes in the y and z (gravity) directions can be written as

$$q_y = -k_y(\theta)\frac{\partial h_m}{\partial y} = -D_y\frac{\partial \theta}{\partial y} \tag{9.15c}$$

$$q_z = -k_z(\theta)\left(\frac{\partial h_m}{\partial z}+1\right) = -D_z\frac{\partial \theta}{\partial z} - k_z(\theta) \tag{9.15d}$$

With eq. (9.14), it can be shown that

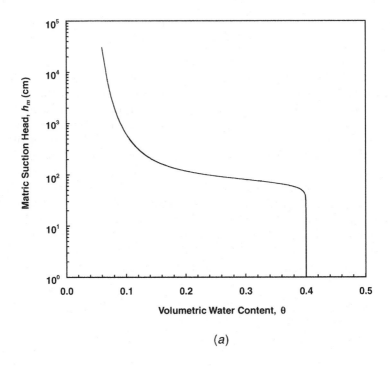

(a)

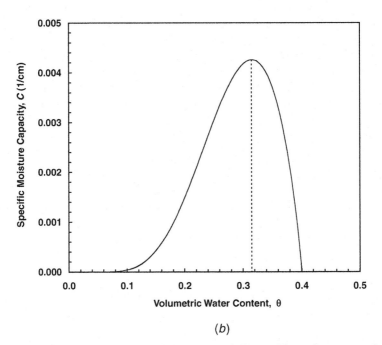

(b)

Figure 9.2 (a) Soil-water characteristic curve and (b) specific moisture capacity function, $C(\theta)$, for typical silty soil.

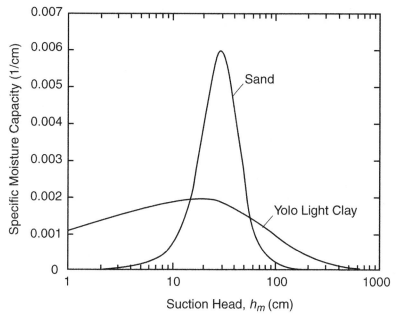

Figure 9.3 Specific moisture capacity functions, $C(h_m)$, for sandy and clayey soils (after Lappala et al., 1993). Relatively sharp peak for sandy material reflects its relatively narrow pore size distribution.

$$D_x = \frac{k_x(h_m)}{C(h_m)} \quad (9.16)$$

where D_x is defined as the ratio of the hydraulic conductivity to the specific moisture capacity and is called hydraulic diffusivity for unsaturated soil. Relationships among hydraulic diffusivity, the hydraulic conductivity function, the soil-water characteristic curve, and specific moisture capacity are illustrated in Fig. 9.4 for a typical silty soil. The hydraulic conductivity varies over five orders of magnitude as the water content varies between 0.08 and 0.4.

Substituting eqs. (9.15b) through (9.15d) into eq. (9.15a) leads to

$$\frac{\partial}{\partial x}\left(D_x(\theta)\frac{\partial \theta}{\partial x}\right) + \frac{\partial}{\partial y}\left(D_y(\theta)\frac{\partial \theta}{\partial y}\right) + \frac{\partial}{\partial z}\left(D_z(\theta)\frac{\partial \theta}{\partial z}\right) + \frac{\partial k_z(\theta)}{\partial z} = \frac{\partial \theta}{\partial t} \quad (9.17)$$

Analytical solutions of eqs. (9.15a) and (9.17) under various initial and boundary conditions constitute a rich body of classical problems in soil physics and groundwater hydrology.

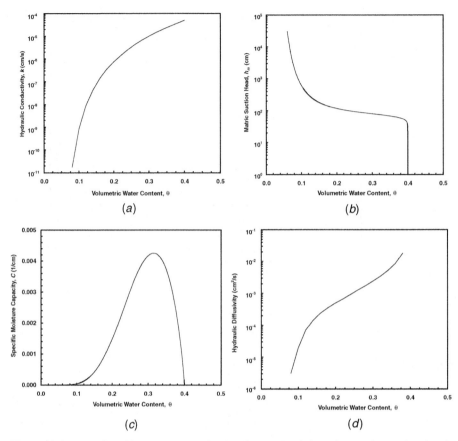

Figure 9.4 Relationships among (a) hydraulic conductivity, (b) matric suction head, (c) specific moisture capacity, and (d) hydraulic diffusivity as functions of water content $[\partial\theta/\partial h_m = C(\theta), D(\theta) = k(\theta)/C(\theta)]$.

9.2 RATE OF INFILTRATION

9.2.1 Transient Horizontal Infiltration

One of the first successful attempts at describing the transient process of water infiltration in unsaturated soil was Green and Ampt's (1911) semianalytical approach to horizontal water movement into an initially dry, uniform column of soil. Under these special conditions, the invasion of water can generally be described in terms of a "sharp wetting front" that initially propagates at a relatively fast rate through the column and gradually slows with time. Green and Ampt (1911) first described this idealized physical process by applying Darcy's law to unsaturated soil. Their analysis is revisited in this section in order to elucidate the basic physics involved in transient unsaturated fluid flow.

Figure 9.5 is a conceptual diagram of an initially dry, one-dimensional soil column undergoing transient horizontal infiltration. The pore fluid in the system, which propagates from the left end of the column to the right end of the column under a constant head, is marked by a distinctly sharp wetting front.

Two assumptions may be made to make the flow problem amenable for analytical solution using Darcy's law: (1) the suction head in the soil beyond the wetting front (in the dry portion of the column) is constant in both space and time and (2) the water content and corresponding hydraulic conductivity of the soil behind the wetting front (in the wet portion of the column) are constant in both space and time. Given these basic assumptions, the total infiltration displacement Q for a unit cross-sectional area at any time t is equal to the product of the change in water content relative to the initial condition and the distance that the wetting front has traveled, or

$$Q = (\theta_0 - \theta_i)x \tag{9.18}$$

where θ_0 is the volumetric water content behind the wetting front, θ_i is the volumetric water content beyond the wetting front, and x is the distance that the wetting front has traveled. The infiltration rate is equal to the infiltration rate at the influent boundary (left end of column), which can be approximated using Darcy's law as

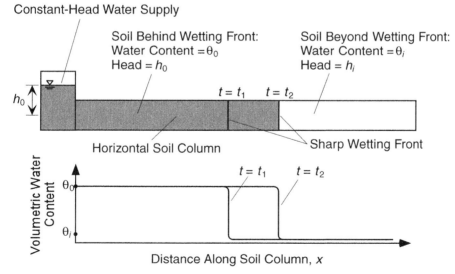

Figure 9.5 Transient infiltration of sharp wetting front in horizontal soil column and corresponding water content distributions with time under the Green-Ampt assumptions.

$$q = \frac{dQ}{dt} = \frac{(\theta_0 - \theta_i)dx}{dt} = -k_0 \frac{dh}{dx} = -k_0 \frac{h_i - h_0}{x} \tag{9.19}$$

where h_i is the suction head at the wetting front, h_0 is the suction head behind the wetting front, and k_0 is the hydraulic conductivity behind the wetting front, which is often assumed to be equal to the saturated hydraulic conductivity, although this may not be the case in general. Integrating with respect to the space and time variables x and t and imposing the initial condition of $x = 0$ at $t = 0$ yields

$$x^2 = \left(2k_0 \frac{h_0 - h_i}{\theta_0 - \theta_i}\right)t \tag{9.20a}$$

or

$$\frac{x}{\sqrt{t}} = \sqrt{2k_0 \frac{h_0 - h_i}{\theta_0 - \theta_i}} = \lambda = \text{const} \tag{9.20b}$$

Since the quantity in parentheses in front of time in eq. (9.20a) has units of length squared over time, it can be considered an effective hydraulic diffusivity D, so that

$$x = \sqrt{2k_0 \frac{h_0 - h_i}{\theta_0 - \theta_i} t} = \sqrt{Dt} \tag{9.21}$$

The above equation predicts that the position of the sharp wetting front advances at a constant rate proportional to the square root of time. The infiltration rate q at the influent boundary and the total infiltration displacement Q may be predicted according to eqs. (9.19) through (9.21) as

$$q = k_0 \frac{h_0 - h_i}{\sqrt{Dt}} \tag{9.22}$$

$$Q = (\theta_0 - \theta_i)\sqrt{Dt} = s\sqrt{t} \tag{9.23}$$

where the parameter s was first called *sorptivity* by Philip (1969). Equation (9.20b) provides a way to physically justify the use of the so-called Boltzmann transformation used to convert the more rigorous governing fluid flow equation (9.17) from a partial differential equation to an ordinary differential equation such that it becomes amenable for analytical solutions for infiltration under various initial and boundary conditions.

For a varying moisture content profile during infiltration, that is, a "nonsharp" wetting front, the sorptivity under the Boltzmann transformation according to Philip (1969) is

$$s = \int_{\theta_i}^{\theta_0} \lambda(\theta)\, d\theta \tag{9.24}$$

The sharp wetting front assumptions (1) and (2) imply that the variable λ is a constant or $h(x) = h_i$. A great uncertainty is inherently involved in eqs. (9.21) and (9.22) in assigning the appropriate value for the constant h_i. For this reason, results based on assumptions (1) and (2) should at most be considered semianalytical in nature. According to Green and Ampt and numerous other experimental works, the matric suction head h_i ranges between -0.5 and -1.5 m. In general, eqs. (9.21) and (9.22) cannot be used to accurately predict the wetting front and infiltration rate unless h_i is predetermined experimentally. If the wetting front is not a sharp one, which may be the case for a fine-grained soil or any soil with moist initial conditions, D is not a constant and an explicit solution of eq. (9.17) is required. Philip's 1957 work represented a classic treatment to such a problem and will be discussed later in this section.

Example Problem 9.1 An infiltration experiment was conducted in the horizontal sandy soil column shown in Fig. 9.5. The cross-sectional area of the column is 100 cm². The cumulative infiltration volume after 20 min of testing was determined to be 200 cm³. Predict the infiltration rate and cumulative infiltration at 1 h and at 6 h.

Solution An advantage of using sorptivity as the controlling parameter for infiltration in this type of problem is that there is no need to quantify k_0, h_i, or θ_0. The cumulative infiltration volume at 20 min may be used to estimate sorptivity from eq. (9.23) as follows:

$$s = \frac{Q}{\sqrt{t}} = \frac{200 \text{ cm}^3}{100 \text{ cm}^2 \sqrt{20 \times 60 \text{ s}}} = 0.058 \text{ cm}/\sqrt{s}$$

From eqs. (9.22) and (9.23), the total infiltration displacement and infiltration rate can then be calculated at 1 h as

$$Q = s\sqrt{t} = (0.058 \text{ cm}/\sqrt{s})\sqrt{3600 \text{ s}} = 3.48 \text{ cm}$$

$$q = \frac{dQ}{dt} = \frac{s}{2\sqrt{t}} = \frac{0.058 \text{ cm}}{2\sqrt{3600 \text{ s}}\sqrt{s}} = 0.00048 \text{ cm/s} = 1.74 \text{ cm/h}$$

and at 6 h as

$$Q = s\sqrt{t} = (0.058 \text{ cm}/\sqrt{s})\sqrt{21600 \text{ s}} = 8.52 \text{ cm}$$

$$q = \frac{dQ}{dt} = \frac{s}{2\sqrt{t}} = \frac{0.058 \text{ cm}}{2\sqrt{21600 \text{ s}}\sqrt{s}} = 0.00020 \text{ cm/s} = 0.71 \text{ cm/h}$$

9.2.2 Transient Vertical Infiltration

Figure 9.6 illustrates a one-dimensional soil column undergoing transient vertical infiltration under the Green-Ampt assumptions. The wetting front advances in this case under the combined effects of suction and gravity gradients. The total head at the wetting front is $h = h_i - z$ such that eq. (9.19) becomes

$$q = \frac{dQ}{dt} = \frac{(\theta_0 - \theta_i)\, dz}{dt} = -k_0 \frac{dh}{dz} = -k_0 \frac{(h_i - z) - h_0}{z} = k_0\left(1 + \frac{h_0 - h_i}{z}\right)$$

(9.25a)

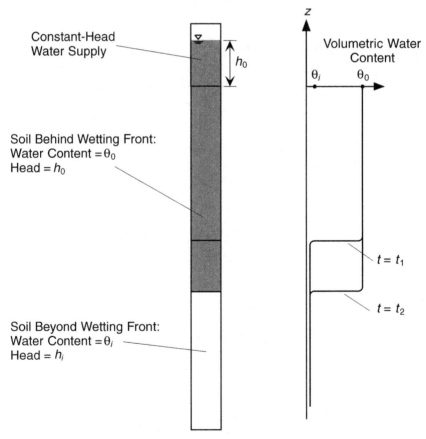

Figure 9.6 Transient infiltration of sharp wetting front in vertical soil column and corresponding water content distributions with time under the Green-Ampt assumptions.

or

$$q = k_0 + \frac{k_0 (h_0 - h_i)(\theta_0 - \theta_i)}{Q} \quad (9.25b)$$

or

$$\frac{dz}{dt} = \frac{k_0}{\theta_0 - \theta_i}\left(1 + \frac{h_0 - h_i}{z}\right) \quad (9.25c)$$

One implication of the above equations is that as the time of infiltration becomes sufficiently large, the depth of infiltration is also large. As such, the infiltration rate q becomes a constant value that is equal to the hydraulic conductivity k_0 corresponding to water content θ_0.

Integrating eq. (9.25c) and imposing the initial condition of $z = 0$ at $t = 0$ yields

$$\frac{k_0}{\theta_0 - \theta_i} t = z - (h_0 - h_i) \ln\left(1 + \frac{z}{h_0 - h_i}\right) \quad (9.26)$$

The above equation predicts the arrival time of the advancing wetting front as a function of the distance traveled from the influent boundary (top of the soil column). The total infiltration displacement Q can also be obtained by imposing eq. (9.18) on eq. (9.25a) to eliminate the spatial variable z and by integrating and imposing the initial condition of $Q = 0$ at $t = 0$:

$$t = \frac{1}{k_0}\left\{Q - (h_0 - h_i)(\theta_0 - \theta_i) \ln\left[1 + \frac{Q}{(h_0 - h_i)(\theta_0 - \theta_i)}\right]\right\} \quad (9.27)$$

The above equation, which relates the total infiltration displacement Q to elapsed time t, can also be obtained directly by substituting eq. (9.20) into eq. (9.26).

The general behavior of horizontal and vertical infiltration problems can be examined by using eqs. (9.20), (9.26), (9.19), and (9.25), as depicted in Fig. 9.7. For horizontal infiltration, where gravity is absent and flow is driven solely by the gradient of the matric suction, the wetting front advances linearly with respect to the square root of time, $\sqrt{t}$. The wetting front for downward infiltration, on the other hand, advances nonlinearly (Fig. 9.7a). Due to the existence of gravity, vertical infiltration advances faster than horizontal infiltration. As shown in Fig. 9.7b, the infiltration rate for both horizontal and vertical infiltration decreases exponentially with respect to the wetting front distance but asymptotically approaches zero for horizontal infiltration and approaches k_0 for downward infiltration. The asymptotic feature of downward infiltration can be used to estimate k_0 from infiltration or ponding experiments.

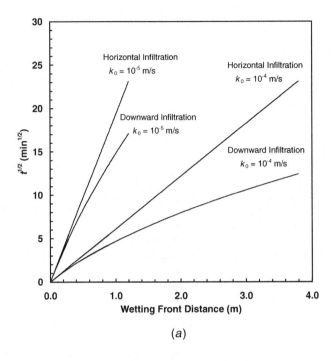

(a)

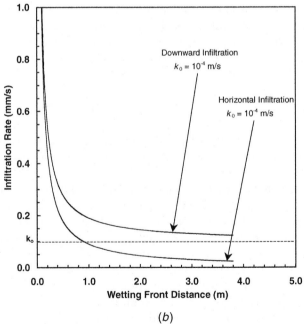

(b)

Figure 9.7 (a) Wetting front arrival time and (b) infiltration rate as functions of wetting front for horizontal and vertical infiltrations for $h_0 - h_i = 0.9$ m and $(\theta_0 - \theta_i) = 0.4$.

However, the test could be costly in water quantity and time due to the fact that the infiltration rate decays slowly in many soils.

Example Problem 9.2 A laboratory vertical infiltration test was conducted on a column of dry silt in order to estimate the time required for infiltration from the ground surface to reach the water table during heavy rainfall events in the field. The saturated hydraulic conductivity of the silt, k_s, is 10^{-6} m/s and the porosity n is 0.45. The column test was conducted such that the top boundary head h_0 was equal to atmospheric pressure, that is, $h_0 = 0$. Results are shown in Table 9.1. Estimate the amount of time required for the wetting front to reach the water table under similar conditions in the field, where the water table is located 1.6 m below the ground surface.

Solution Infiltration distance versus time for the column test is plotted in Fig. 9.8. Equation (9.27) is used to best fit the results. It is found that $h_i = -1.3$ m accords very well with the experimental data. Substituting all the parameters including $z = 1.6$ m into eq. (9.26), the time for a heavy rainfall event at the ground surface to reach to the water table is found to be 123.16 h, or 5.1 days. An upper bound arrival time can also be estimated by the travel distance divided by the saturated hydraulic conductivity and is 22 days. The upper bound value of 22 days far exceeds the more accurate estimate value of 5.1 days predicted by eq. (9.27).

TABLE 9.1 Results of Infiltration Test for Example Problem 9.2

t (min)	Q (mm)
1	8
10	27
20	39
30	45
40	54
50	59
60	63
120	91
180	110
240	130
300	147
360	158
420	168
480	185
540	190
600	205

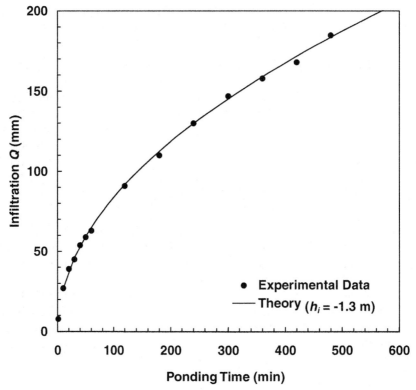

Figure 9.8 Results of infiltration test and theoretical fit to data for Example Problem 9.2.

9.2.3 Transient Moisture Profile for Vertical Infiltration

While the Green-Ampt approach provides insight into the rate of transient horizontal or vertical infiltration, particularly regarding the wetting front arrival time, it offers little information regarding the associated redistribution of water content during infiltration. More realistically, a time-dependent water content profile, or "nonsharp" wetting front, is the case. Philip (1957) provided an explicit and rigorous solution for transient water content profiles during vertical infiltration processes. Thorough understanding of the principles employed by Philip requires complete description of the mathematical form of Philip's transient infiltration approach and is beyond the intended scope of this book. Some highlights are described below for comparison and illustration.

Philip's solution, which takes the form of a power series, includes separate terms to represent gravity and matric suction as functions of both time and the soil properties. If the infiltration time or the distance that the water travels is short, gravity becomes the dominating driving force and the solution for infiltration rate takes the simple form

$$q = k_0 + \frac{s}{2\sqrt{t}} \qquad (9.28)$$

which also can be reduced by substituting x in eq. (9.25a) by eq. (9.21), since eq. (9.21) was derived under horizontal infiltration conditions where the gradient of the matric suction is the dominating driving force.

Philip's solution under both gravity and matric suction potentials offers a rigorous and quantitative way to illustrate the dynamic process of water content redistribution along the wetting front. Philip (1957) illustrated his solution using hydraulic properties of Yolo light clay determined by Moore (1939). Figure 9.9, for example, shows profiles of normalized water content θ^* versus depth for several time increments during a simulated vertical infiltration process. Normalized water content is defined in this case as volumetric water

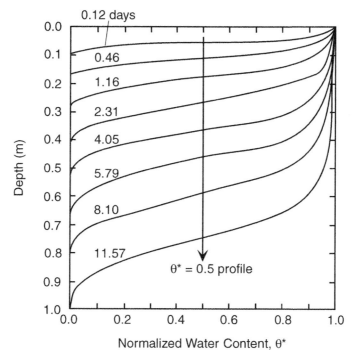

Figure 9.9 Normalized water content profiles during transient vertical infiltration into Yolo light clay (after Philip, 1957). Initial water content of soil located beyond wetting front is $\theta_i = 0.2376$. Final water content of soil located behind wetting front is $\theta_0 = 0.4950$.

content normalized by the initial value for soil located beyond the wetting front θ_i and the final value for soil located behind the wetting front θ_0 [$\theta^* = (\theta - \theta_i)/(\theta_0 - \theta_i)$]. If an "average" wetting front is defined by a value of normalized water content equal to 0.5, the vertical distance between any two profiles along $\theta^* = 0.5$ divided by the corresponding time interval can be used to estimate the infiltration rate. Cumulative infiltration can be obtained by the product of the depth of the average water content and the average water content.

The general behavior of water content redistribution under transient infiltration processes is evidenced in Fig. 9.9 in two distinct aspects: (1) the average slope of the depth-versus-water-content profile increases as infiltration proceeds and (2) the slope of the depth-versus-water-content profile at the dry end increases as infiltration proceeds. Systematic coverage of the transient nature of infiltration requires solution of a well-defined initial and boundary problem for the governing equation (9.12) and is provided next.

9.3 TRANSIENT SUCTION AND MOISTURE PROFILES

9.3.1 Importance of Transient Soil Suction and Moisture

Profiles of matric suction, the effective stress parameter, and suction stress under various steady-state unsaturated flow conditions were described in Chapter 7. These time-invariant profiles were assumed to occur at some depth from the ground surface below the active zone where moisture and temperature variations resulting from fluctuations in ambient environmental conditions are not likely to occur. While steady-state analyses provide an idealized theoretical framework for assessing matric suction and suction stress profiles in unsaturated soil, transient effects that occur closer to the surface within the active zone are indeed important for many geotechnical and environmental engineering problems.

The classic continuum mechanics approach to evaluating time-dependent processes is to cast the physical process into an initial and boundary value problem with an appropriate governing differential equation to be solved analytically, numerically, or graphically. This section illustrates the principles involved in solving initial and boundary value problems for unsaturated fluid flow and clarifies some of the distinct features of transient unsaturated fluid flow via analytical and numerical solutions for one-dimensional infiltration problems involving homogeneous soil.

9.3.2 Analytical Solution of Transient Unsaturated Flow

The governing equation for one-dimensional transient unsaturated fluid flow between the ground surface and the water table in a homogeneous and isotropic soil deposit can be reduced from eq. (9.12) in the form of a nonlinear partial differential equation as

$$\frac{\partial}{\partial z}\left[k(h_m)\left(\frac{\partial h_m}{\partial z}+1\right)\right] = \frac{\partial \theta}{\partial t} \qquad (9.29)$$

The unsaturated hydraulic conductivity function and soil-water characteristic curve can be captured in forms amenable to analytical solution using the following constitutive equations (Gardner, 1958):

$$k(h_m) = k_s \exp(\beta h_m) \qquad (9.30a)$$

and

$$\theta(h_m) = \theta_r + (\theta_s - \theta_r)\exp(\beta h_m) \qquad (9.30b)$$

where β is a soil parameter that captures the reduction in hydraulic conductivity and water content as the suction head increases. Relatively coarse-grained soil (e.g., sands) is typically described by relatively large β values whereas relatively fine-grained soil (e.g., silts, clays) is typically more accurately described by relatively low β values. Equations (9.30a) and (9.30b) may be applied to either infiltration (wetting) or drainage (drying) processes if hysteretic effects are ignored.

Pullan (1990) provides a detailed review of so-called quasilinear approaches that employ exponential models such as eqs. (9.30a) and (9.30b) to linearize and solve the governing unsaturated fluid flow equation. The basic technique is illustrated below through a solution by Srivastava and Yeh (1991). With eqs. (9.30a) and (9.30b), eq. (9.29) can be linearized as

$$\frac{\partial^2 k}{\partial z^2} + \beta \frac{\partial k}{\partial z} = \frac{\beta(\theta_s - \theta_r)}{k_s} \frac{\partial k}{\partial t} \qquad (9.31)$$

Introducing the following dimensionless parameters for distance (Z), hydraulic conductivity (K), ground surface flux (Q_A and Q_B), and time (T), eq. (9.31) can be further simplified:

$$Z = \beta z \qquad K = \frac{k}{k_s} \qquad Q_A = \frac{q_A}{k_s} \qquad Q_B = \frac{q_B}{k_s} \qquad T = \frac{\beta k_s t}{\theta_s - \theta_r}$$

where q_A is the initial steady flux into the ground surface at time equal to zero and q_B is the flux into the ground surface for times greater than zero. Substituting the above parameters into eq. (9.31) leads to the simplified expression

$$\frac{\partial^2 K}{\partial Z^2} + \frac{\partial K}{\partial Z} = \frac{\partial K}{\partial T} \tag{9.32}$$

For a problem of one-dimensional vertical infiltration from the ground surface toward the water table, the initial and boundary conditions in terms of the dimensionless variables are

$$K(Z,0) = Q_A - (Q_A - e^{-\beta h_0})e^{-Z} = K_0(Z) \tag{9.33a}$$

$$K(0,T) = e^{-\beta h_0} \tag{9.33b}$$

$$\left[\frac{\partial K}{\partial Z} + K\right]_{Z=L} = Q_B \tag{9.33c}$$

where h_0 is a prescribed suction head at the water table (typically assumed $h_0 = 0$) and L is the distance from the ground surface to the water table. Equation (9.33a) represents the initial steady state under a constant suction head condition. Equation (9.33b) reflects the constant suction head condition at the water table. Equation (9.33c) reflects the constant flux condition at the ground surface.

Applying the Laplace transform to eq. (9.32) and considering these initial and boundary conditions, Srivastava and Yeh (1991) arrived at the following analytical solution for normalized (dimensionless) hydraulic conductivity:

$$K = Q_B - (Q_B - e^{-\beta h_0})e^{-z} - 4(Q_B - Q_A)e^{(L-Z)/2} e^{-T/4}$$
$$\times \sum_{n=1}^{\infty} \frac{\sin(\lambda_n Z) \sin(\lambda_n L) e^{-\lambda_n^2 T}}{1 + L/2 + 2\lambda_n^2 L} \tag{9.34a}$$

where λ_n is the nth root of the equation

$$\tan(\lambda L) + 2\lambda = 0 \tag{9.34b}$$

The outflow at the water table ($Z = 0$) at any time was derived as

$$Q(T) = k_s \left[\frac{\partial K}{\partial Z} + K\right]_{Z=0} = k_s Q_B - 4k_s (Q_B - Q_A)e^{L/2} e^{-T/4}$$
$$\times \sum_{n=1}^{\infty} \frac{\lambda_n \sin(\lambda_n L) e^{-\lambda_n^2 T}}{1 + L/2 + 2\lambda_n^2 L} \tag{9.35}$$

The following examples demonstrate the analytical solutions given by eqs. (9.34) and (9.35) for a homogeneous unsaturated soil layer with a thickness of 100 cm (i.e., the groundwater table is located 100 cm below the ground surface). Steady-state infiltration profiles are used as initial conditions for two soils designated soil 1 and soil 2. Saturated hydraulic conductivity k_s for both soils is 1.0 cm/h. Soil 1 is described by a pore size distribution index β = 0.1 cm^{-1}, saturated water content θ_s = 0.4, and residual water content θ_r = 0.06. Soil 2 is described by β = 0.01 cm^{-1}, θ_s = 0.45, and θ_r = 0.2. Figure 9.10 shows the corresponding hydraulic conductivity function (Fig. 9.10a) and soil-water characteristic curve (Fig. 9.10b) for each soil modeled using the exponential eqs. (9.30a) and (9.30b). The modeling parameters for soil 1 and soil 2 are selected to represent relatively coarse-grained soil and relatively fine-grained soil, respectively.

Figure 9.11a shows suction head profiles for soil 1 at several times during a transient infiltration process calculated using eqs. (9.34) and (9.30a). The initial suction head profile (at t = 0) is the steady-state profile for a simulated ground surface infiltration rate q_A equal to 0.1 cm/h. For times greater than zero, a larger infiltration rate q_B = 0.9 cm/h is applied. Note that during the early times the suction head significantly increases (becomes less negative) for depths relatively close to the ground surface. This trend gradually progresses downward toward the water table as the wetting front advances. After 100 h, the suction head profile reaches steady state at a value close to zero. Similar patterns in time and space are illustrated in Fig. 9.11b for the calculated volumetric water content profiles.

Suction head profiles for soil 1 during a subsequent drainage process are shown in Fig. 9.12. Here, it is assumed that the hydraulic conductivity function and soil-water characteristic curve are again described by eqs. (9.30a) and (9.30b) (i.e., hysteresis is ignored). Drainage is simulated by allowing the infiltration rate at the ground surface to be suddenly reduced from 0.9 to 0.1 cm/h. Under these conditions, the suction head initially decreases near the ground surface. The increasingly negative suction profile then migrates downward toward the water table with time. After 100 h, the suction head profile returns to the previous steady-state condition for a ground surface flux equal to 0.1 cm/h.

Figures 9.13a and 9.13b show suction head and volumetric water content profiles for soil 2, respectively. Note that the initial steady-state profiles are nearly linear with depth (i.e., where q_A = 0.1 cm/h at t = 0). Compared with the relatively sharp wetting fronts noted for soil 1, the wetting fronts in soil 2 are dispersed to a relatively large distance. The suction head profile for the subsequent drainage process in soil 2 is shown in Fig. 9.14.

The discharge rate to the water table can be calculated by eq. (9.35) and is plotted for both soils during the wetting process in Fig. 9.15. As illustrated, the increase in discharge from the soil with relatively small β (the finer-grained soil) tends to occur earlier and the outflow reaches a new steady state faster than for the soil with the relatively large β value (the coarser-grained soil).

9.3.3 Numerical Modeling of Transient Unsaturated Flow

Numerical solution of the Richards' equation is necessary for many practical problems due to complications in initial and boundary conditions, flow do-

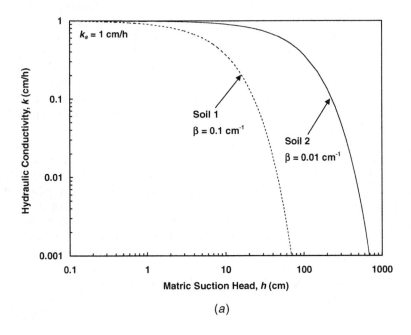

(a)

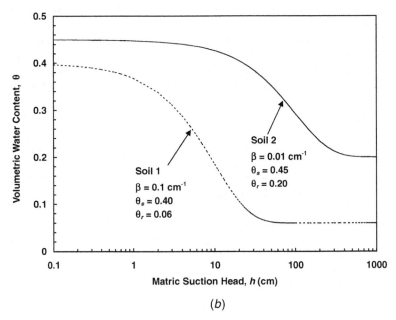

(b)

Figure 9.10 Hydrologic relationships for soil 1 (coarse) and soil 2 (fine) modeled using exponential functions: (a) hydraulic conductivity functions, $k(h)$ and (b) soil-water characteristic curves, $\theta(h)$.

9.3 TRANSIENT SUCTION AND MOISTURE PROFILES   391

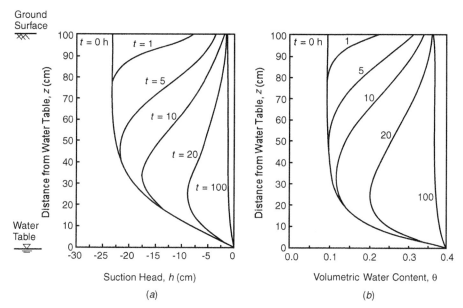

Figure 9.11 Infiltration (wetting) profiles for homogeneous coarse unsaturated soil layer with $\beta = 0.1$ cm^{-1}: (a) suction head profiles and (b) volumetric water content profiles (Srivastava and Yeh, 1991; modified by permission of the American Geophysical Union).

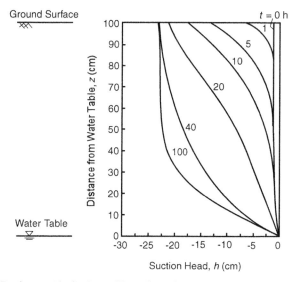

Figure 9.12 Drainage (drying) profiles of suction head for homogeneous coarse unsaturated soil layer with $\beta = 0.1$ cm^{-1} (Srivastava and Yeh, 1991; modified by permission of the American Geophysical Union).

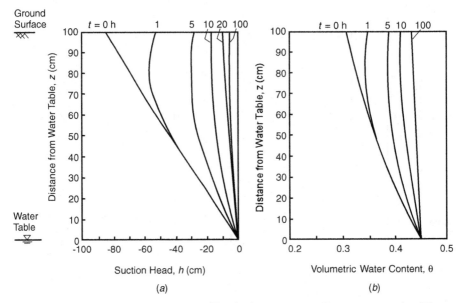

Figure 9.13 Infiltration (wetting) profiles for homogeneous fine unsaturated soil layer with $\beta = 0.01$ cm^{-1}: (a) suction head profiles and (b) volumetric water content profiles (Srivastava and Yeh, 1991; modified by permission of the American Geophysical Union).

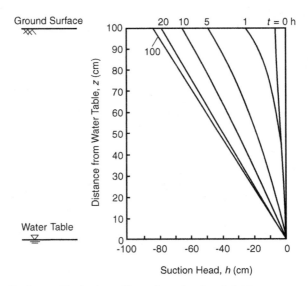

Figure 9.14 Drainage (drying) profiles of suction head for homogeneous fine unsaturated soil layer with $\beta = 0.01$ cm^{-1} (Srivastava and Yeh, 1991; modified by permission of the American Geophysical Union).

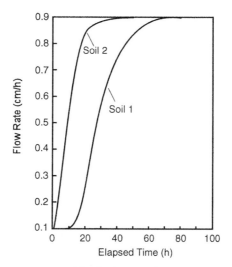

Figure 9.15 Outflow at water table during wetting process for soil 1 (coarse) and soil 2 (fine) (Srivastava and Yeh, 1991; modified by permission of the American Geophysical Union).

main geometry, soil heterogeneity, and the nonlinear hydrologic properties of the soil (e.g., the soil-water characteristic curve and the hydraulic conductivity function). Developing numerical methods that can account for these inherent complexities in typical settings for practical flow problems has been the subject of intense research over the past 30 years in the fields of hydrology, soil science, and environmental and geotechnical engineering. The most common modeling approaches include methods based on finite differences, finite elements, and integrated finite differences. An extensive list of computer software available for numerical flow modeling applications can be found in Tindall and Kunkel (1999).

An example modeling application is described here to illustrate the flexibility that numerical methods can offer in dealing with complicated boundary conditions for unsaturated fluid flow applications. An integrated finite-difference method (Pruess, 1991) is used to simulate vertical transient fluid flow following a sudden precipitation event in a 12-m-thick, one-dimensional deposit of alluvium overlying a fractured rock layer. The example, which was originally developed to model deep infiltration processes at Yucca Mountain, Nevada, during the 1998 El Niño year (Lu and LeCain, 2003), illustrates the large scales in both time and space typically involved in practical transient unsaturated fluid flow problems.

The soil deposit under consideration is a clayey soil with hydrologic functions described by the van Genuchten (1980) model (Chapter 12). The soil-water characteristic curve is described in terms of effective water content Θ or effective degree of saturation S_e as

$$\Theta = S_e = \frac{\theta - \theta_r}{\theta_s - \theta_r} = \left[\frac{1}{1 + (\alpha\psi)^n}\right]^m \quad (9.36a)$$

where specific parameters for the soil under consideration are as follows: $\alpha = 0.2$ kPa^{-1}, $n = 1.3$, and $m = 0.231$. The saturated volumetric water content $\theta_s = 0.4$.

The hydraulic conductivity function for the soil may be written in terms of relative hydraulic conductivity k_{rw} using these parameters as follows:

$$k_{rw} = S_e^{1/2} [1 - (1 - S_e^{1/m})^m]^2 \quad (9.36b)$$

Results from the numerical analysis are shown in Fig. 9.16. Steady-state profiles for effective degree of saturation were initially modeled for a steady infiltration rate of 20 mm/year (labeled February 7, 1998). To simulate a sudden precipitation event, a total flux of 1000 mm was then allowed to percolate into the soil uniformly in time between February 7, 1998, and February 21, 1998. As shown in the results, the sudden precipitation event causes the wetting front to propagate downward toward the underlying fractured rock. By February 14, 1998, the effective degree of saturation in the upper 3 m of the deposit has reached 93%. By March 21, 1998, the wetting front has reached the bottom of the alluvium layer (12 m from the ground surface) and

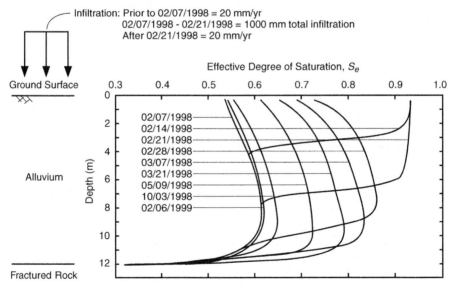

Figure 9.16 Effective degree of saturation profiles modeled using integrated finite-difference method for 12-m-thick alluvium deposit. Saturation profiles show effect of a sudden precipitation event at surface simulated during 1998 El Niño year (Lu and LeCain, 2003).

has started to drain into the underlying fractured rock layer. Note also that by this time the soil located near the ground surface has started to drain back toward the original steady-state condition, requiring another 11 months before the original condition is approached (February 6, 1999).

Figure 9.17 shows results of an alternative simulation where, prior to the major precipitation event occurring between February 7, 1998, and February 21, 1998, an additional total flux of 200 mm was simulated during the month of December 1997. Here, the first precipitation event causes the effective degree of saturation to increase to about 68% in the upper 4 m of the alluvium. Because the first event "prewets" the soil and thus increases its hydraulic conductivity, the wetting front following the major precipitation event propagates to a much greater depth than for the previous simulation.

Numerical solutions of transient unsaturated flow problems using techniques of finite elements, finite differences, and integrated finite differences are becoming common practice in unsaturated soil mechanics and hydrology. The recent shift from analytical to numerical techniques is due to several advantages the numerical methods offer: (1) flexibility in dealing with boundary and initial conditions, (2) capability to deal with complex geometry, and (3) ease in implementing various mathematical models for capturing the nonlinear soil hydrologic properties. Nevertheless, numerical solutions are most suitable for specific problems. In the process of better representing the prob-

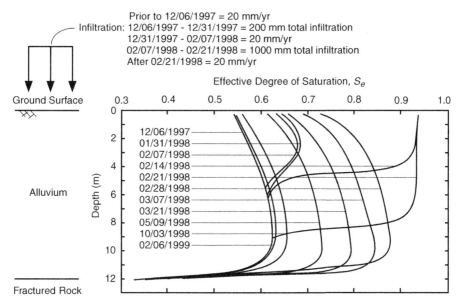

Figure 9.17 Effective degree of saturation profiles modeled using integrated finite-difference method for 12-m-thick alluvium deposit. Saturation profiles show effects of two separated precipitation events simulated during 1998 El Niño year (Lu and LeCain, 2003).

lem geometry and the initial and boundary conditions, as well as unsaturated soil characteristic functions, the generality of the solution is often diminished or lost. In many cases, analytical solutions provide more general and insightful information related to the flow phenomena and their controlling parameters.

9.4 PRINCIPLES FOR PORE GAS FLOW

9.4.1 Principle of Mass Conservation for Compressible Gas

The governing equation for continuous gas flow in soil can be derived by applying the principle of mass conservation, which states that for a given elemental volume the rate of gain or loss of gas is conservative and is equal to the net flux of inflow and outflow. The mass conservation equation for pore airflow in unsaturated soil can be expressed as

$$-\frac{\partial}{\partial x}(\rho_a v_x) - \frac{\partial}{\partial y}(\rho_a v_y) - \frac{\partial}{\partial z}(\rho_a v_z) = \frac{\partial(n_a \rho_a)}{\partial t} \qquad (9.37)$$

where n_a is the air-filled porosity, ρ_a is the density of air, and v_x, v_y, and v_z are the components of flow velocity in the x-, y-, and z-coordinate directions, respectively. For perfectly dry soil, the air-filled porosity n_a is equal to the total porosity n. For saturated soil, the air-filled porosity is equal to zero.

For most practical problems where the airflow velocity is relatively small and within the laminar flow regime, the flow velocities can be effectively described by Darcy's law as

$$v_x = -\frac{K_x}{\mu}\frac{\partial u_a}{\partial x} \qquad v_y = -\frac{K_y}{\mu}\frac{\partial u_a}{\partial y} \qquad v_z = -\frac{K_z}{\mu}\left(\frac{\partial u_a}{\partial z} + \rho_a g\right) \qquad (9.38)$$

where K_x, K_y, and K_z are intrinsic permeability (m²) in the x, y, and z directions, respectively, μ is the viscosity of air (kg/m · s or N · s/m²), and u_a is air pressure (kPa). The second term in the z-coordinate direction reflects the contribution of gravitational acceleration g (m/s²) to the total air pressure.

Assuming that pore air can be considered an ideal gas, air density can be expressed as

$$\rho_a = \frac{\omega_a}{RT} u_a \qquad (9.39)$$

where ω_a is the molecular mass of air (kg/mol), R is the universal gas constant (J/mol · K), and T is absolute temperature (K).

Substituting eqs. (9.38) and (9.39) into eq. (9.37) leads to

$$\frac{\partial}{\partial x}\left(\frac{K_x}{\mu}\frac{\omega_a u_a}{RT}\frac{\partial u_a}{\partial x}\right) + \frac{\partial}{\partial y}\left(\frac{K_y}{\mu}\frac{\omega_a u_a}{RT}\frac{\partial u_a}{\partial y}\right) + \frac{\partial}{\partial z}\left(\frac{K_z}{\mu}\frac{\omega_a u_a}{RT}\frac{\partial u_a}{\partial z}\right) + \frac{\partial}{\partial z}\left(\frac{\omega_a u_a \rho_a}{RT}g\right)$$
$$= \frac{\omega_a}{RT}\left(\frac{\partial(n_a u_a)}{\partial t} - \frac{n_a u_a}{T}\frac{\partial T}{\partial t}\right) \tag{9.40}$$

Assuming that the air-filled porosity does not change, the above equation becomes

$$\frac{\partial}{\partial x}\left(\frac{K_x}{\mu}u_a\frac{\partial u_a}{\partial x}\right) + \frac{\partial}{\partial y}\left(\frac{K_y}{\mu}u_a\frac{\partial u_a}{\partial y}\right) + \frac{\partial}{\partial z}\left(\frac{K_z}{\mu}u_a\frac{\partial u_a}{\partial z}\right) + \frac{\partial}{\partial z}(u_a \rho_a g)$$
$$= n_a\frac{\partial u_a}{\partial t} - \frac{n_a u_a}{T}\frac{\partial T}{\partial t} \tag{9.41}$$

where further simplification leads to

$$\frac{\partial}{\partial x}\left(\frac{K_x}{\mu}\frac{\partial u_a^2}{\partial x}\right) + \frac{\partial}{\partial y}\left(\frac{K_y}{\mu}\frac{\partial u_a^2}{\partial y}\right) + \frac{\partial}{\partial z}\left(\frac{K_z}{\mu}\frac{\partial u_a^2}{\partial z}\right) + \frac{\partial}{\partial z}\left(u_a^2\frac{\omega_a g}{RT}\right)$$
$$= \frac{n_a}{u_a}\frac{\partial u_a^2}{\partial t} - \frac{n_a u_a}{2T}\frac{\partial T}{\partial t} \tag{9.42}$$

Equation (9.42) is the governing equation for transient airflow that includes the gravitational effect and the effects of time-dependent variations in air pressure and temperature. Its explicit solution requires knowledge of the ambient temperature field and its variation with time, which can be obtained by measurement or by solving the appropriate governing equation for heat transport. For some geotechnical engineering problems, however, several simplifications can be made, and fully coupled analysis of temperature and air pressure may not be necessary.

9.4.2 Governing Equation for Pore Airflow

Barometric pressure fluctuation often becomes an important driving force for the flow of pore air in near-surface unsaturated soil. In general problems of this sort, the gravitational and temperature effects captured by eq. (9.42) may usually be ignored. Gravitational effects are usually negligible because the corresponding gradient in air pressure is quite small (less than several pascals) over the relatively shallow depths of interest in most problems (<100 m). Temperature effects are often negligible because most (but not all) subsurface temperature gradients are relatively small and the thermal diffusivity is generally several hundred to several thousand times smaller than the air diffusivity. For problems in which these assumptions are indeed valid, eq. (9.42) may be simplified as

$$\frac{\partial}{\partial x}\left(\frac{K_x}{\mu}\frac{\partial u_a^2}{\partial x}\right) + \frac{\partial}{\partial y}\left(\frac{K_y}{\mu}\frac{\partial u_a^2}{\partial y}\right) + \frac{\partial}{\partial z}\left(\frac{K_z}{\mu}\frac{\partial u_a^2}{\partial z}\right) = \frac{n_a}{u_a}\frac{\partial u_a^2}{\partial t} \qquad (9.43)$$

If the soil is considered to be homogeneous and isotropic and the intrinsic permeability is not affected by either time or the flow process (i.e., if the air-filled porosity remains constant), the above equation becomes

$$\frac{\partial^2 u_a^2}{\partial x^2} + \frac{\partial^2 u_a^2}{\partial y^2} + \frac{\partial^2 u_a^2}{\partial z^2} = \frac{\mu n_a}{K u_a}\frac{\partial u_a^2}{\partial t} \qquad (9.44)$$

or

$$\frac{K\sqrt{p}}{\mu n_a}\left(\frac{\partial^2 p}{\partial x^2} + \frac{\partial^2 p}{\partial y^2} + \frac{\partial^2 p}{\partial z^2}\right) = D_a\left(\frac{\partial^2 p}{\partial x^2} + \frac{\partial^2 p}{\partial y^2} + \frac{\partial^2 p}{\partial z^2}\right) = \frac{\partial p}{\partial t} \qquad (9.45a)$$

where

$$p = u_a^2 \qquad (9.45b)$$

and D_a, or air diffusivity, is as follows:

$$D_a = \frac{K\sqrt{p}}{\mu n_a} = \frac{K u_a}{\mu n_a} \qquad (9.45c)$$

For small changes in air pressure, the air diffusivity according to eq. (9.45c) changes very little. For example, if the maximum barometric air pressure change over one month is less than 3.0 kPa (a typical fluctuation) and the mean barometric pressure is 100 kPa, then the relative error in the air diffusivity calculation using an average air pressure is less than 1.5%. In practice, air diffusivity is often considered as a constant so that eq. (9.45a) can be treated as a linear diffusion equation. Accordingly, the numerous available analytical solutions for the general diffusion-type equation (e.g., Carslaw and Jeager, 1959) become applicable.

9.4.3 Linearization of the Airflow Equation

Consider a flat unsaturated soil layer subjected to transient pore air pressure changes near the ground surface due to natural ambient fluctuations in barometric pressure (Fig. 9.18). The cycling of air pressure at the ground surface causes the pore air pressure in the soil layer to fluctuate at an amplitude that attenuates and lags with increasing depth from the surface.

The governing pore airflow equation for this one-dimensional problem may be reduced from eq. (9.45a) as

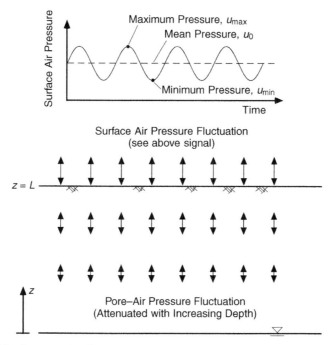

Figure 9.18 Barometric air pressure fluctuation and consequent pore–air pressure fluctuation in near-surface unsaturated soil layer.

$$\frac{\partial p}{\partial t} = D_a \frac{\partial^2 p}{\partial z^2} \tag{9.46}$$

where the coordinate direction z is defined as positive upward from a value of zero at the water table.

Equation (9.46) is a nonlinear partial differential equation and the analytical solution is difficult to find. In typical field settings, however, the amplitude of the barometric pressure fluctuation is usually within 1% of the mean air pressure. Accordingly, linearization of eq. (9.46) becomes possible (Katz et al., 1959; Weeks, 1979). Specifically, if the air pressure defining the diffusion coefficient D_a in eq. (9.45c) can be replaced by the mean air pressure u_0, then eq. (9.46) becomes a linear equation with the dependent variable becoming the air pressure squared, that is, $p = u_a^2$.

If the maximum and minimum pressures in the barometric pressure cycle are u_{max} and u_{min}, respectively, then eq. (9.46) can be rewritten in dimensionless form as

$$\frac{\partial}{\partial t}\left(\frac{u_a^2 - u_{min}^2}{u_{max}^2 - u_{min}^2}\right) = D_a \frac{\partial^2}{\partial z^2}\left(\frac{u_a^2 - u_{min}^2}{u_{max}^2 - u_{min}^2}\right) \tag{9.47a}$$

where

$$D_a = \frac{Ku_0}{\mu n_a} \tag{9.47b}$$

If the maximum amplitude of the air pressure fluctuation is less than $0.01u_0$, it can be shown that

$$\frac{u_a^2 - u_{min}^2}{u_{max}^2 - u_{min}^2} = \frac{u_a - u_{min}}{u_{max} - u_{min}} \frac{u_a + u_{min}}{u_{max} + u_{min}}$$

or

$$\frac{2.00 u_0}{2.01 u_0} \frac{u_a - u_{min}}{u_{max} - u_{min}} \leq \frac{u_a^2 - u_{min}^2}{u_{max}^2 - u_{min}^2} \leq \frac{2.01 u_0}{2.01 u_0} \frac{u_a - u_{min}}{u_{max} - u_{min}}$$

or

$$0.995 \frac{u_a - u_{min}}{u_{max} - u_{min}} \leq \frac{u_a^2 - u_{min}^2}{u_{max}^2 - u_{min}^2} \leq 1.000 \frac{u_a - u_{min}}{u_{max} - u_{min}}$$

which demonstrates that replacement of the squared dimensionless pressure by the nonsquared dimensionless pressure leads to an error of no more than 0.5%. Accordingly, eq. (9.47) or (9.46) may be rewritten as

$$\frac{\partial u_a}{\partial t} = D_a \frac{\partial^2 u_a}{\partial z^2} \qquad D_a = \frac{Ku_0}{\mu n_a} \tag{9.48}$$

The above simplification makes analytical solutions for the diffusion-type equation applicable to barometric pressure–induced airflow in unsaturated soil.

9.4.4 Sinusoidal Barometric Pressure Fluctuation

The following example illustrates the general features of subsurface pore air pressure propagation governed by eq. (9.48). For an idealized daily (diurnal) sinusoidal barometric fluctuation, the initial and boundary conditions are as follows:

$$u_a(z,0) = f_0(z) \tag{9.49}$$

$$\frac{\partial u_a(0,t)}{\partial z} = 0 \tag{9.50}$$

$$u_a(L,t) = f_1(t) = A \sin\left(\frac{2\pi}{24} t\right) \tag{9.51}$$

9.4 PRINCIPLES FOR PORE GAS FLOW

where L is the vertical distance from the water table ($z = 0$) to the ground surface. Equation (9.49) represents the initial condition. Equation (9.50) represents the no-flow boundary condition at the water table. Equation (9.51) is the diurnal barometric variation with amplitude A (Pa) and a period of 24 h.

Assuming the initial air pressure profile is zero [$f_0(z) = 0$], the analytical solution for the initial and boundary value problem defined above is as follows (Carslaw and Jaeger, 1959):

$$u_a(z,t) = C \sin\left(\frac{2\pi}{24} t + \theta\right) \quad (9.52)$$

where

$$C = A\left(\frac{\cosh \sqrt{\pi/6D_a}\, z + \cos \sqrt{\pi/6D_a}\, z}{\cosh \sqrt{\pi/6D_a}\, L + \cos \sqrt{\pi/6D_a}\, L}\right)^{1/2} \quad (9.53)$$

and

$$\theta = \tan^{-1}\left(\frac{\sinh (\sqrt{\pi/24D_a}\, z) \sin (\sqrt{\pi/24D_a}\, z)}{\cosh (\sqrt{\pi/24D_a}\, z) \cos (\sqrt{\pi/24D_a}\, z)}\right)$$

$$- \tan^{-1}\left(\frac{\sinh (\sqrt{\pi/24D_a}\, L) \sin (\sqrt{\pi/24D_a}\, L)}{\cosh (\sqrt{\pi/24D_a}\, L) \cos (\sqrt{\pi/24D_a}\, L)}\right) \quad (9.54)$$

For example, if the thickness of unsaturated soil layer (L) is 40 m, the viscosity of air (μ) is 1.8×10^{-5} kg/m · s, the intrinsic permeability of the soil (K) is 1×10^{-9} m^2, the air-filled porosity (n_a) is 0.25, and the daily pressure fluctuation amplitude (A) at the surface is 300 Pa, then the corresponding profiles of air pressure propagation into the soil layer are illustrated in Fig. 9.19. Note that the surface air pressure amplitude of 300 Pa attenuates to about 134 Pa at 10 m depth, 57 Pa at 20 m depth, 27 Pa at 30 m depth, and 25 Pa at the water table (40 m depth). The peak arrival time also increases as the pressure wave propagates downward into the soil layer. This time lag is about 2.5 h at 10 m, 6.5 h at 20 m, 21.5 h at 30 m, and 24 h (i.e., the entire period) at the water table.

The airflow discharge velocity at any point can be estimated as

$$q = -k \frac{\partial u_a(z,t)}{\partial z} \quad (9.55)$$

The airflow due to atmospheric pressure variation, also called barometric pumping, may be an important mechanism in vapor exchange between the atmosphere and soil.

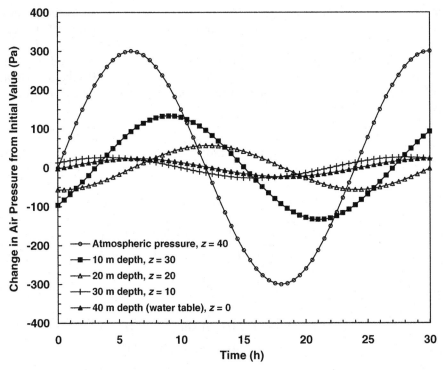

Figure 9.19 Dynamic air pressure propagation into 40-m-thick unsaturated zone showing amplitude attenuation and phase lag with increasing depth from the ground surface.

9.5 BAROMETRIC PUMPING ANALYSIS

9.5.1 Barometric Pumping

Barometric pumping refers to the cycling of subsurface pore air pressure and the associated movement of pore air and pore water vapor in the near-surface unsaturated zone due to natural fluctuations in barometric pressure. The flow of pore gas due to barometric pumping is often an important vapor exchange mechanism between the atmosphere and soil. The net vapor movement is typically upward, thus enhancing evaporation. Under certain atmospheric conditions, however, such as cold weather, the direction of net vapor movement may be reversed. The phenomenon has been recognized in recent years as a potential natural mechanism for removing volatile organic compounds from contaminated soil. As illustrated in this section, analysis of subsurface pore gas pressures measured under naturally induced periodic fluctuations also provides an effective method for determining in situ vertical air permeability in unsaturated soil.

9.5.2 Theoretical Framework

Section 9.4.4 described a relatively simple theoretical framework for analysis of barometric pumping problems with a sinusoidal fluctuation. In reality, barometric pressure fluctuates quite irregularly. The initial and boundary conditions under the more general case of an arbitrary barometric pressure fluctuation function are

$$u_a(z,0) = f_0(z) \tag{9.56a}$$

$$\frac{\partial u_a(0,t)}{\partial z} = 0 \tag{9.56b}$$

$$u_a(L,t) = f_1(t) \tag{9.56c}$$

where, as before, the vertical coordinate z is defined as zero at the water table and as L at the ground surface. The functions $f_0(z)$ and $f_1(t)$ are arbitrary functions in space and time, respectively, that may be directly obtained from field air pressure measurements at the surface and at depth. The general solution for the governing diffusion equation under these initial and boundary conditions is as follows (Carslaw and Jaeger, 1959):

$$u_a = \frac{2}{L} \sum_{n=0}^{\infty} \exp\left(-D_a (2n+1)^2 \frac{\pi^2 t}{4L^2}\right) \cos \frac{(2n+1)\pi z}{2L} [c_1 + c_2] \tag{9.57a}$$

$$c_1 = \int_0^L f_0(z) \cos \frac{(2n+1)\pi z}{2L} dz \tag{9.57b}$$

$$c_2(t) = \frac{(2n+1)\pi D_a (-1)^n}{2L} \int_0^t \exp\left(D_a (2n+1)^2 \frac{\pi^2 \lambda}{4L^2}\right) f_1(\lambda) \, d\lambda \tag{9.57c}$$

Shan (1995) applied this solution to determine vertical air permeability by employing a root-mean-square optimization scheme and demonstrated that the solution can be used to interpret field pneumatic data and to determine the vertical air permeability. However, the integrations in eqs. (9.57b) and (9.57c) are generally difficult to perform.

Alternatively, consider the problem where the initial pressure is at static equilibrium and the barometric pressure fluctuation can be described by a simple harmonic function $f_1(t) = A_a \sin(\omega t + \varepsilon)$, where A_a is the amplitude constant (Pa), ω is the frequency (s^{-1}), and ε is the initial phase (rad). The solution in this case is as follows (Carslaw and Jaeger, 1959):

$$u_a = A \sin(\omega t + \varepsilon + \theta)$$

$$+ 4\pi D_a \sum_{n=0}^{\infty} \left\{ \frac{(-1)^n (2n+1)[4L^2 \omega \cos \varepsilon - D_a (2n+1)^2 \pi^2 \sin \varepsilon]}{16L^4\omega^2 + D_a^2 \pi^4 (2n+1)^4} \right.$$

$$\left. \times \exp\left(-D_a (2n+1)^2 \frac{\pi^2 t}{4L^2}\right) \cos \frac{(2n+1)\pi z}{2L} \right\} \qquad (9.58)$$

where A is the amplitude and θ is the phase lag at depth z defined as

$$A = A_a \left(\frac{\cosh \sqrt{2\omega/D_a}\, z + \cos \sqrt{2\omega/D_a}\, z}{\cosh \sqrt{2\omega/D_a}\, L + \cos \sqrt{2\omega/D_a}\, L}\right)^{1/2} \qquad (9.59a)$$

$$\theta = \arg\left[\frac{\cosh \sqrt{\omega/2D_a}\, z(1+i)}{\cosh \sqrt{\omega/2D_a}\, L(1+i)}\right] \qquad (9.59b)$$

where i denotes the pure imaginary number and arg [ ] is the argument of the complex function in the bracket.

Equations (9.58) and (9.59) have two useful features: (1) the periodic steady-state solution is decoupled from the transient solution and (2) the interdependency among the quantities A, θ, ω, and D_a is explicitly depicted. Physically, the first term in eq. (9.58) represents the pore air pressure response in unsaturated soil to the harmonic excitation presented at the atmosphere. The second term reflects the transient response that is caused by the initial phase jump ε at time equal to zero. This term usually diminishes quickly as time elapses because it is an inverse exponential function of time. The amplitude decay and phase lag are evident in eqs. (9.59a) and (9.59b), respectively.

Equation (9.59) was cited by Weeks (1979), who evaluated the amplitude decay phenomenon to determine the pneumatic diffusivity of an unsaturated zone. As outlined below, Lu (1999) developed an alternative approach using a time series analysis and presented a general analytical solution in terms of the spectrum of pressure amplitudes and frequencies for any initial and boundary barometric pressure fluctuation.

9.5.3 Time Series Analysis

Although natural barometric fluctuation appears to be irregular, periodic patterns exist that are controlled by the diurnal, semidiurnal, and seasonal atmospheric tides. Analysis of surface and subsurface air pressure measurements (e.g., Fig. 9.20) reveals several patterns: (1) the amplitude of the pressure fluctuation usually decreases as the depth increases (i.e., the pressure amplitude is attenuated with depth); (2) progressive phase lags exist for the major tidal modes; (3) higher frequency modes tend to diminish much

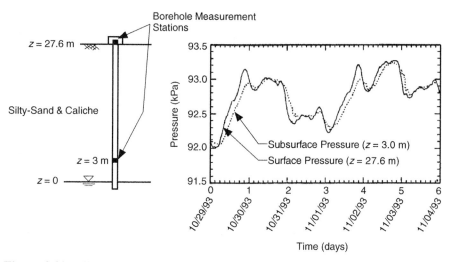

Figure 9.20 Illustration of field pneumatic measurement and typical pressure variations in atmosphere and subsurface (data from Stephens, 1995).

faster than lower frequency modes as the depth increases; and (4) the extent of the these three patterns are greatly controlled by the diffusivity of the porous medium.

In general, any arbitrary atmospheric pressure variation $f_1(t)$ can be represented by a Fourier series as

$$f_1(t) = \sum_{n=0}^{\infty} A_n \sin(n\omega t + \varepsilon_n) \tag{9.60a}$$

where

$$A_n = \sqrt{a_n^2 + b_n^2} \tag{9.60b}$$

$$\varepsilon_n = \arg\left(\frac{a_n}{\sqrt{a_n^2 + b_n^2}}\right) \tag{9.60c}$$

$$a_n = \frac{2}{T} \int_0^T f_1(t) \cos n\omega t \, dt \tag{9.60d}$$

$$b_n = \frac{2}{T} \int_0^T f_1(t) \sin n\omega t \, dt \tag{9.60e}$$

where the frequency $\omega = 2\pi/T$, n is the wave number covering the entire spectrum in the frequency domain, and T is the total observation time (s).

The solution to eq. (9.48) can be obtained by applying the principle of superposition under the zero initial condition $[f_0(z) = 0]$ and the arbitrary boundary condition described by eqs. (9.60a) to (9.60c) as

$$u_a = \sum_{n=0}^{\infty} C_n \sin(n\omega t + \varepsilon_n + \theta_n) \tag{9.61}$$

where the amplitude C_n and the phase lag θ_n are obtained as

$$C_n = A_n \left(\frac{\cosh\sqrt{2n\omega/D_a}\, z + \cos\sqrt{2n\omega/D_a}\, z}{\cosh\sqrt{2n\omega/D_a}\, L + \cos\sqrt{2n\omega/D_a}\, L} \right)^{1/2} \tag{9.62a}$$

$$\theta_n = \arg\left[\frac{\cosh\sqrt{n\omega/2D_a}\, z(1+i)}{\cosh\sqrt{n\omega/2D_a}\, L(1+i)}\right] \tag{9.62b}$$

The subsurface/surface amplitude ratio C_n/A_n describes the attenuation of the barometric pressure wave with depth. Using the properties of complex functions and rewriting D_a in terms of intrinsic permeability K, mean barometric pressure u_0, air-filled porosity n_a, and the viscosity of air μ [eq. (9.47b)], it can be shown that the phase lag θ_n can be rewritten as

$$\theta_n = \tan^{-1}\left(\frac{\sinh\sqrt{n\omega n_a\mu/2u_0 K}\, z \,\sin\sqrt{n\omega n_a\mu/2u_0 K}\, z}{\cosh\sqrt{n\omega n_a\mu/2u_0 K}\, z \,\cos\sqrt{n\omega n_a\mu/2u_0 K}\, z}\right)$$

$$- \tan^{-1}\left(\frac{\sinh\sqrt{n\omega n_a\mu/2u_0 K}\, L \,\sin\sqrt{n\omega n_a\mu/2u_0 K}\, L}{\cosh\sqrt{n\omega n_a\mu/2u_0 K}\, L \,\cos\sqrt{n\omega n_a\mu/2u_0 K}\, L}\right) \tag{9.62c}$$

where $0 \leq \tan^{-1}(\ ) \leq 2\pi$. Defining a dimensionless parameter β as

$$\beta = \left(\frac{2n\omega}{D_a}\right)^{1/2} L = \left(\frac{4n\pi n_a\mu}{TKu_0}\right)^{1/2} L \tag{9.63}$$

the phase lag θ_n and the amplitude ratio C_n/A_n can be treated as functions of two variables β and z/L:

$$\theta_n = \tan^{-1}\left(\tanh\frac{\beta}{2}\frac{z}{L} \tan\frac{\beta}{2}\frac{z}{L}\right) - \tan^{-1}\left(\tanh\frac{\beta}{2} \tan\frac{\beta}{2}\right) \tag{9.64}$$

$$\frac{C_n}{A_n} = \left(\frac{\cosh\beta\,(z/L) + \cos\beta\,(z/L)}{\cosh\beta + \cos\beta}\right)^{1/2} \tag{9.65}$$

The dimensionless parameter β couples the atmospheric pressure parameters (frequency and mean pressure) and the parameters of the unsaturated soil zone (air-filled porosity, pore air viscosity, intrinsic permeability, and thickness of the deposit). The value of β typically ranges between 10^{-4} and 250. For example, for a 10-m unsaturated zone comprised of gravel with an air-filled porosity $n_a = 0.25$, pore air viscosity $\mu = 1.8 \times 10^{-5}$ kg/m·s, and permeability $K = 10^{-6}$ m^2 and if the barometric pressure fluctuation is characterized by period $T = 12$ h, wave number $n = 1$, and mean pressure $u_0 = 92{,}500$ Pa, the β value for the system by eq. (9.63) is 0.0012. On the other hand, for a 100-m unsaturated zone comprised of clay with an air-filled porosity $n_a = 0.25$, air viscosity $\mu = 1.8 \times 10^{-5}$ kg/m·s, permeability $K = 10^{-16}$ m^2, period $T = 336$ h, wave number $n = 1$, and mean pressure $u_0 = 92{,}500$ Pa, the β value for the system is 225. For a similar system soil with a permeability $K = 10^{-12}$ m^2, the β value is 2.25.

9.5.4 Determining Air Permeability

Equations (9.64) and (9.65) can be used to create type curves for determining vertical air permeability from measurements of barometric pressure fluctuation in the unsaturated soil zone. Rojstaczer and Tunks (1995), for example, created type curves portraying either the amplitude ratio C_n/A_n or phase lag θ_n as a function of dimensionless frequency and normalized depth z/L. Air diffusivity could be determined by matching an estimated value of amplitude ratio or phase lag to a specific type curve at known depth. Lu (1999) developed alternative type curves by considering either amplitude ratio or phase lag as a function of the diffusivity-dependent factor β and normalized depth. Figure 9.21 shows type curves in terms of amplitude ratio. If, for example, subsurface pore air pressure measurements are made at some normalized depth z/L to determine the subsurface-to-surface amplitude ratio C_n/A_n for a given wave number n, then the corresponding β value can be determined from the matching type curve. Using the best fitting β value and eq. (9.63), vertical air permeability can be determined given the remaining parameters in eq. (9.63).

Consider the field case shown in Fig. 9.20 where surface and subsurface air pressure measurements were obtained in an unsaturated soil layer. The measurement site, which is located in Tucson, Arizona, consists of a 27.6-m-thick layer of silty sand and caliche soil bounded by the atmosphere at $z = L = 27.6$ m and the underlying water table at $z = 0$. Surface ($z = 27.6$ m) and subsurface ($z = 3.0$ m) air pressure measurements were monitored in a borehole over a period of 6 days from October 29 to November 4, 1993 (Stephens, 1995).

As shown in the figure, the pressure signals at both measurement stations show modes of daily oscillation controlled by the diurnal atmospheric tide. Maximum pressures for any one-day period occur at about 10:00 a.m. and

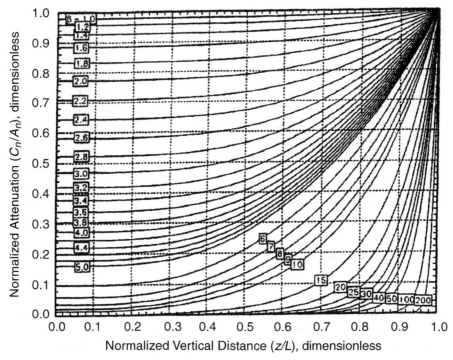

Figure 9.21 Type curves showing dimensionless parameter β as function of normalized depth and subsurface/surface amplitude attenuation ratio (Lu, 1999; modified by permission of ASCE).

10:00 p.m. Minimum pressures occur at about 4:00 a.m. and 4:00 p.m. Amplitude attenuation and phase lag persist in the subsurface during the 6-day period. Fourier time series representations of both the atmospheric and subsurface pressure data are depicted in Fig. 9.22. Four major amplitudes are noted for periods of 3 days ($n = 2$), 1 day ($n = 6$), 12 h ($n = 12$), and 8 h ($n = 18$). Fourier series analysis for different total times of 1, 2, 3, 4, and 5 days shows the same pattern of amplitude dominance at three distinct periods: diurnal, semidiurnal, and 8 h. Amplitudes for frequencies not equal to the three distinct ones follow the decay pattern for the random pressure fluctuation as illustrated by the short-dashed line in Fig. 9.22.

How well do these amplitudes represent the observed air pressure fluctuations both for the atmosphere and subsurface? Figure 9.23 illustrates these distinct modes as well as their superposition. For the atmospheric pressure variation, the amplitude of the diurnal tide is 203 Pa, the semidiurnal tide is 105 Pa, and the 8-h tide is 38 Pa. The combination of these tides has an amplitude of 300 Pa. In the subsurface at a depth of 24.6 m ($z = 3$ m), clear progressive patterns in both the amplitude attenuation and the phase lag are

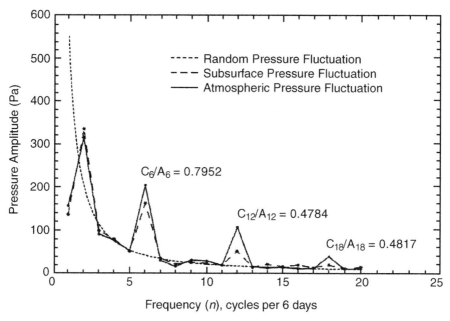

Figure 9.22 Pressure amplitude as function of harmonic frequency number for field data shown in Fig. 9.20 (Lu, 1999; modified by permission of ASCE).

observed. Attenuations for the diurnal, semidiurnal, and 8-h tides are 0.80, 0.48, and 0.48, respectively. The time lags for the diurnal, semidiurnal, and 8-h tides are 1.2, 0.8, and 0.5 h, respectively. The pressure fluctuation represented by the combination of the three modes accurately resembles the two pressure highs occurring at 10:00 a.m. and 10:00 p.m. and the two pressure lows occurring at 4:00 a.m. and 4:00 p.m. Comparison between the combined tides and the observed data can be made by including the longest period mode (3 days), as portrayed in Fig. 9.24.

One amplitude ratio value at the subsurface is mathematically sufficient to determine vertical air permeability using the series of type curves shown in Fig. 9.21. Permeability values determined for other distinct modes may be compared to provide confidence in the results, as shown in Table 9.2 and Fig. 9.25. Air-filled porosity n_a of 0.25 and air viscosity μ of 1.8×10^{-5} kg/m · s were used in the permeability calculations. The air permeability ranges from 1.19 to 4.52 darcies (1 darcy $\approx 10^{-12}$ m²) for different distinct periods (8, 12, and 24 h) and different analysis times (1, 2, 3, 4, 5, and 6 days). A permeability value of 1.47 darcies is obtained using the 6-day analysis time, which agrees well with the value of 1.46 darcies determined by Shan (1995) using a different analytical solution.

410 TRANSIENT FLOWS

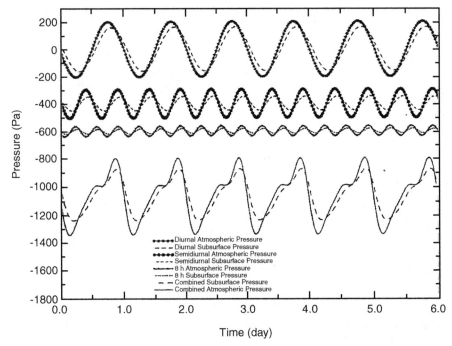

Figure 9.23 Major harmonic tides of diurnal, semidiurnal, and 8 h for field pressure data shown in Fig. 9.20. Pressure magnitudes are shifted to reflect deviations rather than absolute values for comparison (Lu, 1999; modified by permission of ASCE).

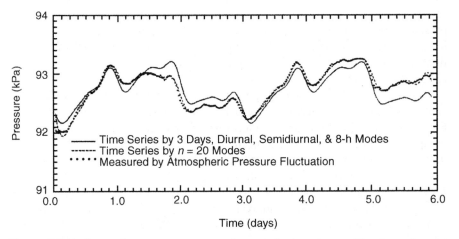

Figure 9.24 Comparison among measured atmospheric pressure, Fourier-series representation for frequency number n up to 20, and four major tidal modes (Lu, 1999; modified by permission of ASCE).

TABLE 9.2 Calculated Permeability Using Field Data of Stephens (1995)

Total Analysis Time (days)	Tide Mode (h)	Amplitude (Pa)		Amplitude Ratio, C_n/A_n	β Parameter (dimensionless)	Permeability (darcy)	Average Permeability (darcy)
		A_n	C_n				
1	8	108	89	0.82	1.85	4.52	2.68
	12	230	143	0.62	2.47	1.69	2.68
	24	480	413	0.86	1.68	1.83	2.68
2	8	47	38	0.81	1.85	4.51	2.58
	12	121	73	0.60	2.53	1.61	2.58
	24	250	208	0.83	1.80	1.59	2.58
3	8	31	18	0.57	2.64	2.22	1.68
	12	95	49	0.51	2.83	1.29	1.68
	24	185	152	0.82	1.83	1.54	1.68
4	8	48	29	0.60	2.58	2.32	1.65
	12	123	63	0.51	2.85	1.27	1.65
	24	235	183	0.78	1.95	1.35	1.65
5	8	44	28	0.62	2.48	2.52	1.76
	12	115	59	0.51	2.85	1.27	1.76
	24	226	183	0.81	1.86	1.49	1.76
6	8	38	18	0.48	2.91	1.83	1.47
	12	105	50	0.48	2.95	1.19	1.47
	24	203	162	0.80	1.93	1.38	1.47
Average	8	—	—	—	—	2.96	1.19–4.52
	12	—	—	—	—	1.38	1.19–4.52
	24	—	—	—	—	1.53	1.19–4.52

Source: From Lu (1999).

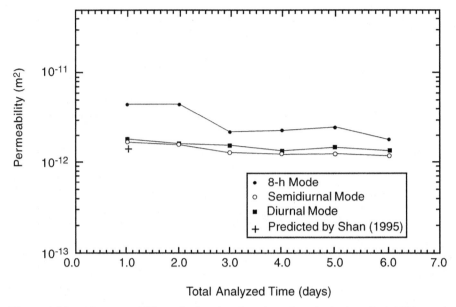

Figure 9.25 Air permeability calculated as function of analysis time (Lu, 1999; modified by permission of ASCE).

PROBLEMS

9.1. Describe the major differences between steady and transient liquid flow in unsaturated soil in terms of boundary conditions, head distribution, and discharge velocity.

9.2. What are the characteristic functions (material variables) required to quantify transient water flow in unsaturated soil? Describe the interdependency among these functions.

9.3. If a soil's hydraulic conductivity and soil-water characteristic function were known, could the moisture field be quantified using eq. (9.17)?

9.4. What is the driving force for fluid flow for a horizontal infiltration problem in initially dry sand when the infiltration just starts? What is the infiltration rate when the infiltration time is sufficiently long? What is the driving force for fluid flow for a vertical infiltration problem into initially dry sand when the infiltration just starts? What is the infiltration rate when the infiltration time is sufficiently long?

9.5. The soil-water characteristic curve for a silty loam can be represented by van Genuchten's (1980) model with the following parameters: $\alpha = 0.0028$ kPa^{-1}, $n = 1.3$, $m = 0.231$, $\theta_r = 0.030$, and $\theta_s = 0.322$. Calculate and plot volumetric water content as a function of matric suction head.

Calculate and plot specific moisture capacity as a function of water content θ.

9.6. A silty soil layer with a 20-m-thick unsaturated zone (from the surface to the ground water table) has the following properties: saturated hydraulic conductivity $k_s = 10^{-7}$ m/s, initial water content $\theta_r = 0.05$, and saturated water content $\theta_s = 0.4$. Estimate the arrival time of a downward infiltration front to the water table after a heavy-rainfall event. Assume that the suction head behind the wetting front is $h_0 = -1.5$ m.

9.7. An unsaturated soil layer has hydrologic properties as follows: $h_0 - h_i = 0.9$ m, $\theta_0 - \theta_i = 0.45$, and $k_0 = 5 \times 10^{-6}$ m/s. If the thickness of the unsaturated zone is 2.0 m, predict the position of the wetting front, the infiltration rate, and the total infiltration as functions of time under a condition of surface ponding due to heavy rainfall. Plot your calculations on three different x-y plots.

9.8. A daily barometric pressure cycle follows a sinusoidal variation with an amplitude of 350 Pa. If the magnitude of pore air pressure at a point 10 m below the ground is measured to be 150 Pa and the thickness of the unsaturated zone is 30 m, what is the pneumatic diffusivity of the soil? If the air-filled porosity $n_a = 0.3$ and the mean air pressure $u_0 = 95,000$ Pa, what is the air permeability? How long will it take for the air pressure wave to propagate to 20 m and to 30 m below the ground surface, respectively?

PART IV

MATERIAL VARIABLE MEASUREMENT AND MODELING

CHAPTER 10

SUCTION MEASUREMENT

10.1 OVERVIEW OF MEASUREMENT TECHNIQUES

Experimental techniques for measuring soil suction and corresponding soil-water characteristic curves vary widely in terms of cost, complexity, and measurement range. Techniques can be generally categorized as either laboratory or field methods and differentiated by the component of suction (matric or total) that is measured. Laboratory methods typically require undisturbed specimens in order to account for the sensitivity of suction to soil fabric, particularly for relatively low values of suction where capillary mechanisms tend to dominate the pore water retention behavior. Disturbance effects generally become less critical at higher values of suction or for highly expansive clays where particle surface adsorption or hydration mechanisms begin to dominate. Table 10.1 summarizes several common suction measurement techniques in terms of their applicable suction component, approximate measurement range, applicability in the laboratory or field, and pertinent references. Figure 10.1 shows a comparison of the approximate suction ranges where the various techniques are practical.

Techniques described in this chapter for measuring matric suction include tensiometers, axis translation techniques, electrical/thermal conductivity sensors, and contact filter paper techniques. Tensiometers are used to directly measure negative pore water pressure. Axis translation techniques rely on controlling the difference between the pore air pressure and pore water pressure and measuring the corresponding water content of soil in equilibrium with the applied matric suction. Electrical or thermal conductivity sensors, often referred to more generally as "gypsum block" sensors, are used to indirectly relate matric suction to the electrical or thermal conductivity of a

TABLE 10.1 Summary of Common Laboratory and Field Techniques for Measuring Soil Suction

Suction Component Measured	Technique/Sensor	Practical Suction Range (kPa)	Laboratory/Field	References
Matric suction	Tensiometers	0–100	Laboratory and field	Cassel and Klute (1986); Stannard (1992)
	Axis translation techniques	0–1,500	Laboratory	Hilf (1956); Bocking and Fredlund (1980)
	Electrical/thermal conductivity sensors	0–400	Laboratory and field	Phene et al. (1971a, 1971b); Fredlund and Wong (1989)
	Contact filter paper method	Entire range	Laboratory and field	Houston et al. (1994)
Total suction	Thermocouple psychrometers	100–8,000	Laboratory and field	Spanner (1951)
	Chilled-mirror hygrometers	1,000–450,000	Laboratory	Gee et al. (1992); Wiederhold (1997)
	Resistance/capacitance sensors	Entire range	Laboratory	Wiederhold (1997); Albrecht et al. (2003)
	Isopiestic humidity control	4,000–400,000	Laboratory	Young (1967)
	Two-pressure humidity control	10,000–600,000	Laboratory	Likos and Lu (2001, 2003b)
	Noncontact filter paper method	1,000–500,000	Laboratory and field	Fawcett and Collis-George (1967); McQueen and Miller (1968); Houston et al. (1994); Likos and Lu (2002)

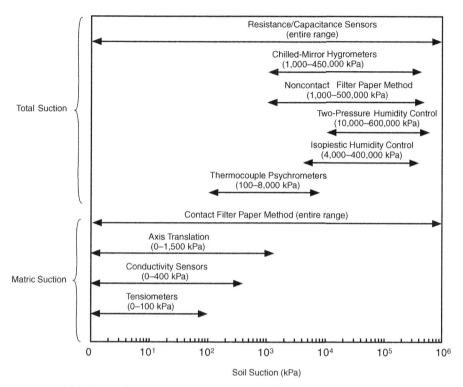

Figure 10.1 Approximate measurement ranges for various suction measurement techniques.

porous medium embedded in a mass of unsaturated soil. Finally, the contact filter paper technique relies on measuring the equilibrium water content of small filter papers in direct contact with unsaturated soil specimens. In each of these cases, water content corresponding to the measured (or controlled) suction is measured to generate data points along the soil-water characteristic curve. The resulting characteristic curve corresponds to either a wetting or drying process depending on the direction of wetting during the measurement.

Techniques described in this chapter for measuring total suction include humidity measurement techniques, humidity control techniques, and the noncontact filter paper method. Humidity measurement devices include thermocouple psychrometers, chilled-mirror hygrometers, and polymer resistance/capacitance sensors. Humidity control techniques described here include isopiestic, or "same pressure," techniques (e.g., humidity control using salt solutions) and "two-pressure" techniques, also known as divided-flow humidity control. The noncontact filter paper method is an indirect humidity measurement technique that relies on determining the equilibrium water content of small filter papers sealed in the headspace (i.e., in communication with the vapor phase) of unsaturated specimens. For each of these techniques, Kelvin's equation (Chapter 3) is used to convert the measured or controlled humidity to total suction. Soil-water characteristic curves are generated by measuring equilibrium water contents corresponding to the suction conditions.

10.2 TENSIOMETERS

10.2.1 Properties of High-Air-Entry Materials

Tensiometers and axis translation techniques rely on the unique properties of high-air-entry (HAE) materials. As introduced in Section 5.4, HAE materials are characterized by microscopic pores of relatively uniform size and size distribution. When an HAE material is saturated with water, the surface tension maintained at the gas-liquid interfaces formed in the material's pores allows a pressure difference to be sustained between gas and liquid phases located on either side. Physically, surface tension acts as a membrane for separating the two phases, thus allowing negative water pressure to be directly measured, as in a tensiometer, or the difference between water pressure and air pressure to be directly controlled, as in axis translation.

The phrase "high air-entry" refers to the fact that relatively high pressure is required for air to break through the membrane formed by surface tension. Figure 10.2, for example, shows an enlarged schematic cross section of a saturated HAE ceramic disk. The maximum sustainable difference between the air pressure above the disk and the water pressure within and below the disk is inversely proportional to the maximum pore size of the material, which is captured by the Young-Laplace equation (Chapter 4) as

$$(u_a - u_w)_b = \frac{2T_s}{R_s} \tag{10.1}$$

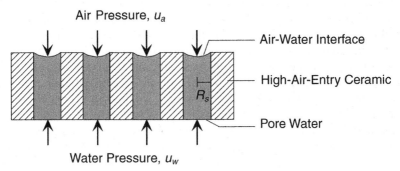

Figure 10.2 Operating principle of high-air-entry ceramic disk showing how air and water pressure are separated by surface tension of water.

where $(u_a - u_w)_b$ is the air-entry, or "bubbling," pressure, T_s is the surface tension of the air-water interface, and R_s is the effective radius of the maximum pore size of the HAE material. Table 10.2 lists the characteristics of several commercially available types of porous ceramics in terms of their approximate average pore diameter, saturated hydraulic conductivity, and air-entry pressure. Note that hydraulic conductivity decreases with increasing air-entry pressure, a reflection of the increasingly smaller pore sizes of the material.

10.2.2 Tensiometer Measurement Principles

A comprehensive description of tensiometer measurement principles, construction guidelines, operating procedures, and applications is provided by Stannard (1992). A standard tensiometer is essentially a water-filled tube with an HAE ceramic tip at one end and some type of sensor for measuring negative water pressure at the other. The ceramic tip, typically in the shape of an inverted cup or small probe, is used to create a saturated hydraulic connection between the soil pore water, the water in the tensiometer body, and the pressure sensor. The pressure sensor may be either a mechanical Bourdon-type gauge or an electronic diaphragm-type transducer. A schematic diagram for a commonly used type of "small-tip" laboratory tensiometer is shown as Fig. 10.3. Tensiometers applicable for field measurements are similar in construction and identical in operating principle.

As illustrated in Fig. 10.4, pore pressure measurements are made by a direct exchange of water between the sensor and the soil. Negative pressure is transmitted through the saturated pores of the HAE ceramic tip such that water is withdrawn from the tensiometer until the internal pressure in the sensor body is equivalent to the matric potential of the soil water. If the soil

TABLE 10.2 Air-Entry Pressure and Hydraulic Conductivity of Several Commercially Available HAE Ceramics

Type of HAE Ceramic	Approx. Pore Diameter ($\times 10^{-3}$ mm)	Saturated Hydraulic Cond. (m/s)	Air-Entry Value (kPa)
1/2 bar high flow	6.00	3.11×10^{-7}	48–62
1 bar	1.70	7.56×10^{-9}	138–207
1 bar high flow	2.50	8.60×10^{-8}	131–193
2 bar	1.10	6.30×10^{-9}	262–310
3 bar	0.70	2.50×10^{-9}	317–483
5 bar	0.50	1.21×10^{-9}	550
15 bar	0.16	2.59×10^{-11}	1520

Source: Soilmoisture Equipment Corp. (2003).

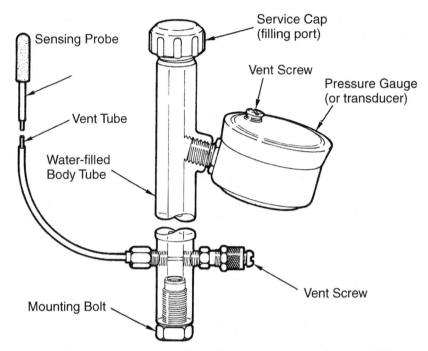

Figure 10.3 Schematic drawing of small-tip laboratory tensiometer (Soilmoisture Equipment Corp., 2003).

is subsequently wetted, water flows in the opposite direction from the soil to the measurement system until a new equilibrium at a new pressure is attained. Because the sensor tip is permeable to dissolved solutes, the osmotic potential of the pore water has no effect on the pressure measurement. The measurement, therefore, becomes a direct measurement of matric suction if gravitational potential is also considered (i.e., corrected for the difference in elevation between the sensor probe and the pressure gauge).

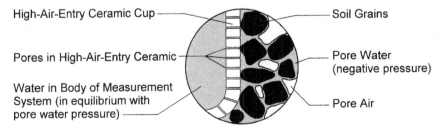

Figure 10.4 Enlarged schematic showing porous ceramic tip in contact with unsaturated soil grains.

The response time for a tensiometer measurement is a function of the system compressibility, the hydraulic conductivity and thickness of the sensor tip, and the hydraulic conductivity of the soil. System compressibility is largely controlled by the presence or absence of air bubbles in the system and the volume of liquid exchange required by the sensor to register an equilibrium pressure change. For measurements below the air-entry pressure of the ceramic, the hydraulic conductivity of the probe is constant. The hydraulic conductivity of the soil is a function of its matric suction. Response times on the order of 1 to 10 min are common.

Good contact is required between the sensor tip and the soil in order to maintain a saturated link between the pore water and the measurement system. Prior to testing, the tensiometer system is saturated with de-aired water, and temporary vacuum may be applied to remove air bubbles from the system. The system must be periodically resaturated through the service port (Fig. 10.3) if measurements over prolonged testing periods are required. Many systems allow the option of automated refilling and flushing during operation.

The range of matric suction obtainable using tensiometers is limited by the air-entry pressure of the porous ceramic tip and the capacity for water to sustain high negative pressures without cavitation occurring. As introduced in Section 2.5, the absolute cavitation pressure for free water at sea level is approximately 1 atm, or about 100 kPa, and decreases proportionally with increasing elevation. In practice, reliable tensiometer measurements using standard testing equipment are limited to about 70 to 80 kPa. Impurities (e.g., dust particles), dissolved gases, and air bubbles that tend to concentrate in tiny crevices on the walls of the sensor body are primarily responsible for this reduction because they may serve as nucleation sites for cavitation to occur.

Alternative types of "high-capacity" tensiometers incorporating extremely small, smooth-walled sensing reservoirs and relatively high air-entry pressure ceramics have been more recently developed (e.g., Ridley and Burland, 1993; Guan and Fredlund, 1997; Tarantino and Mongiovi, 2001). When coupled with specialized operating procedures (e.g., cyclic prepressurization techniques), these types of sensors have been shown to be applicable for matric suction approaching about 1500 kPa. The approach associated with these types of sensors has been to minimize potential sites for nucleation and thus fully realize the tensile strength of water, which although still largely uncertain has been shown experimentally and theoretically to be in the range of megapascals (e.g., Tabor, 1979; Zheng et al., 1991). Comparisons with established measurement systems have shown high-capacity tensiometers to be quite rapid in terms of response time and relatively reliable.

Peck and Rabbidge (1969) and Bocking and Fredlund (1979) also describe a class of "osmotic" tensiometers that rely on confined and prestressed (positively pressurized) aqueous solutions rather than free water for the transmission of negative pore pressure through the measurement system. To date,

424 SUCTION MEASUREMENT

however, difficulties associated with drift and temperature sensitivity have largely precluded the use of osmotic tensiometers in practice.

10.3 AXIS TRANSLATION TECHNIQUES

10.3.1 Null Tests and Pore Water Extraction Tests

As introduced in Section 5.4, axis translation refers to the practice of elevating pore air pressure while maintaining pore water pressure at a reference value through the pores of a saturated HAE material, thus affording direct control of matric suction ($u_a - u_w$). Figure 10.5 shows an enlarged cross section at the boundary between an unsaturated soil specimen and a saturated HAE ceramic disk. For a so-called null measurement of matric suction using such a setup (e.g., Hilf, 1956), the air pressure is elevated and flow of water between the soil and the ceramic disk is not allowed. Matric suction is recorded as the difference between the applied air pressure and the pore water pressure at equilibrium. For a pore water extraction test, the air pressure is increased and drainage from the specimen is allowed to occur through the HAE pores. Drainage continues until the water content of the specimen reaches an equilibrium with the applied matric suction, which is recorded as the difference between the water pressure on one side of the disk, typically atmospheric, and the pore air pressure on the other side of the disk. Several increments of air pressure may be applied to generate several points along the drying loop of the soil-water characteristic curve. Two basic types of extraction systems are commonly used in practice: pressure plate systems and Tempe cell sys-

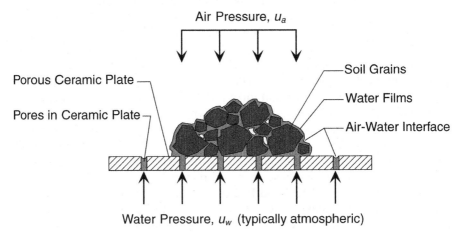

Figure 10.5 Schematic showing enlarged cross section of interface between unsaturated soil and high-air-entry disk for an axis translation measurement.

tems. Pressure plates are applicable for matric suction ranging from about 0–1,500 kPa. Tempe cells are applicable from about 0–100 kPa.

10.3.2 Pressure Plates

Figure 10.6 shows a schematic of a typical pore water extraction testing setup using a pressure plate apparatus. The primary components of the system are a steel pressure vessel and a saturated HAE ceramic plate or cellulose membrane. Ceramic plates are designated by air-entry pressure and are generally available in 1-, 3-, 5-, or 15-bar form (see Table 10.2). Cellulose membranes are generally available in either 15- or 100-bar form. As shown, a small water reservoir is formed beneath the plate or membrane using an internal screen and a neoprene diaphragm. The water reservoir is vented to the atmosphere through an outflow tube located on top of the plate, thus allowing the air pressure in the vessel and the water pressure in the reservoir to be separated across the air-water interfaces bridging the saturated pores of the HAE material.

Several soil specimens are placed on top of the HAE plate such that the pore water is in equilibrium with the water reservoir at atmospheric pressure (Fig. 10.7). For testing remolded samples, "identical" specimens must be prepared (e.g., same dry density and molding water content) to account for the sensitivity of pore water retention to soil fabric. Specimens are initially saturated, typically by applying a partial vacuum to the air chamber and allowing the specimens to imbibe water from the underlying reservoir through

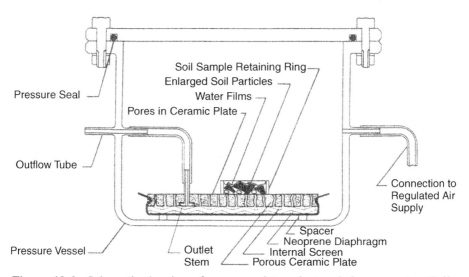

Figure 10.6 Schematic drawing of pressure plate axis translation apparatus (Soilmoisture Equipment Corp., 2003).

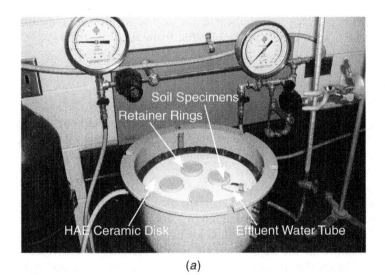

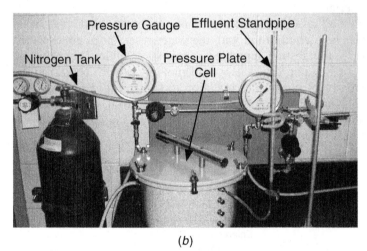

Figure 10.7 Photographs of pressure plate testing in progress: (*a*) initial setup of saturated sand specimens on a 5-bar ceramic plate inside pressure vessel and (*b*) closed pressure vessel with air pressure applied. Nitrogen tank is being used as pressure source and standpipe is being used to monitor effluent water flow to check for equilibrium under applied matric suction.

the ceramic disk. Air pressure in the vessel is then increased to some desired level while pore water is allowed to drain from the specimens in pursuit of equilibrium. The outflow of water is monitored until it ceases, the pressure vessel is opened, and the water content of one or more of the specimens is measured, thus generating one point on the soil-water characteristic curve. Subsequent increments in air pressure are applied to generate additional points on the curve using the other specimens.

Figure 10.8 shows an example of a soil-water characteristic curve determined in this general manner for a poorly graded, fine sand. In general, the pressure plate technique is applicable for relatively coarse-grained soils where the characteristic curve is well defined over a range of suction less than 1500 kPa, which corresponds to the largest air-entry value for most applicable ceramics. Uncertainties arise in testing specimens both near the saturated condition and the dry condition due to uncertainties in the continuity of the pore air and pore water phases, respectively. Detailed analysis of the limitations in the method is provided by Bocking and Fredlund (1980).

10.3.3 Tempe Pressure Cells

Tempe pressure cells are identical in concept to the pressure plate systems described above. As shown in Fig. 10.9, Tempe cells consist of a saturated HAE ceramic disk separating air and water chambers in a closed vessel. Here, however, a single soil specimen is placed in the cell such that several pairs

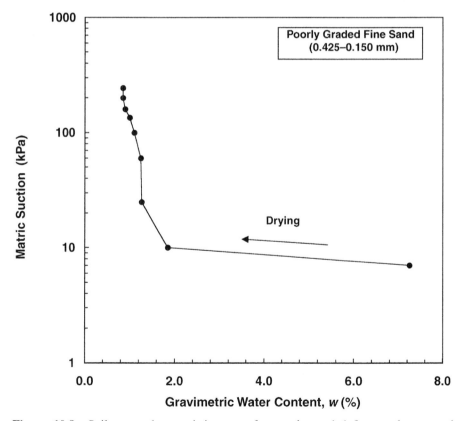

Figure 10.8 Soil-water characteristic curve for poorly graded fine sand measured using pressure plate apparatus.

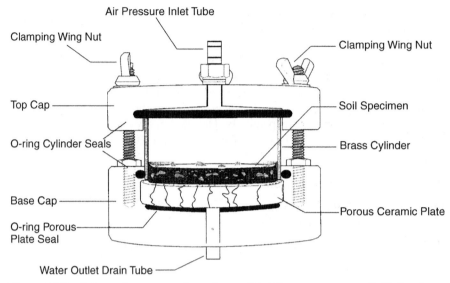

Figure 10.9 Schematic cross section of assembled Tempe pressure cell (Soilmoisture Equipment Corp., 2003).

of data points comprising the soil-water characteristic curve may be determined by applying increasing increments in air pressure. Equilibrium water content is determined for each pressure increment by weighing the entire apparatus and noting the amount of mass lost due to pore water drainage. Once the highest desired level of matric suction is attained, the cell is disassembled and the final water content of the specimen is determined gravimetrically. The final water content may then be considered in light of the incremental changes in mass to back-calculate water content values corresponding to the preceding levels of matric suction.

Figure 10.10 shows the cumulative mass of water expelled from a Tempe cell as a function of time and applied air pressure for tests conducted using a well-graded, fine-sand specimen. As shown, the amount of time required for steady-state ranges from approximately 75 h for relatively low increments in air pressure and increases to as much as 150 h for the latter increments. Equilibrium time generally increases with increasing levels of suction and decreasing water-filled pore size.

Figure 10.11 shows grain size distributions (Fig. 10.11a) and corresponding soil-water characteristic curves obtained using a Tempe pressure cell (Fig. 10.11b) for five sandy soil specimens. Note that the maximum suction determined is less than 100 kPa, a common air-entry pressure for commercially available Tempe cells. The relatively poorly graded and well-graded grain-size distributions of the COR and CHV specimens, respectively, are reflected in their relatively flat and steep characteristic curves.

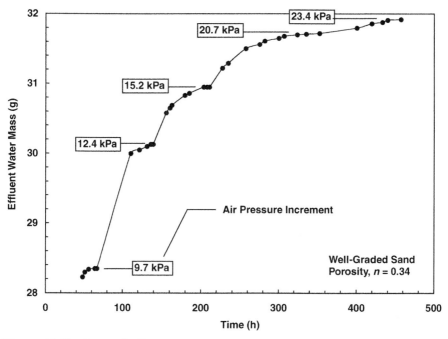

Figure 10.10 Mass of effluent pore water extracted during Tempe cell test for well-graded fine sand.

10.4 ELECTRICAL/THERMAL CONDUCTIVITY SENSORS

The electrical and thermal conductivities of a rigid porous medium are direct functions of water content. If a porous medium is embedded in a mass of unsaturated soil and allowed to reach equilibrium, any subsequent change in the suction of the soil results in a corresponding change in the water content of the porous medium, as governed by its characteristic curve. If the electrical or thermal conductivity of the porous medium is measured, therefore, the matric suction of the soil may be indirectly determined by correlation with a predetermined calibration curve. Modern electrical or thermal conductivity sensors are typically constructed of porous ceramic, polymer synthetics, sintered metal or glass, or gypsum plaster. The more general term *gypsum block* sensor has been commonly adopted in practice.

For thermal conductivity sensors, the thermal conductivity of the porous medium is measured, typically by measuring the rate of internal heat dissipation following an applied heat pulse. For electrical conductivity sensors, the electrical conductivity of the porous medium is measured, typically using two embedded electrodes. One of the major limitations of electrical conductivity sensors, however, is their inherent sensitivity to changes in electrical conductivity that are not related to the moisture content of the porous medium,

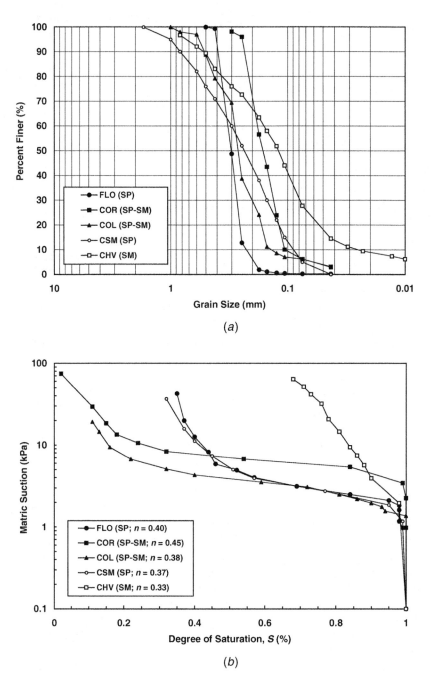

Figure 10.11 (a) Grain size distribution and (b) soil-water characteristic curves obtained using Tempe pressure cell for five sandy specimens (data from Clayton, 1996).

most notably from dissolved solutes. Because this limitation is largely avoided in sensing thermal conductivity, thermal conductivity sensors have found a relatively greater amount of use in geotechnical engineering practice.

Phene et al. (1971a, 1971b) describe the general principles and operating procedures for thermal conductivity sensors in soil suction measurement applications. Most commercially available thermal conductivity sensors are applicable for suction measurements ranging from about 0 to 400 kPa, with the greatest sensitivity existing for suction below about 175 kPa. Fredlund and Wong (1989) describe response characteristics for a group of specific sensors and suggest a technique for their calibration using a modified pressure plate testing system. The applicability of thermal conductivity sensors for both laboratory and field measurements is demonstrated by Picornell et al. (1983), van der Raadt et al. (1987), and Sattler and Fredlund (1989). Advantages include the relative ease with which the sensors may be set up for automated data acquisition and their relatively low cost. Disadvantages include the requirement for a separate calibration curve for each sensor, potential long-term problems associated with drift, and, for many sensors, deterioration in the sensor body over time. Uncertainties for drying and rewetting processes may also arise due to hysteretic effects in the sensor calibration that may or may not be fully accounted for.

10.5 HUMIDITY MEASUREMENT TECHNIQUES

10.5.1 Total Suction and Relative Humidity

As introduced in Chapter 3, the relationship between pore water potential and its partial vapor pressure is described by Kelvin's equation, which can be written in terms of total suction as

$$\psi_t = -\frac{RT}{v_{w0}\omega_v} \ln\left(\frac{u_v}{u_{v0}}\right) \qquad (10.2)$$

where ψ_t is total soil suction (kPa), R is the universal gas constant (8.314 J/mol · K), T is absolute temperature (K), v_{w0} is the specific volume of water (m³/kg), ω_v is the molecular mass of water vapor (18.016 kg/kmol), u_v is the partial pressure of the pore water vapor (kPa), and u_{v0} is the saturated vapor pressure of free water at the same temperature (kPa). Recognizing that relative humidity RH is equal to the ratio u_v/u_{v0}, eq. (10.2) can be written as

$$\psi_t = -\frac{RT}{v_{w0}\omega_v} \ln(\text{RH}) \qquad (10.3)$$

Figure 10.12 shows a plot of eq. (10.3) for $T = 293.16$ K. Total suction is zero when the relative humidity of the pore water vapor is 100%. Relative

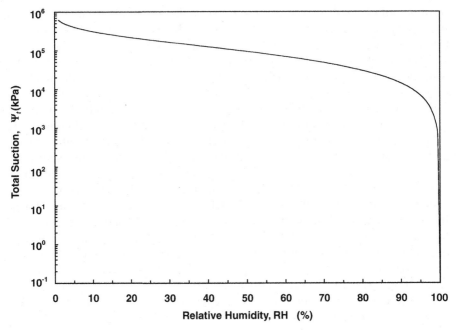

Figure 10.12 Theoretical relationship between total suction and relative humidity ($T = 293.16$ K).

humidity values less than 100% indicate the presence of suction in the soil. Kelvin's equation applies to total suction because all of the mechanisms that act to reduce the potential of the pore water (i.e., dissolved solutes, hydration effects, capillary effects) are accounted for.

Given Kelvin's law, the equilibrium relative humidity of the pore water for an unsaturated soil specimen may be measured to calculate the corresponding total soil suction. Common techniques for measuring relative humidity include thermocouple psychrometers, chilled-mirror hygrometers, polymer resistance/capacitance sensors, and noncontact filter paper techniques. The first three techniques are described in the following three sections. Filter paper techniques are described in Section 10.7.

10.5.2 Thermocouple Psychrometers

Thermocouple psychrometers operate on the basis of the temperature difference between a nonevaporating surface, or reference junction, and an evaporating surface, or measurement junction (Spanner, 1951). Figure 10.13 shows a schematic diagram of a Peltier-type thermocouple psychrometer commonly used for measuring relative humidity in unsaturated soil testing applications. The working component of the psychrometer is an electrical circuit formed

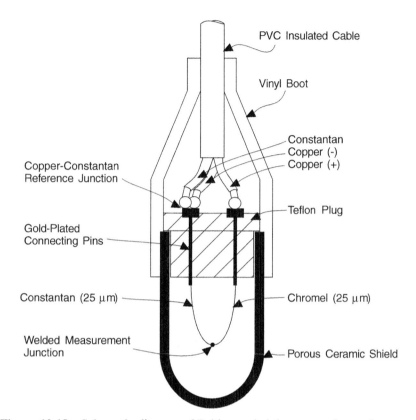

Figure 10.13 Schematic diagram of Peltier-cooled thermocouple psychrometer.

by thin wires of dissimilar metal housed within a shield of porous ceramic (shown) or stainless steel mesh. The metals comprising the circuit are typically constantan (copper-nickel) and chromel (chromium-nickel). A measurement junction is formed by welding the dissimilar wires together in series. A reference junction is formed outside the sensing environment by welding the constantan wire to a heavy-gauge copper heat sink. The heat sink maintains a relatively constant temperature at the copper-constantan junction during operation.

For humidity measurements, the Peltier and Seebeck effects are used to create and measure, respectively, a temperature difference between the measurement junction and the reference junction. A simple electrical circuit demonstrating the Seebeck effect for dissimilar metals A and B is shown as Fig. 10.14a. If the temperature at the two junctions is different by an amount ΔT, a proportional electrical current is generated in the circuit. The corresponding thermocouple output can be measured using a voltmeter in series. The Peltier effect (Fig. 10.14b) is the inverse of the Seebeck effect. Here, the junctions in a circuit comprised of dissimilar metals may be either cooled or warmed

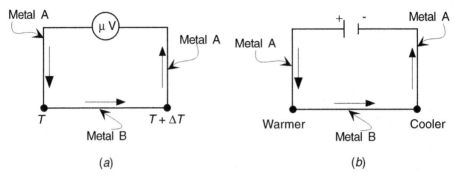

Figure 10.14 Electrical circuits comprised of dissimilar metals for illustrating (a) Seebeck effect and (b) Peltier effect.

by passing an electrical current through the circuit. Depending on the direction of the applied current, the junctions either adsorb or liberate heat in an amount that is a function of the current magnitude. If the temperature of the cooling junction is depressed beyond the dew-point temperature of the ambient environment, then water condenses on the junction at a temperature that is a function of the ambient relative humidity.

Figures 10.15a and 10.15b show a schematic drawing and photograph, respectively, of a typical laboratory setup for measuring total suction using a thermocouple psychrometer. Prior to measurement, a soil specimen is sealed in the sensing chamber and allowed to come to temperature and vapor pressure equilibrium with the psychrometer in the headspace above the specimen. The amount of time required for equilibrium is primarily dependent on the volume and initial relative humidity of the chamber, the soil suction, and the type of thermocouple protective housing (i.e., porous ceramic or stainless steel mesh). Equilibrium time typically varies from several hours to several days. To minimize the effects of temperature fluctuations on the measurement, the sensing chamber should be placed in a controlled-temperature water bath or insulated cooler.

Two modes of operation may be used to determine relative humidity: the "psychrometric" mode and the "dew-point" mode. The physics of each mode is essentially the same, differing primarily in terms of whether or not temperature is actively controlled at the measurement junction. For illustration, a conceptual measurement sequence in terms of thermocouple output as a function of time under psychrometric mode of operation is shown as Fig. 10.16. Initially, current is passed through the thermocouple circuit to cool the measurement junction under the Peltier effect. The applied cooling current is reflected in the thermocouple output from the points in Fig. 10.16 denoted A and B. Current is applied for a length of time sufficient to cool the measurement junction beyond the associated dew-point temperature (see Section 2.2), typically requiring 5 to 30 s. When the temperature of the junction falls below

10.5 HUMIDITY MEASUREMENT TECHNIQUES

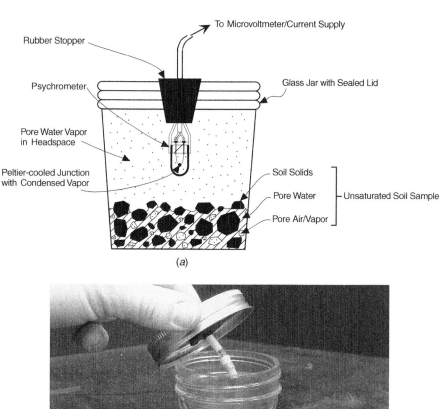

Figure 10.15 (a) Schematic drawing and (b) photograph of typical laboratory set up for measuring total soil suction using thermocouple psychrometer.

the dew point, water vapor in the headspace of the chamber condenses on the junction. The current is then discontinued (point B). As the measurement junction returns to ambient temperature, output generated via the Seebeck effect by the temperature difference between the measurement junction and reference junction is monitored (from point B to C). When the temperature returns to the dew-point temperature (at point C), water begins to evaporate

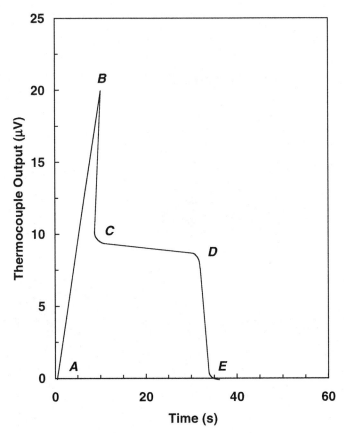

Figure 10.16 Conceptual thermocouple response for Peltier-cooled psychrometer measurement.

from the junction, thus initiating a latent cooling effect that offsets the heat being adsorbed during the concurrent return to ambient. The offsetting cooling is reflected by the relatively flat thermocouple response from point C to D. At point D, the water on the measurement junction has evaporated and the junction's temperature falls back to ambient, finally reaching the same temperature as the reference junction when the output goes to zero.

The thermocouple output at the dew-point temperature (point C) is a function of the ambient relative humidity of the sensor/soil system. Accordingly, psychrometers must be calibrated prior to testing by developing a relationship between sensor output at the dew point temperature and known values of relative humidity. As described in Section 10.6, salt solutions of known concentration may be used to control relative humidity or total suction. Tables 10.3 and 10.4, for example, show suction values corresponding to commonly used NaCl and KCl solutions, respectively, at various temperatures and con-

TABLE 10.3 Total (Osmotic) Suctions for Various NaCl Solutions (kPa)

NaCl Molality	Temperature				
	0°C	7.5°C	15°C	25°C	35°C
0	0.0	0.0	0.0	0.0	0.0
0.2	836	860	884	915	946
0.5	2070	2136	2200	2281	2362
0.7	2901	2998	3091	3210	3328
1.0	4169	4318	4459	4640	4815
1.5	6359	6606	6837	7134	7411
1.7	7260	7550	7820	8170	8490
1.8	7730	8035	8330	8700	9040
1.9	8190	8530	8840	9240	9600
2.0	8670	9025	9360	9780	10,160

Source: From Lang (1967).

centrations. Figure 10.17 shows a typical calibration curve obtained for a Peltier psychrometer using salt solutions. Figure 10.18 shows portions of soil-water characteristic curves obtained using this psychrometer for several mass-controlled mixtures of two expansive clays.

The maximum temperature depression that can be maintained at the measurement junction of a Peltier-cooled psychrometer is limited by a resistive heating effect that increases with the square of the applied current. In practice, the lowest possible humidity measurement is about 94%, corresponding to an upper limit in total suction equal to approximately 8000 kPa. As the suction approaches this upper limit, the scatter in the measurements tends to increase. Similarly, at suctions below about 100 kPa, where the relative humidity approaches 100%, slight temperature fluctuations can result in uncontrolled con-

TABLE 10.4 Total (Osmotic) Suctions for Various KCl Solutions (kPa)

Molality	0°C	10°C	15°C	20°C	25°C	30°C	35°C
0	0.0	0.0	0.0	0.0	0.0	0.0	0.0
0.10	421	436	444	452	459	467	474
0.20	827	859	874	890	905	920	935
0.30	1229	1277	1300	1324	1347	1370	1392
0.40	1628	1693	1724	1757	1788	1819	1849
0.50	2025	2108	2148	2190	2230	2268	2306
0.60	2420	2523	2572	2623	2672	2719	2765
0.70	2814	2938	2996	3057	3116	3171	3226
0.80	3208	3353	3421	3492	3561	3625	3688
0.90	3601	3769	3846	3928	4007	4080	4153
1.00	3993	4185	4272	4366	4455	4538	4620

Source: From Campbell and Gardner (1971).

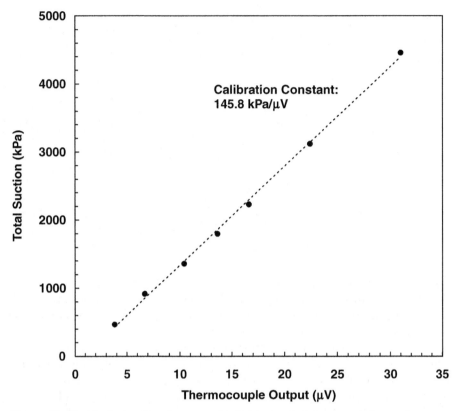

Figure 10.17 Typical calibration curve for Peltier-cooled thermocouple psychrometer ($T = 21°C$).

densation on the measurement junction. Psychrometer measurements below 100 kPa using typical equipment are generally unreliable.

10.5.3 Chilled-Mirror Hygrometers

Chilled-mirror sensing technology has been used since the 1950s for determination of dew-point temperature in a closed, humid environment. Gee et al. (1992) describe the use of a chilled-mirror sensing system for soil suction testing applications. Figure 10.19 illustrates the basic operating principle. Humidity measurement involves thermoelectric chilling of a reflective surface, usually a metallic mirror, to a temperature at which condensation from ambient water vapor in the closed chamber forms on the mirror surface. A beam of light, typically from a light-emitting diode (LED), is directed to the mirror and reflected back to a photodetector. When condensation occurs as the mirror is cooled to the dew-point temperature, the light reflected from the mirror is scattered and the intensity detected by the photodetector is consequently re-

10.5 HUMIDITY MEASUREMENT TECHNIQUES 439

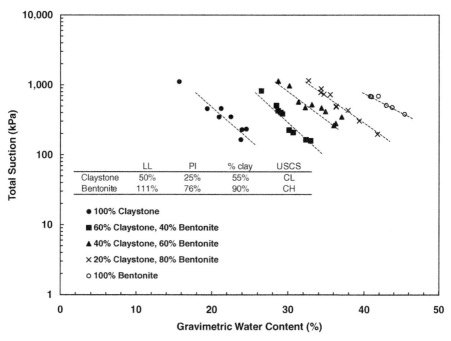

Figure 10.18 Relationships between total suction and water content measured using thermocouple psychrometers for remolded mixtures of natural expansive soils from Denver, Colorado.

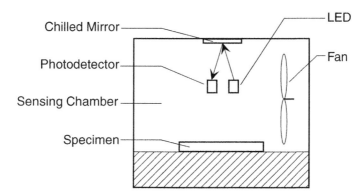

Figure 10.19 Schematic diagram of chilled-mirror sensing technology.

duced. The dew-point temperature is maintained constant by a microprocessor circuit and measured by a resistance thermometer embedded in the mirror. The dew-point temperature may then be related to the ambient relative humidity and corresponding total suction using Kelvin's law. For testing, specimens are placed in the sensing chamber and allowed to come to vapor pressure equilibrium prior to the cooling cycle.

Accuracy on the order of ±0.3% RH is typically reported for most commercially available chilled-mirror sensing systems. The practical measurement range is relatively wide, ranging from about 3% RH ($\approx$450,000 kPa) to about 99.9% RH ($\approx$100 kPa). Scatter has been shown to increase significantly for measurements greater than about 99% RH ($\approx$1000 kPa). Figure 10.20 shows a soil-water characteristic curve obtained for kaolinite clay using a chilled-mirror system. Total suction measurements obtained using the noncontact filter paper method (Section 10.7) are included for comparison. The primary advantages of chilled-mirror sensing technology for soil suction measurement applications are its simplicity and speed. Specimens are simply placed into the sensing chamber and the measurement is automated from that point on. The amount of time required for one measurement can be as little as 5 min.

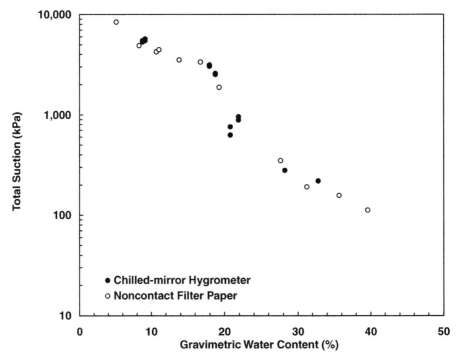

Figure 10.20 Relationship between total suction and water content for kaolinite clay obtained using chilled-mirror hygrometer and noncontact filter paper method.

10.5.4 Polymer Resistance/Capacitance Sensors

Polymer-based sensors are used extensively in the atmospheric, food science, and process industries for measuring relative humidity over a wide range. These sensors consist of a porous probe containing two electrodes separated by a thin polymer film or polymer-coated substrate that adsorbs or releases water as the relative humidity of the gas in equilibrium with the probe changes. As water is either adsorbed or desorbed onto or from the polymer surface, the resistance and capacitance of the electrode-polymer system change. Figure 10.21, for example, shows relationships between relative humidity and sensor capacitance (Fig. 10.21a) and relative humidity and sensor resistance (Fig. 10.21b) for a pair of typical polymer sensors. Measurements of either resistance or capacitance are used along with these types of calibration curves to back-calculate relative humidity. Polymer-based sensors are generally applicable for relative humidity ranging from near 0% RH to near 100% RH.

Wiederhold (1997) describes the major differences between polymer resistance sensors and polymer capacitance sensors. In general, resistance sensors tend to exhibit greater linearity at high relative humidity (>95%) but are not practical below relative humidity of about 20%. Capacitance sensors tend to be linear over a much wider range (<95%), are less sensitive to temperature fluctuations, and are essentially unaffected by most vapor phase or liquid phase contaminants. Both types of sensors are relatively inexpensive, have fast response time (<15 s), exhibit low hysteresis (1 to 6% RH), have excellent long-term stability (± 1 to ± 3% RH/yr), and may be enclosed in small bodies such that they may be directly buried in soil and attached to data logging equipment for remote and continuous monitoring. Sensors for rugged use are usually enclosed by a filtering element such as a plastic or stainless steel screen or a sintered metal cup or tube.

A variety of commercially available polymer-based sensor systems are manufactured to select tolerances, with the most common being accurate to ± 1, ± 2, and ± 3% RH. Accuracy generally reaches a minimum at extremes of very low (e.g., <10% RH) or very high (e.g., >95% RH) relative humidity. Figure 10.22 demonstrates a comparison between "known" relative humidity values and values measured in the laboratory using a thin-film capacitance probe. Here, the known humidity values were controlled using saturated salt solutions (see Section 10.6).

Albrecht et al. (2003) conducted a study to assess the applicability and behavior of a polymer capacitance sensor for use in unsaturated soil mechanics applications. The study includes a comparison between humidity measurements made using the sensor and a chilled-mirror hygrometer, an evaluation of sensor hysteresis, and an analysis of temperature sensitivity. Analysis of scatter in the data showed that the measurement precision was approximately ± 3%. Practical application was demonstrated through a bench-scale study for a dry barrier used as an alternative earthen fill cover.

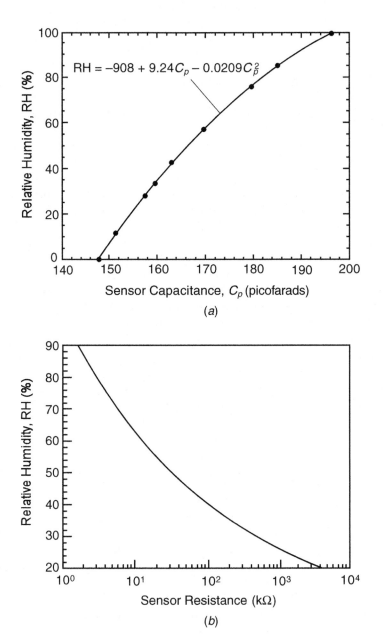

Figure 10.21 Calibration curves for polymer-based relative humidity sensors: (a) relationships between sensor capacitance and relative humidity (Albrecht et al., 2003) and (b) relationship between sensor resistance and relative humidity (Ohmic Instruments Corporation, 2003).

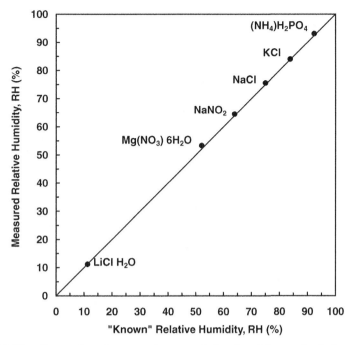

Figure 10.22 Comparison between known relative humidity and relative humidity measured using polymer capacitance probe suspended over various saturated salt solutions.

10.6 HUMIDITY CONTROL TECHNIQUES

Unlike techniques that rely on measurements of total suction for specimens of controlled water content (e.g., psychrometers, noncontact filter paper methods, etc.), humidity control techniques rely on measurement of water content for specimens of controlled total suction. Total suction is controlled by controlling relative humidity in a closed environmental chamber and applying Kelvin's equation. To generate total suction characteristic curves, the water content of specimens placed in the controlled humidity environment is measured as water is adsorbed or desorbed in order to satisfy equilibrium.

Humidity control techniques are applicable for measuring soil-water characteristic curves in the range of relatively high total suction, generally greater than about 4000 to 10,000 kPa. Accordingly, humidity control techniques are often used in conjunction with techniques applicable at lower values of suction to generate combined soil-water characteristic curves over an extremely wide range. Traditional methods for controlling relative humidity include isopiestic, or "same pressure," methods, which rely on attaining vapor pressure equilibrium for salt or acid solutions in a closed thermodynamic environment, and "two-pressure" methods, which rely on active manipulation of relative humidity, either by varying pressure or by mixing vapor-saturated gas with dry gas. Each of these methods are described in the following.

10.6.1 Isopiestic Humidity Control

Using isopiestic humidity control, saturated or unsaturated salt or acid solutions are allowed to come to thermodynamic equilibrium in small sealed containers. Under isothermal conditions, the relative humidity in the headspace above the solution approaches a fixed, reproducible value that is dependent on the solution concentration. The so-called *salt bath* or *osmotic dessicator* is one of the simplest and most well-known strategies for controlling relative humidity in this manner.

Tables 10.3 and 10.4 summarize relative humidity and total suction values corresponding to NaCl and KCl solutions of various concentrations and temperatures. Table 10.5 summarizes humidity and suction values for an additional series of saturated salt solutions. Saturated salt solutions have the practical advantage over unsaturated solutions of being able to liberate or adsorb relatively large quantities of water without significantly affecting the equilibrium relative humidity. The last column of Table 10.5 shows temperature sensitivities for each saturated solution in terms of the change in relative

TABLE 10.5 Summary of Saturated Salt Solution Properties for Relative Humidity Control

Saturated Salt Solution	Temperature (°C)	% RH at 25°C	Total Suction (kPa)	$d(RH)/dT$ (% per °C from 25°C)
$NaOH \cdot H_2O$	15–25	7	365,183	0
$LiCl \cdot H_2O$	20–70	11.3	299,419	−0.01
$MgCl_2 \cdot 6H_2O$	10–50	32.7	153,501	−0.06
$NaI \cdot 2H_2O$	5–35	39.2	128,604	−0.32
KNO_2	20–40	48.2	100,221	−0.18
$Mg(NO_3)_2 \cdot 6H_2O$	0–50	52.8	87,704	−0.29
$Na_2Cr_2O_7 \cdot 2H_2O$	0–50	53.7	85,383	−0.27
$NaBr \cdot 2H_2O$	−10–35	58.2	74,332	−0.28
$NaNO_2$	20–40	64.4	60,431	−0.19
$CuCl_2 \cdot 2H_2O$	10–30	68.4	52,156	0.00
NaCl	5–60	75.1	39,323	−0.02
$(NH_4)_2SO_4$	25–50	80.2	30,300	−0.07
KCl	5–40	84.2	23,617	−0.16
K_2CrO_4	20–40	86.5	19,916	−0.06
$BaCl_2 \cdot 2H_2O$	5–60	90.3	14,012	−0.08
$(NH_4)H_2PO_4$	20–45	92.7	10,409	−0.12
K_2SO_4	15–60	97.0	4,183	−0.05
$CuSO_4 \cdot 5H_2O$	25–40	97.2	3,900	−0.05

Source: From Young (1967).

humidity (%) per unit change in temperature (°C). Solutions with relatively low temperature sensitivity are ideal.

As illustrated in Fig. 10.23a, total suction characteristic curves using isopiestic humidity control may be obtained along either wetting or drying paths by sealing a soil specimen in the headspace of a chamber containing a solution at a concentration corresponding to some desired level of relative humidity. If the initial water content of the specimen is some value less than that required for equilibrium, then water vapor condenses to the soil until equilibrium is reached. In this case, the equilibrium water content of the specimen defines a point along a local wetting path of the soil-water characteristic curve. If the initial water content of the specimen is some value greater than that required for equilibrium, then pore water evaporates until a point corresponding to a local drying path of the characteristic curve is reached. The amount of time required for equilibrium in either case depends on the mass of the soil specimen, the volume of the testing chamber, and the difference between the initial and equilibrium water content of the specimen. As demonstrated in Fig. 10.23b for a specimen of highly expansive Na^+-smectite, equilibrium time varies from as little as several hours to as much as several days when the amount of water that must be adsorbed or liberated is relatively large.

10.6.2 Two-Pressure Humidity Control

Two-pressure methods for controlling relative humidity involve the manipulation of vapor-saturated gas either by varying pressure or temperature (e.g., Hardy, 1992) or by proportioned mixing of vapor-saturated gas with dry gas (e.g., Künhel and van der Gaast, 1993; Hashizume et al., 1996; Chipera et al., 1997). The latter method, also known as the *divided-flow* method, is illustrated in Fig. 10.24 for a system developed by Likos and Lu (2001, 2003b).

Relative humidity for the system in Fig. 10.24 is controlled by proportioned mixing of vapor-saturated, or "wet," nitrogen gas and desiccated, or "dry," nitrogen gas in a closed environmental chamber. Following the figure from left to right, bottled N_2 is split into two separate gas streams. A pair of mass-flow controllers regulates the flow of each gas stream based on an electronic signal from a control computer. One of the gas streams is vapor-saturated by bubbling it through a gas-washing bottle filled with distilled water. The second gas stream is routed through a cylinder filled with drying media. The vapor-saturated and desiccated gas streams are then reintroduced in a three-neck flask where the resulting gas stream has a relative humidity that is a direct function of the wet to dry gas flow ratio (w/d). Electrical heat tape is wrapped around the wet and humid gas lines and connected to a variable voltage transformer to allow the option for elevated temperature testing.

The humid gas stream is routed into an acrylic environmental chamber (Fig. 10.25) containing a soil sample. An effluent gas vent on the top cap of the chamber allows the influent humid gas to escape after flowing around the

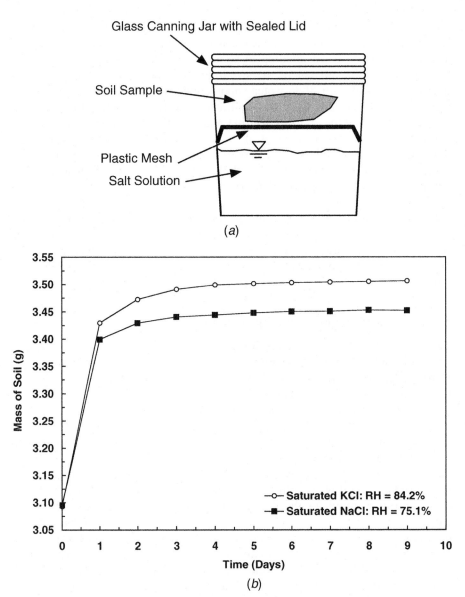

Figure 10.23 Measuring soil-water characteristic curves using isopiestic humidity control: (*a*) example testing configuration and (*b*) response time for a specimen of Na^+-smectite.

10.6 HUMIDITY CONTROL TECHNIQUES   447

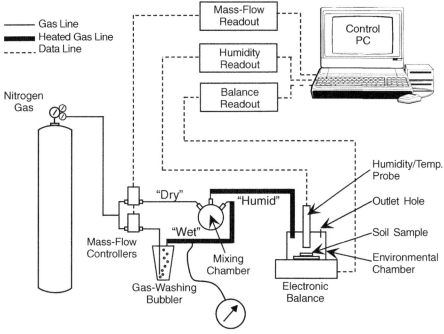

Figure 10.24 Divided-flow humidity control system for measurement of total suction characteristic curves (Likos and Lu, 2003b).

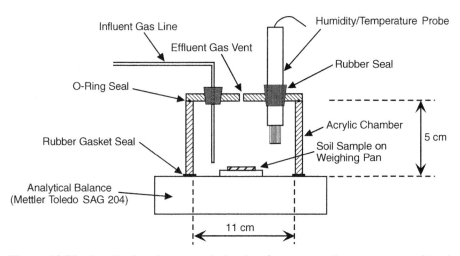

Figure 10.25 Detail of environmental chamber for automated measurement of total suction characteristics (Likos and Lu, 2003b).

soil. Relative humidity and temperature in the chamber are continuously monitored with a polymer capacitance probe (Section 10.5). Signals from the probe form a feedback loop with the control computer for automated regulation of the wet to dry gas flow ratio. An electronic balance forms the bottom plate of the environmental chamber. Soil specimens (typically ranging from 0.5 to 3.0 g) are placed directly on the balance. To develop total suction characteristic curves, the relative humidity in the chamber is incrementally stepped up or down by proportioning the wet to dry gas flow ratio. Soil water content is continuously monitored as water vapor is adsorbed or desorbed at each step in relative humidity. When an equilibrium is reached, the water content is recorded and the humidity in the chamber is stepped (up or down) to the next increment. Typically, humidity is stepped in increments of 10%.

The maximum humidity that may be accurately controlled using the divided-flow system in Fig. 10.24 is approximately 95%, corresponding to total suction of approximately 7000 kPa. Minimum humidity is approximately 0.3%, corresponding to suction of approximately 800,000 kPa. Figure 10.26

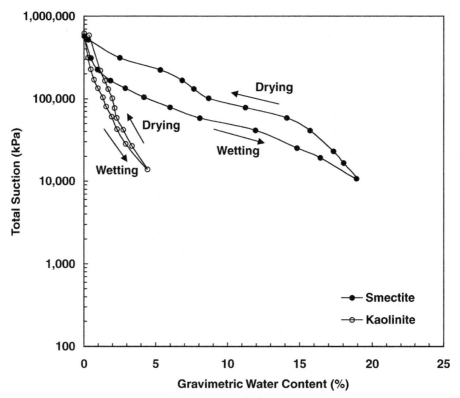

Figure 10.26 Total suction characteristic curves for Wyoming Na^+-smectite and Georgia kaolinite along wetting and drying paths measured using divided-flow humidity control system (Likos, 2000).

shows total suction characteristic curves obtained using the system for powdered samples of Na$^+$-smectite and kaolinite clay. Figure 10.27 compares and combines these results with results obtained using the noncontact filter paper method.

10.7 FILTER PAPER TECHNIQUES

10.7.1 Filter Paper Measurement Principles

Filter paper methods, which were first developed for agricultural and soil science applications, are relatively simple, low-cost, and reasonably accurate alternatives to many of the testing techniques described above. The American Society for Testing and Materials (ASTM) Standard D5298 describes calibration and test procedures for the measurement of either matric suction using the "contact" filter paper technique or total suction using the "noncontact"

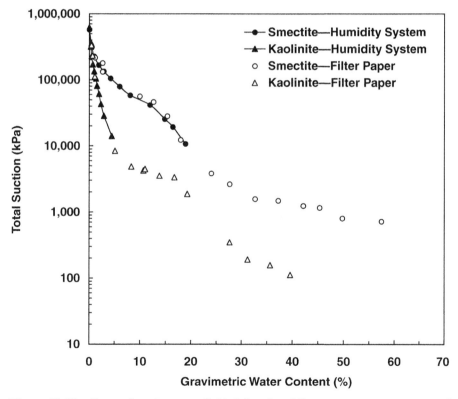

Figure 10.27 Comparison between divided-flow humidity system measurements and noncontact filter paper measurements for Wyoming Na$^+$-smectite and Georgia kaolinite along wetting paths (Likos, 2000).

filter paper technique. Fawcett and Collis-George (1967), McQueen and Miller (1968), Al-Khafaf and Hanks (1974), Hamblin (1981), Chandler and Gutierrez (1986), Houston et al. (1994), and Likos and Lu (2002) all provide additional discussion and analysis.

Both the contact and noncontact filter paper techniques estimate soil suction indirectly by measuring the amount of moisture transferred from an unsaturated soil specimen to an initially dry filter paper. In both cases, the moisture content of the filter paper at equilibrium is measured gravimetrically and related to soil suction through a predetermined calibration curve for the particular type of paper used. Figures 10.28a and 10.28b illustrate general testing setups for total and matric suction measurements using the noncontact

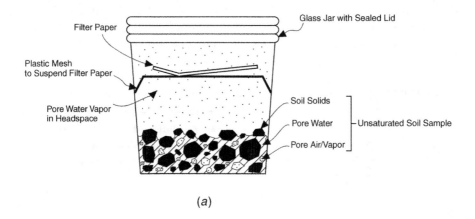

(a)

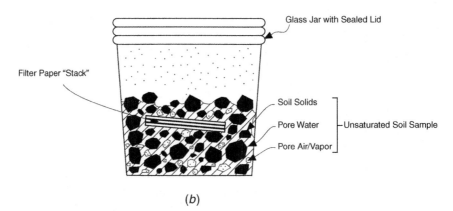

(b)

Figure 10.28 General testing configurations for filter paper testing: (a) "noncontact" method for total suction measurement and (b) "contact" method for matric suction measurement.

and contact techniques, respectively. Following the noncontact technique (Fig. 10.28a), a filter paper is suspended in the headspace above the specimen such that moisture transfer occurs in the vapor phase. The equilibrium amount of water adsorbed by the paper is a function of the pore-air relative humidity and the corresponding total soil suction. Following the contact technique (Fig. 10.28b), filter papers are placed in direct contact with the soil specimen. Accordingly, moisture transfer from the soil to the paper is controlled by capillary and particle surface adsorption forces comprising the matric component of total soil suction. Typically, one paper is sandwiched between two sacrificial papers to prevent fouling or contamination of the internal paper used for the measurement.

10.7.2 Calibration and Testing Procedures

Filter papers used for suction testing should be ash-free, quantitative type II papers. Commonly used types of papers include Whatman #42, Schleicher and Schuell #589 White Ribbon, and Fisher 9-790A. A typically sized paper is circular with a 5.5-cm diameter, weighing on the order of 0.2 g. Prior to "contact" testing, a calibration curve is obtained by measuring the relationship between matric suction and filter paper water content. This can be accomplished by testing representative papers as one normally would test a soil specimen using a pressure plate or pressure membrane device. Alternatively, papers may be buried in moist soil and the corresponding matric suction measured using a tensiometer or other device. Prior to "noncontact" testing, papers are calibrated by determining the relationship between equilibrium water content and relative humidity using salt solutions of known concentration, typically NaCl and KCl (see Tables 10.3 and 10.4). The noncontact method has found greater applicability in geotechnical engineering practice. General calibration and testing procedures for the noncontact method are as follows. Detailed discussion of the contact method can be found in Houston et al. (1994).

Representative filter papers are initially oven-dried to constant mass at 105°C and then allowed to cool to room temperature in a desiccator. Salt solutions are prepared in 30- to 50-mL aliquots at concentrations corresponding to the range of total suction of interest and poured into small testing containers (e.g., a glass jar or equivalent nonreactive container). A thin, perforated sheet of plastic mesh, for example, is cut to fit the inside diameter of the jar and act as a surface on which to suspend one paper above the salt solution. Care is taken such that the paper does not touch the sides or top of the glass jar where liquid water may otherwise be adsorbed. The paper and salt solution are sealed in the jar and placed in an insulated environment for an equilibration period of 7 to 10 days. Ideally, temperature fluctuations during equilibration should be limited to 0.1°C. The paper is then removed and immediately weighed to the nearest 0.0001 g with an electronic balance. The paper is oven-dried and reweighed to determine the filter paper water content. Filter paper water content is plotted versus total suction for each salt solution to define the calibration curve, usually requiring 8 to 10 data pairs.

The procedure for soil testing is essentially identical to that for calibration. A photograph illustrating the basic noncontact setup is shown in Fig. 10.29. Here, 30 to 50 g of soil are placed in the glass jar and one dry paper is suspended above the specimen. The water content of the paper is measured after 7 to 10 days to determine total suction using the calibration curve for the particular type of filter paper in use. The corresponding water content of the soil is determined gravimetrically to develop one point along the soil-water characteristic curve. Specimens prepared or obtained at different water contents may be tested to generate additional points.

Figure 10.30 shows calibration curves according to ASTM Standard D5298 for Whatman #42 and Schleicher and Schuell #589 papers. Both curves are bilinear with an inflection point occurring at a suction value somewhere between 10 and 100 kPa. The inflection has generally been interpreted to indicate transition from an adsorbed film regime at relatively high suction into a capillary adsorption regime at relatively low suction. For practical purposes, the high-suction portions of the calibration curves are applicable for total suction measurements, while the low-suction portions are applicable to matric suction measurements. Ideally, however, independent calibration curves should be directly obtained using lab-specific procedures for the particular type of paper being used. The method of calibration (i.e., control of either matric or total suction) should match the component of suction that is intended for the measurement.

Although far less common, techniques for in situ matric suction measurements using the contact filter paper method and in situ total suction measurements using the noncontact filter paper method have also been described (e.g., Greacen et al., 1989; Fredlund, 1989). As shown in Fig. 10.31, Likos and Lu (2003a) developed a column testing system for measuring transient total suction profiles and the associated moisture transport during one-dimensional evaporation experiments. Figure 10.32 shows an example of results obtained for a column of expansive clay undergoing evaporation at the surface from an initially moist condition over a period of 91 days.

10.7.3 Accuracy, Precision, and Performance

Although noncontact filter paper testing is in theory applicable over the entire range of total suction, the method tends to be impractical for both extremely high and extremely low values of suction. Reliable measurements tend to be limited to a range spanning about 1000 to 500,000 kPa, reasoned as follows. Referring to Fig. 10.12, the relationship between total suction and relative humidity becomes extremely "steep" at suction values less than approximately 1000 kPa. Total suction in this range is highly sensitive to relative humidity and, thus, to measurements of filter paper water content. Slight tem-

10.7 FILTER PAPER TECHNIQUES   **453**

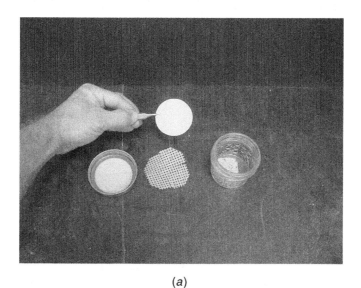

(a)

(b)

Figure 10.29 Photographs showing general test setup for total suction testing using the noncontact filter paper technique.

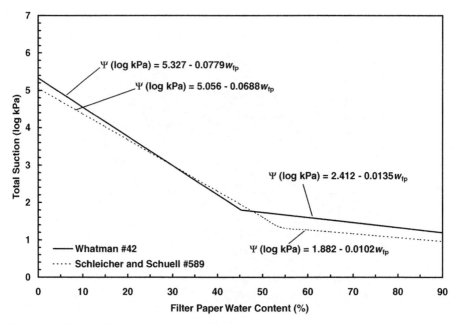

Figure 10.30 Calibration curves for Whatman #42 and Schleicher and Schuell #589 filter papers (ASTM D5298, ASTM 2000).

perature fluctuations may also result in significant changes in relative humidity in this range. At extremely high values of suction, the filter paper adsorbs a smaller and smaller amount of water vapor and the quality of the measurement becomes exceedingly dependent on environmental conditions, operational procedure, and the precision of the equipment used to determine filter paper water content.

Likos and Lu (2002) conducted an analysis to evaluate the accuracy and precision of total suction measurements obtained using the noncontact filter paper technique. As summarized in Table 10.6, twelve NaCl solutions were prepared at concentrations ranging from 354 to 3.3 g/L. Each solution was split into five subsamples, and the standard deviation in measurements of filter paper water content for each was calculated and evaluated in terms of the overall measurement uncertainty.

Figure 10.33a shows 5 measurements of filter paper water content for each of the 12 salt solutions versus the mean measured value of filter paper water content. Similarly, Fig. 10.33b shows the standard deviation of the 5 measurements versus the mean measured value. Both figures indicate that the measurement deviation generally increases as filter paper water content increases (i.e., as suction decreases). Figure 10.34 shows mean values of total suction calculated from the 5 measurements as well as the value based on the known solution concentration. The error bars define the standard deviation

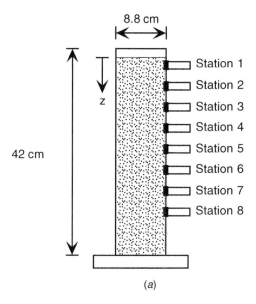

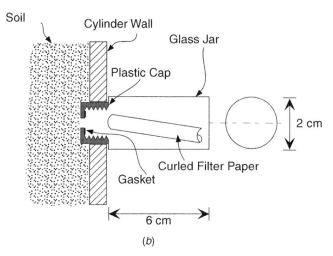

Figure 10.31 Laboratory filter paper column for measuring transient total suction profiles: (*a*) column setup and (*b*) detail of typical measurement station (Likos and Lu, 2003a).

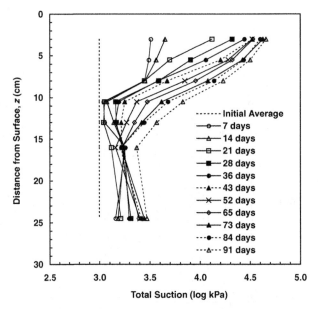

Figure 10.32 Transient total suction profile for Ca^{2+}-smectite during evaporation experiment using noncontact filter paper column (Likos and Lu, 2003a).

TABLE 10.6 Concentration, Relative Humidity, and Total Suction of 12 NaCl Solutions Prepared for Uncertainty Analysis of Noncontact Filter Paper Technique

(1) Solution No.	(2) Concentration (g/L)	(3) RH (%)	(4) Suction (kPa)	(5) Suction (log kPa)
1	354.0	78.8	32,239	4.51
2	265.0	84.1	23,381	4.37
3	178.0	89.3	15,250	4.18
4	87.6	94.7	7,290	3.86
5	70.3	95.8	5,819	3.76
6	52.4	96.9	4,313	3.63
7	35.2	97.9	2,882	3.46
8	17.9	98.9	1,458	3.16
9	14.6	99.1	1,188	3.07
10	10.9	99.3	886	2.95
11	7.2	99.6	585	2.77
12	3.3	99.8	268	2.43

Source: Likos and Lu (2002).

10.7 FILTER PAPER TECHNIQUES   457

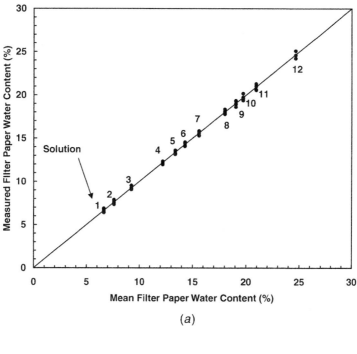

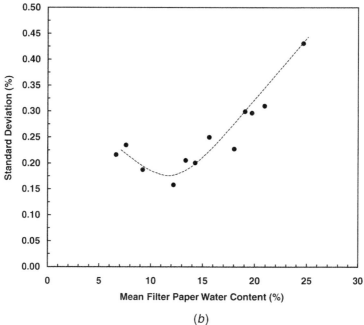

Figure 10.33 Results of filter paper water content measurements for 12 salt solutions of known concentration: (*a*) measured water content versus mean water content and (*b*) standard deviation of measured water content versus mean water content (Likos and Lu, 2002).

458 SUCTION MEASUREMENT

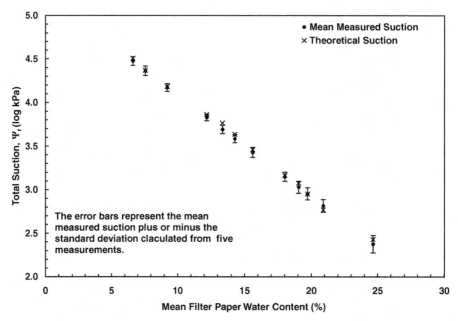

Figure 10.34 Accuracy and precision of total suction measurements for 12 salt solutions (Likos and Lu, 2002).

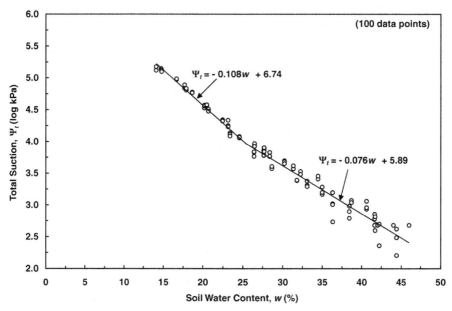

Figure 10.35 Noncontact filter paper measurements for expansive soil showing increased measurement scatter at relatively low values of total suction (Likos and Lu, 2002).

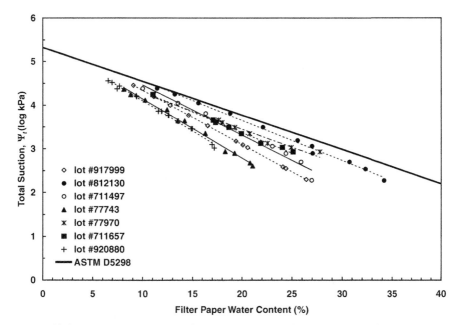

Figure 10.36 Calibration curves for seven batches of Whatman #42 filter paper (Likos and Lu, 2002).

from each mean. Similar deviation in measurements for soil is illustrated by Fig. 10.35, which shows 100 data pairs for an expansive Ca^{2+}-smectite sampled from the Denver, Colorado, area.

It has been shown that filter paper calibration curves can significantly vary among the same type of paper from one researcher to another or among the same type of paper from one "batch" or "lot" to another (e.g., Sibley et al., 1990; Leong et al., 2002; Likos and Lu, 2002). Figure 10.36, for example, shows noncontact calibration curves for seven different batches of Whatman #42 paper. Between the two extreme cases (lot 77743 and lot 812130), measured values of filter paper water content differ by as much as 11% at relatively low values of total suction. At relatively high values of total suction, the difference in measured water content is as much as 4%. In terms of total suction, these discrepancies result in 92% error and 57% error, respectively. As such, batch-specific calibrations are recommended.

PROBLEMS

10.1. Summarize the major techniques to measure matric suction and total suction. Summarize the advantages and disadvantages for each technique including its practical measurement range.

10.2. If a sandy soil were encountered, what would be the appropriate technique(s) for laboratory suction measurement? If a clayey soil were encountered, what would be the appropriate technique(s) for laboratory suction measurement?

10.3. Investigate the sensitivity of total suction to temperature by plotting Kelvin's equation in terms of suction versus relative humidity for three different temperatures. Summarize your findings from this investigation.

10.4. A thermocouple psychrometer is used to measure the relative humidity of an unsaturated clayey soil specimen. At equilibrium, the RH of the pore water vapor is 97.3% and the temperature is 15°C. What is the total suction of the soil in kilopascals? What is the likely water content for the clay (give an approximate range)?

10.5. A Tempe cell test was conducted for a sample of unsaturated sandy silt. The data shown in Table 10.7 was obtained. Column 1 shows values of air pressure that were incrementally applied to the system. Column 2 shows the change in mass of the soil specimen as pore water was expelled at each air pressure increment (e.g., when the pressure was increased from 15 to 20 kPa, 1.15 g of water were expelled). After the system reached equilibrium for the final air pressure increment, the specimen was removed, weighed wet (23.71 g), oven-dried, and then weighed dry (23.00 g). Plot the matric suction characteristic curve for the soil.

TABLE 10.7 Results of Tempe Cell Testing for Problem 10.5

(1) Air Pressure (kPa)	(2) Mass Change (g)
0	
2	0.00
4	0.00
8	0.00
12	0.00
15	1.38
20	1.15
30	0.81
34	0.35
50	0.35
75	0.46
98	0.21

10.6. Concentrations of NaCl and KCl solutions shown in Table 10.8 were used to calibrate a batch of filter papers for total suction testing using the noncontact method. The filter paper water content w_{fp} (%) corresponding to each salt solution is shown. The average temperature during calibration was 25°C. Plot the calibration curve for the filter paper in terms of total suction versus filter paper water content.

TABLE 10.8 Results of Filter Paper Calibration for Problem 10.6

NaCl		KCl	
Concentration (g/L)	w_{fp} (%)	Concentration (g/L)	w_{fp} (%)
40.91	15.60	37.28	16.81
87.66	12.95	44.73	16.21
99.35	12.50	59.64	15.26
105.20	12.29	67.10	14.86
116.89	11.91	74.55	14.51

10.7. An unsaturated soil specimen was tested for total suction using the calibrated filter paper from Problem 10.6. The equilibrium water content of the filter paper after testing was measured as 12.0%. What is the approximate total suction of the soil?

10.8. Evaluate the validity of the van't Hoff approximation [eq. (1.16)] using Tables 10.3 and 10.4.

CHAPTER 11

HYDRAULIC CONDUCTIVITY MEASUREMENT

11.1 OVERVIEW OF MEASUREMENT TECHNIQUES

Techniques for measuring the hydraulic conductivity of unsaturated soil can be generally classified as either laboratory or field methods and as either steady-state or transient methods. Laboratory methods are conducted on disturbed or undisturbed specimens under controlled hydraulic boundary and stress conditions. Field methods are conducted in situ such that the soil fabric and stress conditions are representative, yet often more difficult to quantify. For steady-state testing techniques, the flux, gradient, and water content of the soil-water system are constant with time. For transient techniques, each of these parameters varies with time. Steady-state techniques assume the validity of Darcy's law for unsaturated fluid flow, whereby the hydraulic conductivity corresponding to a specific value of suction or water content is calculated from measurements of flux or hydraulic gradient for a known flow field geometry. Transient techniques rely on solving the governing transient fluid flow equation (Section 9.1) for one-dimensional flow systems under controlled boundary conditions from measurements of flux or moisture content profiles at known locations and times. The majority of techniques and associated analysis procedures assume that the soil matrix does not significantly deform under changes in matric suction or degree of saturation.

Steady-state hydraulic conductivity testing techniques described in this chapter include the constant-head method, the constant-flow method, and the centrifuge method. The "crust" method, which is a steady-state testing technique conducted in the field, is described by Hillel and Gardner (1970) and Bouma et al. (1971). Transient techniques described in this chapter include the horizontal infiltration method, the multistep outflow method, and laboratory and field instantaneous profile methods. Further discussion and analysis of these and other techniques is also available in the literature (e.g., Klute, 1965, 1972; Olson and Daniel, 1981; Klute and Dirksen, 1986; Dirksen, 1991; Stephens, 1994; Benson and Gribb, 1997).

11.2 STEADY-STATE MEASUREMENT TECHNIQUES

11.2.1 Constant-Head Method

One of the oldest and most common laboratory techniques for measuring unsaturated hydraulic conductivity is the constant-head, steady-state method. Much like conventional constant-head testing for saturated soil, measurements are made by maintaining a constant hydraulic head across a soil specimen and measuring the corresponding rate of fluid flow through the specimen at steady state. Matric suction is actively maintained during testing, most commonly by axis translation using an external positive air pressure source. Darcy's law is assumed valid for computing the hydraulic conductivity corresponding to specific levels of applied matric suction.

A number of variations on the basic constant-head testing technique have been described (e.g., Corey, 1957; Klute, 1972; Klute and Dirksen, 1986; Barden and Pavlakis, 1971; Huang et al., 1998). Various systems have been developed for measuring water conductivity, air conductivity, or the conductivity of both phases simultaneously. Figure 11.1 shows a general testing schematic for measuring water conductivity. A constant total head gradient Δh_t is maintained across the specimen using either two Mariotte bottles, a

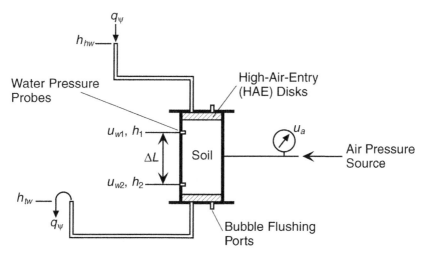

Figure 11.1 Experimental system for measuring unsaturated hydraulic conductivity using constant-head steady-state technique.

pressure panel–burette system, or other equivalent system capable of accurately measuring the flow rate through the specimen, q, and maintaining the head-water head h_{hw} and the tail-water head h_{tw} at constant values. The permeameter may be either rigid-walled (as shown) or flexible-walled, the latter involving a confining cell system to allow for isotropic or triaxial stress control. The specimen is seated in good contact with two high-air-entry (HAE) disks on top and bottom, which are initially saturated with permeant fluid to establish a hydraulic connection between the specimen pore water and the influent and effluent fluid reservoirs. Usually, the specimen is also saturated prior to conductivity testing.

An external pressure supply may be used to maintain the pore air pressure at some value greater than atmospheric such that matric suction may be directly controlled by axis translation (Section 5.4). The air-entry pressure of the HAE disks on either side of the specimen must be at least as large as the maximum suction desired in describing the hydraulic conductivity function. For testing in a rigid-wall setup, air pressure is typically supplied through a side port in the permeameter (as shown) and distributed uniformly throughout the specimen using strips of filter paper situated along the internal wall of the cell. For testing in a flexible-walled setup, special annular end caps comprising both high-air-entry and low-air-entry surfaces are required.

The flux of water through the specimen under the constant-head gradient is measured until steady state is reached. To ensure accurate flux measurements, air bubbles that tend to accumulate in the permeant lines under advective/diffusive transport processes over extended testing periods must be periodically flushed out and accounted for volumetrically. Hydraulic gradient is measured through two or more ports installed along the length of the specimen, separated by a distance ΔL. Positive or negative pore pressures may be maintained, provided that the applied air pressure retains the desired level of matric suction. Tensiometers (Section 10.2) must be used to measure the hydraulic gradient when the pore water pressures are negative. If head loss is not measured inside the specimen, but rather is measured at positions located "outside" the HAE disks, then corrections for their impedance must be made and subtracted from the total system head loss. This impedance may be measured prior to testing by permeating a "blank" specimen having negligible head loss (e.g., a cylinder of free water). For relatively coarse-grained soil, or for many types of soil near saturation, however, the head loss in the end caps may be significantly larger than the specimen head loss, making this correction procedure undesirable.

Hydraulic conductivity is computed from the measured flux, the internal head loss ($\Delta h_s = h_1 - h_2$), and the specimen geometry using Darcy's law:

$$k = -\frac{q}{i} = q\left(\frac{\Delta L}{\Delta h_s}\right) = \frac{Q}{A}\left(\frac{\Delta L}{\Delta h_s}\right) \qquad (11.1)$$

The unsaturated hydraulic conductivity function is usually determined along a drying path by incrementally increasing the matric suction, either by increasing the pore air pressure or decreasing the pore water pressure, and subsequently measuring the steady-state flux for each increment. Although less common, the conductivity function may also be determined along a wetting path for incremental decreases in matric suction. Because the pore pressure in the specimen is not uniform under flowing conditions, the computed hydraulic conductivity is usually referenced to the average matric suction in the specimen determined from the known air pressure and the average of the two internal pore pressure measurements. Water content corresponding to the computed conductivity may be inferred from an independently measured soil-water characteristic curve or may be directly measured by destructive means if the specimen is dissembled after each testing increment. When destructive water content measurements are made, several "identical" specimens (i.e., identical density, molding water content, etc.) are required to obtain more than one point defining the conductivity function. Alternatively, nondestructive water content measurements may be made using a number of techniques [e.g., gamma ray attenuation techniques or time-domain reflectrometry (TDR) probes] or by monitoring the change in mass of the permeameter due to the expulsion of water after each applied suction increment and then back-calculating the water content from a post-test destructive measurement. Gardner (1986) summarizes a variety of nondestructive techniques for water content measurements in unsaturated soil. Topp et al. (1980) and O'Connor and Dowding (1999) describe the use of TDR technology for soil testing applications.

The primary advantage of the constant-head method is its simplicity in both procedure and analysis and its relatively widespread use. Because the test may be conduced in a triaxial cell, hydraulic conductivity may be examined under stress-controlled and simulated in situ conditions. Hydraulic conductivity measurements as low as 10^{-11} m/s are generally possible. Disadvantages associated with the constant-head technique include the requirement for often lengthy amounts of time for steady state to be reached, difficulties associated with accurately measuring extremely low flow rates, and uncertainties in the goodness of contact between the specimen and the HAE end caps, the head loss measurement probes, and the (rigid) walls of the permeameter cell. Because a head gradient is required to induce fluid flow, the suction along the length of the specimen, and thus the hydraulic conductivity, is nonuniform. Balance must be achieved between maintaining a head gradient small enough to minimize this effect but large enough to result in practically measurable flow rates.

Example Problem 11.1 Table 11.1 shows results obtained from a series of rigid-wall constant-head hydraulic conductivity tests conducted on unsaturated specimens of sand. The length and diameter of the cylindrical specimen

TABLE 11.1 Testing and Analysis Results for Constant-Head Hydraulic Conductivity Test Conducted in Example Problem 11.1

Test	(1) h_1 (cm)	(2) h_2 (cm)	(3) Q (cm³/s)	(4) Δh (cm)	(5) k (cm/s)	(6) h_{avg} (cm)
1	−0.51	−15.81	1.30×10^{-1}	15.30	2.40×10^{-3}	8.16
2	−1.02	−25.50	1.42×10^{-1}	24.48	1.65×10^{-3}	13.26
3	−2.55	−44.37	2.30×10^{-1}	41.82	1.56×10^{-3}	23.46
4	−3.06	−66.30	1.64×10^{-1}	63.24	7.32×10^{-4}	34.68
5	−4.59	−83.13	1.03×10^{-1}	78.54	3.72×10^{-4}	43.86
6	−5.10	−109.14	2.36×10^{-2}	104.04	6.41×10^{-5}	57.12
7	−6.12	−120.36	9.83×10^{-3}	114.24	2.43×10^{-5}	63.24
8	−8.67	−134.13	4.54×10^{-3}	125.46	1.02×10^{-5}	71.40

are 10 and 6 cm, respectively. A constant head was maintained across the specimen such that pore water flowed from top to bottom. The first two columns of Table 11.1 show pore water head measured at points located inside the specimen with tensiometer probes inserted near the top and bottom, respectively, separated by a distance of $\Delta L = 8$ cm. The third column shows the volumetric flow rate (Q) measured through the specimen for each test at steady state. Identical specimens were prepared for the eight tests and pore air pressure was maintained at atmospheric for each ($u_a = h_a = 0$). Calculate the hydraulic conductivity function, $k(h_m)$, for the sand.

Solution The head loss between the top and bottom tensiometers for each test ($\Delta h_s = h_1 - h_2$) is shown in the fourth column of Table 11.1. Darcy's law [eq. (11.1)] and the known cross-sectional area of the specimen ($A = 28.27$ cm²) may be used to calculate hydraulic conductivity for each test (column 5). For example, for test 1, $k = Q \Delta L / \Delta h\, A = (0.13 \text{ cm}^3/\text{s})(8 \text{ cm})/(15.3 \text{ cm})(28.27 \text{ cm}^2) = 2.4 \times 10^{-3}$ cm/s. The average value of suction head corresponding to this hydraulic conductivity may be computed as

$$h_{avg} = (h_a - h_{w(avg)}) = h_a - \tfrac{1}{2}(h_1 + h_2) = 0 - \tfrac{1}{2}(-0.51 \text{ cm} - 15.81 \text{ cm})$$
$$= 8.16 \text{ cm}$$

which is shown for each of the eight tests in column 6 of Table 11.1. Figure 11.2 shows a plot of the corresponding hydraulic conductivity function $k(h)$.

11.2.2 Constant-Flow Method

The constant-flow method is a laboratory testing technique quite similar to the constant-head method. In this case, however, the flow rate through the specimen is controlled rather than measured. The applied flow rate can be as

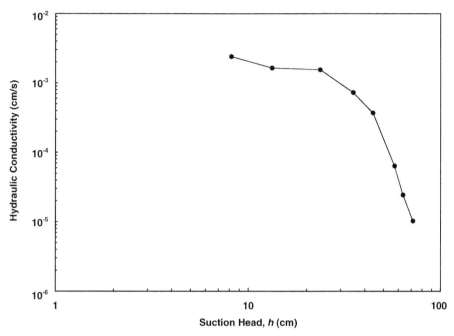

Figure 11.2 Hydraulic conductivity function for sand from Example Problem 11.1.

low as 10^{-7} cm³/s or about 0.01 cm³/day. As such, difficulties associated with measuring extremely low flow rates for low-permeability soils or soils at relatively low saturation are avoided. Motorized flow pumps capable of accurately controlling extremely low flow rates are readily available for geotechnical hydraulic conductivity testing applications (e.g., Olsen et al., 1985, 1988), thus allowing applied hydraulic gradients to be maintained at values that more accurately represent in situ flow conditions and minimize seepage-induced disruption to the soil fabric. If a second flow pump is used to control the water content of the specimen by injecting or extracting known volumes of pore water, the soil-water characteristic curve and the hydraulic conductivity function may be determined simultaneously (Olsen et al., 1994).

Figure 11.3 shows a schematic drawing for one variation of the constant-flow method. A bidirectional flow pump (P) is used to infuse and withdraw identical flow rates at opposite ends of an unsaturated soil specimen (S). Use of the infuse/withdrawal pump effectively cuts the length of the specimen in half, thus reducing the time required for steady state to be reached, which is a diffusion process described by the square of the diffusion path length, by one-fourth. The specimen is seated in a conventional confining cell and isolated from the chamber fluid by a latex membrane. The pedestal and top cap each incorporate HAE ceramic disks through which pore water may be transmitted to or from the specimen. The upper disk has a low-air-entry "hole" comprised of a coarse porous material through which pore air pressure may

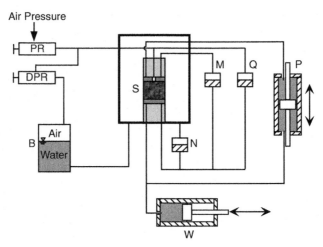

Figure 11.3 Constant-flow permeameter system for concurrent measurement of the soil-water characteristic curve and hydraulic conductivity of unsaturated soil under isotropic stress control (after Olsen et al., 1994).

be controlled by an external pressure regulator (PR). The volumetric water content of the specimen is controlled using a second flow pump (W) to infuse or withdraw water from the base. Isotropic stress can be imposed using a differential pressure regulator (DPR) through an air-water interface (B) to maintain a constant pressure difference between the confining cell pressure and the pore air pressure. One side of each of three differential pressure transducers (M, N, and Q) is connected to one of the pore water lines from the base pedestal. The other side of transducer (M) monitors the water pressure above the top HAE disk to quantify head losses through the specimen. The other side of transducer (Q) monitors the pore air pressure in the center hole of the top cap to quantify matric suction. The other side of transducer (N) monitors the confining fluid pressure to quantify the difference between the confining pressure and the bottom pore water pressure.

For testing, a specimen is placed in the permeameter, saturated under applied backpressure, and consolidated under a desired stress. Saturated hydraulic conductivity may then be measured by inducing a constant flow rate through the specimen using flow pump (P). Output from transducer (M) is recorded to determine the consequent steady-state head loss across the specimen. For determination of unsaturated hydraulic conductivity, matric suction, and the variation of each with moisture content, positive pore air pressure is applied through the hole in the top cap. Water content is decreased from the initially saturated condition by withdrawing a known volume of pore fluid from the base using flow pump (W). The corresponding matric suction is recorded at equilibrium using transducer (Q). Flow is then reintroduced through the specimen and steady-state head loss measured. Hydraulic conductivity corresponding to the applied suction is calculated from eq. (11.1). Testing proceeds by incrementally decreasing (or increasing) the water content of the specimen. Water content, matric suction, and hydraulic conductivity are determined for each increment to simultaneously generate the soil-water characteristic curve and the hydraulic conductivity function.

Figure 11.4a shows a soil-water characteristic curve $\psi(\theta)$ and hydraulic conductivity function $k(\theta)$ obtained using the constant-flow system of Fig. 11.3. Grain size information is shown in the figure. Water content was decreased during the testing series along a drying path from $\theta_s = 0.45$ at saturation to $\theta = 0.37$ at $\psi = 234$ kPa. Hydraulic conductivity at zero suction and 234 kPa are 6.4×10^{-8} cm/s and 2.1×10^{-9} cm/s, respectively. Figure 11.4b shows the hydraulic conductivity function in the form $k(\psi)$. Figure 11.4c shows the amount of time required for steady state to be reached at each increment in suction. In general, response times range from (a) a few minutes or less for materials with k values greater than 10^{-6} cm/s, (b) a few hours for materials with k values in the range of 10^{-7} to 10^{-9} cm/s, and (c) a few days for materials with extremely low k values in the range of 10^{-10} to 10^{-11} cm/s (Olsen et al., 1994).

Example Problem 11.2 Results from a constant-flow hydraulic conductivity test conducted for a cylindrical silty sand specimen are shown as Fig. 11.5. The plot is a time-domain trace of head loss (cm) measured across the specimen as well as the cumulative volume (cm³) of pore water extracted from the specimen. The specimen is initially saturated. A total of 0.63 cm³ of pore water is then extracted (labeled extraction 1) resulting in an equilibrium matric suction of 3.7 kPa. A constant volumetric flow rate of $Q = qA = 0.003$ cm³/s is then imposed through the specimen at this new condition and the corresponding buildup of head loss is measured (labeled HC test 1). The imposed flow is then ceased and the system is allowed to return to the hydrostatic condition. A second extraction results in the cumulative removal of 1.58 cm³ of pore water and a corresponding matric suction of 6.5 kPa. A constant flow rate of 0.003 cm³/s is again imposed and the head loss is measured (labeled HC test 2). Calculate two data pairs on the soil-water characteristic curve and hydraulic conductivity function for the soil based on the results of these tests. The height, diameter, and porosity of the specimen are $L = 2.12$ cm, $d = 5.00$ cm, and $n = 0.324$, respectively.

Solution The total volume of the specimen V_t is 41.63 cm³. Because the soil is initially saturated, the initial volume of water in the voids V_w may be calculated from the total volume and the porosity: $V_w = V_v = nV_t = (0.324)(41.63$ cm³$) = 13.49$ cm³. Volumetric water content after each extraction may be calculated as follows. For extraction 1, $\theta = (13.49 - 0.63)/41.63 = 0.309$, resulting in one pair of points on the drying branch of the SWCC ($\theta = 0.309$, $\psi = 3.7$ kPa). For extraction 2, $\theta = (13.49 - 1.58)/41.63 = 0.286$, resulting in a second pair of points on the SWCC ($\theta = 0.286$, $\psi = 6.5$

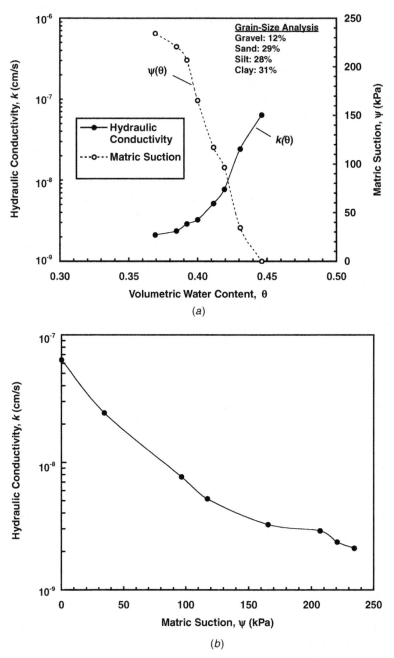

Figure 11.4 Results obtained using constant-flow method for silty soil: (a) soil-water characteristic curve, $\psi(\theta)$, and hydraulic conductivity function, $k(\theta)$, (b) hydraulic conductivity function, $k(\psi)$, and (c) steady-state response time (data from Olsen et al., 1994).

11.2 STEADY-STATE MEASUREMENT TECHNIQUES   471

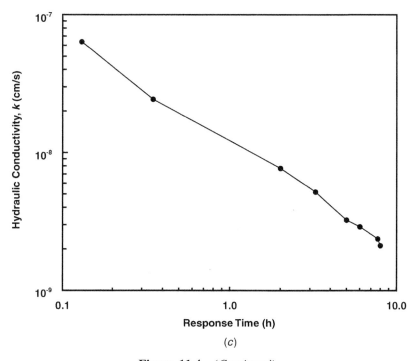

(c)

Figure 11.4 (*Continued*).

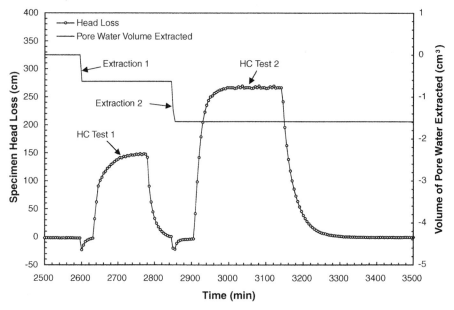

Figure 11.5 Results from constant-flow hydraulic conductivity test conducted for silty sand from Example Problem 11.2.

kPa). From Fig. 11.5, the steady-state head loss measured for the constant flow rate imposed after the first pore water extraction is about 147 cm. The steady-state condition is reached in about 100 min. Given the height of the specimen $L = 2.12$ cm, the corresponding hydraulic gradient is $i = \Delta h_s/\Delta L = 69.34$. Given the specimen area ($A = 19.63$ cm^2) and eq. (11.1), the hydraulic conductivity corresponding to this condition is $k = Q/iA = 2.4 \times 10^{-6}$ cm/s. After the second pore water extraction, where the steady-state head loss is about 270 cm, the hydraulic conductivity is 1.2×10^{-6} cm/s. Thus, two points on the hydraulic conductivity function $k(\theta)$ are defined: $\theta = 0.309$ and $k = 2.4 \times 10^{-6}$ cm/s and $\theta = 0.286$ and $k = 1.2 \times 10^{-6}$ cm/s, or in terms of $k(\psi)$: $\psi = 3.7$ kPa and $k = 2.4 \times 10^{-6}$ cm/s and $\psi = 6.5$ kPa and $k = 1.2 \times 10^{-6}$ cm/s.

11.2.3 Centrifuge Method

The steady-state centrifugation method (SSCM) is a laboratory testing technique that utilizes a spinning centrifuge to quickly establish steady-state fluid flow through an unsaturated specimen. Hydraulic conductivity is calculated by measuring steady-state flow under the elevated gravitational gradient, which significantly reduces the amount of time required for steady state to be reached in relatively low permeability or low degree of saturation materials. Detailed descriptions of various experimental setups and an analysis of the general governing principles are provided by Nimmo et al. (1987, 1992) and Nimmo and Akstin (1988).

The schematic shown as Fig. 11.6 illustrates the basic concept of the centrifuge technique. A cylindrical specimen is placed in a special container located at the end of a centrifuge arm rotating in the horizontal plane at angular velocity ω. The centrifugal gravity field at a distance r from the axis of rotation is equal to the product $\omega^2 r$.

If it is assumed that Earth's gravity field is negligible compared to the centrifugal gravity field, then fluid flow through the specimen may be considered one dimensional and fully described by the hydraulic conductivity and two driving gradients: (1) the suction gradient acting through the specimen in the direction of r and (2) the centrifugal gravity gradient. Fluid flow for a given suction condition may be described in terms of these two gradients as

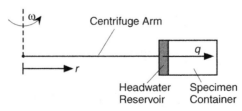

Figure 11.6 Schematic of centrifuge testing technique.

$$q = -k \left(\frac{d\psi}{dr} - \rho_w \omega^2 r \right) \qquad (11.2)$$

where $d\psi/dr$ is the suction gradient along the length of the specimen and ρ_w is the density of the permeant fluid (e.g., water). At sufficiently high rotational speed, Nimmo et al. (1987) showed that the radial suction gradient can be neglected and eq. (11.2) reduces to

$$q = k\rho_w \omega^2 r \qquad (11.3)$$

which may be rearranged to solve for hydraulic conductivity:

$$k = \frac{q}{\rho_w \omega^2 r} \qquad (11.4)$$

Two general approaches are available for delivering permeant fluid to the specimen. The original steady-state centrifugation method (SSCM) uses a self-contained flow delivery system housed within a special specimen testing container. The SSC-UFA method (steady-state centrifuge–unsaturated flow apparatus) uses an external syringe pump to deliver and disperse a precisely controlled constant flow rate to one end of the rotating specimen (Conca and Wright, 1998). ASTM Standard D6527 describes general operating procedures for the SSC-UFA method.

Figure 11.7 shows one variation of specimen container used for SSCM testing. The ceramic disk B is selected to have a saturated hydraulic conductivity and effective cross-sectional area such that the flux through the disk during testing is some value less than the unsaturated conductivity of the specimen. The effective cross-sectional area of the disk is manipulated by selecting various sized O-rings for placement in the groove located immediately below the disk. Given this "impeding" layer, the flow through the specimen is proportional to the conductivity of plate B and the specimen remains unsaturated. Lower conductivity values for the impeding layer result in a drier specimen and consequently lower values of hydraulic conductivity. A constant difference in head is maintained across the specimen using an adjustable overflow port in the sidewall of the head-water reservoir. The height of the overflow port may be changed between tests to provide a variety of head gradients. Hydraulic conductivity is determined using eq. (11.4) by measuring steady-state flow rate at a known angular velocity. The flow rate is determined by periodically stopping the centrifuge and measuring the relative amounts of water in the head- and tail-water overflow reservoirs. Several points comprising the hydraulic conductivity function may be obtained by changing the impedance or effective cross-sectional area of the head-water disk to manipulate the average degree of saturation of the specimen. Typically, specimens are run from an initially saturated condition and then desaturated in a stepwise fashion.

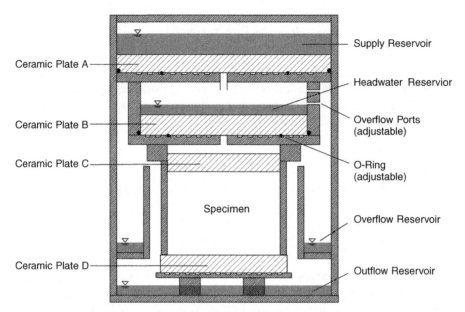

Figure 11.7 Specimen container for SSCM centrifuge testing (adapted from Nimmo et al., 1992).

Figure 11.8a shows results obtained by Nimmo et al. (1987, 1992) for densely packed Oakley sand specimens ($n = 0.333$) using the SSCM method and at normal gravity using the constant-head steady-state technique. A portion of the soil-water characteristic curve for the material is shown as Fig. 11.8b. Figure 11.9 shows results obtained using the SSC-UFA method for specimens obtained from sediment beneath the Hanford Site in Washington State.

The primary advantage of the centrifuge method is the relatively short testing time required, ranging from a few hours for determining relatively large hydraulic conductivities to over 24 h for relatively low conductivity. Nimmo and Akstin (1988) estimated a total measurement uncertainty of $\pm 8\%$. The primary limitations of the method are its relatively high cost (most notably the requirement for a specialized centrifuge) and its limited applicability to incompressible materials such as dense sands or highly overconsolidated sediments. For compressible or highly structured specimens, it has been argued that the high centrifugal forces can alter the soil fabric and thus the hydraulic conductivity. Careful selection of angular velocity is required in these cases. For the type of system shown in Fig. 11.7, difficulties may also be encountered with wet soil or soil having relatively large hydraulic conductivity due the limited volume capacity of the head- and tail-water reservoirs.

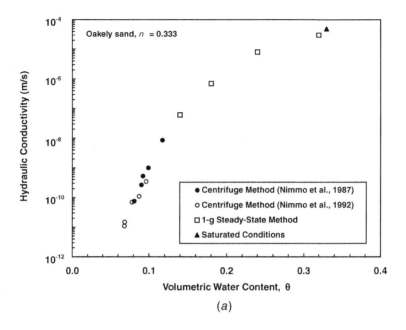

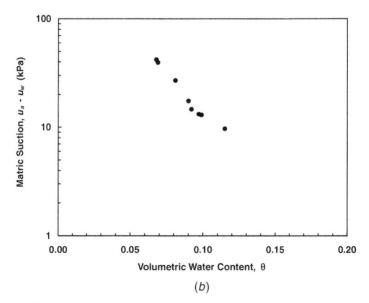

Figure 11.8 Results of hydraulic conductivity and suction tests for Oakley sand: (*a*) hydraulic conductivity function $k(\theta)$ obtained from centrifuge and normal-gravity tests and (*b*) soil-water characteristic curve $\psi(\theta)$ obtained from tensiometer measurements (data from Nimmo et al., 1987, 1992).

476 HYDRAULIC CONDUCTIVITY MEASUREMENT

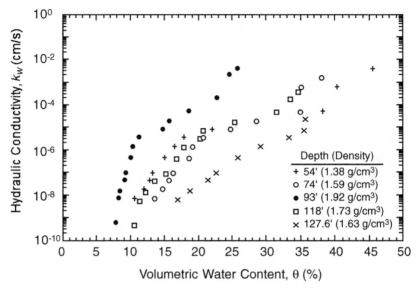

Figure 11.9. Hydraulic conductivity functions $k(\theta)$ obtained from SSC-UFA centrifuge testing (from ASTM D 6527, American Society for Testing and Materials, 2001).

11.3 TRANSIENT MEASUREMENT TECHNIQUES

11.3.1 Hydraulic Diffusivity

As introduced in Chapter 9, the transient flow of water through soil is a diffusion process controlled by the hydraulic diffusivity, which for unsaturated soil is a function of water content, $D(\theta)$. As introduced in Section 9.1.3, hydraulic diffusivity is defined as the ratio of the hydraulic conductivity to specific moisture capacity, or

$$D(\theta) = \frac{k(\theta)}{C(\theta)} \quad (11.5)$$

where the specific moisture capacity $C(\theta)$ describes the slope of the soil-water characteristic curve (θ versus ψ):

$$C(\theta) = \frac{\partial \theta}{\partial \psi} \quad (11.6)$$

The above relation allows eq. (11.5) to be rewritten as

$$k(\theta) = D(\theta)C(\theta) = D(\theta)\left(\frac{\partial \theta}{\partial \psi}\right) \qquad (11.7)$$

Thus, if transient flow experiments are designed to determine hydraulic diffusivity and if the soil-water characteristic curve is measured concurrently or independently to determine specific moisture capacity, then eq. (11.7) may be solved to determine the hydraulic conductivity function. A variety of transient techniques following this general strategy have been developed.

11.3.2 Horizontal Infiltration Method

The horizontal infiltration method was originally developed by Bruce and Klute (1956). Variations on the technique have been explored by Jackson (1964), Cassel et al. (1968), Rose (1968), and Clothier et al. (1983), among others. Testing involves analysis of the distribution of water content in a long horizontal column of soil at some time after a stepwise increase to 100% saturation ($S = 1$, $\theta = \theta_s$) has been introduced at one end of the column. Given the known initial and boundary conditions for the system, a transformed form of the one-dimensional diffusion equation is solved to determine hydraulic diffusivity. The technique is primarily applicable to relatively coarse-grained specimens and typically requires disturbance of the specimen as it is packed into the testing column.

Figure 11.10 shows a schematic of the basic laboratory setup. The testing column is comprised of about 10 segments of glass or acrylic tubing with a diameter of approximately 2 to 3 cm and individual lengths of 2 to 3 cm (Klute and Dirksen, 1986). The segments are held together by a press-fit or external clamping device such that they may be easily separated at the end of the test and the final water content distribution determined. The soil is assumed to have uniform initial water content, typically air dry. At some time t_0, a valve connecting a water supply to one end of the column is opened, creating a stepwise increase from the initial water content, θ_0, to the saturated

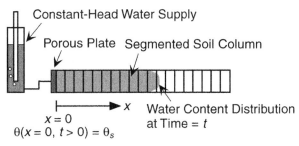

Figure 11.10 Apparatus for horizontal infiltration testing (after Klute and Dirksen, 1986).

water content, θ_s, at the boundary. The supply valve is left open and permeant fluid flows into the system under the imposed suction gradient. Gravity effects are assumed negligible such that the flow process may be considered one dimensional. After some time t, during which the wetting front advances in the direction of x, the supply valve is closed and the column is immediately disassembled into its individual segments so that the average water content of each may be determined.

The governing equation for moisture flow during the transient flow process is as follows (Section 9.1.3):

$$\frac{\partial \theta}{\partial t} = \frac{\partial}{\partial x}\left(D(\theta)\frac{\partial \theta}{\partial x}\right) \tag{11.8}$$

and the initial and boundary conditions for the experimental system are

$$\theta(x,0) = \theta_0 \quad \theta(0,t) = \theta_s$$

The Boltzmann variable $\lambda(\theta)$ is applied to transform eq. (11.8) into an ordinary differential equation from which the diffusivity function can be solved (e.g., Philip, 1957):

$$D(\theta) = -\frac{1}{2}\left(\frac{d\lambda(\theta)}{d\theta}\right)\int_{\theta_0}^{\theta_s} \lambda(\theta)\,d\theta \tag{11.9}$$

where, as described in Chapter 9, the transformed variable $\lambda(\theta)$ is given as

$$\lambda(\theta) = \frac{x}{\sqrt{t}} \tag{11.10}$$

For analysis, volumetric water content measured in each column segment at time t is plotted as a function of λ, as illustrated conceptually in Fig. 11.11. To solve for $D(\theta)$, the derivative and integral terms in eq. (11.9) are determined either graphically or analytically at specific values of water content, θ, from the slope and area under the curve, respectively. The hydraulic conductivity function may then be calculated using eq. (11.7) if the specific moisture capacity (i.e., soil-water characteristic curve) is known. The soil-water characteristic curve may be measured independently for an "identically" prepared specimen (preferably corresponding to a wetting process) or concurrently measured during the infiltration test by installing suction instrumentation (e.g., tensiometers or psychrometers) at several points along the column. Rose (1968) describes a method to determine $D(\theta)$ in the same general manner during an evaporative drying process.

One limitation of the horizontal infiltration method is the often significant scatter in the diffusivity function that propagates from scatter in measurements

11.3 TRANSIENT MEASUREMENT TECHNIQUES

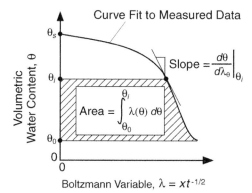

Figure 11.11 Conceptual plot of volumetric water content, θ, versus Boltzmann transform variable, λ, for determining hydraulic diffusivity from horizontal infiltration test (after Klute and Dirksen, 1986).

of the post-test water content distribution. Scatter is particularly evident for portions of the diffusivity function close to saturation where, as shown in Fig. 11.11, the slope of the λ-θ relationship is relatively small (Jackson, 1964). Clothier et al. (1983) describe a mathematical technique for fitting λ-θ with the intent of reducing the propagation of scatter through the analysis. Jackson (1964) used a variation of the method to measure the vapor component of pore water diffusivity for specimens in the relatively dry range.

Example Problem 11.3 The first two columns of Table 11.2 show results from a horizontal infiltration test conducted for a sandy loam soil. Water content measurements (column 2) were obtained after time $t = 1500$ min for several segments of the soil column located at distances x (column 1) from the influent boundary. Saturated water content at $x = 0$ is $\theta_s = 0.39$. Calculate and plot the Boltzmann transform function $\lambda(\theta)$ and the hydraulic diffusivity function $D(\theta)$ for the soil.

Solution The Boltzmann variable λ may be calculated from eq. (11.10) for each segment x and the corresponding analysis time $t = 1500$ min. These calculations are shown in the third column of Table 11.2 and $\lambda(\theta)$ is plotted in Fig. 11.12a. Increments in the area A_i under the Boltzmann variable function (column 4) may be calculated as

$$A_i = (\theta_i - \theta_{i-1}) \left(\frac{\lambda_i + \lambda_{i-1}}{2} \right)$$

Cumulative increments in area ΣA_i (column 5) are

TABLE 11.2 Results and Analysis for Horizontal Infiltration Test from Example Problem 11.3

x (cm)	θ_i	λ_i (cm/min$^{0.5}$)	A_i	ΣA_i	S_i	$D(\theta_i)$ (cm^2/min)
76.00	0.02	1.962				
75.85	0.04	1.958	0.0392	0.0392	−5.1640	0.0038
75.71	0.06	1.955	0.0391	0.0783	−5.5328	0.0071
75.63	0.08	1.953	0.0391	0.1174	−9.6825	0.0061
75.50	0.10	1.949	0.0390	0.1564	−5.9584	0.0131
75.40	0.12	1.947	0.0390	0.1954	−7.7460	0.0126
75.30	0.14	1.944	0.0389	0.2343	−7.7460	0.0151
75.18	0.16	1.941	0.0389	0.2732	−6.4550	0.0212
74.90	0.18	1.934	0.0388	0.3119	−2.7664	0.0564
74.65	0.20	1.927	0.0386	0.3505	−3.0984	0.0566
74.15	0.22	1.915	0.0384	0.3889	−1.5492	0.1255
73.40	0.24	1.895	0.0381	0.4270	−1.0328	0.2067
72.08	0.26	1.861	0.0376	0.4646	−0.5868	0.3959
69.90	0.28	1.805	0.0367	0.5013	−0.3553	0.7054
66.05	0.30	1.705	0.0351	0.5364	−0.2012	1.3330
61.10	0.32	1.578	0.0328	0.5692	−0.1565	1.8187
54.80	0.34	1.415	0.0299	0.5991	−0.1230	2.4364
47.10	0.36	1.216	0.0263	0.6254	−0.1006	3.1086
32.50	0.38	0.839	0.0206	0.6460	−0.0531	6.0880

Source: From Nielson et al. (1964).

$$\Sigma A_i = \Sigma A_{i-1} + A_i$$

Increments in the slope of the Boltzmann variable function $S_i = d\theta_i/d\lambda_i$ (column 6) are

$$S_i = \frac{\theta_i - \theta_{i-1}}{\lambda_i - \lambda_{i-1}}$$

Thus, by eq. (11.9), the diffusivity function $D(\theta)$ (column 7) may be calculated as

$$D(\theta_i) = -\frac{1}{2}\left(\frac{1}{S_i}\right)\Sigma A_i$$

which is plotted in Fig. 11.12b.

11.3.3 Outflow Methods

Outflow methods are relatively widely used transient laboratory techniques that allow concurrent determination of the hydraulic conductivity function and

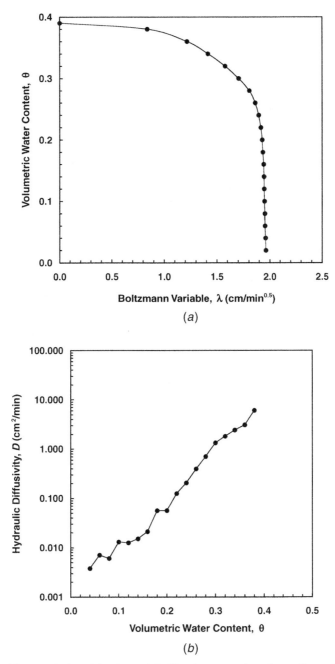

Figure 11.12 Analysis of horizontal infiltration test data from Example Problem 11.3: (a) Boltzmann variable function $\lambda(\theta)$ and (b) hydraulic diffusivity function, $D(\theta)$.

the soil-water characteristic curve. An important advantage of outflow tests is that they are conducted using conventional axis-translation equipment such as pressure plate or Tempe cell systems (Section 10.3). Hydraulic diffusivity is determined by monitoring the time-dependent flow of pore water from specimens subjected to an applied increment or series of applied increments in matric suction. Four general variations on the outflow method have been developed: the multistep method, the one-step method, the multistep direct method, and the continuous outflow method. The original outflow method, the multistep method, was first developed by Gardner (1956) and is summarized here to illustrate the basic testing approach. A more detailed review of each variation on the outflow method is provided by Benson and Gribb (1997).

The multistep outflow method involves subjecting a soil specimen to incremental steps in matric suction in an axis translation cell (e.g., Tempe pressure cell; see Section 10.3). Suction increments are applied by increasing the chamber and pore air pressure while allowing drainage of pore water through a high-air-entry disk or membrane. The rate of the pore water outflow and the total outflow for each suction increment is monitored to calculate the hydraulic diffusivity function. The hydraulic conductivity function is calculated from the hydraulic diffusivity function and the soil-water characteristic curve using eq. (11.7). The soil-water characteristic curve may be obtained directly from the outflow test results by back-calculating the equilibrium water content for each increment.

Several assumptions are made in the analysis of the outflow data for calculating hydraulic diffusivity: (1) the suction increment is small enough such that the hydraulic conductivity of the specimen remains constant, (2) suction is linearly related to water content over the suction increment, (3) the high-air-entry disk has no impedance to the pore fluid outflow, (4) flow is one dimensional, (5) gravity-driven flow is negligible, and (6) the specimen is homogeneous and rigid. Given these assumptions, the governing diffusion equation for the outflow process may be linearized as

$$\frac{\partial \psi}{\partial t} = D \frac{\partial^2 \psi}{\partial z^2} \tag{11.11}$$

where the hydraulic diffusivity D is assumed to be a constant over the small increment in applied suction. The spatial variable z describes the height of the specimen, where $z = 0$ at the bottom of the specimen and $z = L$ at the top ($L \approx 1$ cm $- 5$ cm). The diffusion process described by eq. (11.11) is analogous to the classical one-dimensional consolidation process whereby excess pore pressure resulting from an increment in external loading dissipates as a function of time.

Given initial and boundary conditions, the solution of eq. (11.11) can be written in terms of the pore fluid outflow volume as a linear equation (Gardner, 1956):

$$\ln\left(\frac{V_\infty - V_t}{V_\infty}\right) = \ln\left(\frac{8}{\pi^2}\right) - \frac{D\pi^2 t}{4L^2} \qquad (11.12)$$

where V_∞ is the total volume of pore water expelled for the applied suction increment and V_t is the outflow volume at time t. A plot of t versus $\ln[(V_\infty - V_t)/V_\infty]$ has an intercept of $\ln(8/\pi^2)$ and a slope of $D\pi^2/4L^2$, thus allowing diffusivity D to be calculated if the slope is determined. The hydraulic conductivity corresponding to the average suction during the increment may be calculated using eq. (11.7) in the form

$$k_{\psi_{avg}} = D \frac{\Delta\theta}{\Delta\psi} \qquad (11.13)$$

where $\Delta\theta$ is the change in water content measured for the applied increment in suction, $\Delta\psi$, and ψ_{avg} is equal to $\psi_0 + \Delta\psi/2$, where ψ_0 is the matric suction before the application of $\Delta\psi$. Figure 11.13 shows an example of a hydraulic conductivity function for a silty sand specimen obtained by Gardner (1956) using the multistep outflow method.

Outflow techniques are generally limited to relatively coarse-grained soil (e.g., sand and silt) where drainage occurs relatively rapidly and the hydraulic conductivity function and soil-water characteristic curve are adequately described by a range of suction less than the air-entry value of commercially

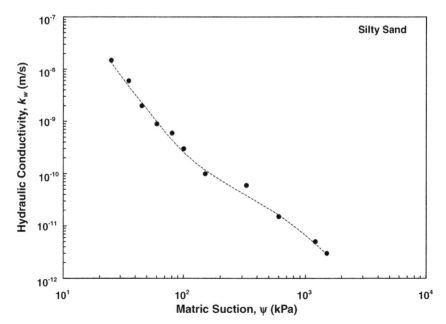

Figure 11.13 Unsaturated hydraulic conductivity function $k(\psi)$ measured using multistep outflow method for silty sand (data from Gardner, 1956).

available ceramics. Accurate measurement of the outflow volume over extended testing periods requires the use of a flushing system to remove air bubbles that tend to accumulate in the outflow system behind the high-air-entry disk (e.g., Kunze and Kirkham, 1962). In many cases, the impedance of the high-air-entry disk is indeed significant compared with the impedance of the soil and must be taken into account (e.g., Miller and Elrick, 1958; Rijtema, 1959; Kunze and Kirkham, 1962). Advantages of outflow methods are that they are relatively rapid and the hydraulic conductivity function and the soil-water characteristic curve may be simultaneously obtained. Both the hydraulic conductivity function and soil-water characteristic curve are measured along drying branches of the functions.

11.3.4 Instantaneous Profile Methods

The instantaneous profile method (IPM) is a transient testing technique applicable for either laboratory or field determination of the hydraulic conductivity function. The name of the technique refers to the fact that profiles of water content and suction at several points along a "column" of soil are obtained at fixed snapshots in time during a transient flow process. For laboratory measurements, this column is a disturbed or undisturbed specimen considered to be representative of the deposit under consideration. For field measurements, the column is defined vertically from the ground surface and fluid flow is assumed to be one dimensional. The volume of water that flows from one point to another over some time is estimated by measuring time-dependent changes in the water content profile. Similarly, the hydraulic gradient responsible for the flow process is estimated by measuring the time-dependent changes in the suction profile. If only one of these profiles is directly measured, the other may be inferred from the soil-water characteristic curve for the column or deposit under consideration.

Fluid flow is allowed to occur under controlled or known boundary conditions at either or both ends of the soil column. This can be either a wetting process, where water flows into the column, or a drying process, where water flows out of the column. Darcy's law is assumed valid to calculate hydraulic conductivity directly from measurements of the fluid flux and hydraulic gradient profiles. Measurements at several locations along the soil profile and at different times during the transient flow process provide multiple and redundant data points comprising the hydraulic conductivity function.

Numerous approaches for controlling the boundary conditions across the soil column have been developed, differing primarily in the technique adopted to add or remove water (Klute, 1972). Examples for tests in the laboratory include: (1) volume-controlled injection of water at one end of a vertically or horizontally oriented soil column, typically using a flow pump or controlled-drip system, (2) volume-controlled withdrawal of water from one end of the soil column, (3) withdrawal of water at a controlled-suction boundary, (4) gravity drainage from an initially saturated condition, and (5) evaporation from an initially saturated condition. The majority of approaches for applications in the field rely on gravity drainage from a ponded water supply at the ground surface.

Laboratory Instantaneous Profile Method The laboratory instantaneous profile method was initially described by Richards and Weeks (1953). Variations on the technique have been developed by Watson (1966), Hamilton et al. (1981), Daniel (1983), Meerdink et al. (1996), and Chiu and Shackelford (1998), among others. The laboratory IPM method can be conducted on either remolded or undisturbed samples. Specimens are typically confined in a rigid-walled column ranging in length from about 10 to 30 cm and oriented either horizontally or vertically. For horizontally oriented columns, gravity-driven fluid flow can generally be assumed negligible in the analysis.

Figure 11.14 shows a typical laboratory IPM setup for a horizontally oriented soil column. Boundary control ports are located at either end of the specimen ($x = 0$, $x = L$) for injection or withdrawal of water. Measurement ports for suction and/or water content instrumentation are located along the column, typically spaced at a distance equal to about 10% of the overall column length. Although only two ports are required to calculate the hydraulic conductivity function, additional measurements are often desired to provide redundancy and smoothness. Instrumentation used for suction measurement most commonly includes tensiometers, thermocouple psychrometers, or a combination of both (e.g., Daniel, 1983; Meerdink et al., 1996). Instrumentation for direct measurement of the water content profile might include a series of time-domain-reflectrometry (TDR) probes, external gamma-ray attenuation techniques, or resistive measurement systems.

To demonstrate the general laboratory testing approach, consider the following analysis for the horizontal column shown as Fig. 11.14. The spatial variable x is defined along the axis of the column from $x = 0$ at the left end to $x = L$ at the right end. Assume that tensiometers and TDR probes are

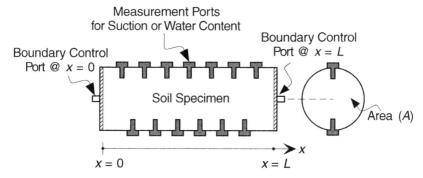

Figure 11.14 Laboratory soil column for measuring hydraulic conductivity function using instantaneous profile method.

inserted through various ports along the specimen for measurement of suction and volumetric water content, respectively.

Initially ($t = t_0$), assume that the soil is uniformly air dry with volumetric water content θ_0. Because the initial water content is relatively low, the corresponding initial suction is relatively high, designated in terms of suction head h_0. Water is then slowly and steadily injected into the column using a flow pump at the left boundary control port located at $x = 0$. A stack of filter papers is used to distribute the influent water over the entire cross-sectional area of the column. The right boundary condition at $x = L$ is open to the atmosphere. The injection of water at the left boundary causes transient changes in the water content and suction profiles along the column. Figures 11.15a and 11.15b illustrate conceptual profiles of suction and water content at t_0 and at snapshots in time t_1, t_2, t_3, and t_4. The suction head gradient i at a point in the column x_i and time t_i is equal to the slope of the suction head profile at that point and time, which may be written as

$$i(x_i, t_i) = \left. \frac{dh}{dx} \right|_{x_i, t_i} \tag{11.14}$$

If the test is terminated before water flows from the right end of the specimen, the total volume of water that passes through any cross-sectional area over a given increment in time is equal to the change in the volume of water occurring between the point under consideration and the right end of the specimen, or

$$\Delta V_w = A \int_{x_i}^{L} \theta_{t=j}(x) \, dx - A \int_{x_i}^{L} \theta_{t=m}(x) \, dx \tag{11.15}$$

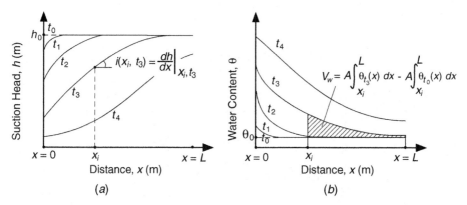

Figure 11.15 Conceptual profiles of (a) suction head and (b) water content measured during laboratory instantaneous profile test.

where ΔV_w is the volume of water that has flowed past point x_i over the time interval Δt from $t = j$ to $t = m$, and A is the cross-sectional area of the specimen. For example, the shaded area shown in Fig. 11.15b multiplied by the cross-sectional area A represents the volume of water that has flowed past point x_i from t_0 to t_3. Similar areas may be evaluated for any location or increment in time. The apparent flow velocity v is equal to the change in the volume of water ΔV_w divided by the area A and the time interval Δt:

$$v = \frac{\Delta V_w}{A \, \Delta t} \tag{11.16}$$

Hydraulic conductivity is calculated using Darcy's law by dividing the flow velocity at a given point by the average hydraulic gradient i at that point over the time interval for which the flow rate was determined:

$$k = -\frac{v}{i} \tag{11.17}$$

Additional calculations may be repeated by determining the flow rate at several points and for different time intervals to develop the hydraulic conductivity function corresponding to a wide range of water content or suction.

Figures 11.16a and 11.16b show soil-water characteristic curves, $\theta(h_m)$, and hydraulic conductivity functions, $k(h_m)$, respectively, determined using the laboratory instantaneous profile method for three mixtures of sand and kaolin clay. Grain size distribution, compaction characteristics, and a series of key water content values for each mixture are summarized in Table 11.3. The soil-water characteristic curves shown in Fig. 11.13a were determined using a special test cell that incorporated measurement ports for tensiometer or psychrometer probes. The normalized volumetric water content is in this case defined as $\Theta = (\theta - \theta_r)/(\theta_m - \theta_r)$, where θ_r is the residual water content of the soil mixture and θ_m is a "maximum" water content similar to the saturated water content.

Figure 11.17 shows hydraulic conductivity as a function of suction head, $k(h_m)$ (Fig. 11.17a), and volumetric water content, $k(\theta)$ (Fig. 11.17b), for Wenatchee silty clay determined using the instantaneous profile method for sorption and desorption tests. Note the distinct difference between the sorption and desorption branches in the k-h_m relationship resulting from hysteretic effects. The differences between the sorption and desorption branches in the k-θ relationship are relatively minor.

Field Instantaneous Profile Method The field instantaneous profile method is identical in principle to the laboratory method described above. Here, transient profiles of water content and/or suction are measured as water

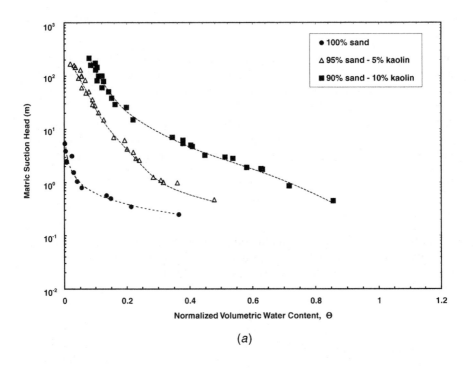

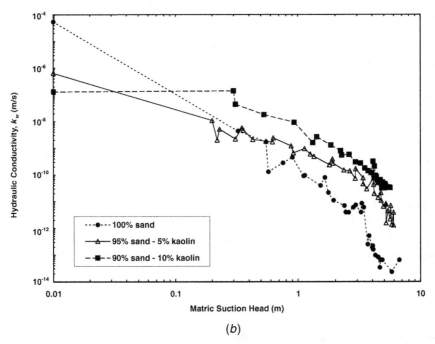

Figure 11.16 Measured properties for sand and sand-kaolin mixtures: (*a*) soil-water characteristic curves based on wetting process and (*b*) unsaturated hydraulic conductivity functions using instantaneous profile method (data from Chiu and Shackelford, 1998).

TABLE 11.3 Physical Properties of Sand and Sand-Kaolin Mixtures

Property	Sand-Kaolin Mixture		
	1	2	3
Sand content (%)	90	95	100
Kaolin content (%)	10	5	0
Sand (0.074–4.75 mm) (%)	90	95	100
Silt (0.002–0.074 mm) (%)	6	3	0
Clay (<0.002 mm) (%)	4	2	0
Classification (USCS)	SP-SC	SP	SP
Maximum dry unit weight, $\gamma_{d,\max}$ (kN/m^3)	17.2	16.5	—
Optimum gravimetric water content, w_{opt} (%)	14.0	13.5	—
Optimum volumetric water content, θ_{opt} (%)	0.246	0.227	—
Saturation at θ_{opt}, S_{opt} (%)	73.7	61.9	—
Residual volumetric water content, θ_r	0.0609	0.0454	0.0284
Saturated volumetric water content, θ_s	0.334	0.367	0.435
Maximum volumetric water content, θ_m	0.282	0.321	0.387
Saturation at θ_m, S_m (%)	84.4	87.5	89.0
Steady-state hydraulic conductivity, k (m/s)	1.3×10^{-7}	7.0×10^{-7}	5.0×10^{-5}

Source: From Chiu and Shackelford (1998).

is added or removed from a vertical "column" of in situ soil. Watson (1966) and Hillel et al. (1972) describe a common procedure whereby the soil column is initially wetted to near saturation and then allowed to drain internally under the gravity gradient. Infiltration and evaporation at the ground surface are prevented to maintain control over the top boundary condition during testing. Meerdink et al. (1996) describe a long-term desorption testing procedure applicable to fine-grained soil that involves evaporation at the soil surface and measurement of the transient moisture content profile. Additional variations on the technique are described by Nielson et al. (1964).

Prior to testing, the soil profile is instrumented with suction and/or moisture content instrumentation. The testing scenario depicted in Fig. 11.18, for example, shows tensiometer probes inserted to various depths for measuring the suction profile and a central access tube for measuring the water content profile by neutron logging (e.g., Gardner, 1986). Ideally, the instrumentation should be concentrated near the center of the soil column to reduce radial boundary effects during testing. If only the suction or water content profile is directly measured, the other may be indirectly estimated if a representative soil-water characteristic curve for the deposit is known. The soil profile is initially saturated by ponding water on the ground surface using a berm structure or infiltration ring. After the deposit is wetted to saturation and steady-state flow has been reached, infiltration is ceased and the deposit is allowed to drain internally under the gravity gradient.

Figure 11.19a conceptualizes the associated system response at three depths (z_1, z_2, and z_3) from the ground surface in terms of volumetric water

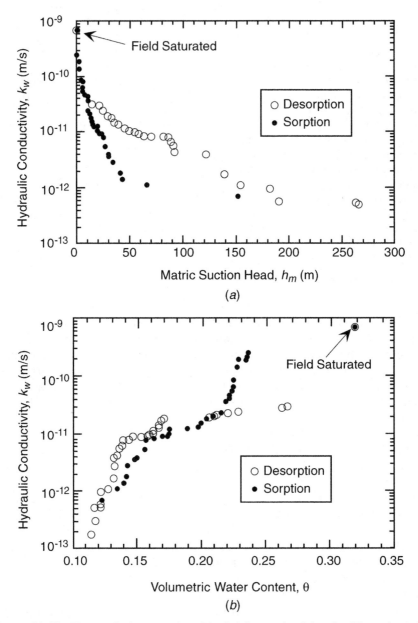

Figure 11.17 Hysteresis in unsaturated hydraulic conductivity for Wenatchee silty clay prepared at field water content and dry unit weight: (a) $k(h_m)$ and (b) $k(\theta)$ (from Meerdink et al., 1996).

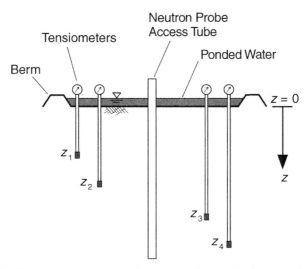

Figure 11.18 Example arrangement for hydraulic conductivity testing using field instantaneous profile method (after Benson and Gribb, 1997).

content as a function of time since drainage began. Similarly, Fig. 11.19*b* shows plots of suction head versus time for each depth. Prior to drainage at $t = 0$, the soil at each depth is saturated and at zero suction. As drainage proceeds, the soil at greater depth maintains relatively high water content for a longer amount of time.

If the soil profile is subdivided into discrete horizontal layers defined by the points were suction is measured, the flux of water through the bottom of any layer at depth z_i and time t_i can be calculated by determining the slope

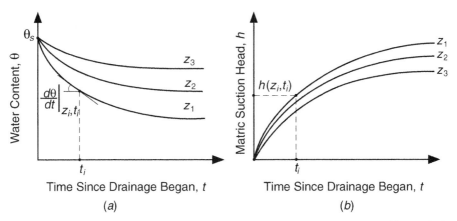

Figure 11.19 Conceptual illustrations of (*a*) transient water content profiles and (*b*) suction profiles measured during field instantaneous profile test.

of the relevant water content versus time curve at the time and depth of interest as

$$q(z_i,t_i) = -dz\left(\frac{d\theta}{dt}\right)_{z_i,t_i} \quad (11.18)$$

where dz is the thickness of the soil layer under consideration. This slope is illustrated on Fig. 11.19a for arbitrary z_i and t_i. The cumulative flux through each layer is obtained by summing the fluxes through all of the overlying layers.

The total hydraulic head h_t at any point in the profile during drainage is equal to the measured matric suction head h_m plus the elevation head z at that point, or

$$h_t = h_m + z \quad (11.19)$$

Thus, the total head profile corresponding to the instantaneous increments in time for which the cumulative fluxes through each layer were calculated may be determined by adding the depth of each suction measurement to the suction value measured at those specific times. This is shown in Fig. 11.20.

The slope of the hydraulic head versus depth curves at any depth z_i and time t_i is the hydraulic gradient acting i at that depth and time:

$$i(z_i,t_i) = \frac{dh_t}{dz}\bigg|_{z_i,t_i} \quad (11.20)$$

Hydraulic conductivity may in turn be calculated using Darcy's law from the calculated flux [eq. (11.18)] and corresponding hydraulic gradient. A se-

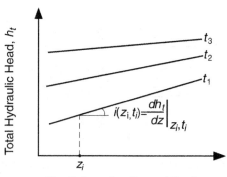

Figure 11.20 Conceptual illustration of variation in total hydraulic head with depth z and time t during gravity drainage for field instantaneous profile method.

ries of hydraulic conductivity values corresponding to various suction heads may be determined by dividing the cumulative flux determined at several points and times by the corresponding gradient.

PROBLEMS

11.1. Summarize the advantages and disadvantages of the various laboratory methods described in this chapter for determining the hydraulic conductivity of unsaturated soil.

11.2. A laboratory constant-flow permeability test was conducted on an unsaturated sandy soil. The following measurements were obtained at steady state: $A = 75$ cm^2, $h_{wt} = 5$ cm, $h_{wb} = 3$ cm, $\Delta L = 10$ cm, $h_a = 76$ cm H$_2$O, and $Q = 4.05 \times 10^{-4}$ cm^3/s, where A is the specimen cross-sectional area, h_{wt} is the pore water head measured at the top of the specimen, h_{wb} is the pore water head measured at the bottom of the specimen, ΔL is the distance between the two head measurements, h_a is the applied air pressure head, and Q is the applied constant flow rate. Calculate the hydraulic conductivity of the sand at the corresponding matric suction.

11.3. Estimate the air-entry pressure (kPa) for the soil from Example Problem 11.1.

11.4. If the hydraulic conductivity test following extraction 2 for the soil from Example Problem 11.2 was conducted using a constant flow rate of 0.005 cm^3/s, estimate the expected head loss.

11.5. Repeat Example Problem 11.3 for an analysis time of $t = 2000$ min using the same data shown in Table 11.2. How do these results compare with the previous analysis where $t = 1500$ min?

11.6. Define hydraulic diffusivity and explain why it is a function of water content.

CHAPTER 12

SUCTION AND HYDRAULIC CONDUCTIVITY MODELS

12.1 SOIL-WATER CHARACTERISTIC CURVE MODELS

Experimental techniques for direct measurement of the soil-water characteristic curve (SWCC) provide a series of discrete data points comprising the relationship between soil suction and water content. Subsequent application of these measurements for predicting flow, stress, and deformation phenomena, however, typically requires that measured characteristic curves are described in continuous mathematical form. Direct measurements also remain a relatively demanding, and often expensive, endeavor. Due to the costs and complexities associated with sampling, transporting, and preparing laboratory specimens or installing, maintaining, and monitoring field instrumentation, the number of measurements obtained for a given site is often too small to adequately capture the spatial variability of soil properties and stress conditions in the field. Available measurements often comprise only a small portion of the soil-water characteristic curve over the wetness range of interest in practical applications. For all these reasons, alternatives to direct measurements are desirable.

Numerous approaches have been proposed for mathematical representation (i.e., fitting) or prediction of the soil-water characteristic curve. This section describes three models commonly adopted for geotechnical engineering applications, specifically: the Brooks and Corey (1964) model, the van Genuchten (1980) model, and the Fredlund and Xing (1994) model. The attributes and limitations of each model are demonstrated through a series of graphical plots and comparisons with experimental data. Detailed reviews and analyses of these and several other models are also provided by Leong and Rahardjo (1997a), Singh (1997), and Sillers et al. (2001). The variety of pedotransfer functions (PTF) and "knowledge-based" systems that have been also developed to indirectly predict soil-water characteristics from measurements or databases of more readily available or easily measured material properties (e.g., grain size distribution, dry bulk density, and porosity) are not described here (e.g., Ahuja et al., 1985; Rawls and Brakensiek, 1985; Bouma and Van Lanen, 1987). Parameter identification methods, which have been developed to estimate modeling parameters from inverse analysis of unsaturated flow systems under known boundary conditions, are summarized by Zachmann et al. (1982) and Durner et al. (1999).

12.1.1 SWCC Modeling Parameters

Parameters used in mathematical models for the soil-water characteristic curve include fixed points pertaining to water content or suction at specific conditions (e.g., saturation, residual saturation, and air-entry pressure) and two or more empirical or semiempirical fitting constants that are selected to capture the general shape of the curve between these fixed points. As illustrated in Fig. 12.1, the saturated water content θ_s describes the point where all of the available pore space in the soil matrix is filled with water, usually corresponding to the desorption branch of the curve. The air-entry, or "bubbling," pressure ψ_b describes the suction on the desorption branch where air first starts to enter the soil's largest pores and desaturation commences. The residual water content θ_r describes the condition where the pore water resides pri-

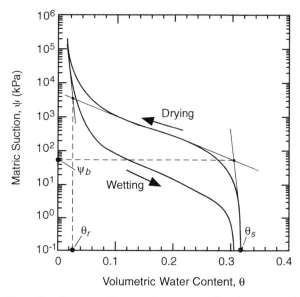

Figure 12.1 Typical soil-water characteristic curve showing approximate locations of residual water content θ_r, saturated water content θ_s, and air entry pressure ψ_b.

marily as isolated pendular menisci and extremely large changes in suction are required to remove additional water from the system. A consistent way to quantify the air-entry pressure and residual water content is to construct pairs of tangent lines from inflection points on the characteristic curve.

For modeling purposes, a dimensionless water content variable, Θ, may be defined by normalizing volumetric water content with its saturated and residual values as

$$\Theta = \frac{\theta - \theta_r}{\theta_s - \theta_r} \tag{12.1}$$

Note that as volumetric water content θ approaches θ_r, the normalized water content Θ approaches zero. As volumetric water content θ approaches θ_s, the normalized water content Θ approaches unity. If the residual water content θ_r is equal to zero, then the normalized water content Θ is equal to the degree of saturation S.

An "effective" degree of saturation S_e may also be normalized by the fully saturated condition ($S = 1$) and the residual saturation condition S_r in a similar manner:

$$S_e = \frac{S - S_r}{1 - S_r} \tag{12.2}$$

where

$$\Theta = S_e \tag{12.3}$$

If the residual saturation S_r is equal to zero, then the effective degree of saturation S_e is equal to the degree of saturation S.

Fitting constants used in the various SWCC models are often related to physical characteristics of the soil such as pore size distribution and air-entry pressure. Models may be differentiated in terms of the number of fitting constants used, most commonly being either two or three. Models incorporating three fitting constants tend to sacrifice simplicity in their mathematical form, but generally offer a greater amount of flexibility in their capability to accurately represent characteristic curves over a realistically wide range of suction. Some form of iterative, nonlinear regression algorithm is typically used to optimize the fitting constants to measured data comprising the characteristic curve (e.g., van Genuchten et al., 1991; Wraith and Or, 1998). Many of the two-constant models may be effectively optimized by visual observation. At least 5 to 10 measured ψ-θ pairs are typically required for a meaningful mathematical representation. The accuracy of the models may be checked by calculating the root-mean-square deviation between the measured and modeled values.

12.1.2 Brooks and Corey (BC) Model

One of the earliest approaches for modeling the soil-water characteristic curve is an equation proposed by Brooks and Corey (1964). Based on observations from a large suite of experimental suction and water content measurements, Brooks and Corey proposed a two-part power law relationship incorporating a "pore size distribution index," λ. The model is nonsmooth or open form about the air-entry pressure, ψ_b, and is written as

$$\Theta = S_e = \begin{cases} 1 & \psi < \psi_b \\ \left(\dfrac{\psi_b}{\psi}\right)^\lambda & \psi \geq \psi_b \end{cases} \qquad (12.4)$$

which, given eq. (12.1), may also be written in the form

$$\theta = \begin{cases} \theta_s & \psi < \psi_b \\ \theta_r + (\theta_s - \theta_r)\left(\dfrac{\psi_b}{\psi}\right)^\lambda & \psi \geq \psi_b \end{cases} \qquad (12.5)$$

or in terms of suction head h and air-entry head h_b:

$$\Theta = S_e = \begin{cases} 1 & h < h_b \\ \left(\dfrac{h_b}{h}\right)^\lambda & h \geq h_b \end{cases} \qquad (12.6)$$

Figure 12.2 shows a series of soil-water characteristic curves modeled using the BC equation to illustrate the relative effects of changes in λ and ψ_b on the behavior of the model. Figure 12.2a shows the effects of changing ψ_b for a constant λ. Figure 12.2b shows the effects of changing λ for a constant ψ_b. In each case, residual saturation S_r is assumed to be equal to zero such that $\Theta = S_e = S$. Note from Fig. 12.2b that relatively large values of λ correspond to characteristic curves where drainage is relatively "rapid," that is the majority of the pores are drained over a relatively narrow range of suction and the SWCC is relatively flat. Physically, large values of λ correspond to soils having a relatively uniform pore size distribution (e.g., poorly graded sand).

Figure 12.3 shows a series of suction-water content measurements obtained using a Tempe cell apparatus and corresponding BC models for three soils, ranging from silty sand to poorly graded sand. The relatively small λ value for the silty sand ($\lambda = 0.15$) reflects its finer grained, less uniform, and relatively dense (porosity, $n = 0.33$) texture.

Overall, the BC model is most appropriate for relatively coarse-grained soils where drainage occurs over a relatively low and relatively narrow range

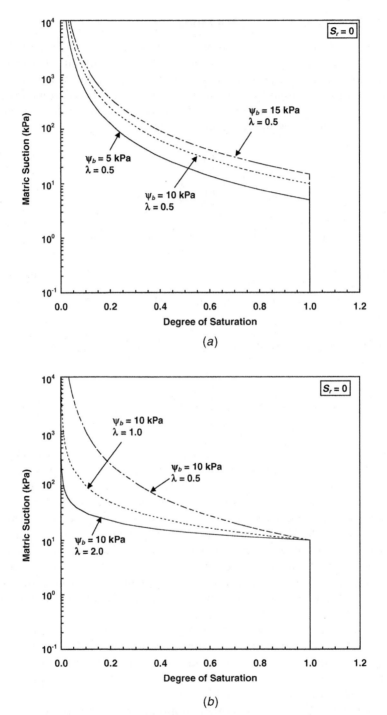

Figure 12.2 Soil-water characteristic curves $\psi(S)$ modeled using the Brooks and Corey (1964) model showing: (a) effect of changing parameter ψ_b for constant λ and (b) effect of changing parameter λ for constant ψ_b.

12.1 SOIL-WATER CHARACTERISTIC CURVE MODELS

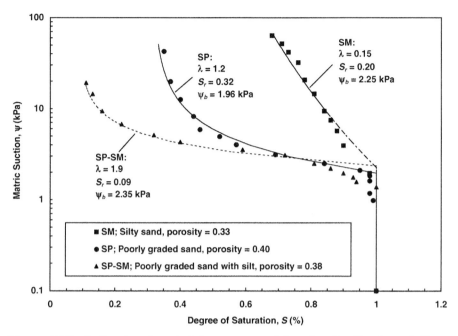

Figure 12.3 Soil-water characteristic curves models using the Brooks and Corey (1964) model (experimental data from Clayton, 1996).

of suction. The model tends to lose applicability at high suctions approaching the residual water content. The absence of an inflection point in the form of the model often results in poor representation of the SWCC over a wide suction range. The nonsmoothness occurring at the air-entry pressure leads to a sharp discontinuity in the specific moisture capacity and hydraulic diffusivity functions [based on the derivative of $\theta(\psi)$], which can often lead to numerical instability when modeling fluid flow behavior near saturation.

Example Problem 12.1 Tempe cell tests were conducted to determine the SWCC of a sandy soil. Results are shown in Table 12.1. Model the SWCC using the Brooks and Corey (1964) equation.

Solution Figure 12.4 shows the Tempe cell data and a best fit to the SWCC using the Brooks and Corey model. The fitting parameters may be optimized by visual observation using a spreadsheet program. The following parameters provide the best fit: $\lambda = 1.0$, $\psi_b = 1.57$ kPa, and $S_r = 0.29$.

12.1.3 van Genuchten (VG) Model

van Genuchten (1980) proposed a smooth, closed-form, three-parameter model for the soil-water characteristic curve in the form

TABLE 12.1 Soil-Water Characteristic Curve Data Obtained from a Tempe Cell Test for Sandy Soil from Example Problem 12.1

$u_a - u_w$ (kPa)	S
0.1	1.00
1.18	0.99
1.86	0.95
2.75	0.77
3.92	0.57
5.00	0.51
7.35	0.45
11.18	0.40
15.78	0.37
35.57	0.32

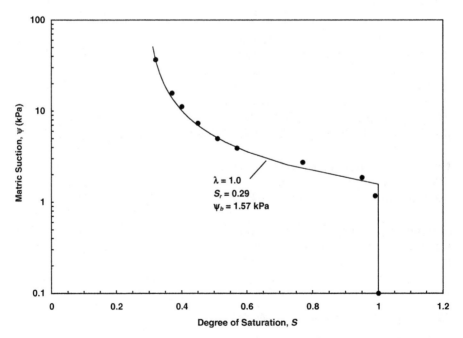

Figure 12.4 Soil-water characteristic curve for sandy soil from Example Problem 12.1 modeled using Brooks and Corey (1964) model.

12.1 SOIL-WATER CHARACTERISTIC CURVE MODELS

$$\Theta = S_e = \left[\frac{1}{1 + (a\psi)^n}\right]^m \tag{12.7}$$

where a, n, and m are fitting parameters. The mathematical form of the VG model, which accounts for an inflection point, allows greater flexibility than the BC model over a wider range of suction and better captures the sigmoidal shape of typical curves. Smooth transitions at the air-entry pressure and for suction approaching the residual condition are more effectively captured.

The suction term appearing on the right-hand side of eq. (12.7) may be expressed in either units of pressure (i.e., ψ = kPa, as shown) or head (i.e., h = m). In the former case, the a parameter is designated more specifically as α, where α has inverse units of pressure (kPa^{-1}). In the latter case, the a parameter is designated β, where β has inverse units of head (m^{-1}). Both α and β are related to the air-entry condition, where α approximates the inverse of the air-entry pressure, and β approximates the inverse of the air-entry head or the height of the capillary fringe. The n parameter is related to the pore size distribution of the soil and the m parameter is related to the overall symmetry of the characteristic curve. The m parameter is frequently constrained by direct relation to the n parameter as

$$m = 1 - \frac{1}{n} \tag{12.8a}$$

or

$$m = 1 - \frac{1}{2n} \tag{12.8b}$$

Both of the above constraints on the m parameter reduce the flexibility of the VG model but significantly simplify it, thus resulting in greater stability during parameter optimization and permitting closed-form solution of the hydraulic conductivity function (van Genuchten et al., 1991).

Figure 12.5a illustrates the effect of changing the pore size distribution parameter n for constant α. Figure 12.5b illustrates the effect of changing α for constant n. In each case, the residual saturation is assumed equal to zero and the m-parameter simplification [eq. (12.8a)] is applied. Much like the BC λ parameter, soils with a "flatter" characteristic curve are most effectively captured by relatively large values of n. Soils with relatively high air-entry pressure are characterized by smaller values of α.

Figures 12.6a and 12.6b show Tempe cell data for three sandy soils modeled using the VG equation. Fitting parameters for the models were optimized by least-squared regression using the RETC (RETention Curve) code described by van Genuchten et al. (1991). The curves in Fig. 12.6a were fit using the constraint on the m parameter [eq. (12.8a)], which clearly places a

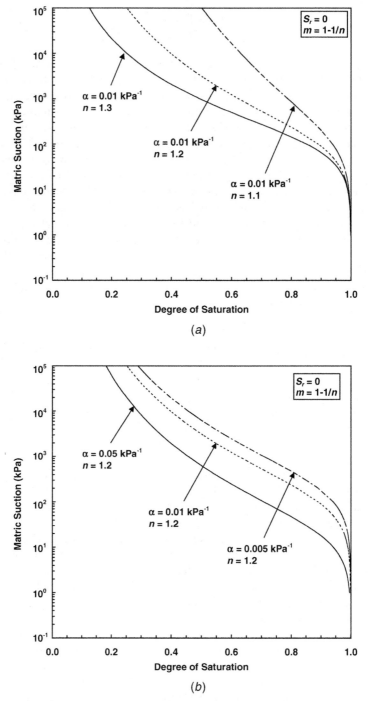

Figure 12.5 Soil-water characteristic curves modeled using van Genuchten equation showing effects of changes in (a) the n parameter and (b) the α parameter. The $m = 1 - 1/n$ constraint is applied in both cases.

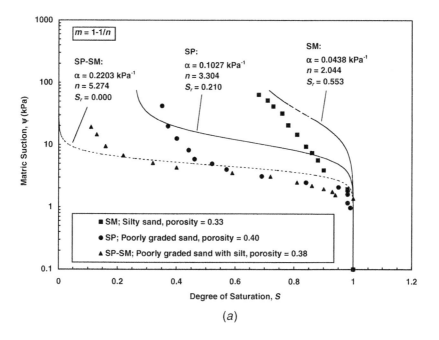

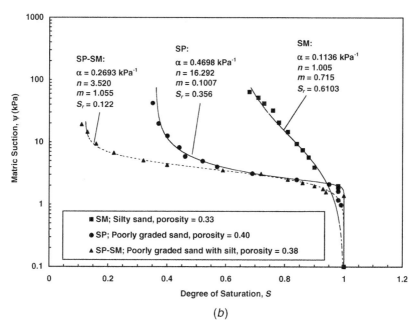

Figure 12.6 Experimental soil-water characteristic curves models using van Genuchten (1980) equation: (*a*) models constrained by $m = 1 - 1/n$ simplification and (*b*) models where m, n, and α are treated as independent parameters (data from Clayton, 1996).

limitation on the flexibility of the model and the accuracy of the best fit. For curves fit by treating each parameter independently (Fig 12.6b), on the other hand, the VG model provides an excellent fit to the experimental data over the entire range.

Tinjum et al. (1997) recognized the similarities between the VG and BC fitting parameters and presented empirical relationships between λ and n and α and ψ_b for a series of compacted clay specimens. Inverse correspondence was found between α (kPa^{-1}) and ψ_b (kPa) for ψ_b ranging from about 1 to 100 kPa as

$$\alpha = \left(\frac{0.78}{\psi_b}\right)^{1.26} \tag{12.9}$$

Others have presented specific methodologies for converting between BC parameters and equivalent VG parameters by investigating the notion of equivalence between the two models. For example, Lenhard et al. (1989) equated the BC and VG expressions at their midpoint ($S = 0.5$) and suggested the following relationship between λ and m:

$$\lambda = \frac{m}{1-m}(1 - 0.5^{1/m}) \tag{12.10}$$

and, for converting between h_b (cm) and β (cm^{-1}), it was suggested that

$$h_b = \frac{S_x^{1/\lambda}}{\beta}(S_x^{-1/m} - 1)^{1-m} \tag{12.11}$$

where it was found empirically that

$$S_x = 0.72 - 0.35 \exp(-n^4) \tag{12.12}$$

Ma et al. (1999) evaluated the influence of three proposed VG-BC conversion methods on the overall prediction of soil-water characteristic curves and water-balance models.

Example Problem 12.2 Model the SWCC from Example Problem 12.1 using the van Genuchten (1980) equation both with and without the $m = 1 - 1/n$ constraint.

Solution Figure 12.7 shows soil-water characteristic curves obtained by least-squared regression using the RETC code. While both curves are fair representations of the data, treating m and n as independent parameters allows the data to be more closely fit over the entire experimental range.

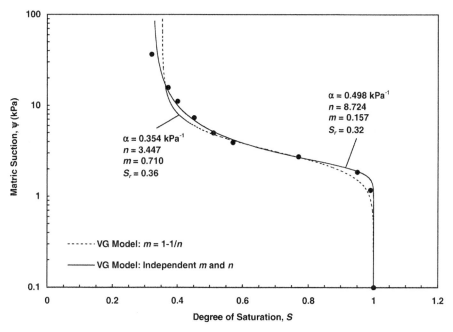

Figure 12.7 Soil-water characteristic curves for sandy soil from Example Problem 12.1 modeled using the van Genuchten (1980) equation.

12.1.4 Fredlund and Xing (FX) Model

Fredlund and Xing (1994) developed a model based on consideration of pore size distribution in a form similar to the VG model as

$$\theta = C(\psi)\theta_s \left[\frac{1}{\ln[e + (\psi/a)^n]} \right]^m \quad (12.13a)$$

where ψ is suction (kPa), a, n, and m are fitting parameters, e is the natural logarithmic constant, and $C(\psi)$ is a correction factor that forces the model through a prescribed suction value of 10^6 kPa at zero water content:

$$C(\psi) = \left[1 - \frac{\ln(1 + \psi/\psi_r)}{\ln(1 + 10^6/\psi_r)} \right] \quad (12.13b)$$

where ψ_r is the suction (kPa) estimated at the residual condition. If the residual water content θ_r is assumed to be zero, eq. (12.13a) can be written in terms of normalized water content Θ or degree of saturation S_e by dividing both sides of the equation by the saturated volumetric water content θ_s. Fredlund and Xing (1994) describe a graphical technique by which the three fitting parameters (a, n, and m) may be estimated from inflection points located on the measured characteristic curve. Comparison with experimental data indicates that the FX model is capable of describing well the characteristic curves over the range of suction from 0 kPa all the way to 10^6 kPa.

Figure 12.8 shows a series of soil-water characteristic curves modeled using the FX equation to illustrate the effects of changes in the a (Fig. 12.8a), n (Fig. 12.8b), and m (Fig. 12.8c) parameters. The a parameter is related to, but generally larger than, the air-entry pressure. For small values of m, the air-entry value can be used as a. The n and m parameters are related to the pore size distribution and overall symmetry of the characteristic curve, respectively. Large n values produce a sharp corner near the air-entry value. The more uniform the pore size distribution, the larger the value of n. The m parameter controls the slope of the characteristic curve in the relatively high suction range, where relatively small m values result in a steeper slope at high suctions.

Example Problem 12.3 Tables 12.2 and 12.3 show soil-water characteristic curve data for a silty loam and glacial till, respectively. Model and compare the SWCC for each soil using the Brooks and Corey (1964), van Genuchten (1980), and Fredlund and Xing (1994) equations.

Solution Figures 12.9a and 12.9b show the experimental soil-water characteristic curves and corresponding BC, VG, and FX models for the silty loam and glacial till, respectively. Table 12.4 summarizes the fitting parameters selected for each. Note that the BC model is incapable of capturing the entire SWCC for the glacial till. The VG and FX models capture the curve reasonably well over the entire range of suction.

12.2 HYDRAULIC CONDUCTIVITY MODELS

A variety of mathematical models have been developed to model the unsaturated hydraulic conductivity function from limited experimental data sets or to predict the hydraulic conductivity function from more routinely obtained constitutive functions, most notably the soil-water characteristic curve. Detailed summaries of several hydraulic conductivity models and modeling techniques include those provided by Mualem (1978), Fredlund et al., (1994), and Leong and Rahardjo (1997b).

Mualem (1986) classifies three types of approaches to modeling the hydraulic conductivity function: empirical models, macroscopic models, and statistical models. Empirical models are typically simple equations that incorporate saturated hydraulic conductivity and one or more fitting parameters optimized to capture the general shape of a given set of data. A sufficient set of experimental measurements is required to optimize the fitting parameters. Macroscopic models assume similarity between laminar fluid flow on a mi-

12.2 HYDRAULIC CONDUCTIVITY MODELS 507

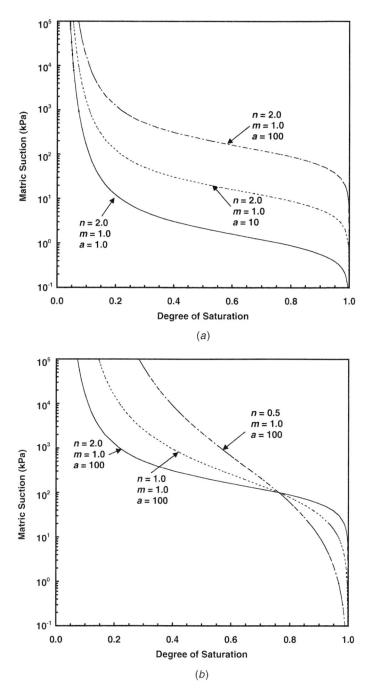

Figure 12.8 Soil-water characteristic curves modeled using the Fredlund and Xing (1994) model showing Effects of changes in (a) the a parameter, (b) the n parameter, and (c) the m parameter.

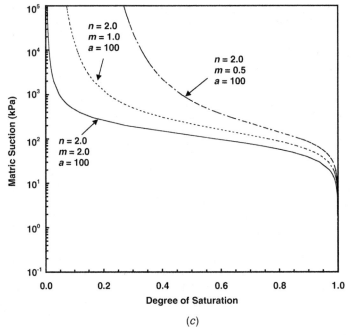

Figure 12.8 (*Continued*).

TABLE 12.2 Soil-Water Characteristic Curve Data for Silty Loam from Example Problem 12.3

$u_a - u_w$ (kPa)	Volumetric Water Content, θ
8	0.463
18	0.463
35	0.458
50	0.449
60	0.435
70	0.384
80	0.329
95	0.278
105	0.255
120	0.231
132	0.208
150	0.190
170	0.181
200	0.162
238	0.157
273	0.139

Source: Brooks and Corey (1964).

TABLE 12.3 Soil-Water Characteristic Curve Data for Glacial Till from Example Problem 12.3

$u_a - u_w$ (kPa)	Volumetric Water Content, θ
0.1	0.322
10	0.322
20	0.322
40	0.322
80	0.305
120	0.299
170	0.289
200	0.283
300	0.273
400	0.260
500	0.257
600	0.251
800	0.241
1,500	0.235
4,200	0.177
40,000	0.080
85,000	0.058
160,000	0.045
300,000	0.032

Source: Fredlund and Xing (1994).

croscopic level (e.g., within the pore throats) and macroscopic fluid flow through the soil. Macroscopic models share a common power law form ($k_r = S_e^\delta$), where a variety of values for the exponent δ have been determined either empirically or theoretically by considering different conceptualizations for the microscale geometry of the pore throats. Variations on the power law model have been developed to account for the important effect of pore size distribution on the hydraulic conductivity function. Statistical models are based on the presumption that the soil matrix can be represented as a network of interconnected capillary tubes of various sizes and that flow through the network occurs only through the liquid-filled tubes. Accordingly, the statistical distribution of tube sizes and their connectivity across a given plane in the soil mass become the controlling parameters for the overall hydraulic conductivity. Because the distribution of fluid-filled pores is dependent on suction, and may be specifically quantified given the SWCC and capillary theory, measurements or models for the characteristic curve become an indirect means to predict the hydraulic conductivity function.

12.2.1 Empirical and Macroscopic Models

Empirical models and macroscopic models for the hydraulic conductivity function are simple mathematical functions incorporating the saturated hy-

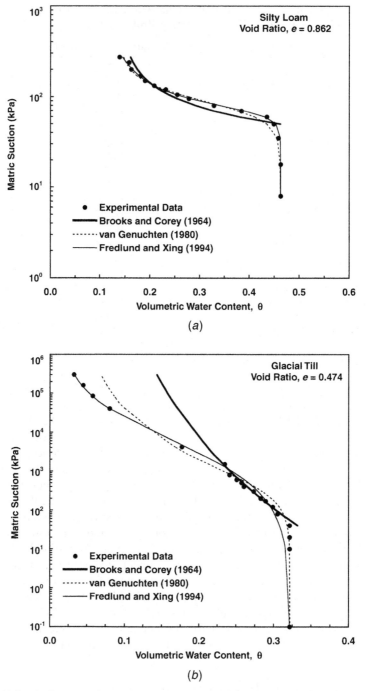

Figure 12.9 Soil-water characteristic curves modeled using the Brooks and Corey, van Genuchten, and Fredlund and Xing equations for (*a*) silty loam and (*b*) glacial till from Example Problem 12.3.

TABLE 12.4 Summary of Fitting Parameters Selected for Soil-Water Characteristic Curves Modeled from Example Problem 12.3

	Curve-Fitting Parameters		
Soil	Brooks and Corey (1964)	van Genuchten (1980)	Fredlund and Xing (1994)
Silty loam $e = 0.862$	$\lambda = 0.11$ $\psi_b = 55$ kPa $\theta_r = 0.030$ $\theta_s = 0.322$	$\alpha = 0.0028$ kPa^{-1} $n = 1.3$ $m = 0.231$ $\theta_r = 0.030$ $\theta_s = 0.322$	$a = 5700$ $n = 0.6$ $m = 2.4$ $\theta_s = 0.322$
Glacial till $e = 0.474$	$\lambda = 1.6$ $\psi_b = 50$ kPa $\theta_r = 0.139$ $\theta_s = 0.463$	$\alpha = 0.012$ kPa^{-1} $n = 4.1$ $m = 0.756$ $\theta_r = 0.139$ $\theta_s = 0.463$	$a = 67.32$ $n = 7.32$ $m = 0.5$ $\theta_s = 0.463$

draulic conductivity and various curve-fitting parameters. The values of the fitting parameters are related to the shape of the soil-water characteristic curve and must be optimized accordingly for various soil types and pore size properties. By assigning different curve-fitting parameters for wetting and drying processes, hysteresis in the conductivity function can be simulated. No single model or set of fitting parameters, however, is valid for all soil types. Numerous models have been proposed to represent hydraulic conductivity in a variety of functional forms [e.g., $k(\theta)$, $k(S)$, $k(\psi)$, or $k(h)$]. Several of these equations are summarized in Table 12.5.

One of the earliest models was proposed by Richards (1931) in the form of a simple linear equation involving two fitting parameters:

$$k(\psi) = a\psi + b \tag{12.14}$$

where the b and a parameters become the intercept and slope of the conductivity function in k-ψ space, respectively, and thus approximate the saturated hydraulic conductivity (i.e., $k_s = k$ at $\psi = 0$) and the subsequent decrease in conductivity with increasing suction.

Averjanov (1950) proposed a model in the form of a power function relating hydraulic conductivity to the saturated conductivity and effective degree of saturation S_e or normalized water content Θ in the form

$$k(S) = k_s S_e^n \tag{12.15a}$$

$$k(\theta) = k_s \Theta^n \tag{12.15b}$$

TABLE 12.5 Summary of Empirical and Macroscopic Equations for Modeling Unsaturated Hydraulic Conductivity Function

Form	Function	Reference
$k(\theta)$ or $k(S)$	$k(\theta) = k_s \Theta^n$	Averjanov (1950)
	$k(S) = k_s S_e^n$	
	$k(\theta) = k_s \exp[a(\theta - \theta_s)]$	Davidson et al. (1969)
	$k(\theta) = k_s \left(\dfrac{\theta}{\theta_s}\right)^n$	Campbell (1973)
$k(\psi)$ or $k(h)$	$k(\psi) = a\psi + b$	Richards (1931)
	$k(\psi) = a\psi^{-n}$	Wind (1955)
	$k(\psi) = \dfrac{k_s}{1 + a\psi^n}$	Gardner (1958)
	$k(\psi) = k_s \exp(-\alpha\psi)$	Gardner (1958)
	$k(h_m) = k_s \exp(\beta h_m)$	
	$k(\psi) = k_s$ for $\psi \leq \psi_b$	Brooks and Corey (1964)
	$k(\psi) = k_s \left(\dfrac{\psi_b}{\psi}\right)^\eta$ for $\psi > \psi_b$	
	$\eta = 2 + 3\lambda$	

where the fitting parameter n is typically equal to about 3.5 for most soils. Numerous derivations of the power function form of eq. (12.15) have been presented for different pore geometry assumptions with the fitting parameter n varying between approximately 2 and 4. In general, the simple form of eq. (12.15) performs relatively poorly for fine-grained soils.

Alternatively, Wind (1955) proposed a two-parameter power function as

$$k(\psi) = a\psi^{-n} \qquad (12.16)$$

Ahuja et al. (1988) suggested a similar but more flexible model in a piecewise form consisting of two equations identical to eq. (12.16) to represent hydraulic conductivity over two subranges of suction.

Gardner (1958) proposed a two-parameter model in the form of a power function and a constant as

$$k(\psi) = \frac{k_s}{1 + a\psi^n} \qquad (12.17)$$

where the a parameter is related to the air-entry pressure and thus controls the point on the conductivity function where the decrease in conductivity with increasing desaturation commences. The n parameter controls the subsequent slope of the decrease in conductivity with increasing suction. Figure 12.10a

12.2 HYDRAULIC CONDUCTIVITY MODELS 513

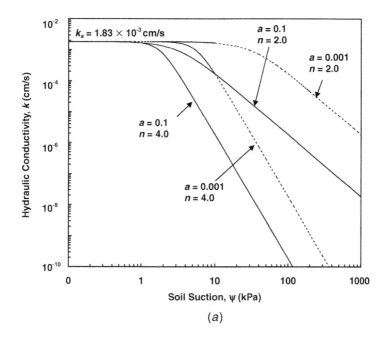

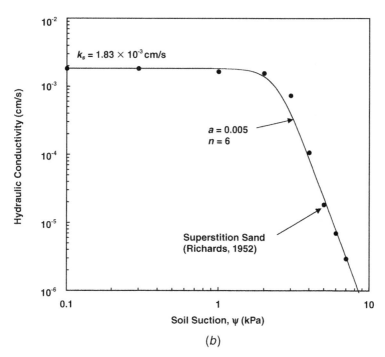

Figure 12.10 Examples of the Gardner (1958) two-parameter hydraulic conductivity model: (*a*) effects of varying fitting parameters *a* and *n* and (*b*) best fit to experimental data for Superstition sand (data from Richards, 1952).

illustrates these effects for a fixed value of saturated hydraulic conductivity. Figure 12.10b shows experimental data for Superstition sand (k_s = 1.83 × 10^{-3} cm/s) modeled using eq. (12.17). The model is generally applicable for relatively coarse-grained soil over a limited range of suction near the air-entry pressure.

Gardner (1958) also proposed a one-parameter exponential function in the form

$$k(\psi) = k_s \exp(-\alpha\psi) \tag{12.18}$$

which may be expressed as a function of matric suction head h_m as

$$k(h_m) = k_s \exp(\beta h_m) \tag{12.19}$$

Davidson et al. (1969) proposed a modified form to express hydraulic conductivity as a function of water content:

$$k(\theta) = k_s \exp[a(\theta - \theta_s)] \tag{12.20}$$

where a is a dimensionless empirical constant.

The α and β parameters in eq. (12.18) and eq. (12.19) are pore size distribution parameters with inverse units of suction pressure (α = kPa^{-1}) and suction head (β = m^{-1}), respectively. Both parameters capture the rate of reduction in hydraulic conductivity as suction increases. Relatively coarse-grained soil is typically modeled by relatively high α or β values, whereas fine-grained soil is more accurately modeled by relatively low values.

Figure 12.11 illustrates the applicability of eq. (12.18) with respect to Richards' (1952) experimental data for Superstition sand. Note that, unlike the case for Gardner's (1958) two-parameter model (Fig. 12.10b), no unique value of α is ideal for fitting the experimental data. An important advantage of the one-parameter exponential model, however, is that its form allows the governing equation for unsaturated fluid flow to be linearized, thus making analytical and quasi-analytical solutions possible. Incorporating either eq. (12.18) or (12.19) into the Richards' equation forms a classical approach to analytical solution of numerous unsaturated fluid flow problems (e.g., Philip, 1987; Pullan, 1990).

Based on their previous work, Brooks and Corey (1964) proposed a relationship between hydraulic conductivity and suction as follows:

$$k(\psi) = \begin{cases} k_s & \text{for } \psi \leq \psi_b \\ k_s \left(\dfrac{\psi_b}{\psi}\right)^\eta & \text{for } \psi > \psi_b \end{cases} \tag{12.21a}$$

where ψ_b approximates the air-entry pressure and the exponent η is related to the BC pore size distribution parameter λ by the equation

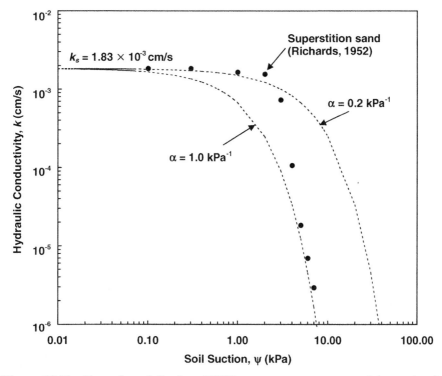

Figure 12.11 Examples of Gardner (1958) one-parameter exponential equation in modeling experimental data for Superstition sand.

$$\eta = 2 + 3\lambda \tag{12.21b}$$

Equation (12.21) may be written in similar form to express hydraulic conductivity $k(h)$ as a function of suction head h and the air-entry head h_b. Campbell (1973) proposed a similar relationship for hydraulic conductivity as a function of volumetric water content:

$$k(\theta) = k_s \left(\frac{\theta}{\theta_s}\right)^n \tag{12.22}$$

A practical advantage of the BC model is that the hydraulic conductivity function may be approximated relatively easily by modeling a given SWCC using the pore-size distribution parameter λ and eq. (12.4). Given eq. (12.21b) and the definition of relative conductivity $k_r = k/k_s$, the BC hydraulic conductivity model may be written in terms of effective water content Θ or effective degree of saturation S_e as

$$k_r = \Theta^{(2+3\lambda)/\lambda} = S_e^{(2+3\lambda)/\lambda} \qquad (12.23)$$

Figure 12.12a, for example, shows a best fit to Richards' (1952) experimental data for Superstition sand using the Brooks-Corey (1964) SWCC model. The air-entry pressure and pore size distribution parameter are $\psi_b = 2.25$ kPa and $\lambda = 1.05$, respectively. If the hydraulic conductivity parameter η is calculated from λ using eq. (12.21b), a close fit to Richards' experimental hydraulic conductivity data is obtained (Fig. 12.12b). It should be noted, however, that the applicability of the BC model and approximation of hydraulic conductivity in this manner is generally limited to relatively coarse-grained soil because the original data from which the model was developed was limited to suction values less than approximately 20 kPa.

Example Problem 12.4 Table 12.6 shows hydraulic conductivity data for an unsaturated clayey soil. Model the relative hydraulic conductivity function $k_r(h)$ using the Brooks and Corey (1964), Gardner (1958) one-parameter, Gardner (1958) two-parameter, and Richards (1931) models.

Solution Figure 12.13 shows $k_r(h)$ for the four models based on visual parameter optimizations in log k_r–log h space. Fitting parameters selected for the BC model are $\lambda = 0.25$ and $h_b = 0.18$ m. Parameters for the Gardner models are $\beta = 3.4$ for the one-parameter model and $n = 1.5$ and $a = 15$ for the two-parameter model. For the Richards model, $a = -3$ and $b = 1$.

12.2.2 Statistical Models

Statistical hydraulic conductivity models may be used to indirectly predict the hydraulic conductivity function from measurements or models of the soil-water characteristic curve. A conceptualization that forms the theoretical basis for statistical hydraulic conductivity modeling is shown in Fig. 12.14. Figure 12.14a shows a plane cut through a finite mass of soil comprising both soil solids and void space. An idealized cross section for the plane is shown in Fig. 12.14b, where the void space is represented by a series of circular pores that are randomly distributed in size and location. The pores are separated into discrete groups, or "pore size classes," of the same radius in order of descending size, for example, $r_1, r_2, \ldots, r_n$.

When any two such cross sections adjoin, as in a continuous soil mass, the hydraulic conductivity across the resulting plane is a function of the probability that water-filled pores of various sizes from each section are connected. The probability that a pore of size r_i is connected to one of size r_j is equal to the product of the probability that each exists at a given location on adjoining cross sections. This can be expressed by

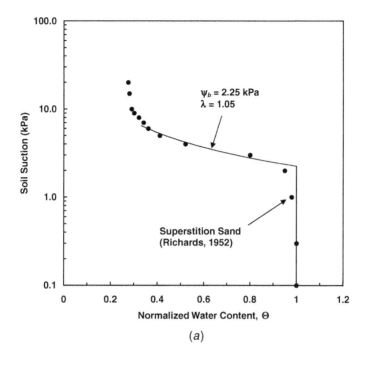

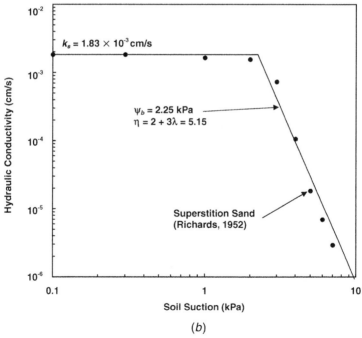

Figure 12.12 (a) Soil-water characteristic curve and (b) hydraulic conductivity function for Superstition sand (Richards, 1952) modeled using the Brooks-Corey (1964) empirical equations.

TABLE 12.6 Hydraulic Conductivity Data for Clayey Soil from Example Problem 12.4

Suction Head (m)	Relative Hydraulic Conductivity, k_r
0.01	1
0.065	0.9
0.13	0.6
0.18	0.4
0.25	0.3
0.3	0.25
0.4	0.18
0.5	0.15
0.55	0.1
0.65	0.095
0.7	0.08
0.8	0.07
1.2	0.06
1.4	0.04
1.5	0.02
2	0.015
2.2	0.009
3	0.006
3.5	0.004
4.5	0.0025
5.5	0.0015
6.5	0.001

$$P(r_i \rightarrow r_j) = f(r_i)f(r_j) \tag{12.24}$$

where $f(r_i)$ is a function describing the probability that a pore with size r_i occurs at a given location on the ith cross section and $f(r_j)$ is the probability that a pore with size r_j occurs at a given location on the jth cross section.

For a *saturated* soil system, the rate of fluid flow through the tube formed by the connection of r_i and r_j may be described by the Hagen-Poiseuille equation, which states that flow velocity is proportional to the square of the tube radius and may be expressed in terms of hydraulic gradient i_h and a characteristic pore radius R as

$$q_{r_i \rightarrow r_j} = \frac{\rho_w g i_h}{8\mu} R^2 f(r_i) f(r_j) \tag{12.25}$$

where ρ_w and μ are the density and absolute viscosity of the permeant fluid, respectively, and g is gravitational acceleration. The characteristic pore radius R is usually considered the smaller of the two connecting pore sizes r_i or r_j,

12.2 HYDRAULIC CONDUCTIVITY MODELS   519

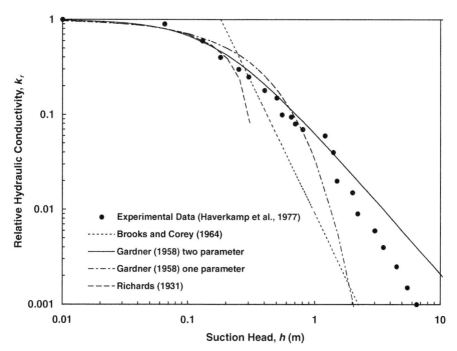

Figure 12.13 Hydraulic conductivity functions for clayey soil from Example Problem 12.4.

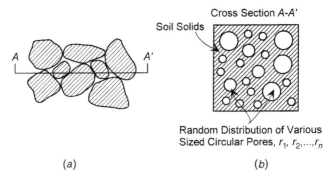

Figure 12.14 Conceptual basis for statistical hydraulic conductivity modeling: (*a*) plane cut through idealized soil mass comprised of solids and voids and (*b*) cross-section *A-A'* idealized as plane of randomly sized and randomly distributed pores.

but may also be considered their mean value or some other function of the two.

The total flow rate across a unit area of the idealized pore system is equal to the sum of the individual flow rates resulting from the connections between each size class:

$$q = \frac{\rho_w g i_h}{8\mu} \sum_{i=1}^{n} \sum_{j=1}^{n} R^2 f(r_i) f(r_j) \quad (12.26)$$

where n is the total number of individual pore size classes.

Applying Darcy's law yields an expression that relates hydraulic conductivity k to the probabilistic pore size distribution:

$$k = \frac{\rho_w g}{8\mu} \sum_{i=1}^{n} \sum_{j=1}^{n} R^2 f(r_i) f(r_j) \quad (12.27)$$

Numerous statistical models have been developed to describe the pore size distribution function $f(r)$ in soil, including those described by Childs and Collis-George (1950), Burdine (1953), and Mualem (1978) among others. Mualem (1986) provides a comprehensive review of several pore size distribution models. Modifications and refinements to these statistical models have resulted in relatively simple predictive equations for use in practice. For example, Marshall (1958) followed the Childs and Collis-George (1950) theoretical development to obtain an equation for saturated hydraulic conductivity in which the average cross-sectional area of connected pores is described in terms of the series

$$k_s = \frac{\rho_w g}{8\mu} \frac{\varepsilon^2}{n^2} [r_1^2 + 3r_2^2 + 5r_3^2 + \cdots + (2n-1)r_n^2] \quad (12.28)$$

where ε is the (fluid-filled) porosity, n is the number of pore size classes, and r_i is the mean radius of pores in a given class i. The parameter r_n represents the smallest of the pore size classes. The term ε^2/n^2 is a *pore interaction term* describing the connectivity of the pores, where the numerator acts as a reduction factor to account for the likelihood of poorer connectivity as porosity decreases.

For an *unsaturated* soil system, liquid flow occurs only within the liquid-filled pores. The unsaturated hydraulic conductivity function can thus be predicted if the relationship between the fluid-filled pore size and suction, that is, the soil-water characteristic curve, is known. Recalling that the radius of the largest water-filled pores r_i under a suction head equal to h_i is given by the Young-Laplace equation (Section 4.1):

$$r_i = \frac{2T_s}{\rho_w g h_i} \qquad (12.29)$$

and substituting this relationship into eq. (12.28) yields an expression for hydraulic conductivity as a function of water content:

$$k(\theta_i) = \frac{T_s^2}{2\mu\rho_w g} \frac{\varepsilon^2}{n^2} [h_1^{-2} + 3h_2^{-2} + 5h_3^{-2} + \cdots + (2n-1)h_n^{-2}] \qquad (12.30)$$

where ε is the fluid-filled porosity at water content θ_i and n is the number of pore classes in the water content range from zero to θ_i. Suction head h_i is determined from the SWCC by subdividing experimental data or a mathematical model for the curve into a series of discrete water content intervals. The soil-water characteristic curve, and thus the hydraulic conductivity function, is most commonly evaluated along a drying (desorption) path.

A matching factor is usually incorporated to provide a more accurate estimate of hydraulic conductivity by accounting for discrepancies between calculated and measured saturated hydraulic conductivities. A general form of eq. (12.30) incorporating such a matching factor can be written as

$$k(\theta_i) = \frac{k_s}{k_{sc}} \frac{T_s^2}{2\mu\rho_w g} \frac{\theta_s^2}{N^2} \sum_{j=i}^{m} [2j + 1 - 2i)h_j^{-2}] \qquad i = 1, 2, \ldots, m \qquad (12.31)$$

where the matching factor k_s/k_{sc} is the ratio of measured saturated hydraulic conductivity k_s (determined independently) to calculated saturated conductivity k_{sc}, j and i are summation indices, and m is the total number of pore size intervals between the saturated water content θ_s and the lowest water content θ_l. The index i increases as volumetric water content decreases. For example, $i = 1$ denotes the pore size class corresponding to the saturated water content and $i = m$ denotes the pore size class corresponding to the lowest water content for which conductivity is calculated. The index N is the number of intervals between the saturated water content and zero water content and is equal to

$$N = m\left(\frac{\theta_s}{\theta_s - \theta_l}\right) \qquad (12.32)$$

Numerous formulations in this family have been developed for predicting the unsaturated hydraulic conductivity function from the soil-water characteristic curve, differing primarily in the form of the pore interaction term and matching factor (e.g., Marshall, 1958; Millington and Quirk, 1964; Kunze et al., 1968; Jackson, 1972). Reviews of several of these formulations are provided by Brutsaert (1967), Green and Corey (1971), and Klute (1972). Agus et al. (2003) provide a quantitative assessment of various statistical modeling approaches by considering the goodness-of-fit between predictions and experimental data. In general, the statistical models perform better for relatively coarse-grained materials such as sands with narrow pore size distributions, presumably because their pore water retention characteristics are more accurately captured by capillary theory and the soil skeleton is relatively rigid. Prediction for fine-grained, structured, or deformable (e.g., expansive) soil remains to a great extent inaccurate.

Example Problem 12.5 The SWCC for a sandy soil is shown in Fig. 12.15. The saturated hydraulic conductivity was determined independently as $k_s = 1.8 \times 10^{-3}$ cm/s. Use the Jackson (1972) formalism to predict the unsaturated hydraulic conductivity function from the SWCC. The Jackson (1972) formalism may be written as

$$k(\theta_i) = k_s \left(\frac{\theta_i}{\theta_s}\right)^c \frac{\sum_{j=i}^{m} [(2j + 1 - 2i)h_j^{-2}]}{\sum_{j=1}^{m} [(2j - 1)h_j^{-2}]} \qquad (12.33)$$

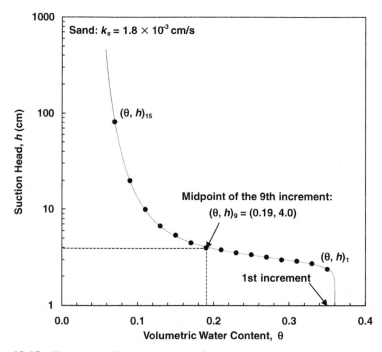

Figure 12.15 Example soil-water characteristic curve for predicting unsaturated hydraulic conductivity function using the Jackson (1972) formalism.

where $k(\theta_i)$ is the hydraulic conductivity at water content θ_i, k_s is the saturated hydraulic conductivity, m is the number of increments of θ subdivided on the characteristic curve, h is the suction head at the midpoint of each water content increment, and j and i are summation indices. The exponent c is a constant that can vary between 0 and 1.33 but is typically set equal to unity.

Solution The characteristic curve is first divided into a series of equal water content increments and the suction head at the midpoint of each increment is estimated. The first increment ($i = 1$), for example, is from $\theta_s = 0.36$ to $\theta = 0.34$. The suction head at the midpoint of the 9th increment is 4.0 cm. Points comprising the conductivity function $k(\theta_i)$ are then calculated for each increment. For example, at $i = 1$, $\theta_1 = \theta_s = 0.36$, and the summations in the numerator and denominator of eq. (12.33) are equal. Thus, $k(\theta_1) = k_s = 1.8 \times 10^{-3}$ cm/s. At $i = 2$, $\theta_2 = 0.34$:

$$k(\theta_2) = 1.8 \times 10^{-3} \left(\frac{0.34}{0.36}\right)^1 \frac{\dfrac{1}{2.75^2} + \dfrac{3}{2.91^2} + \dfrac{5}{3.00^2} + \cdots + \dfrac{27}{82.00^2}}{\dfrac{1}{2.40^2} + \dfrac{3}{2.75^2} + \dfrac{5}{2.91^2} + \cdots + \dfrac{29}{82.00^2}}$$

$$= 1.35 \times 10^{-3} \text{ cm/s}$$

The summation in the denominator is common for all values of i and is equal to 9.387. Thus, at $i = 3$, $\theta_3 = 0.32$ and

$$k(\theta_3) = 1.8 \times 10^{-3} \left(\frac{0.32}{0.36}\right)^1 \frac{\dfrac{1}{2.91^2} + \dfrac{3}{3.00^2} + \dfrac{5}{3.20^2} + \cdots + \dfrac{25}{82.00^2}}{9.387}$$

$$= 9.96 \times 10^{-4} \text{ cm/s}$$

Calculations proceed in this manner to $i = 15$ and $\theta_{15} = 0.08$. Figure 12.16a shows the consequent relationship between water content and hydraulic conductivity. Figure 12.16b shows the corresponding relationship between hydraulic conductivity and suction head.

Two hydraulic conductivity models built upon statistical pore size distributions that have received considerable attention in geotechnical engineering practice are the van Genuchten (1980) and Fredlund et al. (1994) models. Both allow concurrent modeling of the soil-water characteristic curve and the hydraulic conductivity function.

van Genuchten (1980) proposed a flexible closed-form analytical equation for the relative hydraulic conductivity function $k_r(\psi)$ by substituting eq. (12.7)

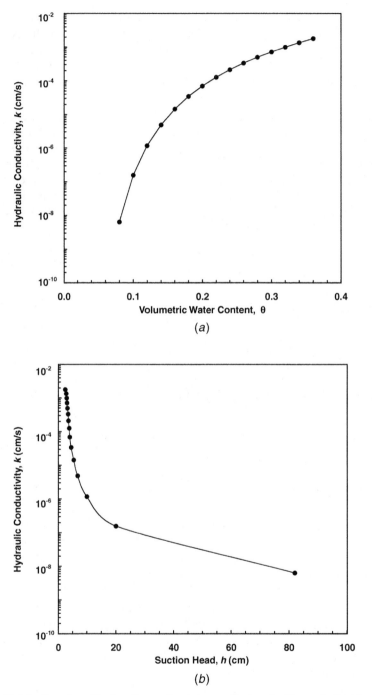

Figure 12.16 Results of hydraulic conductivity prediction for Example Problem 12.5: (a) hydraulic conductivity versus volumetric water content, $k(\theta)$, and (b) hydraulic conductivity versus suction head, $k(h)$.

into the statistical conductivity models proposed by Burdine (1953) and Mualem (1978) as follows:

$$k_r(\psi) = \frac{[1 - (\alpha\psi)^{n-1}[1 + (\alpha\psi)^n]^{-m}]^2}{[1 + (\alpha\psi)^n]^{m/2}} \quad (12.34)$$

which allows the conductivity function to be estimated directly from a corresponding model of the SWCC if the saturated hydraulic conductivity is known. Given eq. (12.7), eq. (12.34) may be written in terms of effective water content Θ (or effective degree of saturation S_e) as follows:

$$k_r = \Theta^{0.5} [1 - (1 - \Theta^{1/m})^m]^2 \quad (12.35)$$

Fredlund et al. (1994) combined eq. (12.13) with the statistical pore size distribution model of Childs and Collis-George (1950) to obtain a model for the relative hydraulic conductivity function as

$$k_r(\psi) = \Theta^q (\psi) \frac{\int_{\ln(\psi)}^{b} \frac{\theta(e^y) - \theta(\psi)}{e^y} \theta'(e^y) \, dy}{\int_{\ln(\psi_{aev})}^{b} \frac{\theta(e^y) - \theta_s}{e^y} \theta'(e^y) \, dy} \quad (12.36)$$

where y is a dummy variable of integration representing $\ln(\psi)$, $b = \ln(10^6)$ kPa, ψ_{aev} is the air-entry pressure, θ' is the derivative of eq. (12.13) with respect to ψ, and Θ^q is a correction factor to take into account tortuosity, with the exponent q typically equal to unity.

TABLE 12.7 Hydraulic Conductivity and Soil-Water Characteristic Curve Data for Soil from Example Problem 12.6

Matric Suction (kPa)	Relative Hydraulic Conductivity, k_r	Volumetric Water Content, θ
0.10	1	0.50
4.47	0.597	0.49
9.50	0.302	0.46
14.71	0.143	0.42
20.00	0.0684	0.38
26.15	0.031	0.34
33.96	0.0127	0.30
44.93	0.00441	0.26
62.34	0.00116	0.22
96.06	0.000183	0.18
195	—	0.14
7843	—	0.11

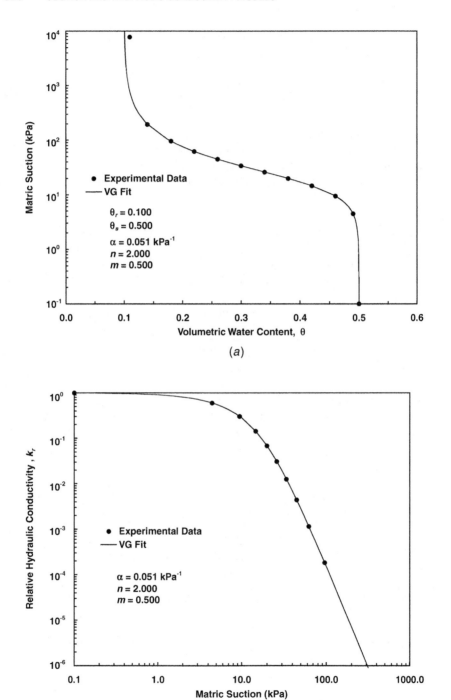

Figure 12.17 van Genuchten (1980) fits to experimental data for soil from Example Problem 12.6: (a) soil-water characteristic curve, $\theta(\psi)$, and (b) relative hydraulic conductivity function, $k_r(\psi)$.

Example Problem 12.6 Table 12.7 shows SWCC and hydraulic conductivity function data for a sandy soil. Represent both curves using the van Genuchten (1980) models, that is, eqs. (12.7) and (12.34).

Solution The RETC code (van Genuchten et al., 1991) may be used to simultaneously fit both curves. Using the $m = 1 - 1/n$ constraint, the optimum fitting parameters for the curves are as follows: $\theta_s = 0.5$, $\theta_r = 0.1$, $\alpha = 0.051$ kPa^{-1}, $n = 2.0$, and $m = 0.5$. The modeled SWCC and hydraulic conductivity functions are shown along with the experimental data in Figs. 12.17a and 12.17b, respectively.

PROBLEMS

12.1. Plot the SWCC for a fine sand specimen using the following Brooks and Corey (1964) parameters: $\psi_b = 3$ kPa, $\lambda = 2.2$, and $S_r = 0.14$.

12.2. Model the hydraulic conductivity function obtained from Example Problem 11.1 by fitting the data using the Gardner (1958) two-parameter power law equation and the Brooks and Corey (1964) equation.

12.3. Estimate the Brooks-Corey (1964) and van Genuchten (1980) fitting parameters for the sandy silt from Problem 10.5. Assume the specimen has void ratio $e = 0.64$ and specific gravity $G_s = 2.65$.

12.4. Hydraulic properties of four different soils are given in Table 12.8 along with fitting parameters for the van Genuchten (VG) and Brooks-Corey (BC) models. Construct and compare the SWCC for each soil using each model.

TABLE 12.8 Hydraulic Properties and Curve-Fitting Parameters for Soils in Problem 12.4

Soil	Residual Water Content, θ_r	Saturated Water Content, θ_s	Initial Water Content, θ_i	VG, n	VG, β (cm^{-1})	BC, λ	BC, ψ_b (cm)
Silty clay loam	0.159	0.496	0.160	5.450	0.014	2.550	56.800
Touchet silt loam	0.102	0.526	0.132	3.590	0.028	1.600	25.100
GE No 2 sand	0.057	0.367	0.083	5.050	0.036	2.340	20.900
Sarpy loam	0.032	0.400	0.045	1.600	0.028	0.506	23.100

12.5. Hydraulic properties of three different clayey soils are given in Table 12.9 along with fitting parameters for the van Genuchten (VG) and Brooks-Corey (BC) models. Construct and compare the SWCC for each soil using each model.

TABLE 12.9 Hydraulic Properties and Curve-Fitting Parameters for Soils in Problem 12.5

Soil Tinjum et al. (1997)	Residual Water Content, θ_r	Saturated Water Content, θ_s	VG, n	VG, α (kPa^{-1})	BC, λ	BC, ψ_b (kPa)
Low plasticity clay B	0	0.354	1.083	0.033	0.068	17.0
Low plasticity clay C	0	0.299	1.063	0.014	0.037	21.7
High plastisity clay F	0	0.407	1.068	0.037	0.054	11.7

12.6. Data comprising the hydraulic conductivity function for a silty loam is shown as Table 12.10. Model the hydraulic conductivity function using as many models described in this chapter as possible.

TABLE 12.10 Data Comprising Hydraulic Conductivity Function for Soil from Problem 12.6

$u_a - u_w$ (kPa)	k_r
0.45	1
1.13	1
1.62	0.95
1.92	0.9
2.94	0.765
4.90	0.595
6.86	0.48
9.80	0.338
13.53	0.2
18.24	0.1
19.61	0.074
25.20	0.03
33.24	0.01

12.7. The SWCC for a silty soil is shown in Fig. 12.18. Use the Jackson (1972) statistical modeling formalism to predict and plot the hydraulic conductivity function. Assume the saturated conductivity $k_s = 5.8 \times 10^{-6}$ cm/s and $c = 1$.

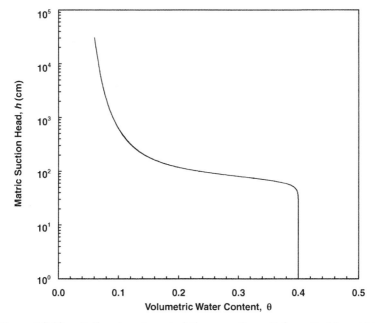

Figure 12.18. Soil-water characteristic curve for soil from Problem 12.7.

REFERENCES

Adamson, A. W., 1976, *Physical Chemistry of Surfaces,* Wiley, New York.

Agus, S. S., Leong, E. C., and Schanz, T., 2003, "Assessment of statistical models for indirect determination of permeability functions from soil-water characteristic curves," *Géotechnique,* **53**(2), 279–282.

Ahuja, L. R., Naney, J. W., and Williams, R. D., 1985, "Estimating soil-water characteristics from simpler properties or limited data," *Soil Science Society of America Journal,* **49,** 1100–1105.

Ahuja, L. R., Ross, J. D., Bruce, R. R., and Cassel, D. K., 1988, "Determining unsaturated hydraulic conductivity from tensiometric data alone," *Soil Science Society of America Journal,* **52,** 27–34.

Albrecht, B. A., Benson, C. H., and Beurmann, S., 2003, "Polymer capacitance sensors for measuring soil gas humidity in drier soils," *Geotechnical Testing Journal,* **23**(1), 3–11.

Al-Khafaf, S., and Hanks, R. J., 1974, "Evaluation of the filter paper method for estimating soil water potential," *Soil Science,* **117**(4), 194–199.

American Society for Testing and Materials (ASTM), 2000, "Standard Test Method for Measurement of Soil Potential (Suction) Using Filter Paper," Designation ASTM D5298-94, *Annual Book of ASTM Standards,* Vol. 4.08, D-18 Committee on Soils and Rock, West Conshohocken, PA, pp. 1082–1087.

American Society for Testing and Materials (ASTM), 2001, "Standard Test Method for Determining Unsaturated and Saturated Hydraulic Conductivity in Porous Media by Steady-State Centrifugation," Designation ASTM D6527, *Annual Book of ASTM Standards,* Vol. 4.09, D-18 Committee on Soils and Rock, West Conshohocken, PA, pp. 868–877.

Anderson, D. M., and Low, P. F., 1958, "Osmotic pressure equations for determining thermodynamic properties of soil water," *Soil Science,* **86,** 251–253.

Averjanov, S. F., 1950, "About permeability of subsurface soils in case of complete saturation," *English Collection,* **7,** 19–21.

Barden, L., and Pavlakis, G., 1971, "Air and water permeability of compacted unsaturated cohesive soil," *Journal of Soil Science,* **22,** 302–317.

Bear, J., 1972, *Dynamics of Fluids in Porous Media,* Dover, New York.

Benson, C. H., and Gribb, M., 1997, "Measuring unsaturated hydraulic conductivity in the laboratory and field," in *Unsaturated Soil Engineering Practice,* S. Houston and D. G. Fredlund, eds., American Society of Civil Engineers Special Technical Publication No. 68, Reston, VA, pp. 113–168.

Berner, E. K., and Berner, R. A., 1987, *The Global Water Cycle: Geochemistry and Environment,* Prentice-Hall, Englewood Cliffs, N.J.

Birkeland, P. W., 1999, *Soils and Geomorphology,* Oxford University Press, New York.

Bishop, A. W., 1954. "The use of pore pressure coefficients in practice," *Géotechnique,* **4,** No. 4, 148–152.

Bishop, A. W., 1959, "The principle of effective stress," *Teknisk Ukeblad I Samarbeide Med Teknikk,* Oslo, Norway, **106**(39), 859–863.

Bishop, A. W., and Blight, G. E., 1963, "Some aspects of effective stress in saturated and unsaturated soils," *Géotechnique,* **13,** No. 3, 177–197.

Bishop, A. W., and Donald, I. B., 1961, "The experimental study of partly saturated soil in the triaxial apparatus," in Proceedings of the 5th International Conference Soil Mechanics and Foundation Engineering, Paris, Vol. 1, pp. 13–21.

Bishop, A. W., Alpan, I., Blight, G. E., and Donald, I. B., 1960, "Factors controlling the shear strength of partly saturated cohesive soils," in ASCE *Research Conference on the Shear Strength of Cohesive Soils,* University of Colorado, Boulder, pp. 503–532.

Blight, G. E., 1961, Strength and Consolidation Characteristics of Compacted Soils, Ph.D. dissertation, University of London, London.

Blight, G. E., 1967, "Effective stress evaluation for unsaturated soils," *ASCE Journal of the Soil Mechanics and Foundations Division,* **93** (SM2), 125–148.

Bocking, K. A., and Fredlund, D. G., 1979, "Use of the osmotic tensiometer to measure negative pore water pressure," *Geotechnical Testing Journal,* **2**(1), 3–10.

Bocking, K. A., and Fredlund, D. G., 1980, "Limitations of the axis translation technique," in *Proceedings of the 4th International Conference on Expansive Soils,* Denver, CO, pp. 117–135.

Bouma, J., Hillel, D., Hole, F., and Amerman, C., 1971, "Field measurement of unsaturated hydraulic conductivity by infiltration through artificial crusts, *Proceedings of the Soil Science Society of America,* **35,** 362–364.

Bouma, J., and Van Lanen, J. A. J., 1987. "Transfer functions and threshold values: From soil characteristics to land qualities," in *Quantified Land Evaluation,* K. J. Beck et al., eds., *Proceedings of the International Society of Soil Science and Soil Science Society of America Workshop,* Washington, DC.

Bouwer, H., 1966, "Rapid field measurement of air entry value and hydraulic conductivity of soil as significant parameters in flow system analysis," *Water Resources Research,* **2,** 729–738.

Brennen, C. E., 1995, *Cavitation and Bubble Dynamics,* Oxford University Press, New York.

Brooks, R. H., and Corey, A. T., 1964, "Hydraulic properties of porous media," Colorado State University, Hydrology Paper No. 3, March.

Bruce, R., and Klute, A., 1956, "The measurement of soil moisture diffusivity," *Soil Science Society of America Proceedings,* **20,** 458–462.

Brutsaert, W., 1967, "Some methods of calculating unsaturated permeability," *Transactions of A.S.A.E.,* **10,** 400–404.

Buckingham, E., 1907, "Studies on the movement of soil moisture," U.S. Dept. Agric. Bur. Soils Bull. 38, U.S. Government Printing Office, Washington, DC.

Burdine, N. T., 1953, "Relative permeability calculation from size distribution data," *Transactions, American Institute of Mining, Metallurgical, and Petroleum Engineers,* **198,** 71–78.

Campbell, J. D., 1973, "Pore pressures and volume changes in unsaturated soils," Ph.D. Thesis, University of Illinois at Urbana-Champaign, IL.

Campbell, G. S., and Gardner, W. H., 1971, "Psychrometric measurement of soil potential: temperature and bulk density effects," *Soil Science Society of America Proceedings,* **35,** 8–12.

Carslaw, H. S., and Jaeger, J. C., 1959, *Conduction of Heat in Solids,* 2nd ed., Oxford University Press, New York.

Cass, A., Campbell, G. S., and Jones, T. L., 1984, "Enhancement of thermal water diffusion in soil," *Soil Science Society of America Journal,* **48,** 25–32.

Cassel, D. K., and Klute, A., 1986, "Water potential: tensiometry," in *Methods of Soil Analysis, Part 1, Physical and Mineralogical Methods,* 2nd ed., A. Klute, ed., Soil Science Society of America, Madison, WI.

Cassel, D., Warrick, A., Nielson, D., and Biggar, J., 1968, "Soil-water diffusivity values based upon time dependent soil water content distributions," *Soil Science Society of America Proceedings,* **32,** 774–777.

Chandler, R. J., and Gutierrez, C. I., 1986, "The filter paper method of suction measurement," *Géotechnique,* **36,** 265–268.

Cheng, L., Fenter, P., Nagy, K. L., Schlegel, M. L., and Sturchio, N. C., 2001, "Molecular-scale density oscillations in water adjacent to a mica surface," *Physical Review Letters,* **87**(15), 156103.

Childs, E. C., 1969, *Soil Water Phenomena,* Wiley-Interscience, New York.

Childs, E. C., and Collis-George, N., 1950, "The permeability of porous materials," *Proceedings, Royal Society, London,* Series A, **210,** 392–405.

Chipera, S. J, Carey, J. W., and Bish, D. L., 1997, "Controlled-humidity XRD analyses: Application to the study of smectite expansion/contraction," in *Advances in X-Ray Analysis,* J. V. Gilfrich et al., eds., Vol. 39, pp. 713–721, Plenum, New York.

Chiu, T.-F., and Shackelford, C. D., 1998, "Unsaturated hydraulic conductivity of compacted sand-kaolin mixtures," *Journal of Geotechnical and Geoenvironmental Engineering,* **124**(2), 160–170.

Cho, G. C., and Santamarina, J. C., 2001, "Unsaturated particulate materials: particle-level studies," *Journal of Geotechnical and Geoenvironmental Engineering,* **127**(1), 84–96.

Clayton, W. S., 1996, "Relative permeability-saturation-capillary head relationships for air sparging in soils," Ph.D. Dissertation, Colorado School of Mines, Golden, CO.

Clothier, B., Scotter, D., and Green, A., 1983, "Diffusivity and one-dimensional absorption experiments," *Soil Science Society of America Journal,* **47,** 641–644.

Coleman, J. D., 1962, "Stress/strain relations for partly saturated soils," *Géotechnique,* **12**(4), 348–350.

Conca, J., and Wright, J. A., 1998, "The UFA method for rapid, direct measurement of unsaturated transport properties in soil, sediment, and rock," *Australian Journal of Soil Research,* **36**, 291–315.

Corey, A. T., 1957, "Measurement of water and air permeability in unsaturated soil," *Proceedings Soil Science Society of America,* **21**(1), 7–10.

Corey, A. T., and Klute, A., 1985, "Application of the potential concept to soil water equilibrium and transport," *Soil Science Society of America Journal,* **49**, 3–11.

Cui, Y. J., and Delage, P., 1993, "On the elasto-plastic behaviour of an unsaturated silt," American Society of Civil Engineers, Special Publication 39, Reston, VA, pp. 115–126.

Cui, Y. J., and Delage, P., 1996, "Yielding and plastic behaviour of an unsaturated compacted silt," *Géotechnique,* **46**(2), 291–311.

Dallavalle, J. M., 1943, *Micrometrics,* Pitman, London.

Daniel, D. E., 1983, "Permeability test for unsaturated soil," *Geotechnical Testing Journal,* **6**(2), 81–86.

Davidson, J. M., Stone, L. R., Nielson, D. R., and Larue, M. E., 1969, "Field measurement and use of soil-water properties," *Water Resources Research,* **5**, 1312–1321.

DeBano, L. F., 2000, "Water repellency in soils: A historical overview," *Journal of Hydrology,* **231–232**, 4–32.

DeCampos, T. M. P., and Carrillo, C. W., 1995, "Direct shear testing on an unsaturated soil from Rio de Janeiro," *Proceedings of the 1st International Conference on Unsaturated Soils,* Paris, pp. 31–38.

Defay, R., Prigogine, I., Bellemans, A., and Everett, D. H., 1966, *Surface Tension and Adsorption,* Wiley, New York.

De Vries, D. A., 1958, "Simultaneous transfer of heat and moisture in porous media," *Transactions of the American Geophysical Union,* **39**, 909–916.

Diamond, S., 1970, "Pore size distributions in clays," *Clays and Clay Minerals,* **18**, 7–23.

Dirksen, C., 1991, "Unsaturated hydraulic conductivity," in *Soil Analysis, Physical Methods,* K. Smith and C. Mullins, eds., Marcel Dekker, New York, pp. 209–269.

Dobbs, H. T., and Yeomans, J. M., 1992, "Capillary condensation and prewetting between spheres," *Journal of Physics: Condensed Matter,* **4**, 10133–10138.

Donald, I. B., 1961, "The Mechanical Properties of Saturated and Partly Saturated Soils with Special Reference to Negative Pore Water Pressure," Ph.D. dissertation, University of London.

Drumright, E. E., 1989, "The Contribution of Matric Suction to the Shear Strength of Unsaturated Soils," Ph.D. Thesis, Colorado State University, Fort Collins.

Durner, W., Schultze, B., and Zurmuhl, T., 1999, "State of the art in inverse modeling of inflow/outflow experiments," in *Proceedings of the International Workshop on Characterization and Measurement of the Hydraulic Properties of Unsaturated Porous Media,* van Genuchten et al., eds., University of California, Riverside, pp. 661.

Dushkin, C. D., Yoshimura, H., and Nagayama, K., 1996, "Direct measurement of nanonewton capillary forces," *Journal of Colloid and Interface Science,* **181**, 657–660.

Edlefsen, N. E., and Anderson, A. B. C., 1943, "Thermodynamics of soil moisture," *Hilgardia,* **15,** 31–298.

Escario, V., 1980, "Suction-controlled penetration and shear tests," in Proceedings of the 4th International Conference on Expansive Soils, Denver, CO, pp. 781–787.

Escario, V., Juca, J., and Coppe, M. S., 1989, "Strength and deformation of partly saturated soils," in Proceedings of the 12th International Conference on Soil Mechanics and Foundation Engineering, Vol. 3, Rio de Janeiro, pp. 43–46.

Escario, V., and Saez, J., 1986, "The shear strength of partly saturated soils," *Géotechnique,* **36**(3), 453–456.

Fawcett, R. G., and Collis-George, N., 1967, "A filter-paper method for determining the moisture characteristics of soil," *Australian Journal of Experimental Agriculture and Animal Husbandry,* **7,** 162–167.

Fisher, R. A., 1926, "On the capillary forces in an ideal soil; correction of formula given by W. B. Haines," *Journal Agricultural Science,* **16,** 492–505.

Fredlund, D. G., 1973, "Volume Change Behavior of Unsaturated Soils," Ph.D. dissertation, University of Alberta, Edmonton, Canada.

Fredlund, D. G., 1989, "Soil suction monitoring for roads and airfields," Symposium on the State-of-the-Art of Pavement Response Monitoring Systems for Roads and Airfields, sponsored by the U.S. Army Corps of Engineers, Hanover, NH, March 6–9, 1989.

Fredlund, D. G., and Morgenstern, N. R., 1977, "Stress state variables for unsaturated soils," *Journal of Geotechnical Engineering Division,* **103,** 447–466.

Fredlund, D. G., and Rahardjo, H., 1993, *Soil Mechanics for Unsaturated Soils,* Wiley, New York.

Fredlund, D. G., and Wong, D. K. H., 1989, "Calibration of thermal conductivity sensors for measuring soil suction," *Geotechnical Testing Journal,* **12**(3), 188–194.

Fredlund, D. G., and Xing, A., 1994, "Equations for the soil-water characteristic curve," *Canadian Geotechnical Journal,* **31,** 521–532.

Fredlund, D. G., Morgenstern, N. R., and Widger, R. A., 1978, "Shear strength of unsaturated soils," *Canadian Geotechnical Journal,* **15,** No. 3, 313–321.

Fredlund, D. G., Rahardjo, H., and Gan, J., 1987, "Non-linearity of strength envelope for unsaturated soils," in *Proceedings of the 6th International Conference on Expansive Soils,* New Delhi, India, Vol. 1, pp. 49–54.

Fredlund, D. G., Xing, A., and Huang, S., 1994, "Predicting the permeability function for unsaturated soil using the soil-water characteristic curve," *Canadian Geotechnical Journal,* **31,** 533–546.

Fredlund, D. G., Vanapalli, S. K., Xing, A., and Pufahl, D. E., 1995, "Predicting the shear strength of unsaturated soils using the soil-water characteristic curve," *Proceedings of the 1st International Conference on Unsaturated Soils,* Paris, pp. 63–69.

Freeze, R. A. and Cherry, J. A., 1979, *Groundwater,* Prentice-Hall, Englewood Cliffs, NJ.

Fung, Y. C., 1965, *Foundations of Solid Mechanics,* Prentice Hall, Englewood Cliffs, NJ.

Gan, J. K., Fredlund, D. G., and Rahardjo, H., 1988, "Determination of the shear strength parameters of an unsaturated soil using the direct shear test," *Canadian Geotechnical Journal,* **25,** No. 3, 500–510.

Gardner, W. R., 1956, "Calculation of capillary conductivity from pressure plate outflow data," *Soil Science Society of America Proceedings,* **20,** 317–320.

Gardner, W. R., 1958, "Some steady state solutions of the unsaturated moisture flow equation with application to evaporation from a water table," *Soil Science,* **85,** No. 4, 228–232.

Gardner, W. H., 1986, "Water content," in *Methods of Soil Analysis, Part 1, Physical and Mineralogical Methods,* Soil Science Society of America, Monograph 9, Madison, WI, pp. 493–544.

Gardner, W., and Widstoe, J. A., 1921, "The movement of soil moisture," *Soil Science,* **11,** 215–232.

Gee, G., Campbell, M., Campbell, G., and Campbell, J., 1992, "Rapid measurement of low soil potentials using a water activity meter," *Soil Science Society of America Journal,* **56,** 1068–1070.

Geiser, F., 1999, "Comportement mécanique d'un limon no saturé: étude expérimentale et modélisation constitutive," Ph.D. Thesis, Swiss Federal Institute of Technology, Lausanne.

Greacen, E. L., Walker, G. R., and Cook, P. G., 1989, Procedure for the Filter Paper Method of Measuring Soil Water Suction, Division of Soils, Report No. 108, CSIRO Division of Water Resources, Glen Osmond, South Australia, Australia.

Green, W. H., and Ampt, G. A., 1911, "Studies on soil physics, Part I. Flow of air and water through soils," *Journal of Agricultural Science,* **4,** 1–24.

Green, R. E., and Corey, J. C., 1971, "Calculation of hydraulic conductivity: A further evaluation of some predictive methods," *Soil Science Society of America Proceedings,* **35,** 3–8.

Guan, Y., and Fredlund, D. G., 1997, "Use of the tensile strength of water for the direct measurement of high soil suction," *Canadian Geotechnical Journal,* **34,** 604–614.

Haines, W. B., 1930, "Studies in the physical properties of soil. V. The hysteresis effect in capillary properties and the modes of moisture distribution associated therewith," *Journal of Agricultural Science,* **20,** 97–116.

Halsey, G. D., 1948, *Journal of Chemical Physics,* **16,** 931.

Hamblin, A. P., 1981, "Filter-paper method for routine measurement of field water potential," *Journal of Hydrology,* **53,** 355–360.

Hamilton, J. M., Daniel, D. E., and Olson, R. E., 1981, "Measurement of hydraulic conductivity of partially saturated soils," *Permeability and Groundwater Contaminant Transport,* ASTM STP 746, T.F. Zimmie and C.O. Riggs, eds., ASTM, Philadelphia, pp. 182–196.

Hardy, B., 1992, "Two-pressure humidity calibration on the factory floor," *Sensors,* July, 15–19.

Hashizume, H., Shimomura, S., Yamada, H., Fujita, T., Nakazawa, H., and Akutsu, O., 1996, "X-Ray diffraction system with controlled relative humidity and temperature," *Powder Diffraction,* **11,** 288–289.

Haverkamp, R., Vauclin, M., Tovina, J., Wierenga, P. J., and Vachaud, G., 1977, "A comparison of numerical simulation models for one-dimensional infiltration," *Soil Science Society of America Proceedings,* **41,** 285–294.

Hazen, A., 1930, "Water supply," in *American Civil Engineers Handbook,* Wiley, New York.

Hilf, J. W., 1956, "An investigation of pore water pressure in compacted cohesive soils," Technical Memorandum No. 654, United States Department of the Interior, Bureau of Reclamation, Design and Construction Division, Denver, CO.

Hillel, D., 1982, *Introduction to Soil Physics,* Academic, New York.

Hillel, D., and Gardner, W., 1970, "Measurement of unsaturated conductivity and diffusivity by infiltration through an impeding layer," *Soil Science,* **109**(3), 149–153.

Hillel, D., Krentos, V. D., and Stylianou, Y., 1972, "Procedure and test of an internal drainage method for measuring soil hydraulic characteristics in-situ," *Soil Science,* **114,** 295–400.

Ho, D. Y. F., and Fredlund, D. G., 1982, "Increase in shear strength due to suction for two Hong Kong soils," in *Proceedings of the ASCE Geotechnical Conference Engineering and Construction in Tropical and Residual Soils,* Honolulu, HI, pp. 263–295.

Houston, S. L., Houston, W. N., and Wagner, A., 1994, "Laboratory filter paper suction measurements," *Geotechnical Testing Journal,* GTJODJ, **17**(2), 185–194.

Huang, S., Fredlund, D. G., and Barbour, S. L., 1998, "Measurement of the coefficient of permeability for a deformable unsaturated soil using a triaxial permeameter," *Canadian Geotechnical Journal,* **35,** 426–432.

Iribarne, J. V., and Godson, W. L., 1981, *Atmospheric Thermodynamics,* 2nd ed., Reidel, Boston.

Israelachvili, J., 1992, *Intermolecular & Surface Forces,* 2nd ed., Academic, San Diego.

Iwata, S., Tabuchi, T., and Warkentin, B. P., 1995, *Soil-Water Interactions: Mechanisms and Applications,* 2nd ed., Marcel Dekker, New York.

Jackson, R. A., 1972 , "On the calculation of hydraulic conductivity," *Soil Science Society of American Proceedings,* **36,** 380–383.

Jackson, R. D., 1964, "Water vapor diffusion in relatively dry soil, I: Theoretical considerations and sorption experiments," *Soil Science Society of America Proceedings,* **28,** 172–175.

Jennings, J. E., and Burland, J. B., 1962 "Limitations to the use of effective stresses in partly saturated soils," *Géotechnique,* **12**(2), 125–144.

Jones, D. E., and Holtz, W. G., 1973, "Expansive soils–The hidden disaster," *Civil Engineering,* ASCE, **43**(8), 49–51.

Katz, D. L., et al., eds., 1959. *Handbook of Natural Gas Engineering,* McGraw-Hill, New York.

Kaye, G. W. C. and Laby, T. H., 1973, *Tables of Physical and Chemical Constants,* Longman, Boston.

Khalili, N., Geiser, F., Blight, G. E., 2004, "Effective stress in unsaturated soils, a review with new evidence," *International Journal of Geomechanics,* **4,** No. 2.

Khalili, N., and Khabbaz, M. H., 1998, "A unique relationship for χ for the determination of the shear strength of unsaturated soils," *Géotechnique,* **48**(5), 681–687.

Klein, C., and Hurlbut, C. S., 1997, *Manual of Mineralogy,* Wiley, New York.

Klute, A., 1965, "Laboratory measurement of hydraulic conductivity of unsaturated soils," in *Methods of Soil Analysis,* Black, C. A. et al., eds., Monograph 9, Part 1, American Society of Agronomy, Madison, WI, pp. 253–261.

Klute, A., 1972, "The determination of the hydraulic conductivity and diffusivity of unsaturated soils," *Soil Science,* **113**(4), 264–276.

Klute, A., and Dirksen, C., 1986, "Hydraulic conductivity and diffusivity: laboratory methods," *Methods of Soil Analysis, Part 1. Physical and Mineralogical Methods,* Soil Science Society of America, Monograph No. 9, Madison, WI, pp. 687–734.

Krahn, J., Fredlund, D. G., and Klassen, M. J., 1989, "Effect of soil suction on slope stability at Notch Hill," *Canadian Geotechnical Journal,* **26**(2), 269–278.

Krynine, D. P., 1948, "Analysis of the latest American tests on soil capillarity," *Proceedings of the 2nd International Conference on Soil Mechanics and Foundation Engineering,* Rotterdam, Vol. 3, pp. 100–104.

Kumar, S., and Malik, R. S., 1990, "Verification of quick capillary rise approach for determining pore geometrical characteristics in soils of varying texture," *Soil Science,* **150**(6), 883–888.

Künhel, R. A., and van der Gaast, S. J., 1993, "Humidity-controlled diffractometry and its applications," in *Advances in X-Ray Analysis,* Vol. 36, Gilfrich, J. V. et al., eds., Plenum, New York, pp. 439–449.

Kunze, R. J., and Kirkham, D., 1962, "Simplified accounting for membrane impedance in capillary conductivity measurements," *Soil Science Society of America Proceedings,* **26,** 421–426.

Kunze, R. J., Uehara, G., and Graham, K., 1968, "Factors important in the calculation of hydraulic conductivity," *Soil Science Society of America Proceedings,* **32,** 760–765.

Lane, K. S., and Washburn, S. E., 1946, "Capillary tests by capillarimeters and by soil filled tubes," *Proceedings of Highway Research Board,* **26,** 460–473.

Lang, A. R. G., 1967, "Osmotic coefficients and water potentials of sodium chloride solutions from 0 to 40 C," *Australian Journal of Chemistry,* **20,** 2017–2023.

Lapalla, E. G., Healy, R. W., and Weeks, E. P., 1993, "Documentation of computer program VS2D to solve the equations of fluid flow in variably saturated porous media," Water Resources Investigations Report 83-4099, U.S. Geological Survey, Denver, CO.

Laplace, P. S., 1806, *Mecanique Celeste,* suppl. 10th vol. English translation reprinted by Chelsea, New York (1966).

Laroussi, C. H., and DeBacker, L. W., 1979, "Relations between geometrical properties of glass bead media and their main $\psi(\theta)$ hysteresis loops," *Soil Science Society of America Journal,* **43,** 646–650.

Lenhard, R. J., Parker, J. C., and Mishra, S., 1989, "On the correspondence between Brooks-Corey and van Genuchten models," *Journal of Irrigation and Drainage Engineering,* **115**(4), 744–751.

Leong, E. C., He, L., and Rahardjo, H., 2002, "Factors affecting the filter paper method for total and matric suction measurements," *Geotechnical Testing Journal,* **25**(3), 321–332.

Leong, E. C., and Rahardjo, H., 1997a, "Reviews of soil-water characteristic curve equations," *Journal of Geotechnical and Geoenvironmental Engineering,* **123,** 1106–1117.

Leong, E. C., and Rahardjo, H., 1997b, "Permeability functions for unsaturated soils," *Journal of Geotechnical and Geoenvironmental Engineering,* **123,** 1118–1126.

Letey, J., Osborn, J., and Pelishek, R. E., 1962, "Measurement of liquid-solid contact angles in soil and sand," *Soil Science,* **93,** 149–153.

Lian, G., Thornton, C., and Adams, M. J., 1993, "A theoretical study of the liquid bridge forces between two rigid spherical bodies," *Journal of Colloid and Interface Science,* **161,** 138–147.

Likos, W. J., 2000, "Total Suction–Moisture Content Characteristics for Expansive Soils," Ph.D. dissertation, Colorado School of Mines, Golden, CO.

Likos, W. J., and Lu, N., 2001, "Automated measurement of total suction characteristics in the high suction range: Application to the assessment of swelling potential," *Transportation Research Record: Journal of the Transportation Research Board,* No. 1755, TRB, Washington, DC, pp. 119–128.

Likos, W. J., and Lu, N., 2002, "Filter paper technique for measuring total soil suction," *Transportation Research Record: Journal of the Transportation Research Board,* No. 1786, TRB, Washington, DC, pp. 120–128.

Likos, W. J., and Lu, N., 2003a, "Filter paper column for measuring transient suction profiles in expansive clay," *Transportation Research Record: Journal of the Transportation Research Board,* No. 1821, TRB, Washington, DC, pp. 83–89.

Likos, W. J., and Lu, N., 2003b, "Automated humidity system for measuring total suction characteristics of clay," *Geotechnical Testing Journal,* **26**(2), 178–189.

Likos, W. J., and Lu, N., 2004, "Hysteresis of capillary stress in unsaturated soils," *Journal of Engineering Mechanics,* ASCE, **130,** No. 6.

Livingston, H. K. 1949, *Journal of Colloid Science,* **4,** 447.

Lowell, S., 1979, *Introduction to Powder Surface Area,* Wiley, New York.

Lu, N., 1999, "Time-series analysis for determining vertical air permeability in unsaturated zones," *Journal of Geotechnical and Geoenvironmental Engineering,* **125**(1), 69–77.

Lu, N., and Griffiths, D. V., 2004, "Profiles of steady-state suction stress in unsaturated soils," *Journal of Geotechnical and Geoenvironmental Engineering.*

Lu, N., and LeCain, G. D., 2003, "Percolation induced heat transfer in deep unsaturated soils," *Journal of Geotechnical and Geoenvironmental Engineering,* **129**(11), 1040–1053.

Lu, N., and Likos, W. J., 2004, "Rate of capillary rise in soils," *Journal of Geotechnical and Geoenvironmental Engineering,* **130,** No. 6.

Lu, N., and Y. Zhang, 1997, "Thermally induced gas convection in mine wastes." *International Journal of Heat and Mass Transfer,* **40**(11), 2621–2636.

Ma, Q. L., Hook, J. E., and Ahuja, L. R., 1999, "Influence of three parameter conversion methods between van Genuchten and Brooks-Corey parameters on water balance predictions," *Water Resources Research,* **35,** 2571–2578.

Maâtouk, A., Leroueil S., and La Rochelle, P., 1995, "Yielding and critical state of collapsible unsaturated silty soil," *Géotechnique,* **45**(3), 465–477.

Malik, R. S., Kumar, S., and Malik, R. K., 1989, "Maximal capillary rise flux as a function of height from the water table," *Soil Science,* **148**(5), 322–326.

Marshall, T. J., 1958, "A relation between permeability and size distribution of pores," *Journal of Soil Science,* **9,** 1–8.

Marshall, T. J., and Holmes, J. W., 1988, *Soil Physics,* Cambridge University Press, New York.

Martin, R. T., 1960, "Adsorbed water on clay: A review" *Clays and Clay Minerals,* **9,** 28–70.

Mason, G., and Clark, W. C., 1965, "Liquid bridges between spheres," *Chemical Engineering Science,* **20,** 859–866.

Maswaswe, J., 1985, "Stress Paths for a Compacted Soil During Collapse due to Wetting," Ph.D. Thesis, Imperial College, London.

McQueen, I. S., and Miller, R. F., 1968, "Calibration and evaluation of wide-range gravimetric method for measuring soil moisture stress," *Soil Science,* **10**(3), 521–527.

McQueen, I. S., and Miller, R. F., 1974, "Approximating soil moisture characteristics from limited data: Empirical evidence and tentative model," *Water Resources Research,* **10**(3), 521–527.

Meerdink, J. S., Benson, C. H., and Khire, M. V., 1996, "Unsaturated hydraulic conductivity of two compacted barrier soils," *Journal of Geotechnical Engineering,* **122**(7), 565–576.

Miller, E., and Elrick, D., 1958, "Dynamic determination of capillary conductivity extended for non-negligible membrane impedance," *Soil Science Society of America Proceedings,* **22,** 483–486.

Millington, R. J., and Quirk, J. P., 1964, "Formation factor and permeability equations," *Nature,* **202,** 143–145.

Mitchell, J. K., 1993, *Fundamentals of Soil Behavior,* Wiley, New York.

Molenkemp, F., and Nazemi, A. H., 2003. "Interactions between two rough spheres, water bridge and water vapour," *Géotechnique,* **53**(2), 255–264.

Monteith, J. L. and Unsworth, M. H., 1990, *Principles of Environmental Physics,* 2nd ed., Hodder Arnold, London.

Moore, R. E., 1939, "Water conduction from shallow water tables," *Hilgardia,* **12**(6), 383–426.

Mualem, Y., 1978, "Hydraulic conductivity of unsaturated porous media: Generalized macroscopic approach," *Water Resources Research,* **14**(2), 325–334.

Mualem, Y., 1984, "A modified dependent domain theory of hysteresis," *Soil Science,* **137,** 283–291.

Mualem, Y., 1986, "Hydraulic conductivity of unsaturated soils: Prediction and formulas," in *Methods of Soil Analysis. Part I. Physical and Mineralogical Methods,* 2nd ed., A. Klute, ed., Agronomy Monograph No. 9, American Society of Agronomy, Madison, WI, pp. 799–823.

Nielson, D. R., Davidson, J., Biggar, Y., and Miller, R., 1964, "Water movement through Panoche clay loam soil," *Hilgardia,* **35**(17), 491–506.

Nimmo, J. R., 1992, "Semiempirical model of soil water hysteresis," *Soil Science Society of America Journal,* **56,** 172–173.

Nimmo, J. R., and Akstin, K. C., 1988, "Hydraulic conductivity of a saturated soil at low water content after compaction by various methods," *Soil Science Society of America Journal,* **52,** 303–310.

Nimmo, J., Akstin, K., and Mello, K., 1992, "Improved apparatus for measuring hydraulic conductivity at low water content," *Soil Science Society of America Journal,* **56,** 1758–1761.

Nimmo, J. R., Rubin, J., and Hammermeister, D. P., 1987, "Unsaturated flow in a centrifugal field: Measurement of hydraulic conductivity and testing of Darcy's law," *Water Resources Research,* **23**(1), 124–134.

Nitao, J. J., and Bear, J., 1996, "Potentials and their role in transport and porous media," *Water Resources Research,* **32**(2), 225–250.

O'Connor, K. M., and Dowding, C. H., 1999, *Geomeasurements by Pulsing Cables and Probes,* CRC Press, Boca Raton, FL.

Office of Arid Lands Studies, University of Arizona, Global humidity index map, 2003.

Ohmic Instruments Corporation, 2003, "Commercial Publications," Easton, MD.

Olsen, H. W., Morin, R. H., and Nichols, R. W., 1988, "Flow pump applications in triaxial testing," in Donaghe, R. T., Chaney, R. C., and Silver, M. L., eds., *Advanced Triaxial Testing of Soil and Rock,* ASTM STP 977, ASTM, Philadelphia, pp. 68–81.

Olsen, H. W., Willden, A. T., Kiusalaas, N. J., Nelson, K. R., and Poeter, E. P., 1994, "Volume-controlled hydrologic property measurements in triaxial systems," in *Hydraulic Conductivity and Waste Contaminant Transport in Soils,* ASTM STP 1142, Daniel, D. E., and Trautwein, S., eds., ASTM, Philadelphia, pp. 482–504.

Olsen, H. W., Nichols, R. W., and Rice, T. L., 1985, "Low gradient permeability methods in a triaxial system," *Géotechnique,* **35**(2), 145–157.

Olson, R. E., and Daniel, D. E., 1981, "Measurement of the hydraulic conductivity of fine-grained soils," in *Permeability and Groundwater Contaminant Transport,* Zimmie, T. F., and Riggs, C. O., eds., ASTM STP 746, ASTM Philadelphia, pp. 18–64.

Olson, R. E., and Langfelder, L. J., 1965, "Pore water pressures in unsaturated soils," *ASCE Journal of the Soil Mechanics and Foundations Division,* **SM4,** 127.

Orr, F. M., Scriven, L. E., and Rivas, A. P., 1975, "Pendular rings between solids: meniscus properties and capillary force," *Journal of Fluid Mechanics,* **67**(4), 723–742.

Pagenkopf, G. K., 1978, *Introduction to Natural Water Chemistry,* Marcel Dekker, New York.

Park, S., and Sposito, G., 2002, "Structure of water adsorbed on a mica surface," *Physical Review Letters,* **89**(1) 85501.

Peck, A. J., and Rabbidge, R. M., 1969, "Design and performance of an osmotic tensiometer for measuring capillary potential," *Soil Science Society of America Proceedings,* **33,** No. 2, 196–202.

Peck, R. B., Hansen, W. E., and Thornburn, T. H., 1974, *Foundation Engineering,* 2nd Edition, Wiley, New York.

Peixoto, J. P., and Kettani, M., 1973, "The control of the water cycle, *Scientific American,* **228**(4), 46–61.

Penman, H. L., 1940, "Gas and vapour movement in soil: I. The diffusion of vapours through porous solids," *Journal of Agricultural Science* (England), **30,** 347–462.

Penman, H. L., 1970 "The water cycle," *Scientific American,* **223**(3), 54–63.

Phene, C. J., Hoffman, G. J., and Rawlins, S. L., 1971a, "Measuring soil matric potential in-situ by sensing heat dissipation within a porous body: I. Theory and sensor construction," *Soil Science Society of America Proceedings,* **35,** 27–33.

Phene, C. J., Hoffman, G. J., and Rawlins, S. L., 1971b, "Measuring soil matric potential in-situ by sensing heat dissipation within a porous body: II. Experimental results," *Soil Science Society of America Proceedings,* **35,** 225–229.

Philip, J. R., 1957, "The theory of infiltration. 1. The infiltration equation and solution," *Soil Science,* **83,** 345–357.

Philip, J. R., 1969, "Theory of infiltration," *Advances in Hydroscience,* **5,** 215–290.

Philip, J. R., 1987, "The quasi-linear analysis, the scattering analog and other aspects of infiltration and seepage," in *Infiltration Development and Application,* Fok, Y. S., ed., Water Resources Research Center, Univ. of Hawaii, Hawaii, pp. 1–27.

Philip, J. R., and de Vries, D. A., 1957, "Moisture movement in porous materials under temperature gradients," *Transactions of the American Geophysical Union,* **38,** 222–232.

Picornell, M., Lytton, R. L., and Steinberg, M., 1983, "Matric suction instrumentation of a vertical moisture barrier," *Transportation Research Record,* **945,** 16–21.

Pruess, K., 1991, *TOUGH2–A General Purpose Numerical Simulator for Multiphase Fluid and Heat Flow,* LBL–29400, Lawrence Berkeley Laboratory, Berkeley, CA.

Pullan, A. J., 1990, "The quasilinear approximation for unsaturated porous media flow," *Water Resources Research,* **26**(6), 1219–1234.

Rawls, W. J., and Brakensiek, D. L., 1985, "Prediction of soil water properties for hydrologic modeling," in *Watershed Management in the Eighties,* Jones, E., and Ward, T. J., eds., ASCE, Denver, CO, pp. 293–299.

Richards, L. A., 1928, "The usefulness of capillary potential to soil moisture and plant investigators," *Journal of Agricultural Research,* **37,** 719–742.

Richards, L. A., 1931, "Capillary conduction of liquids through porous medium," *Journal of Physics,* 318–333.

Richards, L. A., 1952, "Water conducting and retaining properties of soils in relation to irrigation," in *Proceedings of an International Symposium on Desert Research,* Jerusalem, pp. 523–546.

Richards, L. A., and Weeks, L., 1953, "Capillary conductivity values from moisture yield and tension measurements on soil columns," *Soil Science Society of America Proceedings,,* **55,** 206–209.

Ridley, A. M., and Burland, J. B., 1993, "A new instrument for the measurement of soil moisture suction," *Géotechnique,* **43**(2), 321–324.

Rijtema, P., 1959, "Calculation of capillary conductivity from pressure plate outflow data with non-negligible membrane impedance," *Netherlands Journal of Agricultural Science,* **7,** 209–215.

Rojstaczer, S., and Tunks, J., 1995, "Field-based determination of air diffusivity using soil air and atmospheric pressure time series." *Water Resources Research,* **31,** 3337–3343.

Rose, D., 1968, "Water movement in porous materials. III. Evaporation of water from soil," *British Journal of Applied Physics,* **2**(1), 1770–1791.

Ross, B., 1990, "The diversion capacity of capillary barriers," *Water Resources Research,* **26,** 2625–2629.

Ross, B., Amter, S., and Lu, N., 1992, *Numerical Studies of Rock-Gas Flow in Yucca Mountain,* SAND91-7034, Sandia National Laboratories, Albuquerque, NM.

Rossetti, D., Pepin, X., and Simons, S. J. R., 2003, "Rupture energy and wetting behavior of pendular liquid bridges in relation to the spherical agglomeration process," *Journal of Colloid and Interface Science,* **261,** 161–169.

Russell, M. B., 1942, "The utility of the energy concept of soil moisture," *Soil Science Society of America Proceedings,* **7,** 90–94.

Satija, B. S., 1978, "Shear Behavior of Partially Saturated Soils," Ph.D. Thesis, Indian Institute of Technology, Delhi.

Sattler, P. J., and Fredlund, D. G., 1989, "Use of thermal conductivity sensors to measure matric suction in the laboratory," *Canadian Geotechnical Journal,* **26**(3), 491–498.

Schofield, R. K., 1935, "The pF of the Water in Soil," *Transactions of the 3rd International Congress for Soil Science,* **2,** 37–48.

Shan, C., 1995, "Analytical solutions for determining vertical air permeability in unsaturated soils," *Water Resources Research,* **31,** 2193–2200.

Shaw, D. J., 1992, *Colloid and Surface Chemistry,* 4th ed., Butterworth Heinemann, London.

Sibley, J. W., Smyth, G. K., and Williams, D. J., 1990, "Suction-moisture content calibration of filter papers from different boxes," *Geotechnical Testing Journal,* **13**(3), 257–262.

Sillers, W. S., Fredlund, D. G., and Zakerzadeh, N., 2001, "Mathematical attributes of some soil-water characteristic curve models," *Geotechnical and Geological Engineering,* **19,** 243–283.

Singh, V. P., 1997, *Kinematic Wave Modeling in Water Resources: Environmental Hydrology,* Wiley, New York.

Soilmoisture Equipment Corporation (SEC), 2003, "Commercial Publications," SEC, Santa Barbara, CA.

Spanner, D. C., 1951, "The Peltier effect and its use in the measurement of suction pressure," *Journal of Experimental Botany,* **11,** 145–168.

Sparks, A. D. W., 1963, "Theoretical considerations in stress equations for partly saturated soils," *Proceedings of the 3rd Regional Conference for Africa on Soil Mechanics,* Salisbury, Rhodesia, Vol. 1, pp. 215–218.

Sposito, G., 1981, *The Thermodynamics of Soil Solutions,* Clarendon Press, Oxford.

Sridharan, A., Altschaeffl, A. G., and Diamond, S., 1971, "Pore size distribution studies," *Journal of the Soil Mechanics and Foundations Division,* **97**(SM 5), 771–787.

Srivastava, R., and Yeh, T. C. J., 1991, "Analytical solutions for one-dimensional, transient infiltration toward the water table in homogeneous and layered soils," *Water Resources Research,* **27**(5), 753–762.

Stallman, R. W., 1967, "Flow in the zone of aeration," in *Advances in Hydrosciences,* Vol. 4, AGU, Washington, DC, pp. 151–195.

Stallman, R. W., and Weeks, E. P., 1969, "The use of atmospherically induced gas-pressure fluctuations for computing hydraulic conductivity of the unsaturated zone," *Geological Society of America Abstracts with Programs,* **7,** 213.

Stannard, D. I., 1992, "Tensiometers–Theory, Construction, and Use," *Geotechnical Testing Journal,* **15**(1), 48–58.

Steenhuis, T. S., Parlange, J.-Y., and Kung, K.-J. S., 1991, "Comment on 'The diversion capacity of capillary barriers,' by Benjamin Ross," *Water Resources Research,* **27,** 2155–2156.

Stephens, D., 1994, "Hydraulic conductivity assessment of unsaturated soil," in *Hydraulic Conductivity and Waste Contaminant Transport in Soils,* Daniel, D. E., and Trautwein, S., eds., ASTM STP 1142, ASTM, Philadelphia, pp. 169–181.

Stephens, D. B., 1995, *Vadose Zone Hydrology,* CRC Lewis, Boca Raton, FL.

Streeter, V. L., Wylie, E. B., and Bedford, K. W., 1997, *Fluid Mechanics,* McGraw-Hill Science/Engineering/Mathematics, New York.

Tabor, D., 1979, *Gases, Liquids, and Solids,* Cambridge University Press, New York.

Tarantino, A., and Mongiovi, L., 2001, "Experimental procedures and cavitation mechanisms in tensiometer measurements," *Geotechnical and Geological Engineering,* **19,** 189–210.

Terzaghi, K., 1943, *Theoretical Soil Mechanics,* Wiley, New York.

Tetens, O., 1930, "Uber einige meteorologische Begriffe," *Zeitschrift Geophysic,* **6,** 297–309.

Thomson, W., 1871, *Philosophical Magazine,* **42,** 448.

Tindall, J. A., and Kunkel, J. R., 1999, *Unsaturated Zone Hydrology for Scientists and Engineers,* Prentice Hall, Upper Saddle River, N.J.

Tinjum, J. M., Benson, C. H., and Blotz, L. R., 1997, "Soil-water characteristic curves for compacted clays," *Journal of Geotechnical and Geoenvironmental Engineering,* **123**(11), 1060–1069.

Topp, G. C., Davis, J. L., and Annan, A. P., 1980, "Electromagnetic determination of soil water content: Measurement in coaxial transmission lines," *Water Resources Research,* **16**(3), 574–582.

Touloukian, Y. S., Saxena, S. C., and Hestermans, P., 1975, "Viscosity, thermophysical properties of matter," in *TPRC Data Series,* Vol. 11, Plenum, New York.

UNESCO, 1984, "Map of the World Distribution of Arid Regions," Intergovernmental Oceanographic Commission, Paris, France.

Vanapalli, S. K., Fredlund, D. G., Pufahl, D. E., and Clifton, A. W., 1996, "Model for the prediction of shear strength with respect to soil suction," *Canadian Geotechnical Journal,* **33,** 379–392.

Vanapalli, S. K., and Fredlund, D. G., 2000, "Comparison of empirical procedures to predict the shear strength of unsaturated soils using the soil-water characteristic curve," in *Advances in Unsaturated Geotechnics,* Shackelford, C. D, Houston, S. L., and Chang, N. Y., eds., GSP No. 99, ASCE, Reston, VA, pp. 195–209.

van der Raadt, P., Fredlund, D. G., Clifton, A. W., Klassen, M. J., and Jubien, W. E., 1987, "Soil suction measurement at several sites in Western Canada," *Transportation Research Record,* **1137,** 24–35.

van Genuchten, M. T., 1980, "A closed form equation for predicting the hydraulic conductivity of unsaturated soils," *Soil Science Society of America Journal,* **44,** 892–898.

van Genuchten, M. T., Leij, F. J., and Yates, S. R., 1991, "The RETC code for quantifying the hydraulic functions of unsaturated soils," U.S. Department of Agriculture, Agricultural Research Service, Report IAG-DW12933934, Riverside, CA.

Walter, M. T., Kim, J. S., Steenhuis, T. S., Parlange, J. Y., Heilig, A., Braddock, R. D., Selker, J. S., and Boll, J., 2000, "Funneled flow mechanism in a sloping layered soil: laboratory investigation," *Water Resources Research,* **36**(4), 841–849.

Watson, K., 1966, "An instantaneous profile method for determining the hydraulic conductivity of unsaturated porous materials," *Water Resources Research,* **2**(4), 709–715.

Weast, R. C., Astle, M. J., and Beyer, W. H., 1981, *CRC Handbook of Chemistry and Physics,,* 65th ed., Boca Raton, FL.

Weeks, E. P., 1978, "Field determination of vertical permeability to air in the unsaturated zone," U.S. Geological Survey Professional Paper 1051, USGS, Denver.

Weeks, E. P, 1979, "Barometric fluctuations in wells tapping deep unconfined aquifers," *Water Resources Research,* **15,** 1167–1176.

Weeks, E. P., 1991, "Does the wind blow through Yucca Mountain?" *Proc., Workshop V: Flow and Transport through Unsaturated Fractured Rock—Related to High-Level Radioactive Waste Disposal,* Rep. No. NUREGICP-0040, U.S. Nuclear Regulatory Commission, White Flint, MD, pp. 43–53.

Wheeler, S. J., and Sivakumar, V., 1995, "An elasto-plastic critical state framework for unsaturated soils," *Géotechnique,* **45,** 35–53.

Wiederhold, P., 1997, *Water Vapor Measurement Methods and Instrumentation,* Marcel Dekker, New York.

Wilke, C. R., and Chang, P., 1955, *A.E.CH.E. Journal,* **1,** 264–270.

Wind, G. P., 1955, "Field experiment concerning capillary rise of moisture in heavy clay soil," *Netherlands Journal of Agricultural Science,* **3,** 60–69.

Wraith, J. M., and Or, D., 1998, "Nonlinear parameter estimation using spreadsheet software," *Journal of Natural Resources and Life Sciences Education,* **27,** 13–19.

Yeh, T.-C., 1989, "One-dimensional steady state infiltration in heterogeneous soils," *Water Resources Research,* **25**(10), 2149–2158.

Young, J. F., 1967, "Humidity control in the laboratory using salt solutions—A review," *Journal of Applied Chemistry.* **17,** 241–245.

Young, T., 1805, *Philosophical Transactions of the Royal Society,* **95,** 65.

Zachmann, D. W., DuChateau, P. C., and Klute, A., 1982, "Simultaneous approximation of water capacity and soil hydraulic conductivity by parameter identification," *Soil Science,* **134,** 157–163.

Zheng, Q., Durben, D. J., Wolf, G. H., and Angell, C. A., 1991, "Liquids at large negative pressures: water at the homogeneous nucleation limit," *Science,* **254,** 829–832.

INDEX

A

Absolute atmospheric pressure, 82–83. *See also* Atmospheric pressure profile
Absolute humidity, *see* Vapor density
Active conditions, 10
Active earth pressure, 301–313. *See also* Lateral earth pressure
Active limit state, 320
Active zone (unsaturated), 195, 267–268
Adsorbed film, 41, 152
Adsorbed water, 51
Adsorption
   capillary tube model for, 115–118
   mechanisms, 41, 42
Air bubbles, 31, 327, 423
   and axis translation testing, 366, 367
   occluded, 183, 327
Air conductivity, 56, 329, 331–332, 336, 366, 463
Air density, *see* Density
Air diffusion
   in axis translation system, 366–367
   in water, 363–366
Air permeability, determination of, 407–412
Air pressure
   amplitude, 404, 410
   amplitude ratio, 406, 408
   fluctuation, 326
Air viscosity, *see* Viscosity
Air-entry head, 30, 136, 137, 140, 143, 348, 497

Air-entry pressure, 40–42, 199, 201, 203, 205, 241, 243, 252, 333, 340, 423, 495, 496, 497, 499
   of HAE ceramic disk(s), 202–205, 421
Air-expulsion pressure, 243
Air-filled porosity, 396, 406, 409
Air-water interface, 9, 34, 35, 41, 204
Air-water-solid interface, 9, 96, 97, 101–104, 160
Air-water-HAE system, 202–203
Amplitude ratio, air pressure, 406, 408
Angular velocity, 472
Annual precipitation, 13
   net as a function of latitude, 15
   map of global average, 14
Apparent cohesion, 24, 26, 253. *See also* Capillary cohesion
Argon, 58–59
Arid zones, 13
Atmospheric constant, moist, 82
Atmospheric pressure profile, 83, 85
Atmospheric vapor pressure, 66
At-rest or K_0 condition, 21–23, 197. *See also* Lateral earth pressure
Axis translation (technique(s)), 10, 201–207, 417, 424–429
   null tests, 206, 424
   pressure plate(s), 425–427
   Tempe cell(s), 427–429

B

Barometric pressure, 50. *See also* Atmospheric pressure profile

Barometric pressure (*continued*)
  fluctuation, 57, 400–402
Barometric pumping, 10, 50, 402–412
Bearing capacity, 4, 7
  factors, 26
Biological weathering, 13
Bishop's effective stress parameter, *see* Effective stress parameter
Bishop's effective stress, 32, 164, 175, 179, 207, 213, 241, 252, 256
Boiling, definition, 80–81
Boltzmann transformation, 378–379, 478, 479, 481
Book, organization and scope, 8
Bubbling pressure, 203, 495. See also Air-entry pressure
Bundled tubes, *see* Capillary tubes

C

Calcium carbonate (caliche), 18
Capacitance-based humidity sensor(s), 441–443
Capillarity, 7, 34, 128, 174, 175
Capillary adsorption regime, 41–42
Capillary barriers, 10, 341–349
  breakthrough of, 344
  critical head of, 344–345
  diversion capacity of, 346, 347
  diversion width of, 346, 347
  efficiency of, 348
Capillary cohesion
  as a characteristic function, 252–261
  definition, 252–253, 254, 256–261
  determining, 256–261
Capillary condensation, 9, 111–114, 115, 116, 184
Capillary conductivity, 6
Capillary depression, 100
Capillary effects (on chemical potential), 34–36
Capillary finger(s), 136, 137
Capillary flow, 6, 184
Capillary force, 160–161, 176
Capillary fringe, 30, 137, 140, 142
Capillary mechanism(s), 40, 174
Capillary pore size distribution, 147–160
Capillary potential, 6
Capillary radius, 147
Capillary rise, 6, 131, 136, 184, 197
  height of, 9, 133–140, 146, 197
  in soil, 139
  rate of, 9, 140–147
Capillary stress, *see* Suction stress

Capillary tube, 75, 77, 98– 99, 104, 110, 135, 136, 188
Carbon dioxide, 58–59
Cavitation, 80–86, 136, 201, 423
Cellulose membrane(s), 202
Cementation, 5
Centrifuge method, for hydraulic conductivity measurement, 472–476
Ceramic disk(s), *see* HAE ceramic disk(s)
Chemical equilibrium, 9, 107, 108
Chemical potential(s), 7, 34–37, 57, 59, 81, 105–108
Chilled-mirror hygrometer(s), 438–440
Clausius-Clapeyron equation, 60
Climatic factors, 12
Closest packing, 120. *See also* TH packing
Coefficient
  of (unsaturated) earth pressure at rest, 296–298
  of (unsaturated) Rankine's active earth pressure, 305–306, 307
  of (unsaturated) Rankine's passive earth pressure, 314–315
  of permeability, *see* Hydraulic conductivity, Permeability
Cohesion, 20. *See also* Apparent cohesion, Capillary cohesion
Cohesive material(s), 4
Cohesive strength, 24
Collapsing soil, 5, 7
Compaction, 20
Compressibility
  of soil, 371
  of water, 371
Compressible gas, 396
Condensation, 57, 64. *See also* Dew formation
Conductivity, *see* Hydraulic conductivity, Air conductivity
Consolidated-drained (CD) direct shear tests, 227–228
Consolidated-drained (CD) triaxial tests, 223, 224
Constant-flow method, 466–472
Constant-head method, 463–466
Constitutive laws, functions, 26–28, 48
Contact angle, 98, 99–101, 105, 118, 161, 175, 176, 177
Contact angle hysteresis, 184–186
  and effective stress parameter, 187
  and suction stress, 191
  and soil-water characteristic curve, 187

Contact filter paper method, *see* Filter paper method(s)
Continuity principle, *see* Principle of mass conservation
Continuum mechanics, 27, 261
Cracking, *see* Tension cracking
Critical state line, 248–249
Cumulative pore size distribution, 155

D

Dalton's law of partial pressures, 90
Darcy's law, 28, 29, 55, 56, 140, 142, 328, 350, 365, 377, 396, 464
Deformability, 301
Deformation phenomena, 6–8
Density
 of dry air, 47–50, 396
 of moist air, 65–73
 of water, 29, 47, 50–52
 of water vapor, 48. *See also* Absolute humidity
Desaturation, 30, 495
Desert, 19
Desiccation, 20
Desorption, 115
Deviator(ic) stress, 206, 214, 217, 222, 223, 247, 249
Dew formation, 64, 65. *See also* Condensation
Dew point, 64, 65
 mode of operation for thermocouple psychrometers, 434
 temperature, 65, 440
Dielectric constant, 36
Diffusion coefficient
 of free air, 359
 of oxygen, 363
 of water vapor, 359
Diffusivity of water, definition, 375
Dipping capillary barriers, 345–349
Direct shear testing, 226–227, 229, 239, 257, 258, 259
Discharge velocity, 28, 29, 142, 328
Dissolution of gas, 89
Dissolved
 air, 47
 ion mobility, 52
 solutes, 7, 18, 34, 39
Diurnal air tide, 404–405, 409, 410
Diversion of infiltration, *see* Capillary barriers
Droplets, 110

Drying front, 186
Drying loop, 183
DuNouy rings, 75
Dynamic viscosity, 53–55

E

Earth pressure at rest, 10, 294–301. *See also* Lateral earth pressure
Earth pressure profiles, 10, 20–24
Effective angle of internal friction, 220, 229
Effective cohesion, 220, 229
Effective degree of saturation, definition, 496
Effective hydraulic diffusivity, 378
Effective stress, 7, 10, 20, 22, 30–32, 164, 167, 173, 193, 214. *See also* Bishop's effective stress
 due to capillarity, 163
 representation, 215
 validity as a state variable for strength, 247–248
Effective stress parameter, 10, 23, 32, 164–166, 173, 184, 196, 199, 200, 214, 241, 244–247, 248, 251, 256–261
 hysteresis in, 187
 determination of, 242–244
Effective stress profiles, 20–24. *See also* Suction stress profiles
Eight-hour (air) tide, 409, 410
Electrical conductivity sensor(s), 417, 429–431
Electrical displacement, 36
Elevation head, 28, 372
Ellipsoids, interfacial geometry of, 128
Environmental chamber, 445, 447
Environmental factors, 18
Ethyl alcohol, 75
Evaporation, 5, 10, 12, 13, 15, 18, 19, 57, 64
 as a steady flow condition, 352–359
 profile, 195
 rate, 273, 357
Evapotranspiration, 13
Exchangeable cations, 41
 hydration of, 116
Expansive soil(s), 5, 6, 8, 20, 51, 116, 150, 184, 263, 331
 soil-water characteristic curve(s) for, 42, 43, 439, 448, 449, 458
Extended Hooke's law, *see* Hooke's law

Extended M-C (Mohr-Coulomb) criterion, 10, 229–232
and direct shear testing, 229–231
and triaxial testing, 232–233
nonlinearity of, 238–241

F

Fabric, (soil), 175, 332
Failure envelope, 221, 223, 231. *See also* Mohr-Coulomb (M-C) failure criterion
unified representation of, 252–261
Fick's law, 359, 363
Filling angle, 118, 161, 163, 179
Filter paper methods(s), 417–419, 440, 449–459
accuracy, precision, and performance of, 452–459
calibration of, 451–452
column for laboratory testing, 452
principles of, 449–451
Flat capillary barriers, 342–345
Flow phenomena, 6, 10, 323
Fluctuating profiles of suction, 195
Fourier series analysis, 404–406, 408
Free energy formulation, 124, 187
Free energy, 7, 35, 64
Free water, 34–36, 52, 59
Free-stress surface, 305
Freezing point, 52
Friction angle with respect to matric suction, 229
determination of, 233–238
and effective stress parameter, 242–247
Frost formation, 64
Funicular regime, 30
Fusion curve, 80

G

Gamma-ray attenuation technique(s), 485
Gas conductivity, *see* Air conductivity
Gas constant, universal, 49, 154
Gauge cavitation pressure, 85–86
Geo-environmental track, 11
Geomechanics track, 11
Georgia kaolinite, *see* Kaolinite
Gibbs-Duhem equilibrium, 108
Global climatic change, 9
Global humidity index, 13, 16
Governing equation
for transient airflow, 397–400
for transient water flow, 369–378
Green-Ampt assumptions, 376–377
Gypsum block sensor(s), 417, 429–431

H

HAE ceramic disk(s), *see* High-air-entry disk(s)
Hagen-Poiseuille equation, 518
Halsey equation, 152
Hamaker's constant, 37
Head potential, 38
Heat sources, 326
Heat transfer, 27
Heave, 8
Henry's law, 89–91, 365
constant(s), 91, 92
High-air-entry disk(s), 201–206, 222, 226, 366, 420–421, 464. *See also* Cellulose membranes
air-entry pressure of, 202–205, 421
characteristic curve for, 204–205
Hooke's law, 294–295
Horizontal infiltration, 349–351
method for hydraulic conductivity testing, 477–480
Humid zones, 13
Humidity, *see* Relative humidity
Humidity control techniques, 418, 419, 443–449
isopiestic (salt solution), 444–445
two-pressure (divided flow), 445–449
Humidity index, definition, 13
global map for, 16
map for North America, 16
Humidity measurement techniques, 431–443
chilled-mirror hygrometer(s), 438–440
polymer resistance/capacitance sensor(s), 441–443
thermocouple psychrometer(s), 432–438
Hydration, 34, 116, 175
and capillary tube model, 115–118
Hydraulic conductivity, 28, 29, 48, 142, 327, 329–333
and intrinsic permeability, 329–331
relative hydraulic conductivity, 336
Hydraulic conductivity, measurement technique(s), 10, 462–463
steady-state techniques, 463–476
transient techniques, 476–493
Hydraulic conductivity function, 10, 11, 267, 333–341
for HAE ceramic disk(s), 421
hysteresis in, 336, 487
Hydraulic conductivity function, experimental data
for clay, 339
for clayey soil, 518, 519

for sand, 339, 340, 467, 475, 513, 515, 517
for sand-kaolin mixture(s), 488
for silty clay, 490
for silty sand, 483
for silty loam, 340
Hydraulic conductivity function, modeling, 10, 506–527
  Averjanov model, 511, 512
  Brooks and Corey model, 512, 514, 515, 516
  empirical and macroscopic models, 509–516
  Fredlund et al. model, 523–525
  Gardner model(s), 143, 270, 350, 512, 514, 515
  Richards model, 511, 512, 515
  statistical models, 516, 518–524
  statistical model of Jackson, 522
  statistical model of Marshall, 520
  van Genuchten model, 523–525, 526, 527
  Wind model, 512
Hydraulic diffusivity, 372, 476–477, 481
Hydraulic diffusivity function, 478, 479, 481, 485
Hydraulic head, total, 27, 325–326
Hydrologic cycle, 9, 12
Hydrologic parameters representative of clay, sand, and silt, 272
Hydrophilic, 77, 97
Hydrophobic, 97
Hydrostatic condition, 10, 21, 195, 197, 272
Hydrostatic equilibrium, 23, 131, 132
Hyperarid zones, 13
Hysteresis, 10, 117, 174
  in effective stress parameter, 187–189
  in hydraulic conductivity function, 336, 487
  in soil-water characteristic curve, 116–117, 182–184, 187, 188, 337
  in suction stress characteristic curve, 191–192
  mechanism(s), 115, 182–191
  mechanism(s), contact angle effect, 186–187
  mechanism(s), ink-bottle effect, 184–186

I

Ideal gas, 107
  behavior, 58, 107
  law, 48, 60, 360
  volume of, 58

Illite, 19
Illuvial accumulation, 15
Immiscible, 47
Independent state variables, 7
  independent stress state variable(s), 10, 33, 202
Infiltration displacement, 377, 379, 380, 381
Infiltration, 5
  as a steady flow condition, 352–359
  horizontal, 349–351
  horizontal infiltration method for hydraulic conductivity, 477–480
  profile(s), 195
  rate(s), 273, 378–396
  test, 186
  vertical, 352–357
Initial phase of barometric pressure, 403
Ink-bottle effect, 184–186
Instantaneous profile method(s), 484–493
  laboratory, 485–487
  field, 487–493
Interfacial
  equilibrium, 10, 89
  force(s), 163
  physics, 4
  tension, 73
Interparticle
  force(s), 7, 160–163, 175
  stress, 10, 163–168
Intrinsic permeability, see Permeability
Intrusion pressure, 150
Isopiestic humidity control, 418–419, 444–445

K

K_0 condition, see At-rest condition.
Kaolinite, 19, 154, 157, 206, 207
  soil-water characteristic curve for, 43, 155, 440, 448, 449
Kelvin's equation, 9, 104, 105, 109, 110, 111, 147, 431, 432
  derivation of, 106–111
Kelvin's radius, 151, 154
Kinematic viscosity, 53

L

Laminar flow regime, 55, 56
Laplace transform, 388
Lapse rate of atmosphere, 82–83
Lateral earth pressure, 4, 7, 294–321
Latitude, 15
LiCl solution, 64, 444
Limit analysis, 7, 256

552 INDEX

Limit state, 7
Liquid phase, 47
Loess, 5
Low-air-entry porous disk, 222, 467

M

Macrofabric, 332
Mass coefficient of solubility, 95–96
Mass transport, 27
Material properties, 175. *See also* Material variable(s).
Material variable(s), 9, 10, 26, 27, 32, 33, 48, 50
   measurement and modeling, 9, 10, 415
Matric suction, 32, 34, 48, 103, 121, 132, 147, 161, 173, 181, 188, 191, 193, 261, 325. *See also* Suction
   as a state variable, 33
   measurement of, *see* Suction measurement
Matric suction profiles, 268–272. *See also* Suction profiles
Matric suction tensor, 193, 197
Maximum principal stress, 305, 315
Maximum suction stress, 284
M-C criterion, *see* Mohr-Coulomb failure criterion
Mean curvature, 130
Mean effective stress, 214, 217, 247
Mechanical equilibrium, 9, 77, 107, 108, 129, 133, 140
Mechanical stability, 20
Mechanisms for airflow, 326
Menisci, meniscus, 77
   toroidal approximation for, 101, 124, 160, 176, 242
Mercury, 75, 99–100, 150
Mercury intrusion porosimetry, 150
Microfabric, 332
Micromechanical analysis, 21
Minifabric, 332
Minimum principal stress, 305, 315
Miscible, 47
Mohr circle, 24, 207, 209, 210
Mohr-Coulomb (M-C) failure criterion, 23, 301, 302
   extended, 10, 229–232, 256
   for effective stress, 238–248
Moist air, 57
Moisture loading, 195
Moisture profile(s), 8, 23. *See also* Water content profile(s)
Molar concentration, 36, 91

Molar fraction, 58
Molar volume of liquid water, 154
Molecular dynamics, 52
Molecular mass
   of air, 48, 49
   of dry air, 107
   of water vapor, 107
   of water, 38, 66
Montmorillonite, *see* Smectite
Multiphase conditions, 7
Multiphase consolidation, 7
Multiphase system, 47
Multistage testing, 223
Multistep outflow method, *see* Outflow method(s)

N

Negative gauge pressure, 84
Negative pore pressure, 174, 175, 181
Net normal stress, 32, 191, 206, 207, 214, 216
   as a state variable, 33
Net precipitation, *see* Precipitation
Neutral stress, 261
Neutral surface, 100
Neutron logging, 489
Nitrogen gas, 48, 58, 445
Non-contact filter paper method, *see* Filter paper method(s)
Nonwetting interaction, 99
Normal stress tensor, 193, 197
Nucleation, 80, 423. *See also* Cavitation
Null test(s), 33
   for stress state variables, 206
   for matric suction, 424–425

O

Occluded air bubbles, 183, 327
Open-tube column tests, 145
Osmotic effects, 34, 150
Osmotic dessicator, *see* Humidity control techniques
Osmotic mechanisms, 174
Osmotic pressure, 36
Osmotic suction, 34, 39, 325. *See also* Suction
Osmotic tensiometer(s), *see* Tensiometer(s)
Ottawa sand, 139
Outflow method(s)
   for hydraulic conductivity function, 480–484
   principles of, 481–482

Overburden stress, 160, 315
Oxygen, 48, 58, 59, 90, 91, 92

P

Partial molar volume
　of (liquid) water, 38, 107
　of dry air, 107
　of water vapor, 107
　of vapor, 59, 61. *See also* Vapor pressure
Partial pressure, 57–59, 89, 91
Partially saturated soil, definition, 3
Passive earth pressure, 10, 312–319. *See also* Lateral earth pressure
Passive limit state, 313–314, 320
Pedotransfer functions (PTF), 495–496
Peltier effect, 432–436
Pendular
　regime, 30, 160
　state, 327
　water menisci, 137
Permeability, 329–333, 406. *See also* Hydraulic conductivity.
pF, 39
Phase diagram for water, 80, 81
Phase lag of air pressure, 404, 406
Phase transformation, 80
Physical properties of air and water, 47–57
Physical weathering, 13
Piezometer(s), 30
Poise, 53
Poiseuille's law, 330
Poisson's ratio, 21, 22, 196, 295
Polar desert, 19
Polar molecule, 36, 74
Polymer resistance/capacitance sensor(s), 441–443
Pore airflow regime, 326–328
Pore dimension, 128
Pore drainage or adsorption, 183
Pore geometry, 150
Pore pressure regime, 181
Pore radius, 140, 150, 151
Pore size distribution, 9, 40, 42, 154, 497
Pore size distribution parameter, 143
Pore volume, 150
Pore water
　flow regime, 326–328
　potential, 9, 30, 35–38, 39
Porosity, definition, 22. *See also* Air-filled porosity
Potential evaporation, 13

Potential for water flow, 325–326
Potential of soil water, 34–40
Potential, conversion among units, 38–39
Precipitation, 5, 12, 13, 15, 18, 19, 196
Pressure attenuation, 406, 408
Pressure head, 28
Pressure plate(s), 425–427
Pressure potential, 38
Pressure, standard, 66
Principal radii of curvature, 128, 130, 131, 133
Principal stress(es), 191, 208, 209, 210, 221
Principle of mass conservation, 329, 369, 370, 396
Profile(s)
　of active earth pressure, 306–313
　of coefficient(s) of earth pressure, 296–299
　of constant suction stress, 306–308
　of effective stress parameter, 276–282
　of matric suction, 352–359
　of passive earth pressure, 315–319
　of suction stress, 282–292, 296
　of transient moisture, 384–386
　of transient suction, 386–396
　of variable suction stress, 308–310
Psychrometer(s), *see* Thermocouple psychrometers

Q

Quasilinear approach(es), 387

R

Rankine's active state of failure, 302–306
Rankine's passive state of failure, 312–315
Receding front, 186
Recharge rate, 18
Regimes of unsaturated flow, 326–328
Relative conductivity, 336–339
Relative humidity, 9, 20, 30, 57–65, 106, 147, 196
　and total suction, 431
　control of, *see* Humidity control techniques
　definition, 63
　of saturated salt solutions, 444
　measurement of, *see* Humidity measurement techniques
Repellent
　contact angle, 100
　surface, 100, 101

Residual condition(s), 51, 327
Residual (degree of) saturation, 244, 333, 496
Residual water content, 137, 244, 495, 499
Resistance-based humidity sensor(s), 441–442
Reynolds number, 55, 56

S

Saint Venant's principle, 74
Saturated vapor density, 67
Saturated vapor pressure, 60, 61, 62, 63, 113
Saturated water content, 120, 244, 495
Savanna, 19
SC packing, *see* Simple cubic packing
Scanning loop, 183
Seasonal tide, 404–405
Seebeck effect, 433–436
Seepage velocity, 328
Seepage-related problems, 4, 7
Semidesert, 19
Semidiurnal air tide, 404–405, 409, 410
Sesquioxide, 14–15, 17
Settlement, 4
Shear strength, 10, 173, 220–264
  experimental data, 221–228
Shear strength parameters, 10
  for extended M-C criterion, 233–238
  for M-C criterion, 248–252
  for unsaturated soil, 229–230
Shear stress(es), 207, 210
Shear testing, *see* Direct shear testing, Triaxial testing
Short-range adsorption, 34, 39
Short-range particle surface hydration, 116
Short-range physicochemical effects, 7
Shrinkage, 8, 184
Simple cubic (SC) packing, 119, 137, 138, 175–176
Slope stability, 4, 7
Smectite, 19, 52. *See also* Expansive soil(s)
  soil-water characteristic curve for, 42, 43, 439, 448, 449, 458
Soil formation, 9, 13, 19
Soil horizon, 13–18
Soil mechanics, definition, 3
Soil orders, 19
Soil phenomena, classification of, 6
Soil suction, *see* Suction

Soil-water characteristic curve(s), 6, 9–11, 38, 39–43, 48, 114–124, 150, 151, 155, 204, 248, 250, 251, 255, 267, 335, 344, 354, 355, 374
  capillary tube model for, 115–118
  conceptual characteristic curves for sand, silt, and clay, 42, 118
  contacting sphere model for, 118–123
  hysteresis in, 116–117, 182–184, 187, 337
  measurement of, *see* Suction, measurement
  modeling of, *see* Soil-water characteristic curve(s), modeling
Soil-water characteristic curve(s), experimental data
  for expansive soil(s), 42, 43, 439, 448, 449, 458
  for glacial till, 510
  for kaolinite, 43, 155, 440, 448, 449
  for sand-kaolin mixture(s), 488
  for sand(s), 158, 427, 430, 475, 499, 500, 503, 505, 517, 526
  for silty loam, 510
  for silty soil, 470
Soil-water characteristic curve(s), modeling, 494–506
  Brooks and Corey model, 198, 494, 497–499, 500
  Fredlund and Xing model, 494, 505–506, 507, 508
  van Genuchten model, 494, 499–505
Solid mechanics, 7
Solid phase, 47
Solubility
  of air in water, 9, 48, 89–96
  of nitrogen, 91
Solvation, 34
Sorption isotherm(s), 150, 151, 153. *See also* Soil-water characteristic curve(s)
Sorptivity, 378, 379, 385
Specific moisture capacity, 372–376, 476, 477
Specific storage, 371
Specific surface, 153
Spherical
  interface, 101
  particles, 161
  pore geometry, 153
Standard atmospheric pressure, 82
Standard diffusion equation, 371, 372
State variable(s), 7, 9, 10, 26, 27, 28–33, 48

State variable(s)
  deformation, 27, 29
  flow, 27, 29
  stress, 27, 29, 191, 193, 205
Steady evaporation, 349–359
Steady flows, 10, 325–367
Steady infiltration, 349–359
Steady zone, 195, 196, 267–268
Steady-state infiltration, 10
Steppes, 19
Strength characteristic curve, see Suction stress characteristic curve
Strength parameters, 20. See also Shear strength
Stress invariant, 214
Stress phenomena, 6, 7, 10
Stress profile(s), 21, 23
Stress tensor, 191–201
Stress-strain behavior, 4
Strip footing, 24, 26
Sublimation curve, 80
Subsidence, 8
Subsurface-atmosphere interface, 5, 195
Suction, 34–43, 325–326
  of saturated salt solutions, 444
  profile(s), 10, 195, 196, 267–293
  ratio, 243
  regimes of, 39–43, 267–270
  total, matric, and osmotic, 34–35
  units of, 38
Suction, measurement techniques, 417–459
  axis translation, 10, 201–207, 417, 424–429
  chilled-mirror hygrometer(s), 438–440
  electrical/thermal conductivity sensor(s), 429–431
  filter paper techniques, 449–459
  humidity control techniques, 443–449
  polymer-based sensor(s), 441–443
  ranges of, 419
  table of, 418
  tensiometer(s), 80, 84, 86, 417, 420–424, 464, 485
  thermocouple psychrometer(s), 418, 419, 432–438, 485
Suction head, 38, 326, 497
Suction stress, 7, 9, 160–165, 166–167, 173, 187, 191, 214, 216, 242, 252–253, 254, 255, 261
  characteristic curve, 10, 48, 186, 256
  profiles for clay, 289–293
  profiles for sand, 289–294
  profiles for silt, 289–293

regimes, 282–289
stress tensor, 193–194
Surcharge loading, 160
Surface adsorption, 42
Surface charge density, 36, 51
Surface hydration mechanisms, 150
Surface tension, 9, 24, 36, 47, 48, 73–76, 97–101, 105, 161, 187
Surfactants, 75
Suspended water, thickness of, 345
SWCC, see Soil-water characteristic curve
Swelling pressure, 8
Swelling soil(s), see Expansive soil(s)

T

Taiga zone, 19
Tempe cell, 155, 424, 427–429, 499
Temperature
  atmospheric fluctuation, 326
  dew point, 65, 440
  standard, 66
Tensile strength, 195, 297
Tensiometer(s), 80, 84, 86, 417, 420–424, 464, 485
  high capacity, 423
  measurement principles, 421–424
  osmotic, 423
Tension cracking, 8, 297–301, 310–313
Terzaghi's effective stress, 7
Tetens' equation, 62, 360
Tetrahedral (TH) packing, 119, 120, 137, 138, 175–176
TH packing, see Tetrahedral packing
Thermal conductivity sensor(s), 417, 429–431
Thermocouple psychrometer(s), 418, 419, 432–438, 485
  calibration of, 436–438
  principles of, 431–436
Thermodynamic equilibrium, 38, 59, 63
Thermodynamic
  potential, 35
  principles, 4
  properties, 7
Thickness
  of adsorbed water film, 151
  of unsaturated zone, 12, 267, 357
Tightly adsorbed regime, 40–41
Time-domain reflectrometry (TDR), 465, 485
Topographic relief, 326
Toroidal meniscus geometry, 101, 124, 160, 176, 242

Tortuosity factor, 359
Total air potential, 327
Total earth pressure, 310–312, *See also* Lateral earth pressure
Total head, 28, 30, 325, 326
Total suction, 34–38. *See also* Suction
 and relative humidity, 431–432. *See also* Kelvin's equation
 mechanisms, 34–38
Total stress, 21, 174
Trace gases, 48
Transient flows, 10, 369–412
 analytical solution for water flow, 384–389
 gas flow, 10, 396–412
 horizontal infiltration, 376–379
 numerical solution for water flow, 389, 384–396
 vertical infiltration, 380–386
Transpiration, 5
Triaxial testing, 258, 259, 264
 results, 250–252
 setup, 222–223
Triple point of water, 81
Tropical forest zone, 19
Tundra, 19
Turbulent regime, 55
Two-pressure (divided flow) humidity control, 443, 445–449

U

Ultimate bearing capacity, of unsaturated soil, 24, 26
Unit weight of water, 22, 99
Units of potential, head, and pressure, 40
Units of suction, 38–40
Universal gas constant, 48, 49
Unsaturated soil mechanics, definition, 4
Unsteady zone, 195, 267–268

V

van der Waals attraction, 35, 37, 124
van't Hoff equation, 36
Vapor density, 60, 61, 63, 360
 gradient, 361
Vapor enhancement factor, 359
Vapor flow, 359–363

Vapor phase transport, 10, 50, 327, 359
Vapor pressure, 60, 104, 105, 107, 110, 111
 lowering, 9, 104–106, 110
Vaporization curve, 80, 81
Vertical evaporation, 352–359
Vertical infiltration, 352–359
Viral equation, 36
Virtual temperature, 83
Viscosity
 of air, 48, 53–55, 406, 409
 of water, 28, 48, 53–55
Void ratio, 120, 121
Volume of water lens, 118
Volumetric coefficient
 of air solubility, 92–94, 95, 365, 366
 vapor correction for, 94
Volumetric concentration, 92
Volumetric water content, definition, 22

W

Water content profile(s), 10, 267–280
Water droplets, 77
Water meniscus, 103, 161, 177, 181
Water molecules, size of, 113
Water retention characteristics, 136. *See also* Soil-water characteristic curve.
Water retention, 176
Water table, 12
Water-entry pressure, 344
Water viscosity, *see* Viscosity
Wave propagation, 20
Weathering
 front, 18
 depth of, 19
Wetting agent, 75
Wetting contact angle, 100
Wetting front, 142, 377
Wetting interaction, 99
Wetting loop, 183
Wetting process(es), 181, 100
Wilhelmy plates, 75

Y

Young's modulus, 295
Young-Laplace equation, 9, 35, 128, 130, 140, 203, 241, 242, 420

CPSIA information can be obtained at www.ICGtesting.com
Printed in the USA
BVOW08*1442010215
385449BV00013B/167/P